EMBOSS Developer's Guide

The European Molecular Biology Open Software Suite (EMBOSS) is a high-quality, well-documented package of open source software tools for molecular biology. It includes over 200 bioinformatics applications under a consistent command line interface and supported by many graphical interfaces and workflow software. EMBOSS includes extensive and extensible C programming libraries that provide a powerful and robust toolkit for developing new bioinformatics tools from scratch.

- Exploit rich inbuilt functionality with little or no additional coding.
- Comprehensive and consistent application programmer's interface (API).
- Powerful command line interface is handled automatically.
- Extensive qualifiers for controlling program behaviour and specifying IO data.
- Support for both simple and complex (biological) datatypes with easy database configuration.
- Make use of automatic file format handling – no need to code your own.
- Code is well documented and indexed for easy search and navigation; in-depth knowledge of the internals is not required.
- All code is in ANSI standard C with defined standards for coding and documentation.

The *EMBOSS Developer's Guide* is the official and definitive guide to developing software under EMBOSS. It includes comprehensive reference information and guidelines, including step-by-step instructions and real-world code examples:

- Learn how to write fully featured tools guided by the people who developed EMBOSS.
- Comprehensive and up-to-date – all the information you need is at your fingertips.
- Step-by-step guide to writing EMBOSS applications, illustrated with functional, deployed code.
- ACD file development – learn how to customise existing tools without coding, or design and write entirely new application interfaces.
- EMBOSS API programming guidelines – organised into topics to help you quickly master application development.
- Wrapping and porting applications under EMBOSS – learn how to incorporate third-party tools.
- No prior experience with EMBOSS is necessary.

Peter M. Rice is a Group Leader at the European Bioinformatics Institute (EMBL-EBI, Hinxton, UK), a centre for research and services in bioinformatics and part of the European Molecular Biology Laboratory (EMBL). Peter instigated EMBOSS in 1996 when he was based at the Sanger Centre (Hinxton, UK), with Alan Bleasby (SEQNET, Daresbury) and in collaboration with Thure Etzold (EMBL-EBI).

Alan J. Bleasby is a Senior Scientific Officer at EMBL-EBI. He developed the early EMBOSS programming library (AJAX) at Daresbury Laboratory (Warrington, UK) where he was responsible for the SEQNET UK national bioinformatics service. He now works full-time on EMBOSS.

Jon C. Ison has been a developer of EMBOSS since 2000. He too is based at EMBL-EBI and helps coordinate EMBOSS with Peter and Alan. He is currently working on ontologies, data integration and application development.

The EMBOSS team thanks the organisations who have generously supported the project, including the Wellcome Trust, the Medical Research Council, the Biotechnology and Biological Sciences Research Council (BBSRC) and EMBL-EBI.

EMBOSS Developer's Guide: Bioinformatics Programming

Dr Jon Ison
EMBL European Bioinformatics Institute

Mr Peter Rice
EMBL European Bioinformatics Institute

Dr Alan Bleasby
EMBL European Bioinformatics Institute

CAMBRIDGE
UNIVERSITY PRESS

CAMBRIDGE UNIVERSITY PRESS
Cambridge, New York, Melbourne, Madrid, Cape Town,
Singapore, São Paulo, Delhi, Tokyo, Mexico City

Cambridge University Press
The Edinburgh Building, Cambridge CB2 8RU, UK

Published in the United States of America by Cambridge University Press, New York

www.cambridge.org
Information on this title: www.cambridge.org/9780521607247

First published 2011

Printed in the United Kingdom at the University Press, Cambridge

A catalogue record for this publication is available from the British Library

Library of Congress Cataloguing in Publication data
Ison, Jon, 1972–
EMBOSS developer's guide : bioinformatics programming / Jon Ison, Peter Rice, Alan Bleasby.
p. cm.
ISBN 978-0-521-60724-7 (pbk.)
1. Molecular biology – Computer programs. 2. EMBOSS. I. Rice, Peter, 1956–
II. Bleasby, Alan. III. Title.
QH324.2.I8285 2011
572.80285′536–dc22

2010051568

ISBN 978-0-521-60724-7 Paperback

Contents

Acknowledgements

EMBOSS acknowledgements

The EMBOSS developers would like to thank the funding bodies who have supported the project and the many people who have contributed. If you are omitted from the list below please accept our apologies and let us know. Special thanks to all our users who have given valuable suggestions, criticism and praise over the years.

Funding bodies

Biotechnology and Biological Sciences Research Council (BBSRC)

European Molecular Biology Laboratory – European Bioinformatics Institute (EMBL-EBI)

Medical Research Council (MRC)

Wellcome Trust

Contributors

Peter Rice and *Alan Bleasby* were the instigators of the project. Peter was the original project coordinator and, after working for Lion Bioscience and the Sanger Centre, moved to the EMBL-EBI. He runs the EBI Rice Group which houses the EMBOSS project.

Alan Bleasby wrote the original AJAX library while at Daresbury Laboratory and now coordinates the project with Peter Rice. After working at the MRC Human Genome Mapping Project Resource Centre, Alan moved to the EBI to work full-time on EMBOSS in 2005.

Jon Ison has contributed to all areas of the project since 2000 and now helps coordinate the project working full-time with Peter and Alan.

Peter, Jon and Alan wish to thank:

Michael Schuster for contributing code to interface with the ENSEMBL database.

Mahmut Uludag for various contributions.

Gary Williams for writing applications and documentation, providing user support and many other contributions.

Tim Carver for many contributions including the design, implementation and support of **Jemboss**.

Lisa Mullan for many contributions at EMBOSS meetings. She wrote training materials, organised training courses and gave talks.

Ian Longden for work on all aspects of EMBOSS whilst working full-time on a Wellcome Trust grant and in particular, for incorporation of graphics.

David Martin for new applications, bug testing and documentation.

Guy Bottu for wEMBOSS documentation, many valuable bug reports and suggestions.

Nicolas Joly for many bug reports and suggestions.

Damian Counsell for contributions at EMBOSS meetings and work on the documentation.

Hugh Morgan for work on the graphics libraries.

Claude Beazley for work on CORBA integration.

Ranjeeva Ranasinghe, *Waqas Awan* and *Michael Hurley* for work on the protein structure applications.

Bijay Jassal for contributions at EMBOSS meetings and SRS interface support.

Val Curwen for contributing applications, documentation and developing training materials.

Richard Bruskiewich for work on GFF and windows.

Catherine Letondal for integration with PISE.

Kathryn Beal for integration with SPIN.

Thomas Laurent for work on the SRS interface.

Rodrigo Lopez for work on the CpG island applications and for general encouragement.

Sinead O'Leary for contributing applications.

Thon de Boer for ACD documentation

Mark Faller for work on EMMA and other projects.

Andre Blavier for the original windows port.

Martin Sarachu, who has sadly passed away, and *Marc Colet* for **wEMBOSS**.

Luke McCarthy for **EMBOSS Explorer**.

Peter Cock for FASTQ and other sequence format validation.

David Bauer, *Thure Etzold*, *Martin Senger*, *Tom Oinn*, *Don Gilbert*, *Will Gilbert*, *Rodger Staden*, *Bill Pearson*, *Simon Kelley*, *Ewan Birney*, *Susan Jean Johns* and anyone we've missed for their contributions.

Janet Thornton, *Graham Cameron*, *Michael Ashburner*, *Martin Bishop* and *Richard Durbin* for their support.

EMBOSS *Developer's Guide* acknowledgements

The authors would like to thank the following individuals for contributions to the EMBOSS documentation or production of the *EMBOSS Developer's Guide*:

Katrina Halliday from Cambridge University Press for encouragement and suggestions.

Thon de Boer for past contributions to the ACD documentation.

Preface

Introduction to EMBOSS

The European Molecular Biology Open Software Suite (EMBOSS) is a high-quality, well-documented package of open source software tools for molecular biology. It includes over 200 applications for molecular sequence analysis and other common tasks in bioinformatics. It integrates the core applications with a range of popular third-party software packages under a consistent and powerful command line interface. The software has many useful features; for example, it automatically copes with data in a variety of formats and allows for transparent retrieval of sequence data from the web.

EMBOSS includes extensive C programming libraries with a clean and consistent API. There is much useful inbuilt functionality, for example the handling of the command line and common file formats, making it a powerful and convenient platform to develop and release bioinformatics programs. True to the spirit of open source, EMBOSS is free of charge to all and the code is licensed for use by everyone under the GNU General Public Licenses (GPL and LGPL). No one individual or institute 'owns' the code, or ever will. Under the terms of the licenses, it can be downloaded via the internet, copied, customised and passed on, so long as these same freedoms are preserved for others. Contributions are strongly encouraged!

EMBOSS is well established. It is used in demanding production environments reflecting the maturity of the code base. A major new stable version is released each year. For those who need the latest code, the current source code tree can be downloaded via CVS. There have been many thousands of downloads including site-wide installations by administrators across the world, catering for hundreds or even thousands of users. Many interfaces to EMBOSS are available including easy-to-use web interfaces and powerful workflow software, enabling applications to be combined into analysis pipelines.

Developing EMBOSS

Software development and maintenance under EMBOSS is made easy. EMBOSS has powerful inbuilt functionality that any application can make use of with little or no additional coding. This includes support for many simple and complex (biological) data types, common file formats and simple database configuration. Generic mechanisms are in place for sequence and sequence feature specification and for qualifiers controlling program behaviour. Depending on your particular requirements, this might save you a great deal of effort. Furthermore, when, for example, new input and output data formats are added to EMBOSS, your applications will automatically be able to use them; no application code needs to change. Well-defined processes are in place for key aspects such as quality assurance testing, installation, maintenance and support. General aspects are handled by the EMBOSS developers, leaving you to support the parts specific to your own software.

Your application will use the EMBOSS command line which is consistent across the applications. AJAX command definition (ACD) files define the command line interface and the datatype and permissible values for all application parameters. The processing and validation of the command line and user input is handled automatically at startup, before the application proper starts. For example, the production of a sensible prompt and reprompting for values that are out of range. There is clean separation of the user-interface handling aspects from the core functionality of the code: a single function call is used to process the command line and ACD file. The ACD syntax also makes the wrapping of third-party applications under EMBOSS simple.

EMBOSS includes extensive C programming libraries (AJAX and NUCLEUS) for low-level and higher-level tasks respectively. These provide a robust toolkit to develop new bioinformatics applications and workflows and to extend the core library functionality. The application programmer's interface (API) is comprehensive and consistent. A developer needn't know the internals to use the libraries: how to call the functions, the required input data and outputs are all clearly documented. All code is in ANSI standard C and there are defined standards for coding and documentation.

Memory management under EMBOSS is greatly simplified. Memory for all data defined in the ACD file is allocated automatically. Dynamic memory management for programming objects such as strings, sequences and arrays is handled automatically, saving the application programmer a good deal of effort.

The EMBOSS source code is well documented and is indexed as an SRS database to allow easy search and navigation. The documentation is generated automatically from structured comments in the code, which are validated to ensure correctness and consistency, for example, that functions and function parameters have standardised names. The *EMBOSS Developer's Guide* includes programming guides with example code for most library files illustrating their use. Mailing lists for discussions about development and for reporting bugs have a good response time.

AJAX

AJAX is the core low-level library used by all EMBOSS applications and provides a comprehensive set of basic objects and functions. It includes standard data structures for strings, sequences, features, structures, file handles, tables, lists, trees, dynamic arrays, etc. Algorithms for string handling, pattern-matching, sorting, iteration and very fast database indexing are included, and much more besides. AJAX is licensed under the GNU LGPL.

NUCLEUS

NUCLEUS includes higher-level code and algorithms, mostly for common molecular sequence analysis tasks. Functions for sequence comparisons, translation, codon usage and annotation are included. In comparison to the AJAX library and the EMBOSS applications, some parts of NUCLEUS are not as well developed or documented. In future code refactoring, the libraries will be consolidated and the documentation improved. NUCLEUS, like AJAX, is licensed under the GNU LGPL.

About the authors

Peter Rice

Peter Rice is a group leader at the European Bioinformatics Institute (EMBL-EBI, Hinxton, UK), a centre for research and services in bioinformatics and part of the European Molecular Biology Laboratory (EMBL). His group investigates and advises on the e-Science and Grid technology requirements of the EMBL-EBI, through application development plus participation in standards development. His group also houses the EMBOSS project. Peter instigated EMBOSS in 1996 when he was based at the Sanger Centre (Hinxton, UK), with Alan Bleasby (SEQNET, Daresbury) and in collaboration with Thure Etzold (EMBL-EBI). He left Sanger in 2000 to work for LION Bioscience, and in 2003 joined the EMBL-EBI.

Alan Bleasby

Alan Bleasby is a Senior Scientific Officer at EMBL-EBI. Alan developed the early EMBOSS programming library (AJAX) at Daresbury Laboratory (Warrington, UK) where he was responsible for the SEQNET UK national bioinformatics service. He moved to the UK Medical Research Council Human Genome Mapping Project Resource Centre (UK HGMP-RC) when the SEQNET and HGMP-RC services merged in early 1999, where he was Group Leader of the Proteomics Applications Group and coordinated EMBOSS. When the HGMP-RC closed in 2005, he moved to the EBI to work full-time on EMBOSS.

Jon Ison

Jon Ison is a Senior Scientific Officer at EMBL-EBI. He moved from the University of Leeds to the UK HGMP-RC in 1999 to work on the Collaborative Computing Project in Biosequence and Structure Analysis (CCP11), before taking the post of Software Specialist for the Proteomics Applications Group in 2000. He has been a lead contributor and developer of EMBOSS since then, moving in 2005 with Alan to the EMBL-EBI where he helps coordinate the project with Peter and Alan.

How to cite EMBOSS

Please cite EMBOSS where appropriate.

Rice P., Bleasby A. and Ison J. *The EMBOSS User's Guide*. Cambridge University Press.

Ison J., Rice P. and Bleasby A. *The EMBOSS Developer's Guide*. Cambridge University Press.

Bleasby A., Ison J. and Rice P. *The EMBOSS Administrator's Guide*. Cambridge University Press.

Rice P., Longden I. and Bleasby A. EMBOSS: The European Molecular Biology Open Software Suite. *Trends in Genetics* 2000 **16**(6):276–277.

The EMBOSS website. http://emboss.open-bio.org/

Conventions

EMBOSS Guide conventions

Command line sessions and commands

Examples of command line sessions or any other screen output look like this:

```
% seqret
Reads and writes (returns) sequences
Input (gapped) sequence(s): tembl:x65923
output sequence(s) [x65923.fasta]: stdout
>X65923 X65923.1 H.sapiens fau mRNA
ttcctctttctcgactccatcttcgcggtagctgggaccgccgttcagtcgccaatatgc
agctctttgtccgcgcccaggagctacacaccttcgaggtgaccggccaggaaacggtcg
cccagatcaaggctcatgtagcctcactggagggcattgccccggaagatcaagtcgtgc
tcctggcaggcgcgcccctggaggatgaggccactctgggccagtgcggggtggaggccc
tgactaccctggaagtagcaggccgcatgcttggaggtaaagttcatggttccctggccc
gtgctggaaaagtgagaggtcagactcctaaggtggccaaacaggagaagaagaagaaga
agacaggtcgggctaagcggcggatgcagtacaaccggcgctttgtcaacgttgtgccca
cctttggcaagaagaagggccccaatgccaactcttaagtcttttgtaattctggctttc
tctaataaaaaagccacttagttcagtcaaaaaaaaaa
```

In the above example the command, which is typed in on the command line, is given in **`this format`**.

Occasionally you will see values (typically for commands) given in ***`this format`***, indicating a value that must be replaced. For example *`Application Name`* would be replaced by the name of an application. In the command line below *5* and *25* would be replaced

- **`seqret -sbegin`** *5* **`-send`** *25*

Program listings and code

Program listings will look like this:

```
#include "emboss.h"
/* @prog helloworld
***********************************************************
**
** Prints "Hello, World!" to the screen.
**
*******************************************************************************/
int main(int argc, char **argv)
[
         embInit("helloworld", argc, argv);
         ajFmtPrint("Hello, World!\n");
         embExit();
         return 0;
]
```

Occasionally, code is referred to within the text and it is given in `this format`. For example:

- The `main()` function above includes the function `ajFmtPrint`.

Other conventions

Software packages, applications and interfaces, where they are mentioned, appear in **this format**. For example:

- **HMMER** is an EMBASSY package that wraps third-party applications.
- **seqret** is an example of an EMBOSS program.
- **Jemboss** is an example of an EMBOSS interface.

Options to programs appear like this:

- **`-help`**

The name of a database, or parts of a database, appear in *this format*. For example:

- *EMBL* is an example of a sequence database.

Specific EMBOSS system files or directories, environment variables and all other system-specific items appear in `this format`. For example:

- `.embossrc` and `emboss.default` are EMBOSS system files.
- `EMBOSS_ACDROOT` is an EMBOSS environment variable.
- A directory is an example of a system item, for example: `/home/auser/emboss/emboss/ajax/`

Special text blocks

Some special comments are offset from the main body of text.

Caution

A note of caution where there might be undesirable or unexpected consequences of some action.

Important

Comments or notes of special significance.

Note

Important but peripheral information to the main body of text.

Tip

Helpful hints, shortcuts, etc.

Welcome to the EMBOSS *Developer's Guide*

Summary

This manual was written with newcomers to EMBOSS in mind. You will benefit from at least a basic appreciation of molecular biology and some familiarity with UNIX and the C programming language. You should know how to open, use, save and close files using a text editor. It will also help if you've used the EMBOSS programs and are familiar with the command line (see the *EMBOSS User's Guide*).

All of the material in the EMBOSS Guides is available on the EMBOSS website:

http://emboss.open-bio.org

Chapter 1. Getting started

Things you should know about before you start programming, including details of the software license, installation instructions for the CVS (Developer's) release, a summary of developer documentation and instructions on how to navigate the library documentation, details of the mailing lists and how to contribute software.

Chapter 2. Your first EMBOSS application

A step-by-step guide to writing your first EMBOSS application. This is a gentle introduction to application development and illustrates the necessary common steps.

Chapter 3. Software development under EMBOSS

This chapter summarises the inbuilt functionality which any native application can make use of with little or no additional coding. Basic steps to development are explained: project management and planning, ACD file and C source code development, integration and compilation, debugging, quality assurance testing, documentation, distribution, maintenance, support and training. Standards and guidelines for C coding, code documentation and application documentation are included.

Chapter 4. ACD file development

Every EMBOSS and EMBASSY program has an ACD (AJAX command definition) file which describes the application, its options (parameters) and command line interface. The ACD file controls the behaviour of the application at the command line, particularly, all the user input operations. This chapter describes all aspects of ACD file development. It shows how to write application and data definitions, how to perform operations in ACD files and control prompting of the user. Finally, utilities for debugging and testing ACD files are described.

Chapter 5. C programming

An in-depth look at programming and memory management of EMBOSS objects (C data structures), which is essential to master and extend the libraries. The coverage includes basic C datatypes, EMBOSS objects (C data structures), a primer on programming with C pointers and memory management.

Chapter 6. Programming with AJAX

Guides to programming with AJAX that complement the online library documentation. They include the handling of ACD files, arrays, alignments, codon usage tables, exception messages, features, file input and output, graphics, lists, maths, matrices, menus, sequence patterns, phylogenetic data, application reports, sequences, translation, 'simple' ACD datatypes, strings, system code and tables.

Chapter 7. Quality assurance

Various quality assurance tests are performed on the EMBOSS code and documentation to maintain the quality and integrity of the package. You will learn how to write and run quality assurance tests for your applications.

Chapter 8. Application documentation standards

Software without documentation often has little value whereas good documentation can enhance the usefulness of software immensely. You will learn how to write comprehensive documentation for new applications conforming to the EMBOSS style.

Chapter 9. A complete application: seqret

A detailed look at all the steps needed to develop a complete application (**seqret**). Development of the ACD file and C source code illustrate the powerful inbuilt functionality that any EMBOSS program can make use of with little or no additional coding.

Chapter 10. Incorporating third-party applications

There are good reasons for incorporating third-party applications into EMBOSS. This chapter summarises the benefits of wrapping and porting third-party code. Guidelines for the development of application wrappers and porting of code are given. Things are illustrated for the **HMMERNEW** EMBASSY package which wraps Sean Eddy's **hmmer** applications for hidden Markov models. Approaches to porting applications are summarised.

Appendix A. ACD syntax reference

The AJAX command definition (ACD) language was designed for writing ACD files for EMBOSS applications. Here the entire syntax is described in a concise form.

Appendix B. C programming libraries

Reference information for the AJAX and NUCLEUS programming libraries.

Appendix C. C coding standards

To ensure consistency in the EMBOSS code, all C code should conform to a basic style. These standards, which mostly concern code layout, are described in detail.

Appendix D. Code documentation standards

The EMBOSS code documentation standards describe how to comprehensively document your code. Recommendations are provided for general comments and for structured comments documenting application and library source files, functions and datatypes.

Appendix E. Resources

A list of programmatic interfaces to EMBOSS.

Getting started

1

1.1 Licence information

EMBOSS is licensed for use by everyone under the GNU Software licences. The AJAX and NUCLEUS libraries are released under the GNU Lesser General Public Licence (LGPL). The applications are released under the GNU General Public Licence (GPL). If you plan to develop proprietary software using the libraries you should read the full licensing conditions:

```
GPL (http://www.gnu.org/copyleft/gpl.html)
LGPL (http://www.gnu.org/copyleft/lgpl.html)
```

The licences were chosen to provides maximum flexibility and encourage development. They give you freedom in software development, so long as you preserve those freedoms for others.

1.1.1 General Public Licence (GPL)

The GPL allows you to freely modify, copy and distribute the application source code so long as the source code of the derived work is licensed under GPL and made available. This means you can freely extend and improve the EMBOSS applications.

It is important to distinguish a 'derived work' from entirely new code. Typically, a derived work is either a direct modification of the original source code or is linked (statically or dynamically) to the original. For instance, software using a GPL licensed shared library would be a 'derived work', which is why the LGPL is used for the EMBOSS libraries to avoid this restriction. Parsing of ACD files does *not* constitute a 'derived work' and would not in itself be subject to the licence. Many developers have done this, for example, when writing an interface to EMBOSS.

Important

The rules of the GPL licence only take effect when you distribute the code. This means you can use and modify the code for your own internal use without obtaining permission or having to notify anyone.

1.1.2 Lesser General Public Licence (LGPL)

The LGPL requires that all changes to the libraries must be published openly (the source code must be made available). It is possible, however, to develop a completely new application that uses the libraries without the requirement of releasing the source. The libraries can be linked with proprietary software whereas had they been licensed under the GPL the libraries could only be used with free programs.

1.1.3 Licensing under EMBASSY

Developers who do not use the GPL or LGPL licence can still contribute. The EMBASSY collection can include packages that use AJAX or NUCLEUS but under their own licensing conditions. They will be bound by the LGPL of the AJAX and NUCLEUS libraries, but not necessarily by the full GPL.

1.2 Installation of CVS (Developer's) Release

Only very basic information on EMBOSS installation is included here. For complete instructions see the *EMBOSS Administrator's Guide*. The basic installation steps are:

1. Download the EMBOSS source code.
2. Configure EMBOSS. The stable release contains a 'configure' script for this whereas the developer's (CVS) version contains the files necessary to create this script.
3. Compile EMBOSS. This will generate the executable files from the application source code.
4. Set a `PATH` to the executables.
5. Test the installation.
6. Set up your databases.

Caution

To configure and compile EMBOSS the following GNU configuration tools must be installed on your system:

- **autoconf**
- **automake**
- **libtool**

Your OS will often provide these packages; if not then they can be downloaded (from ftp://ftp.gnu.org/gnu/) as source code and installed (in the order shown above and making sure that each newly installed package is being found, in preference to any older version, before installing the next).

You may hit other software dependencies if, for example, you are developing graphics software; see the *EMBOSS Administrator's Guide* for more information.

1.2.1 Downloading via CVS

To download via CVS you must have **cvs** installed on your system. It is there if an application is listed after typing:

```
which cvs
```

You must also use the secure shell SSH for the file transfer. Depending on your shell, type one of the following:

```
setenv CVS_RSH ssh
declare -x CVS_RSH=ssh
export CVS_RSH=ssh
```

To checkout the EMBOSS source code tree, go to the directory where you want the top-level EMBOSS directory created. For example to create EMBOSS in /home/auser/src/ you'd type:

```
cd /home/auser/src
```

To checkout EMBOSS you must first log on to the open-bio.org CVS server by typing:

```
cvs -d :pserver:cvs@cvs.open-bio.org:/home/repository/emboss login
```

You will be prompted for a password which is cvs. You are now logged on to the CVS server. Checkout (download) EMBOSS by typing:

```
cvs -d :pserver:cvs@cvs.open-bio.org:/home/repository/emboss checkout
emboss
```

This will take some time as it's downloading several megabytes of source code and data from the USA. Once the download is complete, terminate your CVS session with:

```
cvs -d :pserver:cvs@cvs.open-bio.org:/home/repository/emboss logout
```

Here is a typical CVS session:

```
% declare -x CVS_RSH=ssh
% cvs -d :pserver:cvs@cvs.open-bio.org:/home/repository/emboss login
Logging in to :pserver:cvs@cvs.open-bio.org:2401/home/repository/emboss
```

```
CVS password:
% cvs -d :pserver:cvs@cvs.open-bio.org:/home/repository/emboss checkout emboss
cvs checkout: Updating emboss
cvs checkout: Updating emboss/emboss
U emboss/emboss/AUTHORS
U emboss/emboss/COMPAT
U emboss/emboss/COPYING
U emboss/emboss/ChangeLog
.
... screen output truncated
.
%
% cvs -d :pserver:cvs@cvs.open-bio.org:/home/repository/emboss logout
Logging out of :pserver:cvs@cvs.open-bio.org:2401/home/repository/emboss
%
% ls
emboss
```

1.2.2 Keeping up to date with CVS

Once you've downloaded the CVS version of EMBOSS, you should keep up to date with the latest changes. To do this, move to the directory which you wish to update (or the root emboss directory to update everything) and type:

```
cvs -d :pserver:cvs@cvs.open-bio.org:/home/repository/emboss update
```

Bear in mind that if you have edited any files then those edits will be merged with the version downloaded via CVS. If you wish to conserve your edits in their original form, you should copy the edited files to a safe place first.

1.2.3 Configuration

You must first build the script and other files used to configure the package. From the second-level EMBOSS directory (e.g. `/home/auser/emboss/emboss`) type:

```
aclocal -I m4
autoconf
automake -a
```

aclocal is part of the **automake** package. These commands build the "configure" script (**./configure**) from the file `configure.in` (in the EMBOSS distribution) and various other files needed for installation. Specifically:

- **aclocal** creates `aclocal.m4` containing m4 macros used by the **auto*** tools.
- **autoconf** reads `configure.in` and creates the "configure" script.
- **automake** reads `Makefile.am` and creates `Makefile.in`

Tip

If you experience any errors at that stage it is possible that your GNU configuration tools are out of date, if so then update them. It is also possible that your version of **libtool** is more recent than that used in the current EMBOSS CVS; you can check that by looking at the version number in the `ltmain.sh` file at the top source level of the checked-out tree. If your installed version of **libtool** is newer then you may have to type **`autoreconf -fi`** and then retype the previous GNU configuration tool commands. **`autoreconf`** is part of the **autoconf** package.

Running the resulting **`./configure`** will

- Check whether your system has the necessary functionality and libraries to compile EMBOSS.
- Read `Makefile.in` and generate platform-specific Makefiles (used later).
- Configure your system. For example, the installation path and various system variables are set and files are flagged to compile (or not).

Note

`./configure` is controlled by command line arguments and by environment variables. Command line arguments are generally used to switch on features **autoconf** was unable to detect. Environment variables are generally used to set build information such as compiler options.

If you intend to compile using **`make install`** (see below) you must specify an installation area for the executables and supporting files. It is good practice to specify these even if you intend to compile with a plain **`make`**. To do this, type:

```
./configure --prefix=/home/auser/emboss
```

Further, if you are using the **gcc** compiler it's a good idea to turn warnings on. You can do this while setting the installation areas:

```
./configure --prefix=/home/auser/emboss --enable-warnings
```

You can also turn on more strict developer warnings by typing:

```
./configure --prefix=/home/auser/emboss --enable-warnings --enable-
devwarnings
```

Important

Note that the warnings switches, such as `--enable-warnings`, are for **gcc** compilers only. There is a further configuration switch, `--enable-devextrawarnings`, which turns on some rather pedantic warnings which are nonetheless useful in some rare circumstances. As that switch can produce compilation noise it is not recommended to specify it in general use.

Caution

To support the Portable Network Graphics (PNG) format, EMBOSS requires the **libgd**, **libpng** and **libz** libraries. On many Linux systems, most support libraries are installed under the directories `/usr/` and `/lib`. For example, `/lib`, `/usr/lib`, `/usr/X11R6/lib`, etc. Whereas Linux distributions include RPMs for these libraries, other operating systems do not.

If you are installing these libraries and include files in somewhere other than `/usr` then you must specify their location when configuring. Assuming you have installed them under `/usr/local` (e.g. `/usr/local/lib` and `/usr/local/include`) you would add the following switch to the configuration command line:

```
--with-pngdriver=/usr/local
```

For more information on the configurability of the build type:

```
./configure --help
```

1.2.4 Compilation

To compile EMBOSS, type:

```
make
```

This will, by using the `Makefiles`, compile all the source files into executable binaries within your chosen checked-out location e.g.:

```
/home/auser/emboss/emboss/emboss
```

Alternatively, compiling with:

```
make install
```

will install the binaries and supporting files into `bin`, `lib` and `share` subdirectories of the directory you specified using **`--prefix`** on the configure line. In the example above, this is the top-level EMBOSS directory level from the CVS checkout, e.g.:

```
/home/auser/emboss
```

Had you not specified **`--prefix=/home/auser/emboss`** they'd be installed to `/usr/local` by default (that isn't recommended).

> **Tip**
>
> If you want EMBOSS installed in somewhere other than **`/usr/local`** then use the **`--prefix`** option of GNU configure to specify the EMBOSS installation directory. This is the recommended method of installation as EMBOSS has hundreds of files which might otherwise obscure other software, particularly under **`/usr/local/bin`**.

To compile and reinstall EMBOSS on subsequent occasions, use the following commands:

```
make clean
./configure
make
make install
```

Though extremely rare, there are circumstances where you may have to type **`rm config.cache`** (on very old installations) or **`rm -rf autom4te.cache`** before doing the above. Those occasions generally happen when trying to update a rather old existing version of EMBOSS.

1.2.4.1 Static and dynamic compilation

> **Note**
>
> A static library (statically linked library) is a library called by an application in which links to external functions and variables are resolved at compile-time by a linker. Static libraries are either compiled into object files and merged with the application during building/linking into a single executable, or are loaded at run-time and are accessible to the executable.
>
> A dynamic library (dynamically linked library) in contrast implements dynamic linking, where libraries remain in separate files on disk and are not copied into the executable or library at compile-time. Instead a loader is used to access the library at load or run-time.
>
> Executables created using static libraries are larger than executables linked to dynamic libraries, because static libraries include the code for the library function(s) within the executable. Static libraries, however, can be useful for debugging purposes and sometimes result in faster executables than when using dynamic libraries.

The above commands would create a dynamically linked EMBOSS. To create a statically linked EMBOSS you would use:

```
./configure --disable-shared
```

1.2.4.1.1 Static compilation via script

There is a script available to compile EMBOSS. If you require a statically linked EMBOSS, simply type:

```
source make-static-developers
```

or

```
./make-static-developers
```

depending on your shell.

This invokes the following commands:

```
aclocal -I m4
autoconf
automake -a
./configure --enable-warnings --disable-shared --enable-debug
make
```

1.2.5 Setting your PATH

Note

Your **PATH** is an environment variable that defines a list of directories that the operating system searches to find executable applications in response to a Unix command. For example, when you type a command such as **cd** or **ls** the invoked application is (typically) **/bin/cd** or **/bin/ls**. Most Unix installations will have put the directory **/bin** in your **PATH** automatically. For EMBOSS, you will likely have to manually set the **PATH** to point to the executables.

You must set the path to the executable applications. Assuming EMBOSS was installed in your home directory and you are using a `csh` style shell then type the following commands (replace the path as required):

```
set path=(/home/auser/emboss/emboss/emboss $path)
rehash
```

Or if you are using an `sh` style shell then type the following commands:

```
export PATH=/home/auser/emboss/emboss/emboss:$PATH
```

If, however, you installed using **`make install`** rather than just a **`make`** then the commands will be different (you must set the `PATH` to where you installed the executables). For a `csh` style shell:

```
set path=(/home/auser/emboss/bin $path)
rehash
```

For a `sh` style shell:

```
export PATH=/home/auser/emboss/bin:$PATH
```

1.2.6 Testing all is well

To test all is well with your installation you can run:

```
embossversion
```

and check that the EMBOSS version number is reported.

You could also try using the **seqret** application to retrieve some sequences from the test databases that come bundled with the EMBOSS distribution. Before you can do this however you need to set up the databases for use with EMBOSS.

1.2.7 Database setup

Note

Database setup is covered in the *EMBOSS Administrator's Guide*; only the bare essentials are covered below.

The EMBOSS distribution comes bundled with some test databases. They are, for example, located in:

```
/home/auser/emboss/emboss/test
```

Any database you want to use must be defined in one of the files:

- `.embossrc` (in your home directory)
- `emboss.default` (in the top-level emboss directory)

These files are used to configure EMBOSS. `.embossrc` is for personal configuration whereas `emboss.default` is used for site-wide configuration. A template file is included in the CVS releases (`.../emboss/emboss/emboss.default.template`). For now, create a file called `.embossrc` in your home directory with the following contents (you'll need to change the paths to the test directories):

```
DB embl
[
type: N
method: direct
format: embl
dir: /home/auser/workspace/emboss/emboss/test/embl/
file: *.dat
comment: "EMBL sequences"
]

DB swissprot
[
type: P
method: direct
format: swiss
dir: /home/auser/workspace/emboss/emboss/test/swiss/
file: seq.dat
comment: "Swissprot sequences"
]
```

Having set up your databases, issue the following commands (and accept the default values at the prompts):

```
seqret embl:x65923
more x65923.fasta
seqret swissprot:UBR5_RAT
more UBR5_RAT
```

If you get output similar to the following then you can rest assured that the installation went well.

```
% seqret embl:x65923
Reads and writes (returns) sequences
Output sequence [x65923.fasta]:
% more x65923.fasta
>X65923 X65923.1 H.sapiens fau mRNA
ttcctctttctcgactccatcttcgcggtagctgggaccgccgttcagtcgccaatatgc
```

```
agctctttgtccgcgcccaggagctacacaccttcgaggtgaccggccaggaaacggtcg
cccagatcaaggctcatgtagcctcactggagggcattgccccggaagatcaagtcgtgc
tcctggcaggcgcgcccctggaggatgaggccactctgggccagtgcggggtggaggccc
tgactaccctggaagtagcaggccgcatgcttggaggtaaagttcatggttccctggccc
gtgctggaaaagtgagaggtcagactcctaaggtggccaaacaggagaagaagaagaaga
agacaggtcgggctaagcggcggatgcagtacaaccggcgctttgtcaacgttgtgccca
cctttggcaagaagaagggccccaatgccaactcttaagtcttttgtaattctggctttc
tctaataaaaaagccacttagttcagtcaaaaaaaaaa

%

% seqret swissprot:UBR5_RAT
Reads and writes (returns) sequences
Output sequence [ubr5_rat.fasta]:

% more ubr5_rat.fasta
>UBR5_RAT Q62671 Q62671 E3 ubiquitin-protein ligase UBR5 (6.3.2.-)
MMSARGDFLNYALSLMRSHNDEHSDVLPVLDVCSLKHVAYVFQALIYWIKAMNQQTTLDT
PQLERKRTRELLELGIDNEDSEHENDDDTSQSATLNDKDDESLPAETGQNHPFFRRSDSM
TFLGCIPPNPFEVPLAEAIPLADQPHLLQPNARKEDLFGRPSQGLYSSSAGSGKCLVEVT
MDRNCLEVLPTKMSYAANLKNVMNMQNRQKKAGEDQSMLAEEADSSKPGPSAHDVAAQLK
SSLLAEIGLTESEGPPLTSFRPQCSFMGMVISHDMLLGRWRLSLELFGRVFMEDVGAEPG
SILTELGGFEVKESKFRREMEKLRNQQSRDLSLEVDRDRDLLIQQTMRQLNNHFGRRCAT
TPMAVHRVKVTFKDEPGEGSGVARSFYTAIAQAFLSNEKLPNLDCIQNANKGTHTSLMQR
LRNRGERDREREREREMRRSSGLRAGSRRDRDRDFRRQLSIDTRPFRPASEGNPSDDPDP
LPAHRQALGERLYPRVQAMQPAFASKITGMLLELSPAQLLLLLASEDSLRARVEEAMELI
VAHGRENGADSILDLGLLDSSEKVQENRKRHGSSRSVVDMDLDDTDDGDDNAPLFYQPGK
RGFYTPRPGKNTEARLNCFRNIGRILGLCLLQNELCPITLNRHVIKVLLGRKVNWHDFAF
FDPVMYESLRQLILASQSSDADAVFSAMDLAFAVDLCKEEGGGQVELIPNGVNIPVTPQN
VYEYVRKYAEHRMLVVAEQPLHAMRKGLLDVLPKNSLEDLTAEDFRLLVNGCGEVNVQML
ISFTSFNDESGENAEKLLQFKRWFWSIVERMSMTERQDLVYFWTSSPSLPASEEGFQPMP
SITIRPPDDQHLPTANTCISRLYVPLYSSKQILKQKLLLAIKTKNFGFV
```

1.2.8 EMBASSY packages

The example below is for installing the CVS (developer's) version of an EMBASSY package.

1. Go to the package directory:

```
cd embassy/PackageName
```

2. Make the configuration file:

```
aclocal -I m4
autoconf
automake -a
```

3. Configure and compile the package. Configuration should be done using the same options you used for the main package. So, if you used:

```
./configure --with-pngdriver=/opt/lib --prefix=/home/auser/emboss
```

for the main package then use the same command for the EMBASSY package. Then:

```
make
make install
```

1.3 Developer documentation

EMBOSS is richly documented. Depending on your experience and requirements you will want to approach it in different ways:

- Application documentation
- Library documentation
- The source code
- Navigate the source code using SRS
- Demonstration applications (for each library file)
- Programming guides on key topics
- AJAX command definition (ACD) documentation
- C coding standards and guidelines
- Quality assurance guidelines
- Code and application documentation standards
- EMBOSS Software Development Course.

1.3.1 Application documentation

Every EMBOSS application is well documented:

CVS (Developer's) Release documentation	http://emboss.open-bio.org/rel/dev/apps
Stable Release 6 documentation	http://emboss.open-bio.org/rel/rel6/apps

1.3.2 Library documentation

AJAX and NUCLEUS contain hundreds of library calls and this can be daunting at first. Documentation is available online for the CVS (Developer's) release and major versions of the Stable Release. The documentation is derived from structured comments in the source code itself (see Appendix D, *Code documentation standards*).

1.3.2.1 AJAX library documentation

AJAX is the core library used by all EMBOSS applications. It covers standard data structures and algorithms:

CVS (Developer's) Release documentation	http://emboss.open-bio.org/rel/dev/libs
Stable Release 6 Documentation	http://emboss.open-bio.org/rel/rel6/libs

1.3.2.2 NUCLEUS library documentation

NUCLEUS provides higher-level functions specific to molecular sequence analysis:

CVS (Developer's) Release Documentation	http://emboss.open-bio.org/rel/dev/libs/
Stable Release 6 Documentation	http://emboss.open-bio.org/rel/rel6/libs

1.3.3 Navigating the libraries

It is easy to navigate the library documentation

- *From the EMBOSS homepage (http://emboss.open-bio.org/), click on "AJAX" or "NUCLEUS".*

This will bring up a table for the AJAX or NUCLEUS library.

Rows in the AJAX or NUCLEUS library tables correspond to an individual library file, e.g. for alignments, array handling, assert functions and so on. There are columns in the table for:

Library documentation	Links to the library file documentation (see below) which references all the available objects (C data structures) and functions.
Short description	A short description of the library file.
Programming guide	Links to a detailed programming guide and usage notes for the library file, if available (see Section 6.2, 'Programming guides').
Example application (C source code)	Links to the C source code for an example application, that illustrates the use of the library, if available (see Section 6.1, 'Demonstration applications').
Example application (ACD code)	Links to the ACD code for an example application (see Section 6.1, 'Demonstration applications').

1.3.3.1 Library file documentation

- *Find "String manipulation" in the table and follow the link under "Library documentation".*

This will bring up the documentation available for string handling (`ajstr.c/h` library files).

The library file documentation includes the following sections:

Summary	A short description of the library file.
Description	A longer description of the library file.
Data structures	Name, short description and links to further information for each object (C data structure).
Description of function categories	Formal description of each function category, organised by object type.
Functions (organised by object and category)	Name, short description and links to formal description for each function, organised by object type and function category.
Functions (alphabetic listing)	Name, short description and link to a formal description for each function in the library, organised alphabetically.

1.3.3.2 Function documentation

The function documentation includes all the critical information.

The sections in the file are as follows:

Function synopsis	The function name, short description and the EMBOSS version number when it was first made available.
Function prototype	Function prototype in standard C form.
Function parameters table	Function parameters are labelled as follows: • `INPUT` parameters are read by the function. • `OUTPUT` parameters are written by it. • `UPDATE` parameters may be read and written.
Returns	Description of return value(s).
Description	Full description of function.
Source code	C source code of function.
Usage example	A typical use of the function, generated automatically.
Notes	Peripheral documentation such as usage notes.
Warnings	Cautionary usage advice, known bugs, etc.
Exceptions	Exception and other messages the function might generate.
Dependencies	External entities the function is dependent upon, for example, environment variables and files.
See also	Links to functions in the same category.

1.3.3.3 Object (C data structure) documentation

The objects are comprehensively described.

Structure synopsis	The C data structure name, short description and EMBOSS version number when it was first made available.
Synopsis	Object synopsis (datatypes and variable names).
Data definitions	Definitions of datatypes for the object.
Description	Full description of object.
Elements	Description of elements in the data structure.
Functions	Functions that operate on the object.
Source code	C source code of the data structure.
Usage example	Typical usage example, generated automatically.
Notes	Peripheral documentation such as usage notes.
Warnings	Cautionary usage advice, known bugs, etc.
See Also	Links to structures in the same library file.

1.3.4 The source code

The source code is a vital reference. A simple method for searching the library or application code is to use the UNIX command **grep** to search the C source files for keywords. This is a convenient and direct way to find objects or functions quickly.

If you are unsure how to do a particular task, for example reading in a data file, then you should quickly be able to find a program that does something similar to what you need. Bear in mind there are many ways to solve a problem and the example you find might not necessarily be the best way.

There are two files (the C source code and the ACD file) to look at for each application. They're kept in the directories:

```
/home/auser/emboss/emboss/emboss/c
/home/auser/emboss/emboss/emboss/acd/
```

1.3.4.1 Navigating the source code using SRS

The source code may be inspected directly and navigated using SRS. The library source code is indexed in SRS at the EBI SRS Server:

```
http://srs.ebi.ac.uk/
```

There are separate SRS databases for objects (C data structures) and functions:

EFUNC: functions in CVS (Developer's) Release	http://srs.ebi.ac.uk/srs7bin/cgi-bin/wgetz?-newId+-page +LibInfo+-lib+EFUNC
EDATA: Objects in CVS (Developer's) Release	http://srs.ebi.ac.uk/srs7bin/cgi-bin/wgetz?-newId+-page +LibInfo+-lib+EDATA
EFUNCREL: functions in Stable Release 6.0.0	http://srs.ebi.ac.uk/srs7bin/cgi-bin/wgetz?-newId+-page +LibInfo+-lib+EFUNCREL
EDATAREL: objects in Stable Release 6.0.0	http://srs.ebi.ac.uk/srs7bin/cgi-bin/wgetz?-newId+-page +LibInfo+-lib+EDATAREL

1.3.4.1.1 Searching EDATA

From http://www.ebi.ac.uk/srs/:

1. *Click on the `Library Page` tab.*
2. *Expand the `Other databases` section by clicking on the + to the left of `Other databases`.*

You will see EDATA, EDATAREL, EFUNC and EFUNCREL listed.

3. *Highlight the check-box next to `EMBOSS Data Structures (CVS)` and then click on the `Query Form` tab.*
4. *Change one of the `AllText` options to `ID` and type a `*` character in its associated box, then click on `Search`.*

You will see a list of every available object. Here is a more specific search:

1. *Return to the query form and replace the `*` by `ajpstr` (the AJAX string object).*
2. *Click on `Search`.*

You'll see that two entries are returned, `AjPStr` and `AjPStrTok`. Click on the link for `AjPStr`.

The documentation here is in several sections. The first three give the name, description and 'aliases' of the object:

- `AjSStr` is the name of the string object.
- `AjPStr` is the datatype for the object pointer.
- `AjPPStr` is the datatype for a pointer to the object pointer.
- `AjOStr` is the datatype for the object proper.

Note

Meaning of `AjSStr`, `AjOStr`, `AjPStr`

`AjSStr` is the formal name of the string object, **`AjOStr`** is the datatype name for the object whereas **`AjPStr`** is the datatype name for the object pointer. In practice **`AjOStr`** (and all other **`AjO*`** datatypes) are never used in EMBOSS. Instead, memory for an instance of the object in memory is dynamically allocated to the pointer **`AjPStr`** (see Section 5.5, 'Programming with objects'). For this reason, **`AjPStr`** is given after "Name" in SRS and for the sake of brevity, "object" is often used to refer to an **`AjPStr`** (for example) when what is really meant is "object pointer". The use of objects and pointers is covered in depth elsewhere (Section 5.5, 'Programming with objects').

EDATA and EDATAREL include links to functions that use each object, which is handy if you want to know what you can do with an object. The functions in EFUNC and EFUNCREL are organised into categories of related functionality that correspond to sections in the C source file (see Appendix D, *Code documentation standards* and below).

After the *`Alias(es)`* section you'll see several more blocks functions within a category. The categories you see will depend upon the library file, but might include:

- Iterators – iteration, e.g. over individual characters in a string.
- Constructors – create new instances of an object (allocate memory).
- Destructors – destroy instances of an object (free memory).
- Assignments – initialise an object, replace contents if necessary.
- Modifiers – change or replace the contents of an object.
- Operators – use, but do not change, the contents of an object.
- Outputs – write the contents of an object to an external file.
- Casts – convert an object into an object or data of another type.

At the bottom of the page you'll see the following section:

- *`Attributes`* lists the elements of the C data structure.
- *`Body`* gives the C code for the object definition.

1.3.4.1.2 Searching EFUNC

The EFUNC database can be searched directly. This is useful if you know the kind of function you want but don't know the name. The function names and names and order of function parameters have been standardised (see Section D.1, 'Code documentation standards') to be intuitive and consistent.

Let's assume you want to search for a function that appends one string to another:

1. *Return to the SRS databases page, uncheck the EDATA database and check the check-box for the EFUNC database.*
2. *Select the query form.*
3. It's often best to limit the search to the description field so as to retrieve more specific matches. So: *change `AllText` to `Description`.*
4. *Type `append & string` into the associated box, then click on `Search`.*

A list of functions will appear. You can only use those functions that begin with `aj` or `emb`, public functions in the AJAX and NUCLEUS libraries respectively. The others are hidden functions, accessed by the internals of EMBOSS and not for general use.

The functions you need are those in the `ajStrAppend*` family. You'll see that some of the functions accept other string objects, character strings or just single characters.

This search method is of course limited by the vocabulary used in the function descriptions. For instance, the term "append" is used rather than "catenate". You can see this for yourself by repeating the above search using `catenate & string`.

To show the advantage of limiting the search:

- *Change the* `Description` *field back to* `AllText` *and repeat the* `string & append` *query.*

You'll see that there is a significant amount of noise in the results list.

Of course you can use SRS if you know the name of a function and need to examine the source code.

1. *Return to the EFUNC page and change* `AllText` *to* `ID`.
2. *Now use* `ajstrappend` *as the search term. Perform the search and then click on* `EFUNC: ajStrAppendS`.

You should see the source code for `ajStrAppendS` on screen. Again, the output is in several sections. The name of the function indicates the source library file in which it is to be found; the `str` of `ajStrAppendS` indicates the `ajstr` library. The description field gives the text you search with a `Description` search.

The most useful information for a user of the library are the `Input`, `Returns` and `Prototype` fields.

The `Input` field shows that this function takes the address of a string object pointer as its first parameter and a string object pointer as its second parameter. The `Returns` field shows the return value of the function (`AjBool`, a boolean value). All this information is given at a glance in the `Prototype` field for the function.

Below the prototype is the body of the function. This patently contains the source code of the function. C language reserved words are highlighted in red. The source code is marked up with any calls to other EMBOSS functions. Unhighlighted function calls are standard C library calls. You could click on, for example, `ajFatal` and see the code for that function.

Clicking on the red arrow on the prototype line will show all the EMBOSS functions that use this particular function. Clicking on the blue arrow will show all the EMBOSS functions that are called by this particular function.

1.3.5 Demonstration applications

EMBOSS includes, for certain AJAX and NUCLEUS library files, an application which illustrates the correct usage of the common functions. Currently, these 'demonstration applications' are kept in the **myembossdemo** package and have the prefix "`demo`". For example the following files illustrates the use of the string library:

```
/home/auser/emboss/emboss/embassy/myembossdemo/emboss_src/demostring.c
/home/auser/emboss/emboss/embassy/myembossdemo/emboss_acd/demostring.acd
```

For information on compiling and using these applications see Section 3.1, 'EMBOSS programming'.

1.3.6 Programming guides

Programming guides (Section 6.2, 'Programming guides') are available for most AJAX sub-libraries. These summarise the available C data structures and functions and give examples of their use. They are very useful if you want to learn all about a particular area of EMBOSS programming.

1.3.7 AJAX command definition (ACD) developer's guide and syntax

Every EMBOSS application has an AJAX command definition (ACD) file which contains a complete definition of the command line interface and defines all the information the application needs to run.

ACD files are written in the ACD syntax (Appendix A, *ACD syntax reference*) which defines a set of datatypes available to the applications, attributes for qualifying the datatypes, and much more besides. To develop new applications you will need to master ACD programming (see Chapter 5, *C programming*).

1.3.8 C coding standards and guidelines

To ensure consistency, all code should conform to a basic style and standards. You should familiarise yourself with these C coding standards (Appendix C, *C coding standards*), most of which concern the layout of code.

1.3.9 Quality assurance guidelines

Various quality assurance tests are performed on the code and documentation to maintain the quality and integrity of the package. This includes application test runs, compilation and memory leak tests and validation of the structured documentation used for objects and functions.

All code should be thoroughly tested and new library code should be documented to the EMBOSS standard (see below) so that checks can be performed. If you develop a new application you should also provide test data for it (see Section 7.2, 'Application quality assurance').

1.3.10 Code and application documentation standards

Software without documentation often has little value whereas good documentation can enhance the usefulness of software immensely. All contributed code should be adequately documented. End-user documentation is also required for any new applications. To ensure consistency, the documentation should conform to a basic style and standards that are defined for the code (Appendix D, *Code documentation standards*) and the applications (Chapter 8, *Application documentation standards*).

1.3.11 EMBOSS software development course

Hands-on courses in 'Bioinformatics software development using EMBOSS' provide a good introduction to programming in EMBOSS, including all the steps to writing a basic bioinformatics application using the EMBOSS programming libraries. If you would like to attend or host a course then get in touch with the EMBOSS developers (emboss-bug@emboss.open-bio.org).

1.4 Project mailing lists

There are three EMBOSS mailing lists which you can join. These provide a forum for discussions about EMBOSS and its future development and are a valuable source of help.

1.4.1 User mailing list

```
emboss@emboss.open-bio.org
```

This is an open list (anyone can join) for general announcements and discussions by end-users. To subscribe to this list, please visit:

```
http://emboss.open-bio.org/mailman/listinfo/emboss
```

1.4.2 Developer mailing list

```
emboss-dev@emboss.open-bio.org
```

This is a closed list for discussions by developers using EMBOSS. To subscribe to this list, please visit:

```
http://emboss.open-bio.org/mailman/listinfo/emboss-dev
```

If you wish to contribute to EMBOSS, please send the code and documentation to:

```
emboss-submit@emboss.open-bio.org
```

1.4.3 Announcements mailing list

```
emboss-announce@emboss.open-bio.org
```

This is a list for major announcements, mostly about new releases and updates. It has far fewer postings than the other groups and is suitable for system administrators who want to keep the software updated but who have little interest in other discussions. To subscribe to this list, please visit:

```
http://emboss.open-bio.org/mailman/listinfo/emboss-announce
```

1.4.4 Mail archives

Archives of previous postings to the discussion mailing lists are available.

1.4.4.1 User mailing list archive

```
http://emboss.open-bio.org/pipermail/emboss/
```

This is the archive of emboss@emboss.open-bio.org.

1.4.4.2 Developer mailing list archive

```
http://emboss.open-bio.org/pipermail/emboss-dev/
```

This is the archive of emboss-dev@emboss.open-bio.org.

1.5 Contributing software to EMBOSS

1.5.1 Please submit your code!

It's likely your in-house developments of EMBOSS will be valuable to others. *Please* submit your code. You can discuss what you've done with the EMBOSS developers:

```
emboss-bug@emboss.open-bio.org.
```

Otherwise, if you have software that's ready to submit, please send the code and documentation to:

```
emboss-submit@emboss.open-bio.org
```

If your additions are of general utility they will be added to the package. The EMBOSS developers will assist you in making any necessary changes to the code, or do this for you.

1.5.2 Types of code submission

The long-term success of EMBOSS relies on code submissions from external developers. Submissions might include:

- Extensions to the existing AJAX or NUCLEUS libraries
- Extensions to the existing EMBOSS or EMBASSY applications
- Code for new libraries or applications
- Code in response to an EMBOSS feature request
- Code fixes in response to EMBOSS bug reports
- Scripts and other software that call the EMBOSS applications
- Something else.

1.5.3 Known required developments

There are many areas that are known to require software development.

```
http://emboss.open-bio.org/wiki/New_Applications
```

The list includes proposals for new applications, requests for new features and other developments that have been submitted by EMBOSS users. Work in these areas will certainly be valuable, so if any of the known requirements interests you, please volunteer to work on it.

If you've a coding project that isn't a known requirement, then your code is still very welcome. Please contact the mailing list with your idea.

1.5.3.1 EMBOSS feature requests and bug reports

A very valuable way to contribute is to write code in response to EMBOSS feature requests and bug reports posted by the EMBOSS users. These are logged on SourceForge:

https://sourceforge.net/projects/emboss/support

Your first EMBOSS application

2

Most programming books begin with a program to write a simple phrase to the screen, and that phrase is invariably “Hello World!”. This section will lead you through the process of writing **helloworld** under EMBOSS. You’ll cover the basic elements common to every EMBOSS program and introduce the steps used to create applications.

Some of the examples below use **myemboss** (Section 3.2, ‘Integration and compilation’), so you must first install and configure this package. **myemboss** is a special EMBASSY package and is found in the normal EMBASSY application directory, e.g.

```
/home/auser/emboss/emboss/embassy/myemboss
```

2.1 helloworld in C

Here is the source code of a C program that prints “Hello World!” to the screen:

```
#include <stdio.h>
int main(void)
{
  printf("Hello World!\n");
  return 0;
}
```

The first line is a preprocessor directive telling the compiler to include the header file `stdio.h`. The angle brackets around the name of the file to be included tell the preprocessor to look in the default location for `stdio.h`. On a Linux system this default location is typically the directory `/usr/include`.

The program consists of a single function (`main`) which has no parameters and an integer return type, in this case it returns `0` to the operating system after printing “Hello World!” to the screen.

If you save the source code to a file called `helloworld.c`, then to get an executable (runnable) version of the program, you have to compile the source by typing one of the following commands.

Using **gcc**, an ANSI C compiler:

```
gcc helloworld.c -o helloworld
```

Using **cc**, the default C compiler:

```
cc helloworld.c -o helloworld
```

If there are no compilation errors you will end up with an executable file called `helloworld`. If you omitted **`-o helloworld`** the executable would be called `a.out`. To run your program you simply type `helloworld` at the UNIX prompt:

```
% helloworld
Hello World!
%
```

People running a `csh` shell may need to type **`rehash`** first.

2.2 helloworld in EMBOSS

You probably write your programs to a standard pattern: you *write* a file of source code, *compile* the source code, then *debug* the resulting executable program. You finally *run* your debugged binary.

There are a couple of additional steps when writing an EMBOSS program. The key difference is that in addition to writing the source code, you must also *write an ACD file* for your new application. An ACD file contains a description of the command line interface. It specifies exactly what input values are required and how to verify them and what is written. It controls the behaviour of the application on the command line, in particular the user input operations.

All of the required parameters are prompted for before the application proper begins. The input values are read and held in memory, files are opened as required and so forth, so that all the parameters are available when the application proper starts. An EMBOSS application cannot ask the user for more information after several hours of processing!

It's good practice to write your ACD file *before* the source code because this forces you to think closely about the application inputs and outputs and exactly what's required from the user. You should then test the ACD file by using an EMBOSS application called **acdc** (see below). In addition, you must *integrate* your application into EMBOSS or EMBASSY.

So, the basic steps to writing your first EMBOSS application are:

1. Write ACD file
2. Test ACD file
3. Write source code
4. Integrate application into EMBOSS or EMBASSY (e.g. **myemboss**)
5. Compile.

Additionally, there are several other considerations with any software project:

- Planning and design – think about the problem and design the software
- Debugging – get it to run without crashing or obvious problems
- Testing – ensure it works correctly under all conditions
- Documentation – describe how it works
- Installation and distribution
- Maintenance, support and training.

2.2.1 Planning and design

You should think carefully about the task at hand and plan or design your software before coding. Think about the inputs and outputs and the major logical steps in the source code. **helloworld** is a very simple program which must:

1. Read and process the ACD file
2. Print “Hello World!” to the screen
3. Exit cleanly

Software planning is covered in Section 3.1, ‘EMBOSS programming’.

2.2.2 Writing the ACD file

The input and output of **helloworld** is trivial. All the program has to do is print “Hello World!” to the screen and so nothing is required from the user. It’s no surprise then that the ACD file is pretty sparse. As a minimum though, all ACD files must contain an *application definition* with a single *documentation:* attribute:

```
application: helloworld
[
    documentation: "Prints 'Hello World!' to the screen."
]
```

This definition must come first in the file. The definition consists of the `application:` token, followed by the application name and a block of attributes held between square brackets. Each attribute is a **`name: value`** pair. The definition above contains a single `documentation:` attribute. The text should be a succinct description of the program and will be printed to screen when the program is run. If the `documentation:` attribute is missing, a warning will be issued when you run the program.

Typically you will develop new code in a special EMBASSY package called **myemboss** that is reserved for applications that are not yet ready to be incorporated into the main EMBOSS or EMBASSY packages (see Section 3.1, ‘EMBOSS programming’). Save your ACD file in the **myemboss** ACD directory:

```
.../myemboss/emboss_acd
```

If the program were to be added to EMBOSS itself then directory would be:

```
.../emboss/acd
```

ACD files have a filename of the form **ApplicationName.acd**, where **ApplicationName** is the name of the application. The file extension .acd is mandatory. It's sensible (but not mandatory) that the filename (without the .acd extension) is identical to the name of the C source code file.

For further information, see the detailed information on the ACD syntax (Appendix A, *ACD syntax reference*) and ACD file development (Section 4.1, 'Introduction to ACD file development').

2.2.3 Testing the ACD file

Special utilities (Section 4.6, 'ACD utilities') are provided to help you test and validate your ACD files. The main one you'll need is called **acdc** (the ACD compiler) which, when given the name of an ACD file as the first argument on the command line, will parse the file, validate it, parse the command line and 'run' the application command line interface as if the application proper was running.

So, testing the ACD file is easy. You simply run **acdc**, giving your application name as an argument:

```
acdc ApplicationName
```

where *ApplicationName* is the name of the application. So, for **helloworld!**:

```
% acdc helloworld
Prints 'Hello World!' to the screen.
%
```

acdc reads helloworld.acd and reads in any required data just as if the application itself was running. It will also test anything on the command line and report errors in exactly the same way as the real application. In this case there is no required data and nothing else on the command line. As **acdc** didn't reported an error in the example above, then we can assume all is well.

2.2.4 Writing the source code

Happy in the knowledge you have a working ACD file you can turn to the C source code itself, which should look something like this:

```
/* @source helloworld Prints "Hello World!" to the screen.
**
** @author: Copyright (C) Arthur Geek (ageek@ebi.ac.uk)
**
** @@
**
** This program is free software; you can redistribute it and/or
** modify it under the terms of the GNU General Public License
** as published by the Free Software Foundation; either version 2
** of the License, or (at your option) any later version.
**
** This program is distributed in the hope that it will be useful,
** but WITHOUT ANY WARRANTY; without even the implied warranty of
** MERCHANTABILITY or FITNESS FOR A PARTICULAR PURPOSE. See the
** GNU General Public License for more details.
**
** You should have received a copy of the GNU General Public License
** along with this program; if not, write to the Free Software
** Foundation, Inc., 59 Temple Place - Suite 330, Boston, MA 02111-1307, USA.
******************************************************************************/
#include "emboss.h"
/* @prog helloworld ***********************************************************
**
** Prints "Hello World!" to the screen.
**
******************************************************************************/
int main(int argc, char **argv)
{
      embInit("helloworld", argc, argv);
      ajFmtPrint("Hello World!\n");
      embExit();
      return 0;
}
```

There are three main parts to the `helloworld.c` file and, indeed, to all EMBOSS programs:

- The standard EMBOSS application documentation header
- `#include` statements
- The application code proper

In C (and EMBOSS code is no exception), every program *must* have a `main()` function. For `helloworld.c` the program itself consists of the `main()` function only, but most programs have other functions besides `main()`.

The source begins with the standard EMBOSS header. This block of comments includes the name of the program and its short description, copyright notice, licence information, disclaimer, author name and contact details. The tags, for instance `@source`, allow EMBOSS to generate documentation automatically from the code. EMBOSS applications are licensed under the GNU General Public License, so these comments must be included in the source.

Next there is the preprocessor directive `#include "emboss.h"`. In contrast to `#include <stdio.h>`, this imports the entire EMBOSS interface, i.e. makes all the EMBOSS library calls available to you. This must be included at the start of every EMBOSS program.

In the EMBOSS version of **helloworld**, the filename `emboss.h` is surrounded by quotes which means that the preprocessor will look in the current directory and any other directories defined in the configuration file `emboss/Makefile.am`.

`emboss.h` is the master include file. It includes all the other header files for the AJAX and NUCLEUS C programming language libraries. If you look inside the header files you'll see that eventually `stdio.h` is itself included:

```
% more nucleus/emboss.h
#ifndef emboss_h
#define emboss_h
#include "ajax.h"
#include "ajgraph.h"
#include "embaln.h"
#include "embcom.h"
#include "embcons.h"
#include "embdbi.h"
.
.


% more ajax/core/ajax.h
#ifdef __cplusplus
extern "C"
{
#endif
#ifndef ajax_h
#define ajax_h
#include "ajarch.h"
#include <stdarg.h>
#include <stdio.h>
#include <stdlib.h>
#include <string.h>
#include "ajassert.h"
#include "ajdefine.h"
#include "ajstr.h"
#include "ajtime.h"
#include "ajfile.h"
.
.
```

Note

The **`#include "emboss.h"`** statement is a directive for the C preprocessor. Any line beginning with a # character is processed by the C preprocessor before the source code is compiled. For example, the line **`#include "emboss.h"`** tells the preprocessor to replace that line with the whole text of the file **`emboss.h`** before compilation.

Following the preprocessor directive there is documentation for the `main()` function. Every function, including `main()`, *must* be documented. Undocumented code often has little value, with the exception of code that is self-explanatory. Even then it's often helpful, especially in larger programs, to document at least the main steps in the program logic.

Note

EMBOSS uses a standard format for function documentation (see Appendix D, *Code documentation standards*). For now, all you need know is that the **@prog** token is used for documenting the **main()** function. You have already seen **@source** and **@author** in the header documentation. These tokens are read by a program that parses the source code and automatically generates the documentation that goes on the web and into SRS.

The source code proper begins with the `main()` function. The `int` indicates that the `main()` function is of type `int`. In other words, when **helloworld** is called, it will return an integer value to the operating system. `main()` functions in C are conventionally of type `int`.

The command line must be available therefore `main` must include it. This is done in the parameter list using `int argc` and `char **argv`. This is the standard way in C of saying that the program should read any input on the command line as an array of character strings. `argc` is the number of arguments (words on the command line separated by whitespace) and `argv` is the array of strings itself.

Note

Let's pretend that **helloworld** took a parameter (it doesn't) and was invoked like this:

```
helloworld "Print this message"
```

then **argv[0]** would have the value **"helloworld"**, **argv[1]** the value **"Print this message"** and **argc** the value of **"2"**.

Three calls to the EMBOSS libraries are made: `embInit`, `ajFmtPrint` and `embExit`.

A big advantage of EMBOSS programming is that you don't need to write your own code to process the command line (the `argv` array); this functionality is built into the function `embInit`. All EMBOSS applications must call this function, which handles all of the user input processing, and do so right at the start of the application. `embInit` does the following:

- Reads in local database definitions
- Finds the right ACD file to use (the application name is `"helloworld"` so it looks for `helloworld.acd` in the ACD directory)
- Reads the ACD file
- Processes the command line (it uses `argc` and `argv` from `main`).

`embInit` handles all prompting of the user for values that are not entered on the command line, including reprompting the user for values that are out of range. If our ACD file was more complicated, and required a sequence as input and a file as output for example, then by the time the call returned it would have read in the sequence and put it somewhere in memory and also opened the output file.

ajFmtPrint is used to print text to the screen. ajFmtPrint is the EMBOSS version of the printf() C function which you'll know from the C stdio (standard input/output) library. embExit calls some internal clean-up and statistical routines.

For further information on C programming see Section 6.2, 'Programming guides'.

2.2.5 Integration (adding the application to EMBOSS)

Once you have your C source code and an ACD file, you must add your application to **myemboss** (or EMBOSS itself) before you compile it. **myemboss** includes two files, both called Makefile.am, which together contain information about every C source file and ACD file known to the package. To add **helloworld** to **myemboss** you must therefore edit these files. Assuming you checked out the CVS version of EMBOSS into /home/auser/emboss you'll have the following directories:

The 'executables directory' for C source files and executables:

```
/home/auser/emboss/emboss/embassy/myemboss/src
```

The 'acd directory' for ACD files:

```
/home/auser/emboss/emboss/embassy/myemboss/emboss_acd
```

The files you have to edit are:

```
/home/auser/emboss/emboss/embassy/myemboss/src/Makefile.am
/home/auser/emboss/emboss/embassy/myemboss/emboss_acd/Makefile.am
```

Were you adding the application to the main EMBOSS package, the files would be:

```
/home/auser/emboss/emboss/emboss/Makefile.am
/home/auser/emboss/emboss/emboss/acd/Makefile.am
```

The Makefile.am in the executables directory contains information about each C source file. Firstly, you must add your program name to the bin_PROGRAMS list. This is usually done in alphabetical order. The before and after editing stages are shown below for EMBOSS but the edits are the same for **myemboss**.

Before editing

```
bin_PROGRAMS = aaindexextract abiview acdc antigenic \
...
garnier geecee getorf helixturnhelix hmoment \
...
```

After editing

```
bin_PROGRAMS = aaindexextract abiview acdc antigenic \
...
garnier geecee getorf helixturnhelix helloworld hmoment \
...
```

Important

When editing **Makefile.am**, the line continuation characters ('\') must be explicitly added to break the entries over more than one line.

Secondly, you must add your application source file to the SOURCES section. The line to add has the following general syntax:

```
ApplicationName_SOURCES = ApplicationName.c
```

where ApplicationName is the name of the application. This line should be added in alphabetic order. So, the appearance of the file would be as follows.

Before editing

```
...
geecee_SOURCES = geecee.c
getorf_SOURCES = getorf.c
helixturnhelix_SOURCES = helixturnhelix.c
hmoment_SOURCES = hmoment.c
iep_SOURCES = iep.c
infoalign_SOURCES = infoalign.c
...
```

After editing

```
...
geecee_SOURCES = geecee.c
getorf_SOURCES = getorf.c
helixturnhelix_SOURCES = helixturnhelix.c
helloworld_SOURCES = helloworld.c
hmoment_SOURCES = hmoment.c
iep_SOURCES = iep.c
infoalign_SOURCES = infoalign.c
```

The Makefile.am in the ACD directory contains information about each ACD file. All that needs to be done for this file is to add the name of the new ACD file. Again, it is usual to do this alphabetically. Here's what the file for the main EMBOSS package looks like:

Before editing

```
pkgdata_DATA = codes.english \
        aaindexextract.acd abiview.acd ajbad.acd ajfeatest.acd ajtest.acd \
...
        garnier.acd geecee.acd getorf.acd helixturnhelix.acd hmoment.acd \
        histogramtest.acd iep.acd infoalign.acd infoseq.acd isochore.acd \
        lindna.acd listor.acd \
        marscan.acd maskfeat.acd maskseq.acd \
        matcher.acd
```

After editing

```
pkgdata_DATA = codes.english \
        aaindexextract.acd abiview.acd ajbad.acd ajfeatest.acd ajtest.acd \
...
        garnier.acd geecee.acd getorf.acd helixturnhelix.acd helloworld.acd \
        hmoment.acd histogramtest.acd iep.acd infoalign.acd infoseq.acd \
        isochore.acd lindna.acd listor.acd \
        marscan.acd maskfeat.acd maskseq.acd \
        matcher.acd
```

Again, line continuation characters ('\') must be added explicitly.

2.2.6 Compilation

You compile the application by typing one of the following from the executables directory (.../myemboss/source):

```
make helloworld
make
```

The latter option may be slower as it will sometimes (when library changes are made) have to compile everything.

The GNU tools will recognise whether the Makefile.am files have been edited and reconstruct the Makefile files when a **make** command is given. It is bad practice to edit the Makefile files themselves.

Here's the example for **helloworld** compiled in the main EMBOSS package:

```
% pwd
/home/auser/emboss/emboss/emboss/
% make helloworld
/bin/sh ../libtool --tag=CC --mode=link gcc -O2 -Wall -fno-strict-aliasing -o
helloworld helloworld.o
../nucleus/libnucleus.la ../ajax/libajaxg.la ../ajax/libajax.la ../plplot/
libplplot.la -
L/usr/X11R6/lib
-lX11 -lm -lgd -lpng -lz -lm
gcc -O2 -Wall -fno-strict-aliasing -o .libs/helloworld helloworld.o
../nucleus/.libs/libnucleus.so
../ajax/.libs/libajaxg.so ../ajax/.libs/libajax.so ../plplot/.libs/libplplot.so-
L/usr/X11R6/lib -lX11
```

```
-lgd -lpng -lz -lm -Wl,--rpath-
Wl,/home/auser/emboss_test_installation_for_course/emboss/lib
creating helloworld
```

Finally, to run the program:

```
% helloworld
Prints 'Hello World!' to the screen.
Hello World!
%
```

2.2.7 Debugging

No debugging should be required in this case but larger programs will invariably contain bugs that need fixing before the application will run to completion, or even run at all. Debugging is covered in greater detail elsewhere (Section 3.3, 'Debugging').

2.2.8 Testing

Thorough testing is an essential part of software development. For EMBOSS this includes formal quality assurance tests that are run on a regular basis by the EMBOSS developers to ensure the applications work as anticipated. If you want to contribute your applications you will need to write these (see Chapter 7, *Quality assurance*).

2.2.9 Documentation

You should ensure that the `main()` function is appropriately documented and the C source file includes the standard documentation block. **helloworld** is so simple it doesn't require end-user documentation other than the basics that are automatically generated from the source and ACD file.

More complex programs should be fully documented. This includes documentation in the code (see Appendix D, *Code documentation standards*), e.g. for datatypes and functions, and end-user documentation (see (Chapter 8, *Application documentation standards*) for the application as a whole.

2.3 Modifying helloworld

EMBOSS allows for very flexible descriptions of an application's command line interface via the ACD language. In this section we'll take a slightly deeper look at writing ACD files and the code required to process them. You'll be introduced to new ACD datatypes and attributes.

Consider this simple task:

- *Modify **helloworld** to print any user-defined string a user-defined number of times to the screen.*

2.3.1 Modifying the ACD file

To replace the hard-coded string "Hello World!" with a user-defined string you must:

- Add a new data definition to the ACD file. The ACD datatype you need is, unsurprisingly, a `string`. Many others are available (Section 4.3, 'Data definition').
- Give your data definition an appropriate label.
- Give the string a default value. You'll need to add a `default:` attribute to the string definition with a sensible default value of `"Hello World!"` (or something else).

Your ACD file should now look something like this:

```
application: helloworld
[
    documentation: "Prints 'Hello World!' to the screen."
]
string: printme
[
    default: "Hello World!"
]
```

2.3.2 Modifying the C source code

To modify your C source code to support the user-defined string you need to:

- Define an appropriate AJAX object, i.e. to hold an ACD `string` datatype.
- Add an appropriate `ajAcdGet*` function (in this case `ajAcdGetString`) to retrieve the value of your new ACD data item. The argument to `ajAcdGetString` must be the same as the data label in the ACD file.
- Call a function to print the string to screen.
- Comment out (or remove) the existing hard-coded `ajFmtPrint` statement.
- Exit cleanly; remembering to free the string object (by calling `ajStrDel`) that will have been allocated (indirectly) by `ajAcdGetString`.

Your source code will now look something like:

```
#include <stdio.h>
int main(void)
{
  AjPStr printme=NULL;
  printme = ajAcdGetString("printme");
  ajFmtPrint("%S\n", printme);
  ajStrDel(&printme);
  embExit();
  return 0;
}
```

You can see that `ajFmtPrint` is used with a conversion specifier (`%S`) to print an EMBOSS string.

Note

Both `ajFmtPrint` and the C equivalent (`printf()`) functions use *conversion specifiers* which control how the arguments to the function are interpreted when they are printed out. The conversion specifiers are given by a single character preceded by a percentage symbol (`%`). For example, a signed decimal integer variable is converted for printing with the conversion **`%d`** and a character string is formatted correctly with the conversion **`%s`**. `ajFmtPrint` supports additional AJAX datatypes including booleans (**`%B`**),date (**`%D`**) and AJAX strings (**`%S`**), etc. (see Section 5.1, 'Basic datatypes')

2.3.3 Running the program

When you run the program, you should get output that looks like this:

```
% helloworld
My first EMBOSS program
Hello World!
```

It should be obvious that it's using the default value. However, the string can be *user-defined*, you just have to specify the appropriate *qualifier* (the label of the data item, `printme` in this example) and an associated string value on the command line when you invoke the program:

```
% helloworld -printme "Hello Sailor!"
My first EMBOSS program
Hello Sailor!
```

As a finishing touch, add a `groups:` attribute to the application definition to assign your application to a functional group. See the list of valid group names (Section 4.2.4.1, 'Application group names file (`groups.standard`)'). This will make it easier to find by end-users.

Also, add a comment line (beginning with #) giving the date your ACD file was last edited.

The `application:` definition should now look something like this:

```
application: helloworld
[
    documentation: "Prints 'Hello World!' to the screen."
    groups: "Test"
#   Last edited 2007.
]
```

2.3.4 Qualifiers and parameters

Having to use the **-printme** option is a bit cumbersome, it would be nice if you could specify the text to print without having to know the label name. For this you use the `parameter:` attribute. If a data item is defined with parameter:, using the flag on the command line is not mandatory, i.e. you can just type:

```
helloworld "Hello Sailor!"
```

To define your string as a parameter, add the following to the ACD string definition:

```
parameter: "Y"
```

You can still use the qualifier name if you want to, but it's not mandatory.

2.3.5 Adding an integer

The next task is to add an integer data item to the ACD file to control how many times the user-defined string is printed to the screen. To add the integer, you repeat the steps you went through for adding the string.

Your ACD file will include the following data definition:

```
integer: n
[
    default: "1"
]
```

The C code will need modifying to look something like this:

```
#include "emboss.h"
int main(void)
{
  AjPStr printme = NULL;
  ajint  n       = 0;
  printme = ajAcdGetString("printme");
  n       = ajAcdGetInt("n");
  while(n)
  {
    ajFmtPrint("%S\n", printme);
    n--;
  }
  ajStrDel(&printme);
  embExit();
  return 0;
}
```

You'll notice a destructor function is not called for the new integer. This is because `ajint` is a fundamental AJAX datatype, not an object (C data structure), and therefore does not require explicit memory management. See elsewhere for further information on programming for fundamental types (Section 5.1, 'Basic datatypes') and objects (Section 5.3, 'Objects (C data structures)').

2.3.6 Parameters and qualifiers revisited

Do not make the integer a parameter just yet – experiment by calling **helloworld** with different command lines. You should notice that it doesn't matter whether the parameter comes before or after the qualifier for your new integer. In other words you can call the application like this:

```
helloworld -n 10 "Hello Sailor!"
```

Or like this:

```
helloworld "Hello Sailor!" -n 10
```

Now make the integer a parameter in your ACD file. Experiment again with calling **helloworld**. You will see for yourself that the order of parameters is important – they have to match the order used in the ACD file!

If, for example, your ACD file looked this:

```
application: helloworld
[
    documentation: "Prints a user-defined string to the screen."
]
string: printme
[
    parameter: "Y"
    default: "Hello World!"
]
integer: n
[
    parameter: "Y"
    default: "1"
]
```

You could invoke the program like this:

```
helloworld "Hello Sailor!" 10
```

But not like this:

```
helloworld 10 "Hello Sailor!"
```

Parameters have to appear on the command line in the order in which they are defined in the ACD file.

2.4 Modifying matcher

The EMBOSS application **matcher** generates pairwise local alignments given either DNA or protein sequences. It has different options depending on the type of input. When wrapping such programs, for instance for incorporation into a graphical user interface, it is often convenient to split them into separate nucleotide and protein applications with the same application code underneath. SRS is a good example of a system which wraps EMBOSS applications in this way.

To deepen your knowledge, use **matcher** as a template to produce new, separate applications for protein (**matcherpro**) and nucleotide (**matchernuc**) sequences.

2.4.1 Planning

You need to change how the **matcher** functionality is presented to the user, therefore you'll need two new ACD files and a copy of the **matcher** source code for each. Both new applications will effectively use the *same* source code with only trivial differences.

The steps to create these applications are as follows:

1. Create the application source code (`matcherpro.c` and `matchernuc.c` files) in:

```
.../embassy/myemboss/src
```

   Copy the source code for **matcher** from `.../emboss/matcher.c`.

2. Create the application ACD files (`matcherpro.acd` and `matchernuc.acd`) in:

```
.../embassy/myemboss/emboss_acd
```

   Use `.../emboss/matcher.acd` as a template.

3. Add the applications to the **myemboss** package by updating the two `Makefile.am` files:

```
.../embassy/myemboss/src/Makefile.am
.../embassy/myemboss/emboss_acd/Makefile.am
```

4. Compile and test the applications.

2.4.2 Editing the ACD files

The original `matcher.acd` is shown below (`help:` attributes are not shown):

```
application: matcher
[
  documentation: "Waterman-Eggert local alignment of two sequences"
  groups: "Alignment:Local"
]
```

```
section: input
[
  information: "Input section"
  type: "page"
]

  sequence: asequence
  [
   parameter: "Y"
   type: "any"
  ]

  sequence: bsequence
  [
   parameter: "Y"
   type: "@($(acdprotein) ? stopprotein : nucleotide)"
  ]

  matrix: datafile
  [
   additional: "Y"
   information: "Matrix file"
   protein: "$(acdprotein)"
 ]

endsection: input

section: additional
[
  information: "Additional section"
  type: "page"
]

  integer: alternatives
  [
    additional: "Y"
    information: "Number of alternative matches"
    default: "1"
    minimum: "1"
  ]

 integer: gapopen
  [
    additional: "Y"
    information: "Gap penalty"
    default: "@($(acdprotein)? 14 : 16)"
    minimum: "0"
    valid: "Positive integer"
    expected: "14 for protein, 16 for nucleic"
  ]

 integer: gapextend
 [
  additional: "Y"
  information: "Gap length penalty"
  default: "@($(acdprotein)? 4 : 4)"
  minimum: "0"
  valid: "Positive integer"
  expected: "4 for any sequence"
 ]
```

```
endsection: additional

section: output
[
  information: "Output section"
  type: "page"
]
  align: outfile
  [
    parameter: "Y"
    aformat: "markx0"
    minseqs: "2"
    maxseqs: "2"
 ]

endsection: output
```

The ACD file introduces several new concepts:

- The `groups:` attribute in the application definition assigns the application to a group (see Section 4.2, 'Application definition').
- The `sequence` ACD datatype is used to define two input sequences; `asequence` and `bsequence`.
- The sequence type of `asequence` is set by the `type:` attribute, in this case to "`any`", i.e. any type of sequence is acceptable.
- For `bsequence`, the sequence type is calculated from the ACD variable `acdprotein`; if `acdprotein` is `true` then `type:` is set to `stopprotein`, otherwise it's set to `nucleotide`.
- `acdprotein` is an 'automatic ACD variable' with a `boolean` type whose value is set automatically when the first sequence is read in. So, if the first sequence is a protein, then `acdprotein` will be `true`. Automatic ACD variables are described in detail elsewhere (Section 4.4, 'Operations').
- The `matrix` datatype is used to define a substitution matrix (called `matrixfile`). EMBOSS will search for this data file in the EMBOSS data directory (see the *EMBOSS User's Guide*).
- `information:` is used to set a user-prompt for some of the data items. It is not needed for the sequence inputs (EMBOSS will automatically generate a suitable prompt) but can be given for the other types used (see Section 4.3, 'Data definition').
- Qualifiers and parameters in the ACD file are organised into sections (`input`, `additional` and `output`). These help to tidy the ACD file and are exploited by user interfaces (see Section 4.3, 'Data definition').
- Options in the `additional` section are defined to be 'additional qualifiers' by the attribute `additional: "Y"`. Values for additional qualifiers are not prompted for (the default value will be used instead) unless **`-options`** is given on the command line, which will turn prompting on for these qualifiers (see Section 4.1, 'Introduction to ACD file development').
- Two gap penalties (`gapopen` and `gapextend`) are defined as `integer` ACD types. The `minimum:`, `valid:` and `expected:` attributes are used to set minimum and expected values and a corresponding message to the user.

- There is a single output, a sequence alignment (outfile) which is defined by the type align. The format (markx0) and minimum and maximum number of sequences (2 in both cases, i.e. a pairwise alignment) are set using the attributes aformat:, minseqs: and maxseqs respectively.

The changes necessary for matcherpro.acd are:

- The application name should be changed to matcherpro.
- The documentation: attribute should state that the application works on protein sequences only.
- The type: attribute of the first input sequence should be changed from any to protein.
- The type of the second input sequence should be stopprotein.
- The residue substitution matrix should be of type protein. Currently this is given as protein: "$(acdprotein)" which means that the protein: attribute will be set to true if the first sequence is a protein. $(acdprotein) should be replaced with y.
- All other occurences of lines containing acdprotein should be replaced as appropriate (see below).

The parts of matcherpro.acd which have been modified and differ from matcher.acd are shown below:

```
application: matcherpro [
 documentation: "Waterman-Eggert local alignment of two sequences"
 groups: "Alignment:Local"
]
... lines omitted

 sequence: asequence [
   parameter: "Y"
   type: "protein"
 ]

 sequence: bsequence [
   parameter: "Y"
   type: "stopprotein"
 ]

... lines omitted


 matrix: datafile [
   additional: "Y"
   information: "Matrix file"
   protein: "Y"
 ]

... lines omitted

 integer: gapopen [
   additional: "Y"
   information: "Gap penalty"
   default: "14"
   minimum: "0"
```

```
    valid: "Positive integer"
    expected: "14"
  ]

  integer: gapextend [
    additional: "Y"
    information: "Gap length penalty"
    default: "4"
    minimum: "0"
    valid: "Positive integer"
    expected: "4 for any sequence"

  ]
… lines omitted
```

2.4.3 Editing the C source file

The `main()` function for **matcher** is shown below. The application includes several functions and macros that are not shown:

```
#include "emboss.h"
/* @prog matcher ***************************************************************
**
** Finds the best local alignments between two sequences
**
*******************************************************************************/
int main(int argc, char **argv)
{
    AjPStr aa0str =0;
    AjPStr aa1str =0;
    const char *s1;
    const char *s2;
    ajint gdelval;
    ajint ggapval;
    ajuint i;
    ajint K;

    AjPAlign align = NULL;

    embInit("matcher", argc, argv);

    seq = ajAcdGetSeq("asequence");
    ajSeqTrim(seq);
    seq2 = ajAcdGetSeq("bsequence");
    ajSeqTrim(seq2);
    matrix = ajAcdGetMatrix("datafile");
    K = ajAcdGetInt("alternatives");
    gdelval = ajAcdGetInt("gapopen");
    ggapval = ajAcdGetInt("gapextend");
    align = ajAcdGetAlign("outfile");

    /*
      create sequence indices. i.e. A->0, B->1 … Z->25 etc.
      This is done so that ajBasecodeToInt has only to be done once for
      each residue in the sequence
```

```
    */

    ajSeqFmtUpper(seq);
    ajSeqFmtUpper(seq2);

    s1 = ajStrGetPtr(ajSeqGetSeqS(seq));
    s2 = ajStrGetPtr(ajSeqGetSeqS(seq2));

    sub = ajMatrixGetMatrix(matrix);
    cvt = ajMatrixGetCvt(matrix);


    aa0str = ajStrNewRes(2+ajSeqGetLen(seq)); /* length + blank + trailing null */
    aa1str = ajStrNewRes(2+ajSeqGetLen(seq2));
    ajStrAppendK(&aa0str,' ');
    ajStrAppendK(&aa1str,' ');

    for(i=0;i<ajSeqGetLen(seq);i++)
        ajStrAppendK(&aa0str,(char)ajSeqcvtGetCodeK(cvt, *s1++));

    for(i=0;i<ajSeqGetLen(seq2);i++)
        ajStrAppendK(&aa1str,ajSeqcvtGetCodeK(cvt, *s2++));

    matcher_Sim(align, ajStrGetPtr(aa0str),ajStrGetPtr(aa1str),
                seq,seq2,
                K,(gdelval-ggapval),ggapval,
                ajSeqGetOffset(seq), ajSeqGetOffset(seq2), 2);

    ajStrDel(&aa0str);
    ajStrDel(&aa1str);

    ajSeqDel(&seq);
    ajSeqDel(&seq2);

    embExit();

    return 0;
}
```

`matcher.c` introduces many new concepts:

- Two AJAX strings (`aa0str` and `aa1str`) are defined using `AjPStr`.
- An alignment object (`align`) is created using `AjPAlign`.
- The values from the ACD file are picked up by calls beginning with `ajAcdGet`. The functions are named to make the types returned obvious: `ajAcdGetSeq` returns sequences, `ajAcdGetMatrix` returns a matrix and so on.
- The next block of code creates the sequences indices. You needn't concern yourself with the details. Functions for this include `ajMatrixGetMatrix`, `ajMatrixGetCvt` and `ajSeqcvtGetCodeK`. The function `matcher_Sim` (not shown) is also called and is a function local to **matcher**.
- Strings and sequences that were allocated are freed by calling `ajStrDel` and `ajSeqDel`.
- The application then exits cleanly by calling `embExit`.

One small edit is needed in the `src/matcherpro.c` file:

- Change the `embInit` call to `embInitP` (which is used for EMBASSY packages) and add `"myemboss"` as the last argument so EMBOSS knows you are in the **myemboss** package
- Change the program name in the `embInitP` call to `"matcherpro"` so that it uses the new ACD file.
- Change the program name in the `main` function comment call to `"matcherpro"` to match the program name.

The parts of `matcherpro.c` which have been modified and differ from `matcher.c` are shown below:

```
/* @prog matcher ***************************************************************

embInitP("matcherpro", argc, argv, "myemboss");
```

2.4.4 Compilation

To compile and install the application from the **myemboss** directory, type:

```
make
```

Or, if you are using **myemboss** as part of a fully installed EMBOSS system then type:

```
make install
```

2.4.5 Testing all is well

matcherpro has just been added to your system, therefore you may need to make it visible in your path, for example, by running the UNIX command **`rehash`**, if you use a `csh` shell.

To check all is well, you can run:

```
acdvalid matcherpro
```

and check that no errors or warnings are reported. As `matcher.acd` has already been validated it should be fine.

You should now have a new application **matcherpro** which should produce the same results as **matcher** with protein input, but will refuse to run with DNA sequences. This is shown below:

```
% matcherpro
Finds the best local alignments between two protein sequences
Input protein sequence: embl:L07770
Error: Sequence is not a protein

Error: Unable to read sequence 'embl:L07770'
Input protein sequence:
...
```

Here it is being used correctly:

```
% matcherpro
Finds the best local alignments between two protein sequences
Input protein sequence: swissprot:UBR5_RAT
Second protein sequence:
...
```

2.4.6 Further developments

From the above it should be obvious what's required to implement `matchernuc.acd` and `matchernuc.c`, so generating two distinct applications, one for nucleotide and one for protein sequences.

If you are familiar with EMBOSS, you might notice that the output format of **matcherpro** is not the same as **water**, another popular alignment program. This could be confusing to users, so as a final modification you can make **matcherpro** consistent with **water**. To do this, set the `aformat:` attribute of the `outfile` data definition, which currently has the value `"markx0"` to the value (`"srspair"`) defined in the `water.acd` file.

A similar effect could be achieved by running **matcherpro** with **`-aformat srspair`** on the command line. If for some reason the old format was required, **matcherpro** could be run with **`-aformat markx0`** on the command line.

2.5 String handling

EMBOSS has a wide range of string handling functions. All of them start with the prefix `ajStr` and are defined in the `ajstr.c` source file. You'll practice using a few of them here by writing a new application that prompts the user for one or more strings, processes them, and writes the results to a plain text output file. This builds upon the string handling code used when modifying **helloworld**.

2.5.1 `stringplay.acd`

To begin, create an ACD file for a new application. We'll call the latter **stringplay**. The steps are as follows:

- Create the ACD file `myemboss/emboss_acd/stringplay.acd`
- Add an `application:` definition with `documentation:` and `group:` attributes
- Add one or more `string:` definitions with appropriate `default:`, `information:` and `help:` attributes

- An output file is needed, so add an `outfile` ACD data definition
- Test the file by running **`acdc stringplay`**
- Fix any warning or error messages generated by **acdc**
- Run **`make`** to compile the application and **`make install`** (depending on your installation) in the `emboss_acd` directory (or above) to install this file.

The ACD file might look something like this:

```
application: stringplay
[
  documentation: "An application for experimenting with basic string handling."
  groups: "Test"
]

string: astring
[
  parameter: "Y"
  default: "ParameterString"
  information: "First input string (parameter)"
  help: "This string is a parameter meaning you needn't specify the label
(-astring) on the command line when specifying a value for it."
]

string: bstring
[
  standard: "Y"
  default: "StandardString"
  information: "Second input string (standard qualifier)"
  help: "This string is a standard qualifier meaning it
 will be prompted for if not specified on the command line."
]

string: cstring
[
  additional: "Y"
  default: "AdditionalString"
  information: "Third input string (additional qualifier)"
  help: "This string is a additional qualifier meaning it will only
 be prompted for if -options is given on the command line."
]

string: dstring
[
 default: "AdvancedString"
 information: "Fourth input string (advanced qualifier)"
 help: "This string is an advanced qualifier meaning it will never be prompted for."
]

outfile: outfile
[
 parameter: "Y"
]
```

You can use **`acd stringplay`** to test the ACD file. **acdc** will 'run' it exactly as if the application source code existed. It will prompt for any string inputs, assuming you defined them with `parameter:` or `standard:` attributes in the ACD file.

If you defined any strings with the `additional:` attribute then they will be prompted for only if you specify **`-options`** on the command line when you run **stringplay**. If you didn't specify `parameter:`, `standard:` or `additional:` then they will default to being an `advanced` qualifier and will never be prompted for.

Regardless of how you define the string inputs, they can be set on the command line by using e.g.:

```
-astring StringValue
```

where **`-astring`** is the label of the data definition and ***`StringValue`*** is the value of the string to set.

If you define the strings with `parameter`: then the flag (**`-astring`**) is optional, although values for parameters must be given in the order they appear in the ACD file.

It is also usually sensible to define a default value with the `default:` attribute. For `additional` and `advanced` qualifiers a default value is essential.

You should experiment by running **`acdc stringplay`** using the above ACD file, specifying values on the command line with and without any data labels.

2.5.2 stringplay.c

The C source code (`myemboss/src/stringplay.c`) must declare variables to hold all the strings and the output file from the ACD file. It must also pick up the values from the ACD file. At a minimum, the application should write the strings to the output file. It should then close the output file and exit cleanly.

If you are using the `stringplay.acd` shown above then the C source code should look something like this:

```
#include "emboss.h"
int main (int argc, char* argv[])
{
  AjPStr astring = NULL;
  AjPStr bstring = NULL;
  AjPStr cstring = NULL;
  AjPStr dstring = NULL;

  AjPFile outf = NULL;
  embInitP("stringplay", argc, argv, "myemboss");
  astring = ajAcdGetString("astring");
  bstring = ajAcdGetString("bstring");
  cstring = ajAcdGetString("cstring");
  dstring = ajAcdGetString("dstring");
  outf = ajAcdGetOutfile("outfile");

  /* functional part of code would go here */
  ajFmtPrintF(outf, "astring: %S\nbstring: %S\ncstring: %S\ndstring: %S\n",
              astring, bstring, cstring, dstring);

  ajFileClose(&outf);
  ajStrDel(&astring);
  ajStrDel(&bstring);
```

```
    ajStrDel(&cstring);
    ajStrDel(&dstring);

    embExit();
    return 0;
}
```

2.5.3 Compilation and testing

To compile your program give the following command from the **myemboss** directory:

```
make
```

And, if you are using **myemboss** as part of a fully installed EMBOSS system, install it using:

```
make install
```

2.5.4 Adding functionality

The EMBOSS string handling library is very extensive. You can inspect `ajstr.h` and `ajstr.c` to get a feel for what's available. The functions are organised into sections for convenience and these include assignment, combination, cut, substitution, query, element retrieval, formatting, comparison, and so on. Just a few functions from these categories are shown below:

```
/* A string "assignment" function to copy a string to a string. */
AjBool ajStrAssignS(AjPStr* Pstr, const AjPStr str);

/* A string "combination" function which appends a strings */
AjBool ajStrAppendS(AjPStr* Pstr, const AjPStr str);

/* A string "combination" function which inserts a text string into a
 string at a specified postion. */
AjBool ajStrInsertS (AjPStr* pthis, ajint pos, const AjPStr str);

/* A string "cut" function which removes a substring from a string. */
AjBool ajStrCutRange(AjPStr* Pstr, ajint pos1, ajint pos2);

/* A string "cut" function which removes the end from a string
reducing it to a defined length. */
AjBool ajStrTruncateLen(AjPStr* Pstr, size_t len);

/* A string "substitution" function which reverses the order of characters
in a string. */
AjBool ajStrReverse(AjPStr* Pstr);
```

```
/* A string "query" function which counts occurrences of a character in a string. */
ajint ajStrCalcCountK(const AjPStr str, char chr);

/* A string "element retrieval" function which returns a single
character at a given position from a string. */
char ajStrGetCharPos(const AjPStr str, ajint pos);

/* A string "formatting" function which converts a string to upper case. */
AjBool ajStrFmtUpper(AjPStr* Pstr);
/* A string "comparison" function which is a simple test for matching
two strings. */
AjBool ajStrMatchS (const AjPStr thys, const AjPStr str);

/* A string "comparison" to test for matching the start of a string against a
given prefix string. */
AjBool ajStrPrefixS(const AjPStr str, const AjPStr str2);

/* A string "comparison" function to find the first occurrence in a string of a
second string. */
ajint ajStrFindS (const AjPStr str, const AjPStr str2);
```

Many of the functions return `AjBool`. This might indicate whether the function was reallocated, was successful or not, whether a test was true or not and so on. You should refer to the documentation in `ajstr.c` for the exact meaning.

`AjPStr* Pstr` indicates a string argument that is (or might be) modified by the function. `const` in front of an argument indicates a variable with a constant value. So here, `const AjPStr str` indicates a string argument that is used but not modified by the function; the pointer cannot be reallocated by the function.

Here are a few examples to illustrate how some of the string functions are called. First, `ajStrAssignS` is used to assign an existing string to a new string. The function should follow construction of `newstr`. `newstr` is modified which is why the address of `newstr` is required:

```
ajStrAssignS(&newstr, astr);
```

Here `ajStrFmtUpper` is used to convert a string to upper-case. Again, the address of the string is required:

```
ajStrFmtUpper(&astring);
```

Here two strings are compared and some action taken if they are identical:

```
if(ajStrMatchS(astring, bstring))
{/* Do something */}
```

2.5.4.1 String memory management

It is good practice to explicitly construct (allocate memory for) each string before it is used, and destroy it (free memory) for the string once you are finished with it. A default constructor function (ajStrNew) and destructor function (ajStrDel) are provided for this purpose. Their use is shown below:

```
/* newstr is NULL, no memory is allocated */
AjPStr newstr=NULL;

/* newstr now points to a string object in memory and is ready for use by
our program */
newstr = ajStrNew();

/* memory for the string object is freed. The program can terminate without
potentially leaking memory. */
ajStrDel(&newstr);
```

You do not need to call the constructor function for ACD variables because memory is allocated for these by the call to embInitP. There is risk of memory error if you do. In the following rather contrived and bugged code the call to ajAcdGetString will overwrite the pointer to memory allocated and returned by the (unnecessary) call to ajStrNewC:

```
/* This code is bugged */
AjPStr astring=NULL;

/* allocates memory for all ACD data items */
embInitP("stringplay", argc, argv, "myemboss");

/* newstr now points to a string object in memory */
astring = ajStrNewC("Hello");

/* this call will break the handle to the memory allocated by ajStrNewC */
astring = ajAcdGetString("astring");
```

The correct code is:

```
AjPStr astring=NULL;

/* allocates memory for all ACD data items */
embInitP("stringplay", argc, argv, "myemboss");

/* Get a string value from ACD */
astring = ajAcdGetString("astring");

/* memory for the string object is freed. The program can terminate without
potentially leaking memory. */
ajStrDel(&astring);
```

A string object was created by embInitP and must be freed once you are done with it. That is what the call ajStrDel(&astring); is for.

Software development under EMBOSS

3

3.1 EMBOSS programming

3.1.1 Introduction

Software development and maintenance under EMBOSS is made easy. EMBOSS has powerful inbuilt functionality which any native application can make use of with little or no additional coding, saving you a great deal of effort. It includes extensive C programming libraries for extending the core functionality and developing new applications. Well-defined processes are in place for key aspects such as quality assurance testing, installation, maintenance and support. General aspects are handled by the EMBOSS developers, leaving you to support the parts specific to your own software.

3.1.2 Inbuilt functionality

Major inbuilt areas of functionality include:

- Support for biological datatypes
- Support for common file formats
- Simple database configuration
- Command line handling
- Command line qualifiers
- Sequence and sequence feature specification.

3.1.2.1 Support for biological datatypes

EMBOSS supports a variety of standard and biological datatypes (Section 4.3, 'Data definition'). Datatypes for application inputs and outputs are defined in the AJAX command definition (ACD) files. Datatype attributes allow an input or output to be specified in greater detail, for example by setting a default value or by defining the permitted type of an input sequence. Calculated attributes, such as the length of an input sequence, are available once the data have been read during ACD file processing but before the application proper starts, allowing for further fine control in the ACD file.

3.1.2.2 Support for common file formats

EMBOSS supports a wide range of file formats for input and output. Biological sequences can be read and written using all common formats. EMBOSS detects the format of an input sequence automatically, but this can be specified for efficiency. Many other formats for sequence features, alignments, data files and so on are handled automatically. Any application you write will support these formats automatically too. Furthermore, when new input and output data formats are added to EMBOSS, your applications will automatically be able to use them; no application code needs to change. For information on supported file formats, see the *EMBOSS User's Guide*.

3.1.2.3 Simple database configuration

Database configuration is simple and flexible. A variety of databases (which can include remote data servers) and diverse access methods are supported. New databases and access methods are easily added. For more information see the *EMBOSS Administrator's Guide*.

3.1.2.4 Command line handling

Your application will use the EMBOSS command line which is consistent across the applications. The ACD file defines the datatype and permissible values for all application options. The ACD file and all input at the command line is validated at startup, before the application proper starts. Command line behaviour, for example the production of a sensible prompt and reprompting for values that are out of range, is handled automatically.

3.1.2.5 Command line qualifiers

EMBOSS includes various command line qualifiers which are not (typically) set in an ACD file. These include *global qualifiers* and various *datatype-specific qualifiers*. Global qualifiers are available to all applications and control general application behaviour. Datatype-specific qualifiers are used to specify particular inputs and outputs, for instance sequences, in more detail.

3.1.2.6 Sequence and sequence feature specification

Sequences are referenced on the command line by their Uniform Sequence Address (USA) which is a powerful and flexible specification for defining the location and format of sequence data. Similarly, sequence features are defined using a Uniform Feature Object (UFO). Any application you write will automatically support USAs and UFOs. For more information see the *EMBOSS User's Guide*.

3.1.3 Basic steps to development

There's more to software than just coding, especially when developing for EMBOSS which is widely distributed and deployed in production environments. All code that is to be integrated into the package must be tested, documented and managed throughout its lifetime. Regardless of whether you intend to develop the libraries or new applications, many of the steps are the same:

1. Project management
2. Planning
3. ACD file development (applications only)
4. C source code development
5. Integration and compilation
6. Debugging
7. Quality assurance testing
8. Documentation
9. Distribution
10. Maintenance, support and training.

It's worth familiarising yourself with these steps, described below, before starting a project. This should help you plan your software effectively, develop it efficiently and deliver your projects on time.

3.1.4 Project management

All software begins with an idea and ends, ultimately, with the software being retired. An appreciation of the steps involved will help the developer avoid pitfalls and problems along the way, or at least be prepared for them, saving much time and effort. Yet many projects are started without this foresight, especially by programmers who are eager to get on coding.

3.1.4.1 Models for releasing software

There is a single major release of EMBOSS on St Swithun's Day (15 July) each year. The version number usually, but not always, increases by 1 each year. Other versions are released in the interim as required. The EMBASSY packages do not have a fixed release date or formal version numbering scheme. To avoid confusion, version numbers for EMBASSY packages that wrap third-party software are set to the same as the version number of the software that has been wrapped. Individual EMBOSS or EMBASSY applications do not have their own version stage or number. New programs are announced on the EMBOSS mailing lists to solicit feedback from users.

You should decide upon a general model for the development and release of your software. No model has been universally adopted; however, all projects begin with a pre-development stage, before coding starts, which is followed by one or more versions of the software reflecting the maturity of the code. The following stages are typical:

1. Pre-development
2. Pre-alpha
3. Alpha
4. Beta
5. Release candidate
6. Release.

In practice, very often there is no formal software release corresponding to every stage. For example, the users might only see a beta version and the final release. Nonetheless it is helpful to bear these stages in mind as it helps when managing the expectations of your users. It should go without saying that, for instance, it would be foolish to distribute broadly code that had not been well tested in a beta version. In practice, early releases often contain many bugs and for this reason people are usually wary of software until it has matured over a period of months or even years.

Pre-development
The pre-development stage occurs before coding starts. Requirements are specified and expected work throughout the software lifetime is planned.

Pre-alpha
Pre-alpha software is still under active design and development. Usually it is a prototype for the version that might eventually be released. From the user's perspective a pre-Alpha release is useful to get the gist of what's to come, from the developer's perspective it provides a means to get early feedback from users in order to inform the design and development.

Alpha
Alpha software is usually reasonably well developed but not ready for general release. There will be very limited, if any, deployment. Such deployment is often restricted to the primary user group or some other select group of users who are known to provide good feedback. Typically, most major features will be implemented and most major bugs removed, but there will be many missing minor features, bugs and other issues. The alpha release might be a complete rewrite or an evolution of the pre-alpha version.

Beta
Beta software is the first significant release. Its main purpose is to get the software tested by as large a group of users as possible and to solicit feedback. Beta software might not be perfect, it might not work under all circumstances, but it should be relatively bug-free. Typically, the majority, if not all, known major bugs should be fixed.

Release candidate
The release candidate is a refinement of the beta version in which all known bugs are fixed. It may have several new features if these were heavily requested during beta testing, but these should not be a major departure from its basic function. Such major requests should (ideally) have been caught earlier on. Major features are now 'frozen', and only bug fixes or minor new features implemented from this point on. The release candidate can be thought of as a halfway house between versions used for testing and that which is deployed. Release candidates should be subject to heavy testing by the developer and in real use-case scenarios.

Release
The release software is fully developed and tested and should be ready for use in anger by your users. It is the final version of the software arising from testing and refinement of the release candidate. The release will normally have a number, for example, Version 1.0.

3.1.4.2 Software life cycle models

The EMBOSS developers use a combination of the 'Code and fix' and 'Synchronise and stabilise' models (see below). This has proved to be the most pragmatic way to keep users happy, especially balancing the requirements for new functionality with that for stability and consistency in the package as a whole.

The general approach is to fix bugs as soon as they are reported and provide bug fix files (see the *EMBOSS User's Guide*) as required. New features or applications are developed in close collaboration with the users, typically as requested, and are developed and released in a scheduled manner. This is done in as timely a manner as possible allowing for all bugs to be fixed and major issues to be addressed. All bug fixes and requests for features that cannot immediately be implemented are logged on the SourceForge website (see Section 1.5, 'Contributing software to EMBOSS'). In this way valuable contributions are not lost even if they cannot be acted upon immediately.

Many software life cycle models have emerged and all tend to focus on the development process itself. They are summarised below. It is up to you to adopt one of these models (or something else) that is most appropriate to your needs and situation.

Code and fix model

The simplest model iterates a cycle of coding and bug-fixing until the software is fit for purpose:

1. Write source code
2. Compile and run program
3. Test and find bugs
4. Fix bugs.

Waterfall model

Development proceeds in a linear series of definite management or development steps. The steps might be followed by testing, which must be passed before proceeding to the next step. Typically, return to a previous step is only allowed if a new problem is identified in the existing step. Thus development proceeds incrementally and logically. The drawback, however, is that such models are rather inflexible to changing requirements.

Spiral model

The spiral model attempts to address the inflexibility of the waterfall model by iterative cycles of development. This introduces more chances for user involvement but the drawback is that the model is rather cumbersome and therefore difficult to stick to in practice. Each cycle has four phases:

1. Specification: user requirements, constraints and possible approaches are identified.
2. Evaluation: alternative approaches are evaluated, typically through development of prototype software.

3. Development: code is designed in detail, implemented and tested.
4. Planning: the software and plans for the next cycle are evaluated in light of user feedback.

Prototyping

The prototyping models try to ensure that software meets user requirements by introducing definite evaluation points in the development process. At these points the user reviews the software, typically a prototype version, via a demonstration or by hands-on testing. The software under review might be a small part of a larger system. As development progresses the developers tend naturally to focus on design and implementation issues, whilst the users remain focused on their requirements. The prototyping models attempt to avoid any divergence of views through regular appraisal.

Rapid application development

This is a general approach to software development and a move away from inflexible formal models. Typically a series of prototypes are reviewed by the user and developer to ensure user requirements are met and fine-tuned in the face of the continuing developments. Analysis, design and implementation at each stage are usually limited to a definite time. The drawbacks to this approach are that, in the absence of a formal statement of a design up front, it can be difficult to agree when a project is finished and the software can evolve in an undesirable or uncontrolled way.

Synchronise and stabilise model

This model coordinates multiple developers who are working in parallel and attempts to quickly produce software that fits most needs, and to improve it in subsequent versions. It relies on frequent depositions of code which are synchronised to ensure compatibility. Testing occurs at all stages and in parallel with development and all reported bugs are fixed immediately. Features tend to be implemented incrementally, starting with those of highest priority to the user. The software is stabilised at key stages in its lifetime through extensive user testing and bug fixing.

Iterative and incremental models

Such models rely on a set of specific use-cases and a detailed software architecture. Each use-case describes what an individual or group of users want from the system, and together describe the complete functionality. The software architecture describes all significant components of the software and environments in which it will run. The use-cases drive the development whereas testing ensures that the use-cases are implemented correctly. Coding is usually divided to cover groups of related use-cases which are tackled iteratively. Each iteration results in an increment of the software, i.e. one with more use-cases catered for. In this model a design may well evolve as required, especially in the early stages, with later iterations tending to add to, rather than modify, what has already been done.

3.1.4.3 User requirements

The EMBOSS mailing lists (Section 1.4, 'Project mailing lists') allow discussions between the user community and developers. Any new software should be discussed on the mailing lists before it's implemented, especially to establish whether it is really needed or already exists.

Software must fulfil the requirements of the intended users and use-cases. It is therefore valuable, essential even, that your users are involved in the development of the software throughout its lifespan, especially these four stages:

1. Pre-development
2. Development
3. Testing and validation
4. Evolution.

Important

There are additional requirements (e.g. software testing and surveying of user requirements) and strategy (e.g. how best to implement for evolving requirements) that are beyond the scope of this manual, but the essential message is simple: *Stay in touch with your users, even if you're the only user, at all stages in the software lifespan.*

`Pre-development`

It's essential to get as complete a picture as possible of the user or use-case requirements at the pre-development stage, before coding starts. Try to pin your users down to exact, specific requirements. The devil is in the detail as any misinterpretations can lead to wasted developments. The requirements must be balanced by the constraints on the developer and it might be that some features are planned for the first release and others are left for a later date, but are considered in the design. It is usually far harder to modify existing code for new features than to code (or at least consider) them from the start, so any effort here will be amply rewarded.

`Development`

Demonstrate your software to its intended users at all stages of development, but especially early on. This allows for confirmation that functionality is developing as planned. As implementation proceeds and becomes fixed on a particular approach, there can be a tendency for the developer's and user's notion of the software to diverge. This is mitigated if good communication is maintained. Inevitably, during implementation, issues arise and ideas evolve. By keeping in touch you avoid unnecessary developments and can fine-tune the user requirements in light of the continuing developments.

`Testing and validation`

A most valuable contribution of users is in testing software. They will find ways to use (and abuse) the software that you did not anticipate. Crucially, this should validate that the software works as required in real-case scenarios and meets the user's needs. Often, new bugs and requirements will arise from testing, which must be fixed or implemented as the beta software is refined. Testing should result in a release candidate in which major features are frozen. From that point on only minor bug fixes and features should be implemented for the release proper. Be organised so that suggestions can be incorporated at the appropriate time: suggestions for major new features are not appreciated the day before the release date!

Try to get a maintain a clear picture of your user's requirements and plan your work accordingly.

`Evolution`
Even software implemented to the very best design can only fulfil the requirements that were known at the time. In practice, your software must evolve to keep pace with your users, whose needs are often evolving. In some cases, when adapting software for changing requirements, it's necessary to move the software back into beta or even to start from scratch where a complete redesign is necessary. In such cases an appreciation of the current stage of the software helps to manage expectations.

3.1.5 Planning

Before you start coding you should think deeply about what exactly the code is supposed to achieve. Do this on at least two different occasions so that you are certain you've considered everything. Once you start coding it'll be much more difficult to incorporate new ideas. The more detailed the planning the better!

Except for trivial cases it is helpful to code to some sort of design. This could be a 'disposable' first version of the code, intended only to test out ideas, that is discarded when coding proper begins. Otherwise use formal and informal methods such as flowcharts, logical steps in prose, etc. Anything here that helps to form a crystal clear idea of the purpose and behaviour of the code will suffice. Give particular attention to inputs and outputs, major logical steps and any tricky steps. The process should result in a design that is fit to adopt for coding.

3.1.5.1 Application development

EMBOSS includes many applications and it's likely that one exists which does something similar to what you need. Before you start coding you should therefore check the application documentation:

- EMBOSS (http://emboss.open-bio.org/rel/dev/apps/)
- EMBASSY (http://emboss.open-bio.org/rel/dev/apps/)

There are three basic types of application development:

- Modify an existing application
- Develop a new application
- Develop a wrapper application to existing third-party software.

If it seems that only a simple modification to an existing application is required, this might easily be achieved by modifying the ACD file without any new C coding at all. See ACD file development (Chapter 5, *C programming*) and ACD syntax (Appendix A, *ACD syntax reference*).

In other cases an entirely new application is required. The AJAX and NUCLEUS libraries (see Section B.1, 'C programming libraries') provide a comprehensive toolkit for developing applications from scratch. The existing application ACD files and C source code might provide a valuable starting point or offer clues as to how to proceed.

There are also cases where you either already have the source code for an application that you want to incorporate into EMBOSS, or you otherwise want to provide an EMBOSS-style command line interface to an existing application. The wrapping and porting of applications under EMBOSS is described later (Section 10.2, 'Wrappers and ports').

When considering a new application there are two basic rules (with a few exceptions):

- An EMBOSS application should perform a single clearly defined operation.
- A new application should only be written if it differs by more than one extra major parameter from an existing program.

Three areas in planning deserve special attention:

- Preliminary documentation
- Inputs and outputs
- Objects and functions.

`Preliminary documentation`

Write some preliminary documentation for the application and its code *before* coding. This might include, but is by no means limited to:

- A single-line description of what the application does
- A more detailed description of what the application does
- A list of the input and output files
- A list of the other application parameters
- A single-line description of each parameter
- A default value for each parameter and whether a value is required from the user
- Numbered comments for all the major logical steps to your program (goes into the C source code file)
- Comments for tricky steps in your program
- Pseudo-code or other structured comments for each major step to capture the program logic, loop structures and so on.

`Inputs and outputs`

Give particular attention to the application parameters; all of its inputs, outputs and any other options. You should make a definitive list of the data definitions (Section 4.3, 'Data definition') you'll need in your ACD file.

`Objects and functions`

You should know what objects (C data structures) and functions you'll need *before* you start coding, so that everything you need is available to you. This is just the same as collecting all your tools together before starting a DIY project. Progress will be slow if you have to repeatedly break off to find functions when what you should be doing is concentrating on the program logic. If any new objects and functions are required then refer to the guidelines for library development (Section 5.5, 'Programming with objects').

3.1.5.2 Library development

The AJAX and NUCLEUS libraries are easily extended, but might already contain the functionality you need. The first thing to do therefore is to check the documentation for AJAX and NUCLEUS (see Appendix B, *C programming libraries*).

Library developments include:

- Extending the functionality of the existing code
- Developing new objects and functions
- Developing entirely new library files.

When developing library code, be careful to avoid redundancy with the existing code. You should, if possible, modify existing objects or functions, or at least use the C source code as a starting point or for ideas on how to proceed.

Where the code is specialised (used by one application only) it should stay in the application C source code file until it is more generally used. Only code that is likely to be of general use should go into AJAX or NUCLEUS.

Where new code is in the area of an existing AJAX or NUCLEUS library file it should be added to that file. Where a new category of functionality is required, a new library file should be created for it.

3.1.6 Before you start coding

Before you start coding, please discuss your ideas with the EMBOSS developers:

```
emboss-bug@emboss.open-bio.org.
```

Also let other EMBOSS users know. Many people may wish to collaborate with you or suggest easier ways of doing things:

```
emboss@emboss.open-bio.org.
```

There are a few basic things to consider, especially if you intend your code to be incorporated into the public package:

Application ACD file Any new application will need an ACD file. They are easy to write once you are familiar with the general layout, syntax (Appendix A, *ACD syntax reference*) and programming methods (Chapter 5, *C programming*).

Linking to the EMBOSS libraries Application code must be linked to the EMBOSS libraries. The minimum requirement is code for processing the application ACD file (see Section 6.3,

'Handling ACD files'). In some cases you will want to leave the original source code untouched and develop an EMBOSS wrapper application to it instead (Section 10.2, 'Wrappers and ports').

Choice of programming language Ideally code should be in ANSI C or C++; however, contributions in other languages will be considered and are welcome.

Coding standards Code should adhere to basic conventions and standards (Appendix C, *C coding standards*). This is not strictly required but is encouraged. Certainly, do not be put off submitting code because you don't think it's up to scratch. It can be improved later if necessary.

Licensing Ideally code will be licensed under a GPL licence; however, code with non-GPL licences can be incorporated under EMBASSY (see Section 1.1, 'Licence information').

Testing All submitted code should be well tested and bug-free. If you've developed new applications then ideally you will provide formal quality assurance tests for it. For further information see Chapter 7, *Quality assurance*.

Documentation All applications and the underlying code should be well documented. The minimum requirement is that others are able to follow what you have done and use the software. Ideally the documentation will conform to basic standards defined for the code (Appendix D, *Code documentation standards*) and the applications (Chapter 8, *Application documentation standards*). Don't be put off if you think your code is not up to scratch. The EMBOSS developers will advise you on how it can be improved.

3.1.7 ACD file development

An ACD file describes in detail the parameters and command line interface for a single EMBOSS application. It specifies exactly what input values are required or permitted, and how user input is prompted for, to ensure that the application can run correctly. It also defines the basic functionality of other interfaces, e.g. graphical user interfaces (GUIs), derived from the ACD files.

An ACD file must contain a single *application definition* (Section 4.2, 'Application definition') and then a *data definition* (Section 4.3, 'Data definition') for each parameter. A definition consists of one or more attributes which are name:value pairs and describe the application or parameter in detail. The application definition is given first followed by the data definitions, which are organised into sections for "Input", "Output" and so on.

It's recommended to write the ACD file *before* the C source code because it defines much of the program's function. Writing the ACD file should be easy if you already have a detailed list of all the application parameters from your design. ACD files are written in the ACD syntax (Appendix A, *ACD syntax reference*); see the programming guidelines (Chapter 5, *C programming*).

Various utilities (Section 4.6, 'ACD utilities') are provided for validating, debugging and processing ACD files. You should make use of these tools to ensure that the ACD file is valid and properly formatted. For example, you can test an ACD file for functionality by using the **acdc** utility. Once you have a functional version you can generate one with standard formatting by using the **acdpretty** utility.

3.1.8 C source code development

3.1.8.1 Coding standards

To ensure consistency in the EMBOSS code, all C code that you write should conform to a basic style. You should familiarise yourself with the standards (Appendix C, *C coding standards*) before coding.

The coding standards mostly concern the layout of code, but includes some general guidelines for C programming in EMBOSS and particularly programming datatypes and functions.

3.1.8.2 Application development

You will need to create (if necessary) a new application C source code (`ApplicationName.c`) file in the appropriate directory. The file should have a short meaningful name which, if possible, bears some resemblance to applications of similar function. Typically the EMBASSY **myemboss** package is used for individual developments. Alternatively the applications directory in another EMBASSY package or EMBOSS itself could be used:

```
/home/auser/emboss/emboss/embassy/myemboss/src
/home/auser/emboss/emboss/embassy/PackageName/src/
/home/auser/emboss/emboss/emboss/
```

A few basic steps are recommended when writing application C source code:

1. Consider the correct layout for the code (see Appendix C, *C coding standards*) including structured comments. Copy the layout of an existing application, or (better still) use the template (.../`myemboss/src/template.c`) file that is provided in the **myemboss**.
2. Add an empty `main()` function.
3. Add prototypes for any new functions. These can be moved to AJAX or NUCLEUS later if required.
4. Add definitions for any new datatypes (see Section 5.5, 'Programming with objects'). Again, these can be moved to the libraries later.
5. Add any comments, pseudo-code, etc. from your preliminary design.
6. Declare variables and implement C code for reading any ACD data items (see Section 6.3, 'Handling ACD files').
7. Add code at the end of `main()` for cleaning up memory for the ACD data items.
8. Write the remaining C source code for the `main()` function.

9. Write the C source code for any functions.
10. Ensure the code in general and functions and datatypes in particular are adequately documented (see Appendix D, *Code documentation standards*).

3.1.8.3 Library development

New C source code (`*.c`) and / or header files (`*.h`) may be created in the core AJAX or NUCLEUS directory as required:

```
/home/auser/emboss/emboss/ajax/core
/home/auser/emboss/emboss/nucleus
```

The files should have a short, meaningful name. AJAX files must begin with the prefix `aj`, for example `ajstr.h` and `ajstr.c` are used for the string handling library. NUCLEUS files must have the prefix `emb`; for example `embaln.h` and `embaln.c` are used for the alignment algorithm code.

To integrate large collections of code, for example third-party libraries, a new directory would usually be created under the main AJAX directory:

```
/home/auser/emboss/emboss/ajax/
```

For example, files in the **zlib** library live in:

```
/home/auser/emboss/emboss/ajax/zlib
```

There are some basic steps when writing library code:

1. Consider the correct layout for the code (see Appendix C, *C coding standards*) including structured comments. Copying the layout of an existing library file is a good strategy.
2. Add function prototypes.
3. Add datatype definitions (see Section 5.5, 'Programming with objects'). This includes private datatypes (in the source file) and public datatypes (in the header file).
4. Write the C source code for the functions. This includes private (static) and public (external) functions, both of which go in the source file.
5. Ensure the code in general and functions and datatypes in particular are adequately documented (see Appendix D, *Code documentation standards*).

3.1.9 Integration and compilation

New applications, regardless of whether they are destined for EMBOSS or an EMBASSY package, will typically be developed first in **myemboss** (Section 3.2, 'Integration and compilation'). This is an almost empty EMBASSY package and provides a way to isolate local application development from the rest of EMBOSS. Applications are built using a `Makefile` file that is generated from a `Makefile.am` file, which must be edited to mention the new applications. An application developed and tested in **myemboss** can be added to the main package or an EMBASSY package using essentially the same process.

To add an entirely new EMBASSY package you must create the directory structure and required files, which includes the application ACD files, C source code files and documentation, and edit the file `configure.in` and several `Makefile.am` files.

New functions and datatypes for AJAX and NUCLEUS must have structured comments to be properly integrated with the rest of the libraries. Entirely new library files are integrated by editing the `Makefile.am` in the AJAX or NUCLEUS directory as appropriate. You must also add any new header files to `ajax.h` and / or `emboss.h`.

For more information see Section 3.2, 'Integration and compilation'.

3.1.10 Debugging

Any contributed code should be as bug-free as possible. EMBOSS includes inbuilt features for debugging applications. In addition, debugging functions are provided for fixing specific modules of code. For further information see Section 3.3, 'Debugging'.

3.1.11 Quality assurance testing

Various quality assurance tests are performed on the EMBOSS code and documentation to maintain quality. All applications are run on test data to ensure they work as advertised. Regular compilation and memory leak tests of the whole package ensure the integrity of the code, for example that applications are not broken by recent changes to the library. The structured documentation for objects and functions is validated. This ensures, for example, that functions and parameters have meaningful and consistent names and functions are listed in the correct section of the file.

Quality assurance testing is handled by the EMBOSS developers but there are ways to help. Your own code should be tested thoroughly in real use-case scenarios. In particular, it should be tested for memory leakage (see Section 3.3, 'Debugging') before it is contributed: if you are unclear how to do this then ask on the EMBOSS mailing lists (Section 1.4, 'Project mailing lists'). Any library code for submission, or at least new public datatypes and functions, should be documented with structured comments (see Appendix D, *Code documentation standards*) so that validation checks can be performed. Any new applications should, ideally, have test cases and test data (see Chapter 7, *Quality assurance*).

3.1.12 Documentation

Software without documentation often has little value whereas good documentation can enhance the usefulness of software immensely. All EMBOSS application and library code for submission should be adequately documented. End-user documentation is also required for any new applications.

The process for generating the application end-user documentation, which is split into different sections, involves combining manually written components with parts that are generated automatically (for example, from parsing the ACD file). The process is largely automated. If you develop a new EMBASSY package then it will require documentation above and beyond that for individual applications.

All library code must be adequately documented using structured comments, which are parsed to generate the online library documentation (Section 1.3, 'Developer documentation'). Comments at the head of application code mostly contain licence and bibliographic information. However, all public functions and datatatypes must also be documented. Functions are organised in the library files under functional sections, which define, amongst other things, naming rules for functions and their parameters. The comments at the head of the library file usually contain extensive documentation. If you develop a new library file then you will need to provide documentation to the same level.

See the instructions on code documentation (Appendix D, *Code documentation standards*) and application end-user documentation (Chapter 8, *Application documentation standards*) for more information.

3.1.13 Distribution

Any contributed code will be widely distributed and installed on many different platforms. EMBOSS is known to work on almost every UNIX system, Microsoft Windows and MacOSX. There is nothing to do here because this is managed by the EMBOSS developers, although it helps greatly if you integrate, test and properly document any new software as summarised above.

3.1.14 Maintenance, support and training

Any contributed code will need maintenance and support. This, to an extent, is covered by the EMBOSS developers who also run EMBOSS training courses. However, it helps greatly if you maintain your own source code and support it through the EMBOSS mailing lists (Section 1.4, 'Project mailing lists').

3.2 Integration and compilation

3.2.1 Using myemboss for application development

The **myemboss** package is an almost empty EMBASSY package where developers can safely build their own EMBOSS applications. It provides a way to isolate local application development from the rest of EMBOSS so preserving the integrity of the package. New applications, regardless of whether they are destined for another EMBASSY package or EMBOSS itself, will typically start their life in **myemboss**.

myemboss requires the GNU tools **autoconf, automake** and **libtool** to be installed (see Section 1.2, 'Installation of CVS (Developer's) Release'). Using the GNU tools you can add new applications very easily. **myemboss** provides a directory structure and `Makefile` stubs for integrating new applications. Application ACD files go in the `myemboss/emboss_acd` directory and source code into the `myemboss/src` directory.

Applications are built using a `Makefile` file, but you do not need to edit this file directly. Instead you edit a file `Makefile.am` which **automake** converts into a file `Makefile.in` and **configure** converts into the `Makefile` file.

The `Makefile.am` file is shown below and has been edited to include two applications:

```
#

SUBDIRS = source emboss_acd emboss_doc

EXTRA_DIST = depcomp ltmain.sh install-sh config.sub config.guess

# tar to pick up the other directories
# then remove any CVS subdirectories

dist-hook:
 tar cBf - emboss_acd | ( cd $(distdir); tar xBf - ; cd emboss_acd; rm -rf CVS)
 tar cBf - emboss_doc | ( cd $(distdir); tar xBf - ; cd emboss_doc; rm -rf CVS; rm -rf
master)
```

To compile applications in **myemboss**, you must configure the package if you have not already done so (see Section 1.2, 'Installation of CVS (Developer's) Release'), then type the following from the **myemboss** applications directory (`myemboss/src`):

```
make
```

`make` will regenerate the `Makefile` file and build your new application, linking it to the EMBOSS libraries (AJAX and NUCLEUS).

This assumes you have installed and configured EMBOSS and the **myemboss** package. If you did a full EMBOSS install using **`make install`** then you must also type:

```
make install
```

to complete the installation of applications developed under **myemboss**.

For more information on installation see Section 1.2, 'Installation of CVS (Developer's) Release' and the *EMBOSS Administrator's Guide*.

3.2.2 Adding new EMBOSS applications

An application that has been developed and tested in **myemboss** can be added to the main package. Adding new applications to EMBOSS is essentially the same as for **myemboss**. EMBOSS has its own `Makefile.am` which, in contrast to **myemboss**, organises applications into two strands:

- Standard EMBOSS applications (the "make" applications) are listed after the `bin_PROGRAMS` line
- Applications that are not yet fully integrated ("make check" applications) are listed after the `check_PROGRAMS` line.

Applications which are not entirely complete should be put into the "make check" area of the `emboss/Makefile.am` file. For example, applications that compile and work correctly, but for which full documentation and quality assurance tests are not yet available, should go here. Only applications that are fully working, tested and documented should go into the "make" area.

An excerpt from the `Makefile.am` file is below and includes a few "make check" applications and standard "make" applications:

```
SUBDIRS = acd data

EXTRA_DIST = emboss.default.template

if ISSHARED
if ISAIXIA64
AM_CFLAGS = -Wl,-bdynamic -Wl,-brtl -L../ajax/.libs -lajax
else
AM_CFLAGS =
endif
endif

check_PROGRAMS = acdlog acdxsd ajtest \
complex corbatest \
entrails entrailsbook entrailshtml entrailswiki giep histogramtest intconv \
origsplitter origunion \
newcoils \
… <lines omitted for brevity>

bin_PROGRAMS = aaindexextract abiview \
acdc acdpretty acdtable acdtrace acdvalid \
aligncopy aligncopypair antigenic \
backtranambig backtranseq banana biosed btwisted \
… <lines omitted for brevity>

water whichdb wobble \
wordcount wordfinder wordmatch wossname \
yank

INCLUDES = -I$(top_srcdir)/nucleus -I$(top_srcdir)/ajax -I$(top_srcdir)/plplot

aaindexextract_SOURCES = aaindexextract.c
abiview_SOURCES = abiview.c
acdc_SOURCES = acdc.c
acdpretty_SOURCES = acdpretty.c
acdtable_SOURCES = acdtable.c

… <lines omitted for brevity>

seqretsingle_SOURCES = seqretsingle.c
seqrettype_SOURCES = seqrettype.c
testplot_SOURCES = testplot.c
treetypedisplay_SOURCES = treetypedisplay.c

# Uncomment to use dual libraries
if ISCYGWIN
LDADD = ../nucleus/libnucleus.la ../ajax/libajax.la ../plplot/libeplplot.la $(XLIB)
else
LDADD = ../nucleus/libnucleus.la ../ajax/libajaxg.la ../ajax/libajax.la ../plplot/
libeplplot.la $(XLIB)
endif

pkgdata_DATA = emboss.default.template
pkgdatadir=$(prefix)/share/$(PACKAGE)
```

```
#if PURIFY
# LINK = purify $(LIBTOOL) --mode=link $(CC) $(CFLAGS) $(LDFLAGS) -o $@
#else
#
#endif
```

Compilation is the same as for **myemboss**. You must configure the EMBOSS package if you have not already done so (see Section 1.2, 'Installation of CVS (Developer's) Release'), then from the applications directory (e.g. `/home/auser/emboss/emboss/emboss/`) type:

```
make
```

This will regenerate the `Makefile` file, build your new application and link it to the libraries. If you installed EMBOSS using **`make install`** then you must also type:

```
make install
```

to complete the installation of the new application.

For more information on installation see Section 1.2, 'Installation of CVS (Developer's) Release' and the *EMBOSS Administrator's Guide*.

3.2.3 Adding new EMBASSY applications

Applications are added to EMBASSY packages in exactly the same way as for **myemboss** (see above).

3.2.4 Adding a new EMBASSY package

To add an entirely new EMBASSY package you must:

1. Create the directory structure and required files for the package under the main EMBASSY directory (.../`emboss/emboss/embassy/`).
2. Write the application ACD and C source code files and populate the appropriate directories.
3. Edit the `Makefile.am` files.
4. Edit the `configure.in` file.
5. Create the application documentation and populate the appropriate directories.

3.2.4.1 Directory structure

The directory structure and file names are mostly the same regardless of the package. For example, the **hmmernew** EMBASSY package looks like this:

```
% cd emboss/emboss/embassy/hmmernew/
% ls
AUTHORS       COPYING    emboss_acd    ltmain.sh     NEWS    Userguide.pdf
ChangeLog     CVS        emboss_doc    m4            README
configure.in  DOCS       INSTALL       Makefile.am   src
```

The required directories are as follows:

src	The "applications directory" holding the application C source code. This should contain a Makefile.am file.
emboss_acd	The "ACD directory" holding the application ACD files. This should contain a Makefile.am file (see below).
emboss_doc	The "documentation directory". This contains the following sub-directories: html" : documentation in HTML format text : documentation in plain text format master : documentation in final format with include files There is a Makefile.am file in this directory and in the html" and text sub-directories, all of which need editing when the new package is created. A semi-automated process (Appendix 8, *Application documentation standards*) is used to create the files of documentation in these directories.
DOCS	This was intended for peripheral documentation but is not currently used.
m4	This contains macro files used in the automake process. The files can be copied from another EMBASSY package.

Additionally, some packages may have the following directories:

include	Header files that are specific to the package.
data	Data files that are specific to the package.

The CVS directory only appears if the package has been committed to the CVS server.

The required files are as follows. If you omit any of these then the GNU software will complain when you come to bundle the EMBASSY package into a distribution file; the GNU software will then usually add a default file:

README	This is the package user guide and normally contains general comments about the package as a whole such as installation instructions, usage notes and so on. It should not normally contain documentation about individual applications. The README is not strictly required but is very highly recommended, especially where the package contains wrappers to third-party applications (see Section 10.2, 'Wrappers and ports') and there are differences between the original applications and the wrappers.
COPYING	Copyright or other information on usage restrictions. This might just be a reference to the README file where such information is given, or contain the GPL licence or something else.
INSTALL	Standard installation instructions for EMBASSY packages. This file can be copied from another EMBASSY package.
NEWS	Package version information. It is used in making a bundled distribution file (via **make dist**).
AUTHORS	Author contact details. Where an EMBASSY package contains wrapper applications, both the authors of the original software and the EMBASSY package should be mentioned.

`ChangeLog`	The log file of changes made to the package. It is helpful to provide one but not strictly necessary.
`configure.in`	The configure file that is processed using **autoconf** to produce a configure script (see Section 1.2, 'Installation of CVS (Developer's) Release'). It should be copied from another EMBASSY package but requires some simple customisation before it can be used (see Section 3.2.4.3, 'Editing the `configure.in` files').
`ltmain.sh`	The **libtool** package front-end script. It can be copied from another EMBASSY package.

Other files can be given. For example, **HMMERNEW** includes documentation for the original **HMMER** application.

3.2.4.2 Editing the `Makefile.am` files

The `Makefile.am` files used in EMBASSY are different from those used in the main EMBOSS package. When creating a new EMBASSY package, there are six `Makefile.am` files which must be edited, in each of the following directories:

- `/packagename/`
- `/packagename/src`
- `/packagename/emboss_acd`
- `/packagename/emboss_doc`
- `/packagename/emboss_doc/html`
- `/packagename/emboss_doc/text`

The required edits to these files should be obvious from inspecting the files below, which are for the **myemboss** package.

`/myemboss/Makefile.am.` The file is shown below:

```
#
SUBDIRS = source emboss_acd emboss_doc

EXTRA_DIST = depcomp ltmain.sh install-sh config.sub config.guess

# tar to pick up the other directories
# then remove any CVS subdirectories

dist-hook:
 tar cBf - emboss_acd | ( cd $(distdir); tar xBf - ; cd emboss_acd; rm -rf CVS)
 tar cBf - emboss_doc | ( cd $(distdir); tar xBf - ; cd emboss_doc; rm -rf CVS; rm -
rf master)
```

The `SUBDIRS` line must reference all top-level subdirectories in the package for which there are `Makefile.am` files.

The `dist-hook` entry must be edited accordingly.

`/myemboss/src/Makefile.am.` The file is shown below:

```
## Process this file with automake to produce Makefile.in

EXTRA_DIST = Makefile.am

bin_PROGRAMS = myseq mytest

if LOCALLINK
INCLUDES = -I../include -I../../../nucleus -I../../../ajax -I../../../plplot
else
INCLUDES = -I../include -I${embprefix}/include -I${embprefix}/include/eplplot
endif


if ISSHARED
if ISAIXIA64
if LOCALLINK
AM_CFLAGS = -Wl,-bdynamic -Wl,-brtl -L../../../plplot/.libs -L../../../ajax/.libs -
L../../../nucleus/.libs -lnucleus -lajax -leplplot
else
AM_CFLAGS = -Wl,-bdynamic -Wl,-brtl -L${embprefix}/lib -lnucleus -lajax -leplplot
endif
else
AM_CFLAGS =
endif
endif


myseq_SOURCES = myseq.c
mytest_SOURCES = mytest.c



if LOCALLINK
LDADD = ../../../nucleus/libnucleus.la ../../../ajax/libajaxg.la ../../../ajax/
libajax.la ../../../plplot/libeplplot.la $(XLIB)
else
LDADD = -L${embprefix}/lib -lnucleus -lajaxg -lajax -leplplot $(XLIB)
endif
```

The names of all the applications must be given on the `bin_PROGRAMS` entry, which may span multiple lines if these are delimited by a backslash character. A reference to the C source code file of each application is given towards the bottom of the file, for example `myseq_SOURCES = myseq.c`.

/myemboss/emboss_acd/Makefile.am.

```
EXTRA_DIST = Makefile.am

pkgdata_DATA = myseq.acd mytest.acd

pkgdatadir=$(prefix)/share/EMBOSS/acd
```

The name of all the application ACD files must be given following the `pkgdata_DATA` entry, which may span multiple lines.

/myemboss/emboss_doc/Makefile.am.

```
SUBDIRS = html text
```

The file does not typically require editing.

/myemboss/emboss_doc/html/Makefile.am.

```
pkgdata_DATA = index.html \
myseq.html

pkgdatadir=$(prefix)/share/EMBOSS/doc/html/embassy/myseq
```

All the application documentation files (Chapter 8, *Application documentation standards*) in HTML format must be listed in the `pkgdata_DATA` entry, which may span multiple lines. Any other documentation files that are automatically generated are listed here also. The line beginning with `pkgdatadir=` must be changed so that it gives the correct package name.

/myemboss/emboss_doc/text/Makefile.am. The file is shown below:

```
pkgdata_DATA =  myseq.txt

pkgdatadir=$(prefix)/share/EMBOSS/doc/programs/text
```

This is essentially the same as `/myemboss/emboss_doc/html/Makefile.am` but references the documentation in plain text format.

3.2.4.3 Editing the `configure.in` files

Two lines only need to be edited here:

- The `AC_INIT` line must be changed so that it references one (any) application C source code file in the package.
- The `AM_INIT_AUTOMAKE` line should give the package name and version number.

Once the application ACD files and C source code are written and integrated, compilation is as for any other EMBASSY package (see Section 3.2.1, 'Using **myemboss** for application development').

3.2.5 Adding new functions and datatypes

Any new functions and datatypes added to AJAX or NUCLEUS should be properly documented (Section D.1, 'Code documentation standards') if they are to be incorporated into the EMBOSS distribution. To compile the new code, type:

```
make
```

from the AJAX or NUCLEUS directory as appropriate. Of course, you will need to recompile any statically linked applications to use the changes. Alternatively, recompile the whole package by typing **make** from the second-level EMBOSS directory, e.g.:

```
/home/auser/emboss/emboss/
```

3.2.6 Adding new library files

If you have added entirely new library files then these must be integrated into AJAX or NUCLEUS by editing the `Makefile.am` in the AJAX or NUCLEUS directory as appropriate.

An excerpt of the the AJAX `Makefile.am` file:

```
## Process this file with automake to produce Makefile.in

if ISCYGWIN
lib_LTLIBRARIES = libajax.la
else
lib_LTLIBRARIES = libajax.la libajaxg.la
endif

if ISSHARED
if ISAIXIA64
if NEEDAJAX
AM_CFLAGS = -Wl,-G -L../plplot/.libs -L.libs -lajax -leplplot -lm
else
AM_CFLAGS = -Wl,-G -L../plplot/.libs -L.libs -leplplot -lm
endif
else
AM_CFLAGS =
endif
endif

if ISCYGWIN
AM_LDFLAGS = -no-undefined -Wl,--add-stdcall-alias -Wl,--export-all-symbols -Wl,--
enable-auto-import -L../plplot -leplplot
endif

## PCRE stuff (version 4.3)

POSIX_MALLOC_THRESHOLD = @POSIX_MALLOC_THRESHOLD@
PCRELIBVERSION = @PCRE_LIB_VERSION@
PCREPOSIXLIBVERSION = @PCRE_POSIXLIB_VERSION@

INCLUDES = -DAJAX_FIXED_ROOT=$(AJAX_FIXED_ROOT) -DPREFIX=\"$(prefix)\" -
I$(top_srcdir)/plplot $(POSIX_MALLOC_THRESHOLD)
#LDFLAGS = -no-undefined
LIBS =

AJAXSRC = ajacd.c ajalign.c ajarr.c ajassert.c \
ajbase.c \
ajcall.c ajcod.c \
ajdan.c ajdmx.c ajdom.c ajdomain.c ajexcept.c \
ajfeat.c ajfile.c ajfmt.c ajgraphxml.c ajindex.c ajjava.c ajlist.c \
ajmath.c ajmatrices.c ajmem.c ajmess.c ajnam.c ajnexus.c \
ajpat.c ajpdb.c ajpdbio.c ajphylo.c \
ajrange.c ajreg.c ajreport.c \
ajseq.c ajseqabi.c ajseqdb.c ajseqread.c ajseqtype.c ajseqwrite.c \
ajsort.c ajstr.c ajsys.c \
ajtable.c ajtime.c ajtranslate.c ajtree.c ajutil.c ajvector.c \
pcre.c pcre_study.c pcreposix.c

AJAXGSRC = ajgraph.c ajhist.c

include_HEADERS = ajax.h ajarch.h ajdefine.h \
ajacd.h ajalign.h ajarr.h ajassert.h \
```

```
ajbase.h ajcall.h ajcod.h \
ajdan.h ajdmx.h ajdom.h ajdomain.h ajexcept.h \
ajfeat.h ajfeatdata.h ajfile.h ajfmt.h \
ajgraph.h ajgraphxml.h ajgraphstruct.h ajhist.h ajindex.h ajjava.h ajlist.h \
ajmath.h ajmatrices.h ajmem.h ajmess.h ajnam.h ajnexus.h \
ajpat.h ajpdb.h ajpdbio.h ajphylo.h \
ajrange.h ajreg.h ajreport.h \
ajseq.h ajseqabi.h ajseqdata.h ajseqdb.h ajseqread.h ajseqtype.h ajseqwrite.h \
ajsort.h ajstr.h ajsys.h \
ajtable.h ajtime.h ajtranslate.h ajtree.h ajutil.h ajvector.h \
pcre.h pcre_chartables.c pcre_config.h pcre_get.c pcre_printint.c \
pcre_internal.h pcreposix.h


if ISCYGWIN
libajax_la_SOURCES = $(AJAXSRC) $(AJAXGSRC)
libajax_la_LDFLAGS = -version-info 5:0:0
else
libajax_la_SOURCES = $(AJAXSRC)
libajax_la_LDFLAGS = -version-info 5:0:0
libajaxg_la_SOURCES = $(AJAXGSRC)
libajaxg_la_LDFLAGS = -version-info 5:0:0
endif
```

The NUCLEUS `Makefile.am` file:

```
## Process this file with automake to produce Makefile.in

lib_LTLIBRARIES = libnucleus.la

if ISSHARED
if ISAIXIA64
AM_CFLAGS = -Wl,-G -lm -L../ajax/.libs -lajax
else
AM_CFLAGS =
endif
endif

if ISCYGWIN
AM_LDFLAGS = -no-undefined -Wl,--add-stdcall-alias -Wl,
--export-all-symbols -Wl,--enable-auto-import -L../plplot -L../ajax -leplplot -
lajax
endif


INCLUDES = -I$(top_srcdir)/plplot $(X_CFLAGS) -I$(srcdir)/ -I$(top_srcdir)/ajax

LIBS =

NUCLSRC = embaln.c embcom.c embcons.c \
    embdata.c embdbi.c embdmx.c embdomain.c \
    embest.c embexit.c embgroup.c \
    embiep.c embindex.c embinit.c \
    embmat.c embmisc.c embmol.c \
    embnmer.c embpat.c embpatlist.c embprop.c embpdb.c \
             embread.c embsig.c \
    embshow.c embword.c

libnucleus_la_SOURCES = $(NUCLSRC)
```

```
libnucleus_la_LDFLAGS = -version-info 5:0:0

include_HEADERS = emboss.h \
   embaln.h embcom.h embcons.h \
   embdata.h embdbi.h embdmx.h embdomain.h \
   embest.h embexit.h embgroup.h \
   embiep.h embindex.h embinit.h \
   embmat.h embmisc.h embmol.h \
   embnmer.h embpat.h embpatlist.h embprop.h embpdb.h \
                 embread.h embsig.h \
                 embshow.h embword.h
```

You must also add any new header files to `ajax.h` and/or `emboss.h`. `ajax.h` looks like this:

```
#ifdef __cplusplus
extern "C"
{
#endif

/*
**
** AJAX master include file
**
*/

#ifndef ajax_h
#define ajax_h

#include <stdio.h>

#include "ajarch.h"

#include <stdarg.h>
#include <stdio.h>
#include <stdlib.h>
#include <string.h>
/*#include <malloc.h>*/

#include "ajassert.h"
#include "ajdefine.h"
#include "ajstr.h"
#include "ajtime.h"
#include "ajfile.h"
#include "ajlist.h"

#include "ajtable.h"
#include "ajnexus.h"
#include "ajfeat.h"
#include "ajseq.h"
#include "ajseqdata.h"
#include "ajseqtype.h"
#include "ajarr.h"
#include "ajseqread.h"
#include "ajseqwrite.h"
#include "ajindex.h"
#include "ajseqdb.h"
#include "ajseqabi.h"
#include "ajreport.h"
#include "ajfmt.h"
```

```
#include "ajmatrices.h"
#include "ajalign.h"

#include "ajrange.h"
#include "ajcod.h"
#include "ajmath.h"

…
… Entries omitted for brevity
…

#endif

#ifdef __cplusplus
}
#endif
```

emboss.h looks like this:

```
#ifndef emboss_h
#define emboss_h

#include "ajax.h"
#include "ajgraph.h"
#include "embaln.h"
#include "embcom.h"
#include "embcons.h"
#include "embdbi.h"
…
… Entries omitted for brevity
…

#ifdef WIN32
#define VERSION "5.0.0"
#endif


#endif
```

3.3 Debugging

Bugs are problems with code that cause it to crash or operate in an unexpected way. They arise through erroneous use of syntax (which is not always caught by compilers) and errors in the code logic. This section gives some practical hints for debugging EMBOSS code.

3.3.1 Direct debugging

Very broadly, debugging proceeds in four stages, where bugs are fixed that:

1. Prevent the program from compiling
2. Cause the program to crash
3. Cause the program to operate incorrectly
4. Manifest under extensive test conditions.

With experience, most bugs are obvious from visual inspection of the code. It is highly recommended that, *before* you compile your code, you read then reread it until you're satisfied it should work as expected. Be scrupulous when writing the code itself. Avoid the temptation to code too quickly; the extra time spent avoiding errors in the first place will be very well rewarded later.

Tip

A simple and powerful debugging method is to use **ajFmtPrint** and **fflush(stdout)** statements to report values of key variables at different stages of execution, allowing you to trace and identify problems. **ajFmtPrint** is used to print variables, and **fflush** is called immediately afterwards to flush the output buffer. This is important as the output buffer might have content at the point of the crash, which will only be printed to screen by calling **fflush(stdout)**. Most bugs are easily squashed using this method.

When writing your code, there will be many stages where you know in advance what value (s) a variable should (or should not) have, particularly when checking a function's arguments or return values. For instance, the value of a pointer used by a function should in most cases not be NULL. At all such places, at least in early versions of the code, code should be added to trap errors and raise appropriate warnings, alerting you to potential bugs before they manifest.

AJAX functions in ajmess.h provide various levels of error handling. Each of the following format and output an exception message (provided as a string):

ajUser	Report an informative message
ajWarn	Report a warning message
ajErr	Report an error message
ajExit	Report a message then exit
ajDie	Report a message then crash (kill) the application
ajDebug	Report a general debugging message to the file programname.dbg if the switch -**debug** was given on the command line

Messages go to stdout or stderr (in both cases usually the screen) or, in the case of ajDebug, to the file programname.dbg.

Important

The EMBOSS code makes extensive use of **ajDebug** so that bugs reported by users can easily be traced. The typical way therefore to debug applications is to produce a debug trace using **-debug**.

In practice, much of the error-trapping code can, for purposes of speed, be commented out or removed once extensive testing of the code is complete. This is especially true for library function code where speed is paramount.

For the vast majority of applications, `ajDebug` and the other functions above will suffice. In special circumstances, however, you might need to write your own exception handling functions. An AJAX library file `ajmess` is provided for this (see Section 6.23, 'Handling exception messages').

3.3.2 AJAX debugging functions

Most AJAX library files include functions for debugging the code in that library file. These usually call `ajDebug`. Typically functions are provided to report on the internal state of data structures defined in that library file. In some cases special debugging files are provided. For example, there is a "debug" output sequence format used when debugging sequence output, and a "trace feature table" used when debugging report formats.

For more information, see the library documentation for AJAX and NUCLEUS. See also the programming guides (Section 3.1, 'EMBOSS programming') for individual library files.

3.3.3 Controlling debugging behaviour

The default behaviour of EMBOSS is *not* to report debugging information generated by calls to `ajDebug`. The global command line qualifier **-debug** can be used to turn debugging on for any EMBOSS application. For example, if you think you have found a bug when the following command is issued:

```
seqret sequence.seq
```

then debugging can be turned on as follows:

```
seqret sequence.seq -debug
```

This will create a debug file called `seqret.dbg`. Debugging could be explicitly turned off by prepending the qualifier with **no**:

```
seqret sequence.seq -nodebug
```

but there's normally no need to do this as the default is false (no debugging) anyway. It could however be useful if debugging was turned on by default in the EMBOSS configuration files or by an environment variable. Debugging can be globally switched on using the EMBOSS environment variable:

```
EMBOSS_DEBUG
```

If this is set `TRUE`, all programs act as if they have **`-debug`** set on the command line. They create a file called `programname.dbg` containing debugging information.

Logging of the processing of the EMBOSS configuration files `emboss.default` and `.embossrc` can be turned on using:

```
EMBOSS_NAMDEBUG
```

This processing takes place *before* the **`-debug`** command line switch is processed. The functions that are called are described in `ajnam.c`. The debugging information includes a report of:

- All defined databases, variables and environment variables.
- Defined attributes for a database definition.
- Defined attributes for a resource definition.

3.3.4 Debuggers

Some bugs might not be obvious from visual inspection or easily traced using `ajFmtPrint` and `fflush(stdout)`. For these, it is worth using specialised debugging software. A debugger executes the bugged program and traces its internal state to allow problems with the code to be rapidly identified and fixed. The functions available depend on the debugger used. Most give control over the program, allowing it to be executed in a stepwise manner, variables to be given values and so on, providing a quicker and more reliable means of stepping through the logic of the code than doing this mentally. Most debuggers should provide at least the following information:

- The line of code and statement the program crashed on.
- If the error occurred within a function, the line the function was invoked from and the arguments.
- The values of variables (local to a function or global variables) at a particular point during execution of the program.
- The result of a particular expression in a program.

The most popular UNIX debugger is **GDB**, the GNU debugger (see http://sourceware.org/gdb/). It includes powerful features for tracing and altering the execution of a program, but you'll only need to use the very basics for it be extremely useful.

If you intend using **GDB** to debug EMBOSS code, it's necessary to configure the package using:

```
--enable-debug
```

before you build the package (see Section 1.2, 'Installation of CVS (Developer's) Release').

Note

The output of debuggers cannot always be entirely trusted.
Rare cases can arise where the behaviour of the executable compiled for debugging is subtly different to the standard executable. For this reason, `ajFmtPrint` and `fflush` statements, which are less invasive as far as the executable is concerned, remain the tool of last resort.

3.3.5 Tracing memory problems

Some bugs evade identification by direct means or debuggers. Such bugs are usually caused by serious programming errors including the invalid use of pointers, memory corruption, memory leaks, or invalid memory access, for instance, an out-of-bounds array. These errors can be extremely difficult to trace but can easily arise in C programming, especially where C pointers are used extensively. Fortunately, there are several excellent programs available that can trap and identify such problems. Even if your code appears to be working correctly it is well worth using a memory checker to check that memory is not being violated. Their use can avoid hours of frustration later on and will help ensure your code remains stable in different use cases. Popular programs include:

Valgrind	(http://valgrind.org/) A suite of tools for debugging and profiling Linux programs. It detects many memory management and threading bugs. It also performs profiling which helps you to identify ways to speed up and reduce memory usage of your programs.
PurifyPlus	Commercial software for memory corruption detection, memory leak detection, application performance profiling and code coverage analysis.
Insure++	Powerful commercial software checking the integrity of memory usage and detecting potential defects and inefficiencies in memory usage.

ACD file development

4

4.1 Introduction to ACD file development

4.1.1 Introduction to ACD files

Every EMBOSS and EMBASSY program has an ACD (AJAX command definition) file which describes the application, its options (parameters) and command line interface. The ACD file controls the behaviour of the application at the command line, particularly, all the user input operations. It includes:

- An application definition describing the application itself, for instance its name, documentation text and functional grouping.
- Attributes defining exactly what options are permitted or required under different conditions, how option values may be specified on the command line and how the user is prompted for values.
- A data definition for every application option. These describe the program parameters, i.e. the data the program can accept or requires to run. This includes input and output files and all other parameters.
- A datatype for each application option and attributes describing the permissible type or value in detail, for example default values and whether data has to be within limits. These are used for user input validation to ensure the application parameters are set correctly.
- Dependencies between application options, such as when a value is only required if another is specified, or the permitted type of value depends on the value or the presence of another. If for example the input sequence for an alignment program is DNA it should not accept a protein comparison matrix.

When an application is run, the ACD file is read and the command line processed automatically by an ACD file parser and command line processor that is internal to EMBOSS. Together these ensure that the required information is available at startup. All of the required option values are prompted for as necessary before the application starts. Input values that are incorrect or out of range are reprompted for automatically. The input values are read and held in memory, files are opened as required and so forth, so that all the required data are available when the application proper runs. An EMBOSS application cannot ask the user for more information after several hours of processing!

ACD files are written in the AJAX command definition (ACD) language which was designed specifically for EMBOSS. The ACD syntax allows for very flexible descriptions of an application's options and its command line interface. It specifies everything that can appear on the command line or in another interface such as a web page. Metadata specifically for web, GUI, and other types of interface are provided. For example, whether graphical elements such as radio buttons are appropriate for setting option values. These metadata and the syntax in general make it easy for interface developers to create friendly and intuitive interfaces. Many different interfaces are available (see the *EMBOSS User's Guide*).

The description of an application's options and interface frequently involves calculations, conditional statements and the use of variables and menus. The ACD syntax supports all these operations and also lets you probe the properties of input datatypes such as sequences. ACD files support various biological datatypes and many different file formats, saving the application developer huge effort. For instance, biological sequences can be read from many different formats and using a variety of different access methods. Usage of these features is transparent to an application and they do not have to be specified in the ACD file.

4.1.2 ACD general syntax

ACD (Appendix A, *ACD syntax reference*) is a forgiving language that does not restrict the available syntax or layout any more than is strictly necessary. You don't need to know the full syntax to start writing ACD files; all the major features you'll need are summarised below.

Various utility applications (Section 4.6, 'ACD utilities') are provided for validating, debugging and processing ACD files. These can be used to validate and cleanly reformat an ACD file once it has been written.

4.1.2.1 ACD file names

ACD files are plain ASCII text files. They *must* have the extension `.acd` and typically have the same name as the application:

```
ApplicationName.acd
```

`ApplicationName` is the name that's passed to the function embInit which invokes ACD file parsing from the application C source code (see Section 4.2, 'Application definition').

4.1.2.2 Whitespace

During ACD file parsing, the entire file content is effectively treated as a single string which is parsed into tokens delimited by *space* characters. Only a single space between individual tokens is required, extra whitespace is ignored. All the tokens are treated as strings until they are finally resolved to their appropriate datatype. There is only one pass through when parsing, so any element (for example a datatype definition) must be declared before it can be referred to.

4.1.2.3 Comments

Comment lines can be added and begin with "#" and continue to the end of the line:

```
# This is a valid comment line and will be ignored during ACD parsing.
```

4.1.3 ACD definitions

An ACD file contains a definition of the application and all its options. An ACD file must contain a single *application definition* and then a *data definition* for each option. A definition contains one or more attributes, each of which is an `AttributeName: AttributeValue` pair. The application definition must be given first, followed by the data definitions which are organised into sections (Section 4.1.5, 'ACD file sections').

The general format for the application definition (Section 4.2, 'Application definition') is:

```
application: ApplicationName
[
   ApplicationAttribute1Name: "ApplicationAttribute1Value"
   ApplicationAttribute2Name: "ApplicationAttribute2Value"
]
```

The general format for a data definition (Section 4.2, 'Application definition') is:

```
Datatype: OptionName
[
   Attribute1Name: "Attribute1Value"
   Attribute2Name: "Attribute2Value"
]
```

Attribute values are normally enclosed in double quotes, although this is only mandatory for values (typically strings) which include whitespace. For example, here is a valid attribute definition:

```
string: "A String Value"
```

Here is an invalid one:

```
string: A String Value
```

The `application:` token and tokens representing datatypes can be abbreviated so long as they remain unambiguous. For example:

```
appl:
```

is sometimes given instead of:

```
application:
```

Similarly the sequence datatypes:

```
sequence:
seqset:
seqall:
```

could in principle be abbreviated to:

```
sequ
seqs
seqa.
```

Attributes can also be abbreviated. For example, the attribute to set a default value for a data definition:

```
default:
```

can be abbreviated to:

```
def:
d:
```

Such abbreviations are *not* recommended, however, because they tend to make the ACD file more difficult to read.

4.1.4 Parameters and qualifiers

Each application option, i.e. every data definition in the ACD file, can be defined via the appropriate ACD attribute to be one of the following:

- Parameter
- Standard qualifier
- Additional qualifier

with the default being:

- Advanced qualifier

The attributes to use are as follows:

```
parameter:    "Y"
standard:     "Y"
additional:   "Y"
```

`parameter: "Y"` means that the data definition is a *parameter*. `standard: "Y"` and `additional: "Y"` mean that the data definition is a *standard qualifier* or an *additional qualifier*. If none of `parameter: "Y"`, `standard: "Y"` or `additional: "Y"` are specified then the data definition defaults to being an *advanced qualifier*. Parameters are commonly used for input and output files and other primary input and output data, whereas the different types of qualifier are used for other application options.

Caution

Only one of **`parameter: "Y"`**, **`standard: "Y"`** or **`additional: "Y"`** should ever be given in a data definition. If more than one is used an error message will be generated during ACD processing.

Note

You should never explicitly specify **`Parameter: "N"`**, **`standard: "N"`** or **`additional: "N"`**. An error will be generated during ACD processing if you do. In practice, calculated values of "**`N`**" are in fact supported and can be used, in exceptional circumstances, to override the default behaviour of these attributes (see Section 4.5, 'Controlling the prompt').

Note

In earlier versions of EMBOSS, **`required:`** and **`optional:`** were used.

`standard:` now replaces **`required:`** and **`additional:`** replaces **`optional:`**.

The difference between parameters and qualifiers is that you don't have to use a flag to specify a value for a parameter on the command line. Values are specified like this:

```
ApplicationName Option Value.
```

In contrast, you *do* have to use a flag (the option name) to specify a value for a qualifier. Values for standard, additional and advanced qualifiers are specified like this:

```
ApplicationName -OptionName OptionValue.
```

The flag (option name) can also be given before parameter values but it is not required. If flags for parameters are not used it is, however, necessary to give the values in the *same* order as the corresponding data definitions appear in the ACD file.

This gives a simplified view of how to assign a value to an application parameter. There are more possibilities not described here, particularly for handling multiple options of the same type. For more information see the *EMBOSS User's Guide*.

4.1.4.1 Prompting behaviour and default values

Values for parameters and standard qualifiers are always prompted for if not specified on the command line. If a default value is specified in the ACD file, the user may accept this by pressing `<Return>`.

Values for additional qualifiers are not prompted for (the default value will be used instead) unless **`-options`** is given on the command line, which will turn prompting on for these qualifiers. A default value should *always* be given in the ACD file for additional qualifiers by specifying the `default:` attribute.

An advanced qualifier is never prompted for, even if **`-options`** is used, therefore a default value should be given in the ACD file.

4.1.4.2 Help information

Help information is generated when a program is run with the **`-help`** inbuilt qualifier. Help information for parameters and standard qualifiers appears in the 'Required' section of the help display. Information for additional and advanced qualifiers appears in the 'Advanced' section.

4.1.4.3 Examples

The following ACD file defines an application with a sequence parameter (`sequence`) and standard qualifier (`thresh`) to set a scoring threshold:

```
application: seqdemo
[
  documentation: "Demo application"
]

sequence: asequence
[
   parameter: "Y"
]

float: thresh
[
  standard: "Y"
```

```
  default: "2.5"
  minimum: "0.0"
  maximum: "100.0"
  information: "Score threshold"
]
```

The `asequence` parameter can be set without the qualifier name (flag), but one is needed for `thresh`. The application could be invoked as follows:

```
seqdemo filename.seq -thresh 5
seqdemo filename.seq
```

In the first case, `thresh` is set directly. In the second case, it is not set and the user would be prompted for a value.

If, in contrast, `thresh` was defined as an additional qualifier:

```
float: thresh [
  additional: "Y"
  default: "2.5"
  minimum: "0.0"
  maximum: "100.0"
  information: "Score threshold"
]
```

the user would only be prompted for a value if **`-options`** was set:

```
seqdemo filename.seq -options
```

The following ACD definition would define an advanced qualifier (`abool`):

```
boolean: abool
[
  default: "Yes"
]
```

The user would never be prompted for a value for `abool`, so if it was not set on the command line, the application would proceed with the default value (`"Yes"`). `abool` could be set on or off as follows:

```
acddemo -asequence filename.seq -abool
acddemo -asequence filename.seq -noabool
```

4.1.5 ACD file sections

4.1.5.1 Sections in an ACD file

EMBOSS processes ACD definitions in the order in which they appear in an ACD file. This order is insufficient, however, to allow interface developers to build clean groupings of options for user interfaces. To help interface developers, all ACD data definitions are grouped into five major sections, which must appear in the following order:

1. Input
2. Required
3. Additional
4. Advanced
5. Output

A section appears like this:

```
section: input
[
  information: "Input section"
  type: "page"
]
#
# Input data definitions go here
#

endsection: input
```

All data definitions in an ACD file *must* be contained within an appropriate section and the sections must be given in the order indicated. An error will be generated during ACD processing otherwise (see Section A.1.6, 'ACD file sections').

Each major section may include nested within it one or more subsections, which may contain further nested subsections as required although this is seldom necessary. Subsections, in contrast to major sections, have arbitrary names and can be given in any order.

For a detailed description of these sections and what goes in them, see Section A.1.6, 'ACD file sections'.

4.1.5.2 Input section

Input datatypes and other data definitions related to input. See Section A.2.2, 'Description of input ACD datatypes'.

4.1.5.3 Required section

Options which always require a value from the user but which are not in the input or output section. This includes all data definitions that are defined as *parameters* and all those defined as *standard qualifiers*.

4.1.5.4 Additional section

Additional qualifiers which are occasionally set by the user.

4.1.5.5 Advanced section

Advanced qualifiers for which a value may be set but is never prompted for. They are defined by the *absence* of the attributes:

```
parameter: "Y"
standard: "Y"
additional: "Y"
```

4.1.5.6 Output section

Output datatypes and other parameters related to output. See Section A.2.3, 'Description of output ACD datatypes'.

4.1.5.7 Example

The `search` input string of the **wossname** application is contained within an input section. The output section contains all the parameters and qualifiers related to output and also contains a subsection called `htmlsection`:

```
section: input
[
  information: "Input section"
  type: "page"
]

  string: search
  [
    parameter: "Y"
    default: ""
    information: "Text to search for, or blank to list all
                  programs"
    help:"Enter a word or words here and a case-independent search
          for it will be made in the one-line documentation, group names and
          keywords of all of the EMBOSS programs. If no keyword is
          specified, all programs will be listed."
    knowntype: "emboss keyword"
  ]

endsection: input
.
.
.
section: output
[
  information: "Output section"
  type: "page"
]
```

```
  outfile: outfile
  [
    additional: "Y"
    default: "stdout"
    knowntype: "wossname output"
  ]

  section: htmlsection
  [
    information: "Html section"
    type: "frame"
  ]

    toggle: html
    [
      additional: "Y"
      default: "N"
      information: "Format the output for HTML"
      help: "If you are sending the output to a file, this will format
             it for displaying as a table in a WWW document."
      outputmodifier: "Y"
    ]

  endsection: htmlsection

  boolean: groups
  [
    additional: "Y"
    default: "N"
    information: "Output only the group names"
    help: "If you use this option, then only the group names will
           output to the file"
    outputmodifier: "Y"
  ]

  boolean: alphabetic
  [
    additional: "Y"
    default: "N"
    information: "Output an alphabetic list of programs"
    help: "If you use this option, then you will get a single list of
           the program names and descriptions instead of the programs being
           listed in their functional groups."
    outputmodifier: "Y"
  ]

endsection: output
```

4.1.5.8 Standard file sections (`sections.standard` file)

All sections and subsections are defined in the file `sections.standard` which is stored in the EMBOSS application ACD file directory, e.g.:

```
.../emboss/emboss/emboss/acd/sections.standard
```

An excerpt of this file is shown below:

```
#
# ACD sections
#
# top level (page) sections
# ========================
additional page Additional section
advanced page Advanced section
input page Input section
output page Output section
required page Required section
#
#
# sub level (frame) sections
# ==========================
afeaturesection input First feature options
bfeaturesection input Second feature options
consensussection additional Consensus section
datasection output Data output section
fastalignsection additional Fast align options
```

The file defines the permissible order of the top level (page) sections ('Input', 'Required', 'Additional', etc.) The file also defines the permissible subsections (given under `# sub level (frame) sections`), which have been defined and used for individual applications as required. The first token on a line is the subsection name. The second token is the top-level section to which the subsection belongs: a subsection cannot appear in any other section other than that listed. The remaining text on the line is a description of the section.

4.2 Application definition

4.2.1 Application definition format

The first definition in an ACD file must be the application definition. This consists of the `application:` token, the application name with which this ACD file is associated, and a block of application attributes in square brackets. Each attribute is a `name: value` pair:

```
application: ApplicationName
[
   ApplicationAttribute1Name: "ApplicationAttribute1Value"
   ApplicationAttribute2Name: "ApplicationAttribute2Value"
]
```

For example, the **wossname** application definition has two attributes (`documentation:` and `groups:`):

```
application: wossname
[
    documentation: "Finds programs by keywords"
    groups: "Display"
]
```

The application name (ApplicationName) is arbitrary but is typically the same as that used for the ACD file name. During program initialisation (see Section 6.3, 'Handling ACD files') the embInit function is called to associate the C source code with an ACD file and invoke ACD file processing. embInit is passed the name of the ACD file to parse, e.g.:

```
embInit("ApplicationName", argc, argv)
```

This allows multiple ACD files (and therefore command line interfaces) to be developed for a single file of application C source code.

Note

embInit does *not* compare **ApplicationName** to the string given after the **application:** token in the application definition.

4.2.2 Application attributes

The application attributes are used for metadata describing the application itself. They let you set such things as documentation text and the group to which the application belongs. See Section A.3, 'Application attributes'.

Note

The **documentation:**, **keywords:** and **groups:** attributes are worthy of special mention. Their string values are searched by the **wossname** program which, given a keyword, will return all the applications containing that keyword in the **groups:**, **keywords:** or **documentation:** text.
The **groups:** string is also used by the **seealso** application which takes the name of an EMBOSS program and returns a list of the programs which share some functionality with it.

4.2.3 Application documentation

The documentation: attribute text is a succinct description of application function, printed to the screen when the program is run or when help is displayed (when the program is run with the **-help** qualifier). Here's the application definition for **seqret**:

```
application: seqret
[
    documentation: "Reads and writes (returns) a sequence"
    groups: "Edit"
]
```

The ACD file above would give the following behaviour at the command line:

```
% seqret
Reads and writes (returns) a sequence
Input sequence :
```

If the `documentation:` attribute is missing then a warning will be issued when you run the program. The string value must conform to certain conventions (Section A.3, 'Application attributes').

4.2.3.1 Application keywords file (`keywords.standard`)

The `keywords:` attribute is used to attach keywords to an attribute. The permissible keywords that may be given are defined in the file `keywords.standard` which is installed in the same directory as the ACD files, e.g.:

```
.../emboss/emboss/emboss/acd/Keywords.standard
```

The file is shown below:

```
#
# Application keywords
#
# Keywords are additional search terms to locate applications
# Valid group and subgroup names are not allowed. Keyword searches are
# expected to include the groups. Keywords defined here and in the
# application keywords: attribute are a few terms to be used where the
# short documentation string is not sufficient to identify an
# application by function.
```

You will see that no keywords are currently defined. If you add any, one keyword should be given per line following the comment block.

4.2.4 Application groups

The `groups:` attribute places the application into groups or sub-groups based on its functionality. Programs in the same sub-group generally do similar things on operate in the same general area. The string value of the `groups:` attribute must obey certain conventions (Section A.3, 'Application attributes').

The general format is as follows:

```
application: ApplicationName
[
    groups: "GroupName1, GroupName2, … : SubgroupName1, SubgroupName2, … "
]
```

For example:

```
application: syco
[
  documentation: "Synonymous codon usage Gribskov statistic
                 plot"
  groups: "Nucleic:Codon usage, Nucleic:Gene finding"
]
```

Typically just a single group and sub-group is specified. For example, the following ACD file indicates that the application **pscan** is in the group `Protein` and sub-group `Motifs`:

```
application: pscan
[
  documentation: "Scans proteins using PRINTS"
  groups: "Protein:Motifs"
]
```

A single group is also commonly sufficient, for example:

```
application: trimest
[
  documentation: "Trim poly-A tails off EST sequences"
  groups: "Edit"
]
```

4.2.4.1 Application group names file (`groups.standard`)

The application groups have become widely used elsewhere so EMBOSS now uses a standard set of group names which all group names in an ACD file are validated against. The standard is defined in the file: `groups.standard` which is installed in the same directory as the ACD files, e.g.:

```
…/emboss/emboss/emboss/acdgroups.standard
```

`groups.standard` contains one line for each valid `group:` attribute value. Group and sub-group (if specified) are delimited with a '`:`' (colon) character, and spaces in group or sub-group names are replaced by an underline character ('_'). There is also a short description of what might be expected to be placed in the group. An excerpt from the file is shown below:

```
#
# Application groups
#
Acd ACD file utilities
Alignment:Consensus Merging sequences to make a consensus
Alignment:Differences Finding differences between sequences
Alignment:Dot_plots Dot plot sequence comparisons
Alignment:Global Global sequence alignment
Alignment:Local Local sequence alignment
Alignment:Multiple Multiple sequence alignment
Display Publication-quality display
Edit Sequence editing
Enzyme_Kinetics Enzyme kinetics calculations
Feature_tables Manipulation and display of sequence annotation
.
.
.
```

If a group and sub-group are specified, for example:

```
Alignment : Consensus
```

then the group and sub-group names must not be given singly after the `groups:` attribute. For example, this is valid:

```
groups: "Alignment : Consensus"
```

but this is not:

```
groups: "Alignment"
```

If the group string has one level only, i.e. a top-level group, for example:

```
Display
```

then please don't add sub-groups to it. This, for example, is not valid:

```
Display : Features
```

You are *strongly* encouraged to use the groups structure defined in `groups.standard`. When, however, a completely new category of program is written, please discuss the creation of the new group name in the developers' mailing list (emboss-dev@emboss.open-bio.org).

Note

The group names in **`groups.standard`** are reported by the EMBOSS programs **wossname** and **seealso**.

4.3 Data definition

4.3.1 Data definition format

Every application option must be defined in the ACD file. The definition of these options (or 'data definitions') follow the application definition in the file. All data definitions *must* be contained within an appropriate ACD file section (see Section 4.1.5, 'ACD file sections'). An error will be generated during ACD processing otherwise.

The general format for data definitions is as follows. The first token is an ACD datatype, the second token the option name, followed by data attributes given between square brackets. Each attribute is a `name: value` pair:

```
Datatype: OptionName
[
   DataAttribute1Name: "DataAttribute1Value"
   DataAttribute2Name: "DataAttribute2Value"
]
```

`Datatype` must be a valid ACD datatype (Section A.2, 'Datatypes'). These are predefined and include simple types such as `integer` and `float`, and biological types such as `sequence`. Other types are available to define such things as menus (e.g. `list`) and control whether values of other data definitions are prompted for (e.g. `toggle`).

`OptionName` is the name of the option and is synonymous with the *qualifier name*, *parameter name* or *command line flag*. It is used to refer to the data definition from the command line. The value of any option can be set on the command line if the flag is specified before it:

```
-OptionName OptionValue
```

The programmer must have a handle on each application option from within the C source code and the flag is used for this purpose too (see Section 6.3, 'Handling ACD files'). The option name can be any string you like within reason, but conforming to certain conventions (Section 4.3.2, 'Parameter naming conventions').

The attributes (Section 4.3.4, 'Types of data attributes') give you control over the application parameters. They allow you to specify such things as 'help' documentation, default values and the requirements for a correct value, such as a permissible ranges of values (maxima and minima) can be set. They also allow the application programmer to control how the user is prompted for values, if at all (Section 4.5, 'Controlling the prompt').

The attribute values are the strings enclosed in double quotes and may be parsed during ACD file processing into some other type depending on the data definition in which they occur. For example, this defines a default value:

```
default: "5.0"
```

If the above appeared in the definition for a floating point number (`float:`) then it would be held internally as a floating point number.

Example

wossname has a string input parameter (`search`) and the string definition has five attributes (`parameter:`, `default:`, `information:`, `help:` and `knowntype:`):

```
string: search
[
    parameter: "Y"
    default: ""
    information: "Text to search for, or blank to list all
                  programs"
    help: "Enter a word or words here and a case-independent search
          for it will be made in the one-line documentation, group names and
          keywords of all of the EMBOSS programs. If no keyword is
          specified, all programs will be listed."
    knowntype: "emboss keyword"
]
```

4.3.2 Parameter naming conventions

The qualifier or parameter name (the command line flag) of an ACD data definition must be a lowercase alphanumerical string without whitespace characters. In theory it can be of any length. In practice, there are conventions (Section A.1, "Introduction to ACD syntax"), and it is strongly recommended that you follow them. In some cases a warning or error message will be generated during ACD processing (i.e. when the ACD file is parsed, typically after the application is run) if you do not.

For example, the conventional name for the `sequence` datatype (input sequence) is `sequence` or `*sequence` and for `seqout` (output sequence) is `outseq` or `*outseq`. In other words, where more than one instance of a datatype is specified in an ACD file, then the characters `a`, `b`, etc. can be prepended to the flag. For example, use `asequence`, `bsequence`, etc. where more than one sequence is specified, or `aoutseq`, `boutseq`, etc. where more than one sequence output stream is specified. You can see this in the **matcher** application, which takes two input sequences:

```
sequence: asequence
[
    parameter: "Y"
    type: "protein"
]
```

```
sequence: bsequence
[
    parameter: "Y"
    type: "stopprotein"
]
```

4.3.3 ACD datatypes

4.3.3.1 Groupings of ACD datatypes

For convenience the available ACD datatypes are organised into five groupings reflecting similar properties or modes of usage as follows:

- Simple datatypes
- Input datatypes
- Selection datatypes
- Output datatypes
- Graphics datatypes.

The groupings are described below. For a complete description of the datatypes including their data value, default value and key attributes see Section A.2, 'Datatypes'.

4.3.3.2 Simple ACD datatypes

The simple datatypes (Section A.2.1, 'Description of simple ACD datatypes') include primitive and some derived datatypes:

`boolean`	Simple boolean value
`float`	Simple floating point number
`integer`	Simple integer number
`string`	Simple string
`toggle`	Simple boolean switch for controlling other parameters
`array`	List of either integer or floating point numbers
`range`	Range of sequence positions
`regexp`	Regular expression pattern
`pattern`	Sequence pattern

4.3.3.3 Input ACD datatypes

The input datatypes (Section A.2.2, 'Description of input ACD datatypes') cater for input of sequences, sequence features, files and directories, inputs specific to the **phylipnew** EMBASSY package, data files and other files of biological data.

4.3.3.3.1 Sequence input

For biological sequences input use:

`sequence`	A single sequence for reading
`seqall`	A set of single sequences that are addressed one after another
`seqset`	A set of single sequences that can be used all at the same time
`seqsetall`	One or more sets of single sequences that can be used all at the same time

The data value in all cases is the Uniform Sequence Address (USA) of an input sequence stream. The USA specification is described in the *EMBOSS User's Guide*.

4.3.3.3.2 Feature input

For biological sequence features input use:

`features`	Sequence feature annotation in any supported feature format

The data value in all cases is the Uniform Feature Object (UFO) of an input sequence feature stream. The UFO specification is described in the *EMBOSS User's Guide*.

4.3.3.3.3 Files and directories

For general input files and directories use:

`directory`	A directory that can be used for input or output
`dirlist`	A list of file names that are read from a directory
`filelist`	A list of input files
`infile`	Non-sequence-related input file

In all cases, the data value is any valid file or directory name.

4.3.3.3.4 Data files

For input data files use:

`datafile`	A formatted data file read from the standard EMBOSS data search path
`matrix`	Comparison matrix file (integer values)
`matrixf`	Comparison matrix file (floating point values)

In all cases, the data value is the name of a file in the EMBOSS data search path (see the *EMBOSS User's Guide*).

Typically where a comparison matrix is specified, gap penalties will also be required. These must be specified separately in one or more other data definitions.

4.3.3.3.5 **phylipnew** EMBASSY package

Input datatypes specific to the **phylipnew** EMBASSY package are given below. They allow detailed type checking, and the detection and validation of the various alternative formats that **phylip** supports without the need for complex extra command line options.

`discretestates`	Discrete states file
`distances`	Distance matrix
`frequencies`	Frequency value(s)
`properties`	Property value(s)
`tree`	Phylogenetic tree

In all cases, the data value is any valid file name.

4.3.3.3.6 Other biological inputs

Other biological input datatypes include:

`codon`	Codon usage table file
`cpdb`	Protein coordinate data
`scop`	SCOP and CATH domain classification information

In all cases, the data value is any valid file name.

4.3.3.4 Output ACD datatypes

The output datatypes (Section A.2.3, 'Description of output ACD datatypes') cater for output of sequences, sequence features, alignments, files and directories, outputs specific to the **phylipnew** EMBASSY package, data files, other files of biological data and formatted application output files (reports).

4.3.3.4.1 Sequence output

For biological sequence output use:

`seqout`	Output file for single sequence
`seqoutall`	Output file for multiple sequences
`seqoutset`	A set of single sequences stored in memory together, to be written to file

The behaviour of these datatypes is identical but they are provided for consistency with the input sequence datatypes (see above). The data value in all cases is the Uniform Sequence Address (USA) of an output sequence stream (see the *EMBOSS User's Guide*).

FASTA format is used by default for the output sequence(s). The format is normally set at the command line but a default may be hard-coded with `osformat:` (see Section 4.3.6.2.2, 'Feature output').

4.3.3.4.2 Features

For biological sequence feature output use:

`featout`	Output file for sequence feature annotation

The data value is the Uniform Feature Object (UFO) of an output sequence feature stream (see the *EMBOSS User's Guide*).

4.3.3.4.3 Alignments

For alignment output use:

`align`	Output file for sequence alignments output use:

The data value is any valid file name.

4.3.3.4.4 Output files and directories

For general output files and directories use:

`outdir`	Output directory for writing of multiple output files
`outfile`	General output file
`outfileall`	Multiple general output files

`outfile` and `outfileall` are used for data not catered for by some other output ACD datatype. For example, the output file would not normally contain sequence data. They are suitable for general application output in plain text.

The data value is any valid file name.

4.3.3.4.5 Output data files

For output data files use:

`outdata`	Output file for data formatted cleanly as a table or list
`outmatrix`	Output file for integer comparison matrix data
`outmatrixf`	Output file for floating point comparison matrix data

In all cases the data value is any valid file name.

4.3.3.4.6 **phylipnew** EMBASSY package

Output datatypes specific to the **phylipnew** EMBASSY package allow EMBOSS to provide detailed type checking, and automatically detect and validate the various alternative formats that **phylip** supports without the need for complex extra command line options.

`outdiscrete`	Output file for phylogenetic discrete characteristics data
`outdistance`	Output file for phylogenetic distance matrix data

`outfreq`	Output file for phylogenetic character frequency data
`outproperties`	Output file for phylogenetic property data
`outtree`	Output file for phylogenetic tree data

In all cases, the data value is any valid file name.

4.3.3.4.7 Other biological outputs

Other biological output datatypes include:

`outcodon`	Output file for codon usage data
`outcpdb`	Output file for protein coordinate data
`outscop`	Output file for SCOP and CATH domain classification information

The data value is any valid file name.

4.3.3.4.8 Report output

For formatted application output use:

`report`	Output file for sequence annotation

The data value is any valid file name.

4.3.3.5 Selection ACD datatypes

The selection datatypes cater for menus (Section A.2.4.2, '`selection`').

The user is presented with a limited list of options they can choose from. An option consists of a label and descriptive text. One or more options may be selected. Selection datatypes include:

`list`	A list of options (typically terse text descriptions) with text labels
`selection`	A list of options (typically verbose text descriptions) with automatically generated numerical labels

The data value is one (or more) of the valid options. An option is specified by label (whether text or numerical) or by a non-ambiguous part of the descriptive text itself given after the label. If multiple selections are allowed then the user must supply a comma-separated list of options.

4.3.3.6 Graphics ACD datatypes

The graphics datatypes cater for graphical output (Section A.2.5, 'Description of graphics ACD datatypes').

`graph` Graphical output of any general kind

`xygraph` Graphical output as a simple two-dimensional (2D) XY plot with the sequence along the *x*-axis

The data value is a supported graphics device (Section A.2.5, 'Description of graphics ACD datatypes').

4.3.4 Types of data attributes

There are three basic types of attributes that may be defined for a data definition:

- Global attributes (Section A.4, 'Global attributes')
- Datatype-specific attributes (Section A.5, 'Datatype-specific attributes')
- Calculated attributes (Section A.6, 'Calculated attributes').
- Datatype-associated qualifiers defined as attributes (Section A.2, "Datatypes").

Global attributes are available for all datatypes and can be defined in any ACD data definition.

Datatype-specific attributes, in contrast, are available for certain datatypes only. Each ACD datatype has its own set of specific attributes. The values of global and datatype-specific attributes are set explicitly in the ACD file.

Calculated attributes are a type of datatype-specific attribute that are assigned values *after* the data definition has been processed (read and validated), for example, once a sequence has been read in from file. This allows the data definition to refer to attributes whose value is not known up front but depends on the input data, and is calculated automatically during ACD file processing.

Datatype-associated qualifiers (or simply *associated qualifiers*) are command live qualifiers that are inbuilt for certain ACD datatypes. They may, however, also be 'hard-coded', i.e. defined as attributes in the appropriate data definition.

4.3.5 Global attributes

Global attributes are available for all datatypes. They may be defined in any ACD data definition and their value is set explicitly in the ACD file (Section A.4, 'Global attributes'). Global attributes are as follows:

```
GlobalAttributeName: "Value"
```

Most global attributes have string or boolean values. The booleans have a hard-coded default value which can be overridden by `"Y"`, `"Yes"`, `"N"` or `"No"` (the strings are case-insensitive). All of the following attributes are therefore valid:

```
parameter: "YES"
parameter: "Yes"
parameter: "Y"
parameter: "NO"
parameter: "No"
parameter: "N"
```

The global attributes are summarised below. For convenience they are grouped by function as follows:

- Parameters and qualifiers
- User prompting
- Datatype definition
- Help information and documentation
- Hints for GUIs
- For use by SOAPLAB.

4.3.5.1 Parameters and qualifiers

Each application parameter, i.e. every data definition in the ACD file, can be defined via the appropriate global attribute to be one of a 'parameter', 'standard qualifier' or 'additional qualifier' with the default of 'advanced qualifier'.

```
parameter:   "Boolean"
standard:    "Boolean"
additional:  "Boolean"
```

The default value is `"N"`; however, this is never specified explicitly in the ACD file.

4.3.5.1.1 When to use parameters

Reports

The `report` datatype, where used, is typically a primary output of an application and, as such, should be defined as a parameter using the `parameter:` attribute. The first report file *must* be defined as `parameter:` (an error will be generated during ACD processing otherwise). It is recommended that subsequent report definitions (second, third and so on) are also defined as parameters (a warning will be generated during ACD processing if they are not). The exception is if the `default:` or `nullok:` attributes are set, in which case no warning or error messages are generated as the application can run with a default or without any value for the definition respectively.

Sequences

Sequence inputs (`sequence`, `seqall`, `seqsetall` or `seqset` ACD datatypes) and outputs (`seqout`, `seqoutall` and `seqoutset` ACD datatypes) are typically the primary input or output of an EMBOSS application, and as such should be defined as parameters in the same way as for the `report` datatype above.

Features

Sequence features, where used, are typically the primary input (`feature` ACD datatype) or output (`featout`), as such, should be defined as parameters in the same way as for the `report` datatype.

Alignments

The `align` datatype, where used, is typically the primary output and should be defined as a parameter. The first and subsequent alignment outputs should be defined as `parameter:` types, a warning will be generated during ACD processing if they are not. The exception is if the `default:` or `nullok:` attributes are set, in which case no warning or error messages are generated as the application can run with a default or without any value for the definition respectively.

Files

File datatypes, where there are used, are typically the primary input (`infile`, `filelist`, `directory` or `dirlist` ACD datatypes) or output (`outdir` ACD datatype) and should be defined as parameters in the same way as for the report datatypes described above.

4.3.5.2 User prompting

Attributes are used to provide the text that will be used on the screen to prompt the user for values. Whether a prompt will appear at all depends on whether the option was defined as a parameter or one of the qualifier types (see above) and on the command used to invoke the application (i.e. what options were specified). In some cases, the prompt may or may not appear depending on the value of other inputs. For more information see Section 4.5, 'Controlling the prompt').

The attributes used are:

```
information: "String"
code: "String"
prompt: "String"
```

Only one of `code:`, `prompt:` or `information` should ever be defined. The use of `information:` (with a standard name, see below) is preferred. In practice, `prompt:` is only ever required in the rare cases where the `information:` string might be misleading.

A default value for the `information:` string is defined for most common datatypes (see Section 4.3.5.2.1, 'Standard prompts file (`codes.english`)') and should be used. A warning will be generated during ACD processing if either the `information:` attribute is missing or if the value used is not the standard value (where a standard value is available).

If a non-standard prompt is used, the text given after the `information:` attribute should conform to certain conventions (Section A.4.4.1, '`information: "String" (" ")`').

In the example below a non-standard prompt is defined for `asequence`. A warning will be generated if you try to run this:

```
sequence: asequence
[
    standard: "Y"
    information: "Enter filename"
]
```

In the next example, no `information:` string is specified, therefore the standard prompt of `"Read sequence from"` (from `codes.english`) would be used instead:

```
sequence: asequence
[
    standard: "Y"
]
```

4.3.5.2.1 Standard prompts file (**`codes.english`**)

To provide standard prompts, a default value for the `information:` string is defined for most common datatypes. The defaults are in the EMBOSS system file `codes.english` in the application ACD file directory, e.g.:

```
.../emboss/emboss/emboss/acd/codes.english
```

Prompts are given on lines beginning with `DEF`*`DatatypeName`* where `DatatypeName` is the name of the ACD datatype. The file also contains some additional standard prompts for specific instances of individual datatypes, identified by a code (not beginning with `DEF`). These prompts can be used with the `code:` attribute, for example `code: "GAP"`.

An excerpt of the file is shown below:

```
#
# This is the EMBOSS coded prompts file. Any messages found here will
# override those in the .acd file, and can be translated into other
# languages. The file extension (default "english") is set by the
# variable "emboss_language"
#
# Messages are referred to by .acd files as code: NAME
#
# DEFXXXX codes are automatically searched for ACD type XXXX if there
# is no information (or prompt) defined in the ACD file of as a
# standard prompt for the ACD type.
#
# HELPXXXX codes are automatically searched for ACD type XXXX if there
# is no help text in the ACD file.
#

# Default prompts: these are used where no prompt for a data type
# has been provided.

DEFALIGN      "Write output alignment to"
DEFREPORT     "Write output report to"
DEFINTEGER    "Enter a number"
DEFFLOAT      "Enter a number"
DEFBOOL       "Yes or No"
.
.
.
# Gap penalties
GAP           "What gap penalty"
GAPEXT        "What gap extension penalty"
```

4.3.5.3 Datatype definition

Five global attributes concern the data:

```
knowntype: "String"
default: "Value"
relations: "String"
outputmodifier: "Boolean"
missing: "Boolean"
```

`knowntype:` describes the type of data and should be defined where the type is not already clear from the datatype itself. It is typically defined for `string:`, `infile:`, `outfile:` and `outfileall` datatypes but not, for example, for a `sequence:`.

Valid known type strings are listed in the file `knowntypes.standard` (The other attributes are used to set a default value (`default:`), describe relations to other data definitions (`relations:`), control what fields are written (`outputmodifier:`) and whether a NULL value can be specified (`missing:`). A few other values are accepted, for example "`ApplicationName` output" for an `outfile` datatype. These are documented with the datatypes (Section A.2, 'Datatypes').

Note

If a value is given that is not a standard known type or other accepted value, a warning message will be generated during ACD processing. The **acdvalid** utility will check all **`knowntype`** values in an ACD file, and report any missing values for data definitions that require a known type.

4.3.5.3.1 Application data known types file (`knowntypes.standard`)

The standard values (known types) are read from the EMBOSS system file `knowntypes.standard` which can be found in the application ACD file directory:

```
.../emboss/emboss/emboss/acd
```

```
# Known Types
# Knowntype_string              Type        Comment
aaindex_data                    file        AAINDEX entry
aaindex_database                file        AAINDEX database
abi_trace                       file        ABI sequencing trace
ajint_ajlong_data               file        Standard format info on ajint and
ajlong
alistat_input                   file        HMMER alistat program input
alistat_output                  file        HMMER alistat program output
amino_acid_classification       file        Amino acid chemical classes data
.
.
.
```

4.3.5.4 Help information and documentation

Every EMBOSS application accepts the global qualifiers **-help** and **-verbose**, the latter used in combination with the former (if at all). When an application is run with **-help**, helpful information is printed to the screen, including all the program parameters and qualifiers with explanatory text alongside.

The attributes below define the text that's used:

```
help: "String"
valid: "String"
expected: "String"
```

`help:` is usually only defined if a deeper explanation of an application parameter is needed. If `help:` is not defined, the value of the `information:` attribute (if available) or the default help string (see below) will be used instead. The `help:` string must conform to certain conventions (Section A.4.6.1, '`help: "String" ("")`'). `valid:` and `expected:` are used to describe the allowed and expected values for the online documentation. They are not usually required as, in most cases, a reasonable value is obvious from other attributes.

An example where `help:` is helpful:

```
integer: window
[
    standard: "N"
    default: 10
    minimum: 5
    maximum: 100
    information: "Window size"
    help: "Number of residues used to calculate the value for each point"
]
```

4.3.5.4.1 Standard help strings file (**codes.english**)

Default help strings are given for each datatype in the EMBOSS system file `codes.english` file (see Section 4.3.5.2.1, 'Standard prompts file (`codes.english`)'). They have the general form:

```
HELPDatatypeName
```

where `DatatypeName` is the name of the ACD datatype.

For example:

```
HELPSEQUENCE    "Sequence USA"
HELPDIRECTORY   "Directory name"
```

4.3.5.5 Hints for GUIs

The `needed:` global attribute indicates whether a parameter is expected to be included in a GUI form. It is used to help GUI developers. Certain datatype-specific attributes, for example `button:` which can be defined for selection lists, are also used to provide such hints. See Section A.5, 'Datatype-specific attributes'.

The attribute is defined as follows:

```
needed: "Boolean"     ("Y")
```

4.3.5.6 For use by SOAPLAB

The following attributes provide metadata for the SOAPLAB project. They are not used by EMBOSS directly and should not appear in standard EMBOSS application ACD files.

The attributes are defined as follows:

```
qualifier:   "String"
template: "String"
comment: "String"
```

4.3.6 Datatype-specific attributes

Datatype-specific attributes given within an ACD data definition define the characteristics of application options. They control such things as:

- The name and location of input and output files
- Detailed type-checking of inputs to ensure data is as required by the application
- Exactly what data is generated on output
- The appearance and behaviour of the user interface at the command line
- Validation of user input, e.g. to ensure user input is within permissible ranges
- Validation of application output, e.g. to ensure it is as expected
- Suggestions for graphical interfaces.

Datatype-specific attributes are defined for individual datatypes or for groups of related datatypes. The key attributes for each datatype are summarised below. For descriptions of all the available attributes see Section A.5, 'Datatype-specific attributes'.

4.3.6.1 Attributes for simple ACD datatypes

These are used to validate user input.

4.3.6.1.1 Validation of user input

`minimum:` and `maximum:` can be defined for several datatypes and restrict the value within a given range. Attributes to control the length (`minlength:` and `maxlength`) and case (`upper: "Y"` or `lower: "Y"`) of a regular expression or sequence pattern are available for `regexp` and `pattern` datatypes.

A regular expression can be used to validate a `string` if necessary.

4.3.6.2 Attributes for input ACD datatypes

These are used to specify the input data and to validate user input.

4.3.6.2.1 Sequence input

`sask:` is available for all sequence inputs and sets the default for the **`-sask`** qualifier. If set to `"Y"` it specifies that a sequence begin and end position, and the reversing of a nucleotide sequence, will be prompted for.

The sequence input datatypes (`sequence`, `seqall`, `seqset` and `seqsetall`) share a common set of attributes used to define and validate the sequence input. The type of sequence can be restricted with `type:`, so that the program accepts, for example, only DNA sequences. The type must be a standard sequence type, for example `dna`, `pureprotein`, `gaprna`, etc. See Section A.7, 'Sequence types'.

4.3.6.2.2 Feature input

For all the sequence inputs (`sequence`, `seqall`, `seqset` and `seqsetall` datatypes), sequence features (Section 6.9, 'Handling features') can be read if the `features:` attribute is set.

The type of features can be restricted by setting `type:`. For example, the program can be made to accept only DNA features. The feature type must be one of `protein` or `nucleotide`. There is a default based on the type of an input sequence (where used), but a value should be specified. If no type is specified for input features and there is no sequence input from which to take a default type, then an error will be generated during ACD processing.

4.3.6.2.3 **phylip** EMBASSY package

Attributes of the **phylip** package datatypes provide detailed type checking and can automatically detect and validate the various alternative formats that **phylip** supports, without the need for complex extra command line options. See Section A.5, 'Datatype-specific attributes'.

- `length:` specifies the number of property values per set (`properties` datatype) or the number of frequency loci / values per set (`frequencies` datatype).
- `size:` specifies the number of discrete state sets (`discretestates` datatype), the number of frequency sets (`frequencies` datatype) or the number of trees (`tree` datatype).

4.3.6.2.4 Naming of input files

The `name:`, `extension:` and `directory:` or `fullpath:` attributes are defined for several datatypes and define the name and location of input file and directory datatypes. In some cases a default naming scheme or a hard-coded default is available but can be overridden by using these attributes. A default value for an input file name can also be set by defining the `default:` global attribute. Data files, for example, often have a hard-coded filename and you are free to define this using `name:` or `default:`.

In cases where a default file naming scheme is available, but a default value is also specified in the ACD file, then the default value in the ACD file can be overridden and the naming scheme used if a null value (`""`) for the parameter is given on the command line. It's necessary to set the global attribute `missing:` for such data definitions.

4.3.6.2.5 Missing input files

`nullok:` is available for most input datatypes. It is a boolean attribute and specifies whether a missing input file is acceptable. If the application can accept a null value for this definition and can run without the corresponding input file, the `nullok:` attribute must be set to `"Y"`. This allows a default value to be omitted or, if a default value is specified, for the application to be run with **`-no`*`Flag`*** on the command line (where `Flag` is the data definition flag) to specify the input file is *not* available.

4.3.6.2.6 Input file datatype identification

See Section 4.3.5.3, 'Datatype definition'.

4.3.6.3 Attributes for output ACD datatypes

These are used to specify the output file and in some cases exactly what data is generated and how it is validated.

4.3.6.3.1 Sequence and sequence feature output

For sequence and feature output datatypes (`featout`, `seqout`, `seqoutset` and `seqoutall`), if the `name:` attribute is not defined in the ACD file, it will default to the name of the *first* sequence that is read in (if available).

The type of sequence can be restricted by setting `type:`. The application will then validate that the output is of the specified type. The type must be a standard sequence type (see Section A.7, 'Sequence types').

If no type is specified for an output sequence, and there is no sequence input from which to take a default type, then an error will be generated during ACD processing.

4.3.6.3.2 Feature output

Sequence features can be written alongside an output sequence (`seqout`, `seqoutall` and `seqoutset` datatypes) if their `features:` attribute is set. For all these datatypes, the default data format can be specified with `osformat:` which the **`-osformat`** associated qualifier can override.

The type of features can be restricted by setting `type:`. There is a default based on the type of an input sequence (where used), but it should be specified for validation purposes. The feature type must be one of `protein` or `nucleotide`. If no type is specified for output features, and there is no sequence input from which to take a default type, then an error will be generated during ACD processing.

4.3.6.3.3 Alignment output

Any `seqset` or `seqsetall` datatype *must* have the `aligned:` attribute set: an error will be generated during ACD processing otherwise.

Alignment format can be set with the `aformat:` attribute.

For an `align` data definition, `minseqs:` and `maxseqs:` set the expected minimum and maximum number of sequences. The `multiple:` boolean attribute should be set to `"Y"` if the output can contain more than one alignment from the same input.

4.3.6.3.4 Report output

Attributes for the `report` datatype include:

`rformat:`	specifies the report format to use, which must be one of the supported report formats (see the *EMBOSS User's Guide*)
`type:`	is defined as one of `"protein"` or `"nucleotide"` where the report format is one of the standard feature table formats (see the *EMBOSS User's Guide*)
`taglist:`	defines the tag / value pairs from the internal feature table to be reported in the output.

4.3.6.3.5 Naming of output files

Output filenames are constructed from the `name:` and `extension:` values and have the format:

```
name.extension
```

The name and extension may have default values. For example, for the `outcodon` datatype, `name:` defaults to `outfile` and `extension:` defaults to the format name, with `cut` defined as the default output format.

In cases where a default file naming scheme is available, but a default value is also specified in the ACD file, the default value can be overridden and the naming scheme used if a null value (`""`) for the parameter is given on the command line. It's necessary to set the global attribute `missing:` for such data definitions.

Datatype-specific command line qualifiers (see the *EMBOSS User's Guide*) may also be used to name the output file, either by specifying the qualifier on the command line or by hard-coding it as an attribute in the ACD file. For example, an alignment filename with the format `aname.aextension` is constructed if the qualifiers **`-aname`** and **`-aextension`** are specified. Values may be hard-coded with the corresponding `aname:` and `aextension:` attributes.

A default value for the output file name can also be set by defining the `default:` global attribute.

4.3.6.3.6 Missing output files

`nullok:` and `nulldefault:` are available for all output datatypes. If the application can accept a null value for a data definition and can run without generating the corresponding output file, the `nullok:` attribute must be set to `"Y"`. This allows a default value to be omitted or the application to be run with **`-noFlag`** (where `Flag` is the data definition flag) to specify the output file is *not* to be generated.

`nulldefault:` overrides the default name generation, and uses an empty string (no output file) as the default for programs where an alignment file is only occasionally required. If an empty string is specified on the command line, the standard default value will be generated instead.

4.3.6.3.7 Output datatype identification

The type of data is identified using `knowntype:`. See Section 4.3.5.3, 'Datatype definition'.

4.3.6.3.8 Specification of output data

Attributes may be used to control exactly what data is generated and to validate that data. For example an alignment output file is defined in the same way as a plain output file (`outfile` datatype) but has extra attributes to allow a choice of alignment formats and to specify, for validation purposes, the expected number of aligned sequences. Similarly, a report output file is defined in the same way as a plain output file (`outfile`) but has extra qualifiers to allow a choice of report formats.

The default data format of `codon`, `cpdb` and `scop` datatypes can be specified with `oformat:` which the **`-oformat`** associated qualifier can override.

4.3.6.4 Attributes for selection ACD datatypes

These control the aspects of the menus (lists of options) the user is presented with. For example, the text appearing above the menu, and how user input is validated, the minimum and maximum number of selections and whether the options are case-sensitive.

For example, consider this ACD list definition:

```
list: matrix
[
     default: "blosum"                   # default value
     minimum: 1 maximum: 1               # must select exactly 1
     header: "Comparison matrices"       # printed before list
     values: "B:blosum, P:pam, I:id"     # 3 valid values
     delim: ","                          # delimiter default ";"
     codedelim: ":"                      # label delimiter default ":"
     prompt: "Select one"                # prompt after list
     button: Y                           # use radio buttons rather than checkboxes in HTML,
ignored by ACD.
]
```

What you get on screen is:

```
Comparison matrices
      B : blosum
      P : pam
      I : id
Select one [blosum] : PAM
```

With this ACD list definition:

```
select: matrix
[
     default: "blosum"                # default value
     minimum: "1" maximum: "1"        # must select exactly 1
     header: "Comparison matrices"    # printed before list
     values: "blosum, pam, id"        # valid values
     delimiter: ","                   # delimiter default ";"
     information: "Select one"        # prompt after list
     button: "Y"                      # use radio buttons rather than checkboxes in HTML,
ignored by ACD
]
```

you get:

```
Comparison matrices
      1 : blosum
      2 : pam
      3 : id
Select one [blosum] : PAM
```

4.3.6.5 Attributes for graphics ACD datatypes

These specify the graph data that are output.

`gtitle:` specifies the graph title for a `graph` or `graphxy` datatype. Many other graphical elements can be set.

`multiple:` specifies the number of multiple graphs in a single `graph` or `graphxy` datatype output.

4.3.7 Attributes for datatype-associated qualifiers

Various command line qualifiers are inbuilt for certain ACD datatypes. These *datatype-associated qualifiers* are normally specified only on the command line. They may, however, also be 'hard-coded' as attributes in the appropriate data definition and it is sometimes desirable to do this. Some examples for setting file format are shown below:

Input sequence	**-sformat**
	sformat:
Output sequence	**-osformat**
	osformat:
Output alignment	**-aformat**
	aformat:
Report output file	**-rformat**
	rformat:
Sequence feature input	**-fformat**
	fformat:
Sequence feature output	**-offormat**
	offormat:

The qualifiers are described in more detail in the *EMBOSS User's Guide*.

4.3.8 Introduction to calculated attributes

Calculated attributes are datatype-specific attributes that are assigned values *after* an input file has been read during ACD file processing. The values are calculated or extracted from the actual data that an ACD data definition refers to.

Most calculated attributes are for datatypes for sequence input (`sequence`, `seqall`, `seqset` and `seqsetall`) and sequence feature input (`features`). For instance, for a sequence datatype, the length and type of sequence are available once the sequence file or an entry from it has been read.

For a complete description of the available calculated attributes see Section A.6, 'Calculated attributes'.

4.3.8.1 Retrieving values of calculated attributes

Values of calculated attributes are retrieved by an operation (Section 4.4, 'Operations') from within the ACD file. The operation uses a `ParameterName.AttributeName` term surrounded by parentheses with a dollar sign (`$`) at the front:

```
$(ParameterName.AttributeName)
```

The `$` syntax means 'get the value of', in this case the term enclosed by parentheses.

This is typically done from within a data definition that relies on the value of a calculated attribute from some other data definition. For example, the following ACD file excerpt defines a `sequence` parameter and an `integer` window:

```
sequence: sequence
[
  parameter: "Y"
  type: pureprotein
]

integer: window
[
  standard: "Y"
  default: "10"
  maximum: "$(sequence.length)"
]
```

The maximum window size (`maximum` attribute of the `window` datatype) is being set to the length of the sequence by using `maximum: "$(sequence.length)"`.

4.3.8.2 Sequence calculated attributes

All of the sequence input datatypes:

```
sequence
seqall
seqset
seqsetall
```

have six calculated attributes that are properties of an input sequence which can be queried within ACD:

`begin`	Start residue (**`-sbegin`** value)
`end`	End residue (**`-send`** value)
`length`	Length
`protein`	Boolean, `True` if sequence is protein
`nucleic`	Boolean, `True` if sequence is nucleic
`name`	Name

Let's assume we've defined a sequence input called `asequence` and have an ACD file looking something like:

```
sequence: asequence
[
  parameter: "Y"
  type: protein
]
```

The attributes would be referred to as follows:

- `asequence.begin`
- `asequence.end`
- `asequence.length`
- `asequence.protein`
- `asequence.nucleic`
- `asequence.name`

Note

The value of the **`begin`** and **`end`** attributes can be set on the command line by specifying the **`-sbegin`** or **`-send`** which are inbuilt qualifiers for the sequence types.

The `seqset` and `seqsetall` datatypes have the following calculated attributes :

`totweight`	Total sequence weight for the set(s) of sequences
`count`	Number of sequences in the set(s)

`seqsetall` also has :

`multicount`	Number of sets of sequences

To retrieve calculated attribute values from the definition of `asequence` shown above you can use:

```
$(asequence.begin)
$(asequence.end)
$(asequence.length)
```

Example
When writing a program to insert one sequence into another, one way to make sure that the insertion position isn't greater than the length of the first sequence is to use code like the following:

```
if(position > ajSeqGetLen(seq))
    ajFatal("Insertion position out of bounds");
```

The problem is that the program can prematurely terminate after the user has gone to all the effort of configuring the application (entering all the inputs). What would be better is if the interface forced the correct input, and there is a way to achieve that by using calculated attributes in the ACD file itself.

Now, instead of hard-coding the check for the sequence insertion position you just need to add:

```
maximum: $(asequence.end)
```

to the integer definition of the insert position. An ACD file with two input sequences and an integer insert position might look something like:

```
sequence: targetsequence
[
  parameter: "Y"
  type: "protein"
]

sequence: insertsequence
[
  parameter: "Y"
  type: "protein"
]

integer: position
[
  parameter: "Y"
  maximum: $(targetsequence.end)
]
```

You may of course retrieve the value of calculated attributes from both sequences. In the following example, a sequence window is defined which cannot be any longer than the sum of the lengths of the two sequences. A calculation is used, which therefore has to be enclosed by parentheses and started with the 'at' symbol ('@'):

```
integer: window
[
```

```
  parameter: "Y"
  maximum: @($(targetsequence.length) + $(insertsequence.length))
]
```

These calculated attributes are also useful for conditional statements. In the example below the sequence input is of the any type. A ternary conditional (see Section 4.4, 'Operations') is being used to set a default substitution matrix based on the automatically determined type of sequence:

```
sequence: sequence
[
  parameter: "Y"
  type: "anY"
]

integer: penalty
[
  parameter: "Y"
  default: "@($(sequence.protein) ? EBLOSUM62 : EDNAFULL)"
  etc
]
```

4.3.8.3 Feature calculated attributes

The sequence feature input datatype:

```
features
```

has seven calculated attributes:

`fbegin`	Start of the features to be used (**`-fbegin`** value)
`fend`	End of the features to be used (**`-fend`** value)
`flength`	Total length of sequence
`fprotein`	Boolean, `True` if feature table is protein
`fnucleic`	Boolean, `True` if feature table is nucleotide
`fname`	Name of the feature table
`fsize`	Number of features

These are properties of an input feature that can be queried within ACD.

Note

The value of the **`fbegin`** and **`fend`** attributes can be set on the command line by specifying **`-fbegin`** or **`-fend`** which are inbuilt qualifiers for the **`features`** datatype.

Assuming you've defined a feature input called `afeatures` and have an ACD file looking something like:

```
features: afeatures
[
  parameter: "Y"
]
```

the attributes would be referred to as follows:

- `afeatures.fbegin`
- `afeatures.fend`

and so on.

To retrieve calculated attribute values from your definition of `afeatures` above you can use:

```
$(afeatures.begin)
$(afeatures.end)
$(afeatures.length)
```

Example

When writing a program to compare the features of two sequences, which might be protein or nucleotide sequences, you need to ensure that both sequences are of the same type. One way to do this from the application code is to call an AJAX function to return the type of both, and exit if they are not the same:

```
if((ajFeattableIsProt(features1)&& ajFeattableIsNuc(features2)) ||
   (ajFeattableIsNuc(features1) && ajFeattableIsProt(features2))
    ajFatal("Input feature tables are not of the same type");
```

The problem with this code is that the application will terminate once it is running if the feature tables are not of the same type. Far better if the interface forced the correct input before the application proper started, which can be achieved by using calculated attributes.

Assume the application takes two feature table parameters as input and the ACD file therefore looks something like:

```
features: afeatures
[
   parameter: "Y"
]

features: bfeatures
[
   parameter: "Y"
]
```

Instead of hard-coding the check for feature type, it can be enforced on the second of the features that is read: the first feature table can be of the `any` type but the second must match the first. This is achieved by declaring the `type:` attribute for the feature tables. The `type:` of the second feature table is set to that of the first by retrieving the `fprotein` calculated attribute of the first feature table in a ternary conditional (see Section 4.4, 'Operations'):

```
features: afeatures
[
   parameter: "Y"
   type: "anY"
]

features: bfeatures
[
   parameter: "Y"
   type: "@($(afeatures.fprotein) ? protein : nucleotide)"

]
```

4.4 Operations

4.4.1 Types of operation

Operations are used within ACD files to achieve flexible and precise control over the command line interface. Currently the following types of operation are supported:

- To retrieve the value of a data definition
- To retrieve the values of attributes and calculated attributes of a data definition
- To retrieve the values of variables
- Arithmetic operations
- Tests for equality
- Boolean logical operations
- Calculations that use the above operations
- Unary and ternary conditional statements using the above operations.

4.4.2 General operation syntax

An operation must be enclosed in a pair of parentheses and preceded by the 'at' symbol ('@'):

```
@(operation)
```

If the operation contains whitespaces, the whole token should be enclosed by double quotes (" "):

```
"@(operation with white space)"
```

Operations can be nested, in which case each operation must be enclosed in parentheses:

```
"@(operation @(nested operation))"
```

An operation typically contains a reference to the value or attribute of a data definition or variable. They may also contain arithmetic, logical and conditional operators.

4.4.2.1 Arithmetic operators

- `@(a+b)` [Addition]
- `@(a-b)` [Subtraction]
- `@(a*b)` [Multiplication]
- `@(a/b)` [Division]

4.4.2.2 Logical operators

- `@(!a)` [Not boolean]
- `@(a|b)` [Or]
- `@(a&b)` [And]

4.4.2.3 Equality operators

- `@(token1==token2)` [Equality]
- `@(token1!=token2)` [Non-equality]
- `@(token1<token2)` [Less-than]
- `@(token1>token2)` [Greater-than]

4.4.2.4 Conditional operators

- `@(boolval ? iftrue : iffalse)` [If]
- `@(testval = A : 1 B : 2 else : 0)` [Case]

4.4.3 Retrieving data values in ACD files

The attribute values for a given data definition can depend on the values from other data definitions. It is possible to retrieve the value of:

- A data definition (application parameter)
- An attribute of a data definition
- A calculated attribute of a data definition
- An ACD variable definition.

Such values are retrieved using the ACD "get the value of" syntax which consists of a term `ParameterName`.AttributeName term surrounded by parentheses with a dollar sign ($) at the front:

```
$(ParameterName.AttributeName)
```

If only the value of the data definition is to be retrieved then the `AttributeName` component is omitted:

```
$(ParameterName)
```

Of course a variable name may also be given. Variables do not have attributes therefore the syntax is:

```
$(VariableName)
```

For example, in the following ACD file excerpt two integer inputs are defined:

```
integer: gappenalty
[
    standard: "Y"
    default:  "10"
]

integer: gapextpenalty
[
    standard: "Y"
    default:  "$(gappenalty)"
]
```

The default for the standard qualifier `gapextpenalty` is set to the value of qualifier `gappenalty`.

As another example assume a sequence input and a window over that sequence which should not be longer than the sequence itself. There is also a sub-window which (by default) is one residue shorter than the main window, and an array of numbers that is the same size as the sub-window. The following ACD code would suffice:

```
sequence: sequence
[
  parameter: "Y"
]

integer: window
[
  standard: "Y"
  default: "10"
  maximum: "$(sequence.length)"
]

integer: subwindow [
  standard: "Y"
  default: "@($(window)-1)"
  maximum: "@($(window.maximum)-1)"
```

```
]

array: values
[
  standard: "Y"
  size: "$(subwindow)"
]
```

`sequence.length` is a calculated attribute and is the length of the sequence. The `$(sequence.length)` value is calculated during ACD processing once the input sequence has been read in. In contrast `window.maximum` simply refers to the `maximum:` attribute of the `window` data definition.

Of course you need to be careful that the type of retrieved value matches that of the attribute which is being calculated. For instance, if you have defined a `toggle` datatype:

```
toggle: dovalues
[
  standard: "Y"
]
```

then the following definition is perfectly valid because the `standard:` global attribute has a boolean value, which is the same type as the `toggle` datatype.

```
array: values
[
  standard: "$(toggle)"
  size: "$(subwindow)"
]
```

The following definition is *not* valid and an error will be generated during ACD processing:

```
array: values
[
  standard: "Y"
  size: "$(toggle)"     /* This is not valid ! */
]
```

It is invalid because the `size:` attribute of the `array` datatype has an integer value, whereas the `toggle` datatype has a boolean value.

4.4.4 Calculations and tests

Calculations can be performed in ACD using the @ syntax. A rather silly, but legal, calculation would be:

```
@(5 + 9)
```

which equates to the value 14. Calculations can be used to add, subtract, multiply or divide, or test for equality, inequality, "greater than" or "less than". The test values can be integers, floats and strings. You can use calculations with most attributes of datatypes where they make sense. In the following example a start condition is set to have a maximum value of the sequence length minus a window size value plus 1:

```
sequence: sequence
[
  parameter: "Y"
  type: pureprotein
]

integer: window
[
  standard: "Y"
]

integer: start
[
  standard: "Y"
  maximum:   "@(@($(sequence.length) - $(window)) + 1)"
]
```

Note that there are two separate calculations here so each needs to be surrounded by the @() syntax. Long calculations can get messy. If you need to use them then you possibly need to rethink your ACD logic. If they can't be avoided then they can be tidied up with the use of variables (see below).

4.4.4.1 Arithmetic operations

The supported arithmetic operations are addition, subtraction, multiplication and division. The standard characters for the arithmetic operations are used: +, -, * and /.

```
@(a+b) (Addition)
@(a-b) (Subtraction)
@(a*b) (Multiplication)
@(a/b) (Division)
```

The operands a and b *must* resolve to a numeric value (integer or floating point); the result is undefined otherwise but will most probably result in an error during ACD processing.

Only a single arithmetic operation is allowed per set of parentheses, for instance, the code below is *not* valid:

```
integer: a []
integer: b []
integer: c []
integer: n
[
  default: "@(a + b + c)"     /* This is not valid ! */
]
```

It could however be rewritten using nested operations. The following is perfectly valid:

```
integer:  n
[
default: "@(@(a + b)+c)"
]
```

Note that two sets of parentheses are required: The first around the addition of `a` and `b` which gives `@(a + b)`. The second around the addition of `@(a + b)` to `c`, which gives `@(@(a + b)+c)`.

Where more than one arithmetic operation is required, however, one would typically use an internal ACD variable to hold the intermediate results.

4.4.4.2 Tests for equality

The supported equality tests (symbols in parenthesis) are "equality" (==), "non-equality" (!=), "less than" (<) and "greater than" (>):

```
@(token1 == token2)    (Equality)
@(token1 != token2)    (Non-equality)
@(token1 < token2)  (Less-than)
@(token1 > token2)  (Greater-than)
```

The above equality tests can be used on strings in which case the lexicographical sorting order of the string is used.

4.4.4.3 Boolean tests

The supported boolean operations are logical `AND`, logical `OR` and logical `NOT`. Again, the standard characters are used: `&`, | and `!`:

```
@(!a)  (NOT)
@(a|b) (OR)
@(a&b) (AND)
```

In the following ACD code snippet:

```
integer: fubar
[
   standard: "Y"
   default: 5
   etc
]

integer: rtfm
[
   standard: "@(@($(fubar)==3) | @($(fubar)==7))"
   etc
]
```

The integer `rtfm` will only be prompted for if the value of `fubar` is either 3 or 7. Each of the equality tests is a calculation and the boolean test is another calculation. There are therefore three instances of `@()`.

4.4.5 Conditional statements

There are three kinds of conditional statements in ACD: unary, ternary and case-type.

4.4.5.1 Unary conditional statements

A typical use for unary conditionals is to switch prompts on or off. Assume that a window size should only be prompted for if the sequence turns out to be a protein. The ACD to achieve this would look as follows:

```
sequence: sequence
[
  parameter: "Y"
  type: gapany
]

integer: window
[
  standard: "$(sequence.protein)"
  etc
]
```

If the sequence is a protein then the required statement is equivalent to:

```
standard: "Y"
```

and the prompt is switched on. If the sequence is nucleic the statement is equivalent to:

```
standard: "N"
```

This will effectively disable the prompt. Controlling prompting is described in detail elsewhere (Section 4.5, ‘Controlling the prompt’).

4.4.5.2 Ternary conditional statements

Ternary conditional statements have the general form:

```
@(conditional ? value-if-true : value-if-false)
```

They are useful when setting up the application for two distinct modes of usage, for example when setting gap penalty values differently for proteins and nucleic acids in alignment programs. The example below will set the penalty to 14 for proteins and 16 for nucleic acids:

```
integer: penalty
[
  standard: "N"
  default: "@($(sequence.protein) ? 14 : 16)"
  etc
]
```

4.4.5.3 Case conditional

In the case-type operation, the test value is compared with a list of possible values. If a match is found then the operation resolves to the result associated with that possible value. The test value, which is parsed as a string, is followed by an equals sign (=), which in turn is followed by one or more pairs of possible and associated values separated by a colon (`:`). If none of the possible values match then the operation will resolve to the default result that is associated with the keyword `else`.

The `else :` default value pair is not mandatory and if none of the possible values match in a operation without the default value then the operation will resolve to a `NULL`.

This is formalised as follows:

```
@(testval = poss_valA : ass_valA
            poss_valB : ass_valB
            else : default_val)
```

For example:

```
string: matrix
[
    default: "@($(sequence.type) =
          protein : BLOSUM62
          dna : dnamat
          rna : rnamat
          else : unknown)"
]
```

The `$(sequence.type)` variable is a string value that holds the sequence type of the ACD data item named `sequence`. If the type is `protein`, the operation resolves to `BLOSUM62`, if the type is `dna` it resolves to `DNAMAT`. If the type is not in this list, the operation resolves to `unknown`.

If the test value cannot unambiguously be assigned to a single associated value then the operation will resolve to the LAST associated value that matches its possible value.

4.4.6 Use of variables

Variables are useful for holding partial calculations or values. The general syntax for them is:

```
VariableName : Variable value
```

Note that, as a variable only has a single value and no attributes, square brackets are not used.

As an example, here is a calculation to determine the maximum size of a sequence window:

```
integer: start
[
  standard: "Y"
  maximum: "@(@($(sequence.length) - $(window)) + 1)"
]
```

This can be tidied by storing one of the calculations in a partial result as follows:

```
variable: lminusw "@($(sequence.length) - $(window))"
integer: start
[
  standard: "Y"
  maximum: "@($(lminusw) + 1)"
]
```

In the following ACD code, an internal ACD variable `protlen` is used to store an intermediate result. The value of the variable `$(protlen)` is calculated from the length of the input sequence (`sequence` datatype) and used in the definition of the maximum size of the `window` parameter:

```
variable: protlen "@( $(sequence.length) / 3 )"
integer: window
[
    maximum: "@($(protlen)-50)"
    default: 50
]
```

The same result could be achieved using nested operations as shown below:

```
integer: window
[
    maximum: "@( @( $(sequence.length) / 3) - 50)"
    default: 50
]
```

The `window` parameter is calculated directly from the `sequence.length` variable (calculated attribute) by using the divide arithmetic operation nested within a separate subtraction operation.

Variables may be used to simplify the ACD file making it easier to read and parse. An ACD file can use a variable definition to define a result once only, and then to refer to the variable by name in all later ACD data type definitions.

Caution

The use of variables might indicate that there is some complexity in the ACD definitions. When a variable is used, or when a conditional operation refers to another ACD value, the application might logically be regarded as two or more separate applications forked by the conditions resolved.

4.4.6.1 Automatic variables

Currently there is just one of these (`acdprotein`) which is set to true or false depending upon the type of the first sequence read.

4.5 Controlling the prompt

4.5.1 Introduction

Reliable prompting of the user for input values is a critical part of any application interface. EMBOSS allows you precise control over how application parameters are prompted for and thereby saves you a great deal of development time. This is achieved by using ACD data attributes:

- `parameter: "Y"` means that the data item is a parameter, i.e. you do not have to use the data label to specify a value for it on the command line, e.g. `myprog 10`.
- `standard: "Y"` and `additional: "Y"` mean that the data item is a qualifier, i.e. you *do* have to use the data label to specify a value for it on the command line, e.g. `myprog -somevalue 10`. If a data definition is not defined to be a parameter, or a standard or additional qualifier then it will default to an advanced qualifier.

Values for parameters and standard qualifiers are always prompted for (with their default value) if not specified on the command line. Values for additional qualifiers are not prompted for (a default value will be used) unless **`-options`** is given on the command line. A default value for additional qualifiers should always be given in the ACD file. Values for advanced qualifiers are never prompted for.

4.5.2 Parameters

Consider the following ACD file, for an application called **helloworld**, which defines two parameters, namely a string input (`string` datatype) and an output file (`outfile` datatype):

```
application: helloworld
[
    documentation: "Prints something arguably uninteresting"
]

string: message
[
    parameter: "Y"
]

outfile: outfile
[
    parameter: "Y"
]
```

Values for parameters must be specified on the command line in the order they appear in the ACD file. So to specify both parameters for **helloworld** you'd have to type something like:

```
helloworld "Hello World!" message.dat
```

That would cause *Hello World!* to be printed to the output file `message.dat`. Alternatively it's also perfectly acceptable to use one of:

```
helloworld -message "Hello World!" -outfile message.dat
helloworld -outfile message.dat -message "Hello World!"
helloworld -message "Hello World!" message.dat
```

For parameters you don't have to use the name flag whereas for qualifiers you do. If you omit the flag for parameters then their values *must* appear on the command line in the order in which they appear in the ACD file.

However, you might not want to force the user to specify a message, instead relying on a default message ("Hello World!") defined in the ACD file to be printed if nothing else is given. So by typing:

```
helloworld message.dat
```

you want "Hello World!" (the default message) to be printed to the file message.dat.

The above ACD file will not do that. All the data definitions are parameters (parameter:) and therefore will be prompted for if they are not given on the command line. Typing the command above would result in message.dat being taken as the string to be printed, and you'd then be prompted for an output file name. In short, not the desired behaviour. This is where qualifiers come in.

4.5.3 Qualifiers

Values for qualifiers (whether "standard", "additional" or "advanced") can appear anywhere on the command line but you must always refer to them by their flag. So if your message is defined as a qualifier, rather than a parameter, the reference to it has to look like this:

```
-message "Hello World!"
```

To specify a data definition as a qualifier, you use standard: or additional: rather than the parameter: attribute that's currently specified. Alternatively, if you don't specify any of these attributes it will default to being an advanced qualifier. For example:

```
application: helloworld
[
    documentation: "Prints something arguably uninteresting"
]

string: message
[
    standard: "Y"
]

outfile: outfile
[
    parameter: "Y"
]
```

As you now have only one parameter, both of the following command lines are valid:

```
helloworld -message "Hello World!" message.dat
helloworld message.dat -message "Hello World!"
```

Now if you don't specify the message text on the command line, i.e. you just type:

```
helloworld message.dat
```

then `message.dat` will be treated as a parameter, and taken to be the output file name. The program will then prompt you for a string to print out. This is close to what we wanted, but still not quite the desired behaviour. What we wanted was for it to go ahead and run with a default string.

4.5.4 Defaults

You can associate a default value for most data definitions using the `default:` attribute. The ACD can be modified as follows:

```
application: helloworld
[
    documentation: "Prints something arguably uninteresting"
]

string: message
[
    standard: "Y"
    default: "Hello World!"
]

outfile: outfile
[
    parameter: "Y"
]
```

Now if you typed:

```
helloworld message.dat
```

then although you specified a default it's still prompting for a value. Remember that all values for `standard` data definitions are always prompted for, regardless of whether a default is specified or not. You need to specify `message` as being an 'additional' or 'advanced' qualifier, which are not normally prompted for.

The ACD file will look like this:

```
application: helloworld
[
    documentation: "Prints something arguably uninteresting"
```

```
]

string: message
[
    additional: "Y"
    default: "Hello World!"
]

outfile: outfile
[
    parameter: "Y"
]
```

This ACD finally does what is needed. Typing:

```
helloworld message.dat
```

will print `Hello World!` to the file `message.dat` i.e. the desired behaviour.

You can of course override the default message by specifying the message on the command line:

```
helloworld -message "Goodbye World!" message.dat
```

This will print a rather morbid message to the output file. You should usually supply a default value for `additional` and `advanced` ACD data definitions as EMBOSS will not prompt you for a value if you don't specify one on the command line. EMBOSS would generate an error if you tried, from within your C source code, to access the value of an unspecified data item. In contrast, values for `standard` and `parameter` data definitions are always prompted for if they're not specified on the command line, therefore a default is useful but not essential.

> **Note**
>
> Note that some datatypes such as **`boolean`** have an inbuilt default value.

4.5.5 Maxima and minima

It is often either useful or vital to be able to set limits on the maximum and/or minimum values to be associated with an ACD datatype definition. This is done in an intuitive way using the `minimum` and `maximum` attributes:

```
integer: window
[
    standard: "Y"
    default:  "10"
    minimum:  "5"
```

```
    maximum:   "100"
]
```

4.5.6 Setting the prompts

EMBOSS will always provide default prompt text. Consider the following ACD file:

```
integer: window
[
    standard: "Y"
    default: "10"
    minimum: "5"
    maximum: "100"
]
```

The user would be prompted as follows:

```
-window : Enter a number [10] :
```

Though adequate this is not entirely friendly. You can set the prompt for a datatype definition by using the `information:` attribute. The ACD file would then look like this:

```
integer: window
[
    standard: "Y"
    default: "10"
    minimum: "5"
    maximum: "100"
    information: "Window size"
]
```

This will give the following as the prompt.

```
Window size [10] :
```

which is much more meaningful.

4.5.7 Negation

So far you have only described the value of `"Y"` after a parameter or qualifier definition. However, negation (specifying, indirectly, a value of `"N"`) often finds a use in ACD files. Take a simple example:

```
sequence: sequence
[
    parameter: "Y"
]
```

```
integer: n
[
    standard: "@($(sequence.length) > 100)"
]
```

The `standard:` attribute is being set to `"Y"` if the sequence length is greater than 100 or "N" otherwise. Although a value of `"N"` should never be specified explicitly after `parameter:`, `standard:`, or `additional:`, calculated values that evaluate to `"N"` are in fact supported for the qualifiers (but not parameters). In such cases, the `"N"` overrides the default behaviour of the attributes such that prompting for a value will be turned off. This is useful in some situations. In this case, the calculation will switch a prompt on only if the sequence length is greater than 100.

Now let us assume that your application can produce both graphic and textual output. Assume further that you only want textual output if the user hasn't selected graphical output. First you would set up a `toggle` ACD datatype definition as follows:

```
toggle: plot
[
    standard: "Y"
    default: "N"
    information: "Plot a graph"
]
```

The value of `$(plot)` will be `"Y"` if the user adds **`-plot`** to the command line. The value is `N` if either the user doesn't add anything to the command line or if the user adds **`-noplot`** to the command line.

The output file can now be defined as:

```
outfile: outfile
[
    standard: "@(!$(plot))"
]
```

This becomes equivalent to `standard: "Y"` only if `plot` is *not* true. The negation operator (`!`) is a calculation so the term must be surrounded by `@()`.

The only sad thing about this is that it doesn't work as written but not for any reason involving the logic. The reason is because EMBOSS handles file input/output operations in a different manner to other datatypes. If it sees one of the file (e.g. `outfile`) or sequence (e.g. `seqout`) definitions it will *always* try and open it.

If the term equates to `standard: "N"`, and no filename has been specified on the command line or as a default (and you wouldn't normally specify a default name for an output file) then ACD parsing will try and open a file with no name. That would cause an error.

There is a way around this and that is to use the `nullok:` attribute. So, a definition of `outfile` that works is:

```
outfile: outfile
[
```

```
    standard: "@(!$(plot))"
    nullok: "Y"
]
```

The `nullok:` attribute above means that it's OK to continue (do not generate an error) if no filename is given.

Note

If you run an ACD file (for instance, testing it by using **acdc** (Section 4.6.4.1, '**acdc**') with calculated values for **`standard:`**, you will see a warning message. The use of calculations in this way is supported but is only recommended if absolutely necessary. This is because it can confuse the comprehension of the ACD file, for instance, by third party interfaces.

Caution

Calculations should *not* be used to set the value of a **`parameter:`**: an error will be generated during ACD processing if you do. Parameters are taken to be essential inputs to the application. In contrast, calculations can be used to control the prompting for *qualifiers* (whether standard, additional or advanced).

4.5.8 Boolean datatypes

The ACD syntax provides two datatypes (`boolean` and `toggle`) that have boolean values. `boolean` is a standard boolean datatype whereas a `toggle` is a special type of boolean datatype that is used exclusively to control the prompting of other attributes.

In the ACD file below, an application with two parameters and an advanced qualifier (`abool`) of `boolean` datatype is shown. `abool` might conceivably be used to set verbose or terse text in the output file, though it would be given a more intuitive name.

```
application: seqdemo
[
    documentation: "Demo applicatioN"
]

sequence: asequence
[
    parameter: "Y"
]

boolean: abool
[
    default: "Y"
]

outfile: outfile
[
    parameter: "Y"
]
```

The application could be invoked in either of the following ways:

```
seqdemo filename.seq filename.out -abool
seqdemo filename.seq filename.out -noabool
```

In the first example the boolean qualifier `abool` is set to `True`. In fact, the value needn't have been set explicitly since the default value is `True`. In the second command line `abool` is set to `False` using the prefix **no**.

4.5.8.1 Prompting for boolean values

Care should be taken over the definition of the `information:` and `help:` global attributes for `boolean` datatypes. These are used to prompt the user (interactively or via a GUI) and to provide help text. The text provided in each case should reflect the expected default value of the boolean option, which may be the opposite of what the name implies. For example, if set to "Y" by default, then the command line option would typically be **-noFlag** where *Flag* is the qualifier. If set to "N" by default, then the default action may be the opposite of what the information or help text implies. If the value is calculated then the user may need some extra guidance.

4.6 ACD utilities

4.6.1 Introduction

EMBOSS includes six utilities that are used to run, test and document an ACD file without the need to write the C source code for the program itself:

acdc Reads and processes an ACD file, prompting the user as necessary in exactly the same way as if the application was running. If it encounters an error it will terminate. This allows the behaviour of the command line interface to be tested under different conditions, even if the application itself has not yet been written. It is the most widely used utility and is the basic tool for testing that an ACD file works correctly.

acdtrace Works the same as **acdc** but reports the processing of lines in an ACD file. It also prints the results of evaluating any variables and operations (see Section 4.4, 'Operations').

acdvalid Processes an ACD file and reports any errors or warnings caused by incorrect, deprecated or obsolete syntax. The tests go above and beyond those of **acdc** and **acdtrace**. It enforces recommendations for parameter and qualifier naming, and highlights ACD syntax that might cause problems to third-party parsing software.

acdlog Processes an ACD file and user input while logging the contents of internal data structures. For the time being its use is deprecated and the **entrails** application (see below) should be used instead.

acdpretty Generates a correctly formatted ACD file from one which is parsable but which is partly or incorrectly formatted.

acdtable Creates the table of parameters, allowed values and defaults, used in the application documentation and to provide user help.

4.6.2 Using the utilities

The utilities do not have options of their own. They are all invoked in the same way, by specifying the name of the application to be tested along with any command line options after the name of the utility application itself:

```
UtilityName ApplicationName Options
```

For example, this would test the ACD file of an application called **helloworld**:

```
acdc helloworld
```

acdc, **acdtrace** and **acdlog** will run the command line interface of any EMBOSS program by reading its ACD file and processing the command line and user responses in exactly the same way as if the true program itself were running. The user is prompted for any required values for application parameters and any input files are read. The application proper is *not* run, it is only the ACD file that is read.

Any command line options can be specified. In the following, **seqret** is being called to retrieve region 50 – 100 from a sequence (`sequence.in`) and write the fragment to the output file (`sequence.out`):

```
acdc seqret sequence.in sequence.out -sbegin 50 -send 100
```

Note

All EMBOSS applications call the NUCLEUS function **`embInit`** (or for EMBASSY applications **`embInitP`**) which makes further calls to ACD file processing functions. The utility applications pass any arguments (i.e. the command line for the application being tested) to **`embInit`** or **`embInitP`** and so run exactly like the native application. Output files can be specified in any available format however they will be empty (the application code is not called so there will be no output).

4.6.3 Global qualifiers

The utilities can use all the global qualifiers (see the *EMBOSS User's Guide*):

`-auto` Turns off any prompting of the user. It is always useful to test an application with the **`-auto`** option because this emulates the condition of the application being called by a script or some other way that doesn't involve interactive user input.

-debug Turns on debugging. This outputs the parameters associated with calls to the AJAX function `ajDebug`.

-filter Reads from `stdin` and writes to `stdout` and implies **-auto**.

-stdout Writes by default to `stdout`, but still prompts the user.

-help Will give usage information for the utility.

-verbose When used with **-help** also gives the associated qualifiers and the global qualifiers.

-warning Program will generate warning messages if necessary.

-error Program will generate error messages if necessary.

-fatal Program will generate fatal error (termination) messages if necessary.

None of the utilities themselves will generate debugging, warning, error or fatal messages. However, because input files can be read, any debug calls or messages made by the input functions *will* be reported.

4.6.4 Description of utilities

4.6.4.1 acdc

acdc reads and processes an ACD file, prompting the user as necessary in exactly the same way as if the application was running. If it encounters an error it will terminate. This allows the behaviour of the command line interface to be tested under different conditions, even if the application itself has not yet been written. It is the most widely used utility and is the basic tool for testing that an ACD file works correctly.

acdc does not have its own options, but takes a single parameter which is the name of the EMBOSS application that is being tested. It is invoked by specifying the name of the application to be tested along with any command line options after the name of the utility application itself:

```
acdc ApplicationName Options
```

Any command line options for the application can be specified. The user is prompted for any required values for application parameters and any input files are read. The application proper is *not* run, it is only the ACD file that is read.

acdc will use any data files specified in the ACD file of the specified program, but will ignore any data files that are only used directly by the program's code. EMBOSS data files are distributed with the applications and stored in the standard EMBOSS data directory (see the *EMBOSS User's Guide*).

Output files can be specified in any available format however they will be empty (the application code is not called so there will be no output).

Here is a sample session with **acdc**:

```
% acdc antigenic
Finds antigenic sites in proteins
Input protein sequence(s): tsw:actl_fugru
```

```
Minimum length of antigenic region [6]:
Output report [act1_fugru.antigenic]:
```

4.6.4.2 acdtrace

acdtrace works in the same way as **acdc** but reports the processing of lines in an ACD file. It also prints the results of evaluating any variables and operations in the ACD file (see Section 4.4, 'Operations').

It is invoked as follows:

```
acdtrace ApplicationName Options
```

acdtrace will run the command line interface of any EMBOSS program by reading its ACD file and processing the command line and user responses in exactly the same way as if the true program itself were running. Any command line options for the application being called can be specified. The user is prompted for any required values for application parameters and any input files are read. The application proper is *not* run, it is only the ACD file that is read.

Output files can be specified in any available format however they will be empty (the application code is not called so there will be no output).

acdtrace will use any data files specified in the ACD file of the specified program, but will ignore any data files that are only used directly by the program's code. EMBOSS data files are distributed with the applications and stored in the standard EMBOSS data directory (see the *EMBOSS User's Guide*).

Here is a sample session with **acdtrace**:

```
% acdtrace antigenic
Finds antigenic sites in proteins
Input protein sequence(s): tsw:act1_fugru
Minimum length of antigenic region [6]:
Output report [act1_fugru.antigenic]:
```

The output on screen (not shown) can look a little confusing but is by far the best way to see how variables and operations work in your ACD file.

4.6.4.3 acdvalid

acdvalid processes an ACD file and reports any errors or warnings caused by incorrect, deprecated or obsolete syntax. The tests go above and beyond those of **acdc** and **acdtrace**. It highlights ACD syntax that might cause problems to third party parsing software which wrap the EMBOSS applications.

acdvalid validates an ACD file, testing many features that will not prevent an application from running, but which may create problems for any user interfaces, whether on the command line or through use of a wrapper. Most of the messages from **acdvalid** are warnings and do not prevent a program from running.

It is invoked as follows:

```
acdvalid ApplicationName Options
```

acdvalid will, like **acdtrace**, run the command line interface in exactly the same way as if the true program itself were running.

It will use any data files specified in the ACD file of the specified program, but will ignore any data files that are only used directly by the program's code. EMBOSS data files are distributed with the application and stored in the standard EMBOSS data directory (see the *EMBOSS User's Guide*).

Output files can be specified in any available format; however, they will be empty (the application code is not called so there will be no output).

Here is a sample session with **acdvalid**:

```
% acdvalid antigenic
```

acdvalid generates both error and warning messages. If the message is an 'Error' then the ACD file will *not* work and requires fixing. If the message is a 'Warning' then the ACD file will work, although it is worth trying to fix the problem.

> **Note**
>
> Warning messages are only advisory, but as many warnings as possible are fixed for any application in the main release.
> It is worth running **acdvalid** on all local ACD files with each new version of EMBOSS, because extra validation tests are added with each new release.

4.6.4.3.1 Features tested by acdvalid

These include:

- Sections are in the correct order of 'input', 'required', 'additional', 'advanced' and 'output'.
- Section and subsection names (Section 4.1.5, 'ACD file sections') are defined in the EMBOSS system file `sections.standard`.
- Section definitions use a standard `information:` attribute.

  This ensures that interface wrappers will have a consistent description when using sections to group application parameters.
- All data definitions are contained in a section or subsection.
- Parameters that are restricted to one section are in the correct section.

  All input ACD datatypes (Section A.2.2, 'Description of input ACD datatypes') must be in the "input" section and all output ACD datatypes (Section A.2.3, 'Description of output ACD datatypes') are in the "output" section.
- Other parameters are valid in the section in which they appear.

For example the "advanced" section cannot include parameters that might be prompted for, which excludes data definitions with a `parameter:`, `standard:` or `additional:` attribute set (parameters, and standard and additional qualifiers). The "additional" section can only have qualifiers whose `additional:` attribute is set, or `toggle` objects which control them.

- Calculations are *not* used to set the value of a `parameter:` attribute.
  Parameters are taken to be essential inputs to the application. In contrast, calculations can be used to control the prompting for *qualifiers* (whether standard, additional or advanced).
- If a value for the `group:` attribute in the application definition of the ACD file is used, that it is defined in the file `groups.standard`.
- Calculated values are not used for `standard:`.
  If an ACD file has calculated values for `standard:`, a warning message is reported. The use of calculations in this way is supported but is only recommended if absolutely necessary, because it may confuse comprehension of application inputs by, for instance, third-party interfaces.
- Calculated values (see Section 4.5, 'Controlling the prompt') for the `standard:` and `additional:` attributes use `toggle` datatype values but not `boolean`.
  This allows interface wrappers can use `toggle` values to split an application into two or more simpler definitions.
- `parameter:`, `standard:` and `additional:` attribute values are not set explicitly to `"N"` (the default).
  This confuses parsers. You should never explicitly specify `Parameter: "N"`, `Standard: "N"` or `Additional: "N"`. In practice calculated values of `"N"` are in fact supported and can be used, in exceptional circumstances, to override the default behaviour of these attributes (see Section 4.5, 'Controlling the prompt').
- Datatypes which are typically the primary input or output of an EMBOSS application are defined as a parameter by using the `parameter:` attribute. The datatypes include:
  - Alignment output (`align:` datatype)
  - Sequence features for input (`features` datatype) and output (`featout` datatype)
  - File datatypes for input (`infile`, `filelist`, `directory` or `dirlist` datatypes) and output (`outdir`)
  - Application report output (`report:` datatype)
  - Sequence inputs (`sequence`, `seqall`, `seqsetall` or `seqset` datatypes)
- The first data definition for any of the datatypes above *must* be defined as a parameter. It is recommended that subsequent definitions of that datatype (second, third and so on) are also defined as parameters. The exception is if the `default:` or `nullok:` attributes are set, in which case no warning or error messages are generated as the application can run with a default or without any value for the definition.
- Parameter and qualifier names (data definition flags) match an accepted standard (Section 4.1, 'Introduction to ACD file development').

For example all sequence input should end in "sequence" so that short forms (e.g. -**aseq**, **-bseq**, etc.) can be used on the command line. A warning message will explain what the standard requires if an error is made.

- ACD definitions for alignment output, feature input and output, file inputs and outputs, report output, and sequence inputs and outputs (seqout, seqoutall, seqoutset datatypes) have standard names (Section 4.1.4, 'Parameters and qualifiers').
- Where the knowntype: attribute is specified, the value matches a standard name defined for the appropriate datatype in the EMBOSS system file knowntypes.standard, or matches an accepted standard format for the datatype.
  A few values not in knowntypes.standard are accepted, for example "ApplicationName output" for an outfile: datatype. These are documented with the datatypes (Section A.5, 'Datatype-specific attributes').
- Data definitions that require a value for the knowntype: attribute have one defined.
  These include any output files (outfile datatype) or directories (outdir), and the string datatypes.
- EMBOSS must be able to discern the type of the feature input and output. If no type is specified (using the type: attribute) for a feature or featout definition and there is no sequence input from which to take a default type, then an error will be generated.
- EMBOSS must also be able to discern the type of the sequence output. If no type is specified (using the type: attribute) for an output sequence (seqout, seqoutset or seqoutall definition) and there is no sequence input from which to take a default type, then an error will be generated.
- Datatypes with standard prompts should not have information: or prompt: attributes (see Section A.4, 'Global attributes').
  To provide standard prompts, a default value for the information: string is defined for most common datatypes. The defaults are in the EMBOSS system file codes.english.
- The text given after the help:, information: and prompt: attributes conforms to the following conventions:

  - The string should start with an uppercase, alphabetic character and should not end with a full stop

- Any seqset or seqsetall datatype *must* have the aligned: attribute set.

4.6.4.4 acdlog

acdlog processes an ACD file and user input, while logging the contents of internal data structures. For the time being its use is deprecated and the **entrails** application (Section 4.6.5, 'Use of **entrails**') should be used instead.

It is invoked as follows:

```
acdlog ApplicationName Options
```

Like **acdtrace**, **acdlog** will run the command line interface in exactly the same way as if the true program itself were running.

Output files can be specified in any available format; however, they will be empty (the application code is not called so there will be no output).

4.6.4.5 acdpretty

acdpretty will generate a correctly formatted ACD file from one which is parsable but which is partly or incorrectly formatted.

acdpretty reads an ACD file and rewrites it with clean formatting to the file:

```
(ProgramName).acdpretty
```

which can be used to overwrite the original ACD file. All sections and attributes will be indented. All values will be quoted. Lines will be wrapped at a reasonable width. The output file is intended as a standard format for parsers used by GUIs and other interface wrappers, rather than using the original ACD file directly.

It is invoked as follows:

```
acdpretty ApplicationName Options
```

Any command line parameters that might normally be specified for the application are essentially ignored as processing stops once the reformatted ACD file is printed. A few of the global qualifiers have the potential to conflict with the function of **acdpretty**.

4.6.4.6 acdtable

acdtable is used to create the table of parameters, allowed values and defaults used in the application documentation and to provide user help.

It is invoked as follows:

```
acdtable ApplicationName Options
```

Acceptable expected and default values can be generated automatically for most ACD datatypes. In other cases the allowed value is defined with the `valid:` global attribute, and the default value by the `expected:` global attribute (see Section A.4, 'Global attributes').

4.6.5 Use of entrails

entrails will report the internals of data structures used to parse the ACD file. It reports:

- Details of all known ACD datatypes
- Details of all known ACD attributes
- Details for all calculated attributes

- Internal data structures for sequence access methods
- Internal data structures for sequence reading and writing
- Internal data structures for feature handling
- Internal data structures for alignments
- Internal data structures for application reports
- Internal data structures for codon usage tables
- Report of the database attributes available
- Report of the resource attributes available.

Note

entrails is not compiled in a standard EMBOSS installation because it is in the 'make check' section of the **`Makefile.am`** file.
To compile the **`"check_PROGRAMS"`** applications, type:

```
make check
```

from the applications directory. If for some reason that fails (the section is after all reserved for applications that are in active development) edit the **`Makefile.am`** files accordingly so that **`entrails.c`** is treated as a normal application, and compile it as usual. For further information see Section 1.2, 'Installation of CVS (Developer's) Release'.

C programming 5

5.1 Basic datatypes

5.1.1 Fundamental C datatypes

The fundamental datatypes that the C programming language supports include two basic arithmetic types. *Integer* types represent characters or whole numbers whereas *floating* types represent floating point numbers and have a sign bit, mantissa and exponent.

The types may be prepended with modifiers to specify whether or not the type is signed (integer types only) and the size (not for `char` which is usually always 8 bits) and precision (floating types only). Integer types on most operating systems (e.g. IRIX is an exception) are all signed unless specified otherwise: unsigned integers can hold a greater range of positive values. The exact types available and their sizes depend on the compiler. The common types and nominal sizes are given in the table below.

Type	Description	Size (bytes)
`char`	Character	1
`short int`	Short integer	2
`int`	Integer	4
`long int`	Long integer	8
`unsigned char`	Unsigned character	1
`unsigned short int`	Unsigned short integer	2
`unsigned int`	Unsigned integer	4
`unsigned long int`	Unsigned long integer	8
`float`	Floating point number	4
`double`	Double precision floating point number	8
`long double`	Extra precision floating point number	12

Others may be available, for example the `long long` integer type. The standard C header `limits.h` defines the size of a `char` (e.g. 8 bits) and the largest and smallest values of the other integer types that are permitted in any implementation, such that each type is guaranteed to be of a minimum size and no larger than the following type. For example, an `int` is never longer than a `long int`. Similarly, the header `float.h` (or equivalent) defines constants such that a floating type is always at least as precise as the previous type. For example, a `double` is always at least as precise as a `float`.

In addition to the fundamental types described above, C defines the `void` type which specifies an empty value and is used, for example, for the return type of functions that do not return a value.

There are also *enumerations*, unique types that are associated with a set of named constant integer values.

5.1.2 Fundamental AJAX datatypes

5.1.2.1 Integer types

The exact size and precision of the fundamental C datatypes (excluding `char`) is implementation-dependent. To avoid any potential problems with these system-dependent limits, AJAX defines in the AJAX file `ajarch.h` a new set of fundamental datatypes as follows:

Type	Description
`ajshort`	Short integer
`ajint`	Standard integer
`ajlong`	Long integer
`ajushort`	Unsigned short integer
`ajuint`	Unsigned standard integer
`ajulong`	Unsigned long integer

There are some differences between the systems listed in `ajarch.h` but the typical definitions are as follows:

```
typedef short ajshort;
typedef int ajint;
typedef long long ajlong;

typedef unsigned int ajuint;
typedef unsigned short ajushort;
typedef unsigned long long ajulong;
```

An `ajint` is typically 32 bits and an `ajlong` typically 64 bits. Use `ajint`, if 32 bits is enough, instead of `int`. Use `ajlong` instead of `long` or `long long`. That said, standard C `int` and `long` *should* be used in some circumstances, for example as parameters to C system library functions.

You should match your datatype to what you need. If, for example, you are using an Alpha box then both your `int` and `long` variables will be 64 bits. In this case don't use only `ajlong` out of laziness as your code may run more slowly on other platforms.

5.1.2.2 Other AJAX types

For convenience ajdefine.h also defines a few datatypes given in the table below:

Type	Description
AjBool	Boolean
AjStatus	Status code
AjIntArray	Integer array (int*)
AjFloatArray	Float array (float*)

AjBool Used to store true (ajTrue, AJTRUE) and false (ajFalse, AJFALSE) values. On output, the conversion code %b writes Y or N while conversion code %B writes Yes or No.
There is also a macro for testing boolean values:

```
#define AJBOOL(b) (b ? "TRUE" : "FALSE")
```

AjStatus Intended as a general return code for functions, but currently unused because AjBool is enough. Has a constant value to indicate success (ajStatusOK), message (ajStatusInfo), warning (ajStatusWarn), error (ajStatusError) and fatal error (ajStatusFatal).

AjIntArray A simple C-type array of integers:

```
typedef int* AjIntArray;
```

AjFloatArray A simple C-type array of floats:

```
typedef float* AjFloatArray;
```

To use these AJAX datatypes you must include the files ajdefine.h and ajarch.h in your code. All applications must include at the start of the code the preprocessor directive #include "emboss.h" (see Chapter 2, *Your first EMBOSS application*).

emboss.h is the master include file and imports the entire EMBOSS interface: it includes all the header files in the AJAX and NUCLEUS C programming libraries making all the code available to you. If you inspect the file you'll see that ajax.h is included:

```
% more nucleus/emboss.h

#ifndef emboss_h
#define emboss_h

#include "ajax.h"
#include "ajgraph.h"
.
.
```

which itself includes `ajdefine.h`, which includes `ajarch.h`:

```
#ifdef __cplusplus
extern "C"
{
#endif

#ifndef ajdefine_h
#define ajdefine_h

#include "ajarch.h"
.
.
```

If you develop library code that uses the fundamental types, you must include `ajdefine.h` explicitly. For example from `ajstr.h`:

```
#ifdef __cplusplus
extern "C"
{
#endif

#ifndef ajstr_h
#define ajstr_h

#include "ajdefine.h"
#include "ajtable.h"

.
.
```

5.1.3 Derived types

A potentially infinite number of other types may be derived from the fundamental C datatypes as follows:

- Arrays of objects of a single type
- Functions returning objects of a single type
- Pointers to objects of a given type
- Structures of objects of various types
- Unions capable of holding one of several objects of different datatypes.

'Object' here might refer to a variable with a primitive datatype, to a function, to a data structure and so on. In general these methods can be applied in a compound manner. It is possible, for instance, to have a data structure that includes an array of functions which all return a pointer to an array of `float` variables. Pointers provide a handle on objects of a particular type and are used when managing memory for objects.

When programming under EMBOSS things are, for most intents and purposes, simplified:

- There is a standard way for defining new data structures and pointers to them. Data structure and pointer types are referred to as 'objects' and 'object pointers' respectively.
- Non-void functions typically return either a primitive datatype or an object pointer.
- Structures are passed to functions by reference (object pointer): the structure itself is never passed.
- Constructor and destructor functions handle object memory management.
- Macros are provided for general-purpose memory management.
- AJAX implements dynamic arrays of common fundamental datatypes for which memory management is handled automatically. Memory management is also handled automatically for some other datatypes, for example strings.

5.1.4 Storage class and linkage

There are two storage classes in C, automatic and static. Automatic objects are initialised whenever the code block in which they are declared is entered (excluding jumps into the code), and in the order in which they are declared. In contrast, static objects are initialised only once before the program proper starts. The storage class of an object depends on the context of its declaration and the keywords used.

Automatic objects are local to a block and are discarded when the block is exited. Declarations in a block are automatic by default although this may be made explicit with the `auto` keyword. Objects declared with `register` are automatic and, where possible, are handled in the fastest available memory register.

Static objects might be local to a block or external to all blocks at the same level as the function definitions. In either case they retain their value when the block is exited and re-entered. They are declared with the keyword `static`.

Objects that are external to all blocks are always static. In such cases the `static` keyword gives them *internal linkage* which means they are only visible in the local file. Otherwise, they have *external linkage* which means they will be global to the entire program or other compiled unit.

There are a few implications here when programming for EMBOSS. Any unions and C data structures (objects) that are private to a library file or application should be declared `static` in the library or application C source code file. Any public (external) unions and structures are given in appropriate library header files and should not include the `static` keyword. Similarly, all application functions and private functions in the libraries should be declared `static`. Public functions in the library should not include this keyword.

Avoid exporting names outside individual C source files; i.e. declare as `static` (in the library header file or application code) every function that you possibly can. Where code is specific to an application only, it should stay in the application C source code file and not be moved to the libraries until it is of more generally use.

All datatypes should be defined in the EMBOSS style and functions must be prototyped using the full ANSI C style (see Appendix C, *C coding standards*).

5.2 C pointers primer

5.2.1 Pointer basics

Pointers can be a difficult aspect of C programming and their misuse leads to more problems than any other part of the language. With a proper understanding of the principles, however, things are much easier. To become good at EMBOSS programming you *must* master at least the basics of pointers.

With the exception of register variables, every variable you declare in your program resides somewhere in memory, that 'somewhere' is the *memory address* of the variable. A pointer is merely a variable whose value is such a memory address. On most C implementations the longest unsigned integer that the hardware supports is used for pointers, for example an `unsigned long int`. The value of this integer is the memory address of another object, such as another variable, C data structure etc.

So, when this line of a program is executed:

```
ajint x=0;
```

sufficient memory to hold an integer (usually 4 bytes) is reserved for use by the program. The value of those bytes is set to zero.

To declare a pointer variable, you use the pointer operator (`*`) in the declaration. So when this line is called:

```
ajint *ptr=NULL;
```

memory for a pointer variable is reserved and the value of the variable set to `NULL`.

To return the memory address of a variable, you use the `&` (address) operator:

```
ajint x=10;
ajint *ptr=NULL;

ptr = &x;
```

To get to the value held at a particular memory address you use the `*` (pointer) operator; this is called *dereferencing* the pointer or getting a value by *indirection*:

```
ajint x=10;
ajint y=0;
ajint *ptr=NULL;
ptr = &x;
y = *ptr; /* y is set to 10 */
```

From the above code it's clear that `x` and `y` hold integer values or, put more simply, 'x and y are integers'. Similarly, `ptr` holds a memory address or, more simply, '`ptr` is a pointer'. The pointer is made to point to `x` in the above code.

If you don't complicate the basic idea that a pointer is a variable whose value is a memory address then you have the foundation for understanding and using pointers in EMBOSS.

Example
Consider the following code:

```
int main()
{
                ajint x=0;
    /* 1. */    printf("Value of x : %d\n", x);
    /* 2. */    printf("Memory address of x : %p\n", &x);
    /* 3. */    printf("Value of x by indirection : %d\n", *(&x));
}

/* Output will look like:

Value of x : 0
Memory address of x : #1
Value of x by indirection : 0

(In reality, a hexadecimal number would be printed instead of '#1',
but '#1' is easier to follow).
*/
```

The variable name `x` is our handle on the reserved memory. It refers to an integer value that resides at memory address `#1`. In the code the following is done:

1. First print the value of `x`.
2. Then use `&` to get the memory address of variable `x` and print it.
3. Then use the `*` operator to dereference this address and print the value stored there.

5.2.1.1 Declaring pointers

In practice a pointer holds the memory address of a specific object such as an integer, C data structure or even another pointer. The type of data pointed at must be specified when the pointer is declared. This is not because the memory address of an `int` is any different to that of a `float`, it's so that the compiler knows how the pointer can be used in the source code. For instance the compiler must know the type of data pointed at to be able to access the memory pointed at in order to, for example, print a value correctly. This is why in C pointers are declared in the manner shown:

```
ajint *ptr=NULL;
```

The `*` in the declaration means that `ptr` is a memory address and the `ajint` tells us that it's the address of an integer, i.e. `ptr` is a pointer to an integer.

When that line of the program is executed, sufficient memory to hold a memory address will be reserved for use by our program. This, like an integer, is normally 4 bytes on 32-bit machines. The value of these 4 bytes is set to `NULL`.

It's important to appreciate that the meaning of the * operator is context-dependent. It's only in the context of a variable declaration that it declares a variable as being a pointer, for example:

```
ajint *ptr=NULL
```

means 'ptr is a pointer variable, set its value to NULL'.

If * was found outside of a variable declaration it would indicate a variable being dereferenced as mentioned above, for example:

```
ajint *ptr = NULL;

*ptr =0;
```

Here the second instance of *ptr = 0; means 'dereference ptr and set the value stored there to 0' or in other words 'set the value held at memory address ptr to 0'. Note that the code, as written, would crash – but it is only illustrative.

The final thing to mention is that we've assigned the value of NULL to the pointer in the declaration and the integer stored at that memory address to 0 in the code. They *cannot* be used interchangeably as they are not of the same type: you should only ever use NULL for pointers.

You can see that in the code below:

```
int main()
{
/* 1. */  ajint x=0;
/* 1. */  ajint *ptr=NULL;

/* 2. */  printf("Value of x : %d\n", x);
/* 3. */  ptr = &x;
/* 4. */ *ptr=5;
/* 2. */  printf("Value of x : %d\n", x);
}

/* Output will look like this:
Value of x : 0
Value of x : 5
*/
```

In the code:

1. Declare an integer and a pointer to an integer.
2. Print the value of x.
3. Give ptr the value of the address of x.
4. Set the value of x to 5 by indirection.

In the above example, you would normally say that 'ptr holds the address of x' or simply 'ptr points to x'.

5.2.2 Pointers to pointers

It was mentioned above that a pointer can hold the memory address of another pointer. This is obvious when you realise that a pointer, like any variable, resides somewhere in memory. So if a pointer that holds the memory address of an integer is a 'pointer to an integer', then a pointer that holds the memory address of another pointer is, of course, 'a pointer to a pointer'.

This bit of code shows how you declare a pointer to a pointer-to-an-integer:

```
ajint **ptrto=NULL;
```

For better comprehension let us write the declaration as follows:

```
(ajint *) *ptrto=NULL;
```

The second `*` means that `ptrto` is a memory address. The `ajint *` tells us that it's the address of a pointer-to-an-integer. When the code is executed, enough memory to hold an address is reserved for our use and the value of the bytes is set to `NULL`.

Of course, the `&` (address) and the `*` (pointer) operators still work with pointers to pointers. Where you have multiple levels of pointers you can use multiple `*` (pointer) operators for dereferencing. `*ptrto` would dereference once and retrieve an address (a pointer to an integer). `**ptrto` would dereference twice and retrieve an integer.

You can see that in the code below:

```
main()
{
/* 1. */ ajint x=0;          /* an integer */
/* 1. */ ajint *ptr=NULL;    /* a pointer to an integer */
/* 1. */ ajint **ptrto=NULL; /* a pointer to a pointer-to-an-integer */

/* 2. */ printf("Address of x : %p\n", &x);
/* 2. */ printf("Address of ptr : %p\n", &ptr);
/* 2. */ printf("Address of ptrto : %p\n", &ptrto);

/* 3. */ ptr = &x;
/* 3. */ ptrto = &ptr;

/* 4. */ printf("Value of x : %d\n", x);
/* 4. */ printf("Value of ptr : %p\n", ptr);
/* 4. */ printf("Value of ptrto : %p\n", ptrto);

/* 5. */ printf("Value of x by dereferencing ptr : %d\n", *ptr);
/* 5. */ printf("Value of x by dereferencing ptrto : %d\n", **ptrto);
}
/* Output will look like this:
Address of x : #1
Address of ptr : #2
Address of ptrto : #3

Value of x : 0
Value of ptr : #1      /* i.e. the address of x*/
```

```
Value of ptrto : #2    /* i.e. the address of ptr*/

Value of x by dereferencing ptr : 0
Value of x by dereferencing ptrto : 0
*/
```

There are no new concepts in the above code, it's merely an extension of what you already know about pointers:

1. Declare three variables called x, ptr and ptrto. x is the integer, ptr is a pointer-to-an-integer and ptrto is a pointer to a pointer-to-an-integer.
2. Print the address of each variable; x lives at #1, ptr at #2 and ptrto at #3.
3. Assign the address of x to ptr. The address of ptr is assigned to ptrto.
4. Print the value of each variable. x has a value of 0, ptr has a value of the address of x, i.e. #1 and ptrto has the value of the address of ptr, i.e. #2.
5. Print x out by indirection.

You already know what *ptr means. Further on ptrto is dereferenced twice, which is what you've got to do if you want to get to the integer from it. The first time you dereference ptrto you get to ptr, the second time you are effectively dereferencing ptr, which takes you to x

This, and in fact all operations with pointers, is very easily understood if you sketch what's happening on a piece of paper: *draw a diagram* if you're not sure what's happening with your pointers.

5.3 Objects (C data structures)

To program effectively using the AJAX and NUCLEUS libraries you need to understand how the EMBOSS derived datatypes (objects) are defined and used. These include simple objects such as dynamic AJAX strings and arrays and more complex biological datatypes, such as sequences and alignments. To extend the functionality of the libraries with new datatypes and functions you'll need a deeper understanding of C pointers and memory management, and their particular implementation in EMBOSS.

EMBOSS borrows the concept of objects from C++. An object can be thought of as a 'black box' with clearly defined inputs and outputs, but possibly more opaque internals with which the developer need not be concerned. An object stores its own (member) data and knows how to perform certain actions via member functions. From the perspective of the developer it doesn't matter what is going on inside so long as the interface, i.e. the inputs and outputs, remains the same. The use of objects allows the programmer to model their code on the problem more closely, breaking it down into small easily managed pieces.

In EMBOSS the objects are the C data structure definitions. Elements in the structures are the member data. There are no member functions as such; however, all the functions that use an object are documented, along with the data elements, in the structured documentation in the C source file. The consistent structuring of code sections and their documentation enforces the naming and classification of all functions in sections for each datatype. This documentation (see Appendix D, *Code documentation standards*) is used online and is accessible via SRS (see Section 1.3, 'Developer documentation'). EMBOSS maintains this link between datatypes and the functions that act upon them so that it's easy to find the objects and functions you need.

5.3.1 Object definition

Objects in EMBOSS are defined (Appendix C, *C coding standards*) and documented (Appendix D, *Code documentation standards*) in a standard way.

A typical definition, for the public AJAX string object, is shown below and includes the standard documentation:

```
/* @data AjPStr ***************************************************************
**
** Ajax string object.
**
** Holds a null terminated character string with additional data.
** The length is known and held internally.
** The reserved memory size is known and held internally.
** The reference count is known and held internally.
** New pointers can refer to the same string without needing
** to duplicate the character data.
**
** If a string has multiple references it cannot be changed. Any
** instance to be changed is first copied to a new string. This
** means that any function which can change the character data must
** pass a pointer to the string so that the string can be moved.
**
** A default null string is provided. New strings are by default
** implemented as pointers to this with increased reference counters.
**
** AjPStr is implemented as a pointer to a C data structure.
**
** @alias AjPPStr
** @alias AjSStr
** @alias AjOStr
** @iterator AjIStr
**
** @attr Res [ajuint] Reserved bytes (usable for expanding in place)
** @attr Len [ajuint] Length of current string, excluding NULL at end
** @attr Ptr [char*] The string, as a NULL-terminated C string.
** @attr Use [ajuint] Use count: 1 for single reference, more if several
**                    pointers share the same string.
**                    Must drop to 0 before deleting. Modifying means making
**                    a new string if not 1.
** @attr Padding [ajint] Padding to alignment boundary
** @@
******************************************************************************/

typedef struct AjSStr
{
    ajuint  Res;
    ajuint  Len;
    char   *Ptr;
    ajuint  Use;
    ajint   Padding;
}   AjOStr;
#define AjPStr AjOStr*
typedef AjPStr* AjPPStr;
```

The declaration defines:

- The object name (`AjSStr`)
- A datatype for the string object proper (`AjOStr`)
- A datatype for the string object pointer (`AjPStr`)
- A datatype for a pointer to the string object pointer (`AjPPStr`).

Note

The object pointer (**`AjPStr`**) is the datatype you'll commonly use and for this reason an **`AjPStr`** is often referred to as a 'string object', rather than the more cumbersome 'string object pointer'. Of course an **`AjPStr`** points to a string object in memory.

The datatype naming conventions are supposed to make the names intuitive:

- `Aj` indicates that the object belongs to the AJAX library
- `P` indicates that the datatype is a pointer
- `Str` gives a clue to the name of the AJAX library file in which the object is defined, i.e. the string-handling library `ajstr.h`.

Many other basic datatypes are available. For example the basic file object `AjPFile` is defined in the AJAX file `ajfile.h` whereas the input (`AjPSeq`) and output (`AjPSeqout`) sequence objects are in `ajseqdata.h` and `ajseqwrite.h` respectively. Different naming conventions apply for public NUCLEUS objects and for private objects including those listed in the application code (see Appendix C, *C coding standards*).

5.3.2 Object functions

You should never access the elements of an object directly, that is what the library functions are for. Objects should always be accessed by calling the appropriate functions. These are fully described by structured comments in a similar way to the objects themselves (see Appendix D, *Code documentation standards*). Functions are organised by the datatype they act upon and, for easier navigation, into sections of related functionality.

The function sections are supposed to help you to quickly find the functions you need. Functions in the same section tend to have similar names and return types, and similar number, order and type of parameters. As functions in the same section are all used in a similar way then programming with the libraries is reasonably intuitive.

Most of the sections are unique to a given library file however there are some common sections. For example, many of the library files have an 'element retrieval' section for functions which return a data element of an object and an 'element set' section for functions that set the value of a data element directly. Most of the complex biological datatypes have 'input' and 'output' sections for reading or writing the data to file in a formatted way. This includes input (`AjPSeq`) and output (`AjPSeqout`) sequence objects, application reports (`AjPReport`) and sequence alignments (`AjPAlign`).

The common sections are described in more detail in Appendix D, *Code documentation standards*.

The main thing to be aware of when using objects with functions is that object pointers (for example an `AjPStr`) are always used: a data structure proper is *never* passed to or returned from a function for reasons of efficiency. Furthermore, for consistency, all functions in EMBOSS should obey the following rules:

- If a function changes the pointer (so that it points to a new object) or changes the data pointed to in any way then the *address of the object pointer* is passed.
- If the function merely reads the data pointed to and does not change the pointer itself then the plain *object pointer* is passed.

Consider, for example, two functions in the AJAX string library `ajstr.c`: `ajStrMatchS` and `ajStrAssignS`. `ajStrMatchS` compares two strings and returns `ajTrue` if they are the same whereas `ajStrAssignS` copies one string value into another.

You can deduce that `ajStrMatchS` merely reads two string values and therefore object pointers will be passed. In contrast, `ajStrAssignS` must change the value of the destination string, possibly allocating a new string (as a failsafe in case a `NULL` pointer is passed to it) or reallocating a new bigger string in cases where a destination string is passed but is too small to hold the new value. In either case the string value will be changed and possibly the pointer too, therefore the address of the object pointer for the destination string is required.

The prototypes show that this is indeed the case:

```
AjBool ajStrMatchS(const AjPStr str, const AjPStr str2);
AjBool ajStrAssignS(AjPStr* Pstr, const AjPStr str);
```

The two strings (`str` and `str2`) passed to `ajStrMatchS` are only read from, therefore the parameters are object pointers (`AjPStr`). The source string (`str`) of `ajStrAssignS` is also read-only, whereas the destination string (`Pstr`) is modified and therefore the address must be passed (`AjPStr* Pstr`).

Tip

When you are using the existing library functions you needn't worry about function internals. The documentation describes whether an object pointer or the address of it is required. So long as you pass to functions what is shown in their prototype you will be fine. Furthermore, the EMBOSS application code is a rich source of examples of how the functions are used in practice.

5.4 Memory management in EMBOSS

5.4.1 Introduction to memory management

Memory management in C can be a difficult area to master, especially if you are used to scripting or object-oriented languages where this aspect might be automatically taken care of.

It requires a sound knowledge of pointers and discipline in coding but is one of the most powerful aspects of the language, allowing very memory-intensive code to be written in an efficient way that might not be feasible in other languages. Curiously many C programming books omit a detailed coverage of pointers and memory management, yet these areas account for most of the time spent debugging C programs. For this reason great effort has been made to make handling memory in EMBOSS as simple as possible.

Memory management when using the libraries is greatly simplified, at least when programming with the existing datatypes. As far as possible the developer is shielded from low-level C calls to allocate and free memory. This is achieved in the following ways:

- General memory management macros
- Object memory management macros
- Object constructor and destructor functions
- Failsafe object construction
- Dynamic objects
- EMBOSS is free of arbitrary limits.

5.4.1.1 General memory management macros

General memory management macros (Section 5.4.2.1, 'General macros') are provided to wrap the C `malloc`, `calloc` and `free` functions. `malloc` allocates memory with uninitialised content, `calloc` allocates memory and initialises it to zero, and `free` frees allocated memory. When programming using the libraries you should use the objects provided (or create new ones) and therefore will seldom need to call these C functions. In some cases, however, it is necessary or desirable to do so and you should use these associated macros.

5.4.1.2 Object memory management macros

Macros (Section 5.4.2.2, 'Object macros') are provided to simplify the memory allocation for single objects and arrays of objects of any type, and for freeing that memory. Bear in mind these macros only allocate memory for a basic object (or an array of them) as defined by a single object definition. Where the object itself includes pointers for nested data structures then memory for these nested objects is *not* allocated. That is what the constructor functions are for (see below).

5.4.1.3 Object constructor and destructor functions

A *constructor* function (memory allocation) (Section 5.4.3.2, 'Object construction') and *destructor* function (freeing memory (Section 5.4.3.3, 'Object destruction')) are provided for every type of object. Their function goes beyond that of the object macros, which merely allocate or free a single block of memory referenced by a single pointer. Constructors will allocate the object and all nested objects and initialise the memory, possibly with values passed in by argument. Similarly, destructors ensure that all pointers nested within the structure are freed correctly and that the main object pointer passed is reset to `NULL` so that it's ready for reuse in the calling code.

5.4.1.4 Failsafe object construction

Many (but currently not all) functions that modify an object passed by argument will allocate memory for it if an unallocated (`NULL`) pointer is passed. This is provided as a safety measure against sloppy programming. You should not rely on it. It's recommended that, where appropriate, all object pointers are explicitly allocated in your code before they are used, and of course freed later once you are done with them.

5.4.1.5 Dynamic objects

Memory for most objects is dynamically reallocated (and freed) as needed by the library functions. This means for example you can append text to a string object without worrying whether there is sufficient space available, or write to an array element without first checking the array is big enough (new elements will be created as necessary). Similarly, memory is freed automatically when it is no longer needed. The most commonly used dynamic objects are the strings and arrays mentioned, but most of the object functions show this behaviour.

5.4.1.6 EMBOSS is free of arbitrary limits

There are no arbitrary hard-coded limits in the code. For example there is no hard-coded maximum to the length of a sequence or the number of sequences in a sequence alignment, and no upper limits to the size of a matrix you can create. The only restrictions come from the hardware you are using.

5.4.2 General memory management

5.4.2.1 General macros

Instead of calling `malloc`, `calloc` or `free` directly you should use the macros provided:

`AJALLOC(nbytes)`	Allocates `nbytes` of uninitialised memory. This is equivalent to using `malloc`.
`AJALLOC0(nbytes)`	Allocates `nbytes` of memory and initialises the memory to zero. Equivalent to using `calloc`.
`AJCALLOC(count, nbytes)`	Allocates an array of `count` elements of `nbytes`. The array elements are uninitialised. Equivalent to using `malloc`.
`AJCALLOC0(count, nbytes)`	Allocates an array of `count` elements of `nbytes` and initialises the memory to zero. Equivalent to using `calloc`.
`AJRESIZE(ptr, nbytes)`	Resizes previously allocated memory (referenced by `ptr`) to a new size of `nbytes`. Initialises new additional reserved memory (if any) to zero. The original memory contents are preserved regardless of whether the block is moved or not. If a `NULL` pointer is passed then a new block of memory is allocated automatically. Equivalent to using `realloc`.
`AJFREE(ptr)`	Frees memory using `free` and sets the pointer to `NULL`. Ignores `NULL` pointers.

Most of these macros allocate (or reallocate) some memory and return a pointer to it. In case of failure a `NULL` pointer is returned and an exception raised. The exception message is printed to the standard error stream and the code exits. The exception message includes the source file name and source line number.

For most applications, you would use AJALLOC0 or AJCALLOC0 as it's safer to initialise the memory to zero by default. If you are certain the memory will be overwritten after it's allocated then AJALLOC or AJCALLOC should be used instead for efficiency.

5.4.2.2 Object macros

The macros below are used to create a single object or an array of objects of any type, and for freeing that memory. They allocate memory for a basic object (or array of them) as defined by a single object definition. Where the object itself includes pointers for nested data structures then memory for these nested objects is *not* allocated (or freed) by these macros. They are normally called from within object constructor and destructor functions.

AJNEW(ptr)	Allocates memory to an object pointer (p) for a single object of the correct type. The memory is not initialised. This is equivalent to using malloc.
AJNEW0(ptr)	Allocates memory to an object pointer (p) for a single object of the correct type. The memory is initialised to zero. This is equivalent to using calloc.
AJCNEW(ptr, count)	Allocates memory to an object pointer (p) for an array of c objects of the correct type. The memory is not initialised. This is equivalent to using malloc.
AJCNEW0(ptr, count)	Allocates memory to an object pointer (p) for an array of c objects of the correct type. The memory is initialised to zero. This is equivalent to using calloc.
AJCRESIZE0(ptr, count)	Resizes a previously allocated array of objects (referenced by p) such that it becomes an array of c objects of the correct type. Initialises new additional reserved memory (if any) to zero and preserves the original memory content.
AJFREE(ptr)	Frees a previously allocated object or array (referenced by ptr). Tests that the memory pointer has a non-NULL value to protect against twice freeing, or freeing unallocated memory.

5.4.2.3 Arrays of fundamental C-type datatypes

Two datatypes are defined in ajdefine.h for handling arrays of C-type integers (int) and floats (float):

```
/* @datatype AjIntArray *******************************************************
**
** Array of integers
**
** @attr typedef [ajint*] Value
** @@
******************************************************************************/

typedef ajint* AjIntArray;

/* @datatype AjFloatArray *****************************************************
**
```

```
** Array of floats
**
** @attr typedef [float*] Value
** @@
*****************************************************************************/

typedef float* AjFloatArray;
```

They may be used with the macros above to allocate memory for such arrays. A typical use is shown below:

```
AjIntArray  integers = NULL;
AjFloatArray floats    = NULL;
ajint        dim       = 10;

AJCNEW0(integers, dim);
AJCNEW0(floats, dim);

/* Do something with arrays */

AJFREE(integers);
AJFREE(floats);
```

5.4.2.4 Memory leaks

Memory leaks occur when, in your source code, you lose a reference to an allocated block of memory. This usually happens by accidentally making a pointer point somewhere else, without first freeing the memory or copying the pointer and freeing it later. They are one of the most common sources of error in C programming, accounting for much of the time spent debugging code. Leaks are easily avoided if you have a strong grasp of pointers, particularly their implementation in EMBOSS, and take a disciplined approach when coding.

It is vital that you keep track of exactly what objects you have in memory and what references (pointers) you have to this memory. Most memory leaks can be avoided if you explicitly allocate memory for objects before they are used and free this memory later once you are done with the object. Be careful to match calls to destructor functions with calls to constructors. If you rely on a function's failsafe memory allocation mechanism then the allocation is hidden from you and it's no longer obvious from the code that memory has been allocated and needs freeing.

Consider the following code.

```
int main(int argc, char **argv)
{
    AjPStr mystring=NULL;

    embInit("noleaks", argc, argv);

    mystring = ajAcdGetString("astring")
    ajStrDel(&mystring);
    embExit();

    return 0;
}
```

An AJAX string object (`AjPStr`) is declared and `embInit` called to invoke ACD file processing. The ACD file is read and the user prompted for input values. The call to `embInit` also allocates memory for all ACD data items and initialises the objects (see Section 6.3, 'Handling ACD files').

When retrieving, for example, a string (`AjPStr`) data item by using a call to `ajAcdGetString` then the function will return a pointer to the string created by `embInit`. This means that you *do not* have to allocate memory for the string first, which is why the above code does not call explicitly a string constructor function.

Nonetheless, a string object was created by `embInit` and must be freed once you are done with it. That is what the call `ajStrDel(&mystring);` is for. Had you omitted this then you would be relying on the operating system to free the process memory when the application exits. This is not strictly a memory leak but would be bad practice.

In the following code, the string constructor function `ajStrNewC` is called unnecessarily:

```
int main(int argc, char **argv)
{
    AjPStr  mystring=NULL;

    embInit("leaks", argc, argv);

    mystring = ajStrNewC("Hello");
    mystring = ajAcdGetString("astring")          /* Memory leak */
    ajStrDel(&mystring);

    embExit();

    return 0;
}
```

A string object pointer (`mystring`) is defined as before and is made to point to a new object allocated by `ajStrNewC()`. Then, by calling `ajAcdGetString`, the same pointer is made to point to the string allocated by `embInit` instead. You have lost the handle on the memory allocated by `ajStrNewC`. In other words you've created a memory leak.

Note

There are other ways to create memory leaks than described here. Details of how to avoid leaks are described for individual datatypes in the programming guides (see Section 6.2, 'Programming guides'). All code submitted to EMBOSS should be appropriately tested and debugged so that it does not leak memory (see Section 3.3, 'Debugging').

5.4.3 Object memory management

5.4.3.1 Introduction

In C++ an object is a definition or a template for *instances* of that object. The *instance* is the actual thing that can be manipulated. If you want to do anything you must create an instance in memory, i.e. *instantiate* the object. When programming with EMBOSS objects it's important to make the distinction between the object pointer and the object proper (or instance) residing in memory. An object pointer is merely a variable which holds the memory address of a certain type of object. The object proper is a particular instance of an object residing in memory.

In principle, it would be possible to instantiate (allocate memory for) an object in this way:

```
#include "emboss.h"
int main(void)
{
    AjOStr my_structure;

    /* Do something with my_structure */

    AJFREE(my_structure);
}
```

The declaration of `my_structure` would create a single string object in memory, which is later freed by calling `AJFREE`. However, this approach is *not* taken in EMBOSS because, as already explained, object pointers are always used for reasons of efficiency and convenience. The above definition does not give the programmer the freedom to manage the memory of the object. Even if you only need one structure you should never use (for example) `AjOStr` because it would be inconsistent with the rest of EMBOSS. The above code is almost certain to fail anyway owing to the way the library handles string objects. The `AJFREE` would also not free any required internal memory allocation in `my_structure`.

Note

The standard way to instantiate an object is to dynamically allocate memory to the object pointer. It's for this reason and for brevity that, for example, an **`AjPStr`** may be referred to as an 'object' even though 'object pointer' is more accurate. The terms are not important as long as you understand whether you are dealing with a pointer or a structure in memory.

All objects should be allocated dynamically and freed once you're done with them. This is easy because a *constructor* function (for memory allocation) and *destructor* function (for freeing memory) are provided for every type of object.

Consider for example the following code:

```
#include "emboss.h"
int main(int argc, char **argv)
{
    AjPStr my_string=NULL;

    embInit("helloworld", argc, argv);

    my_string = ajStrNew();

    ajStrAssignC(&my_string, "Hello, World!\n");
    ajFmtPrint("%S", my_string);

    ajStrDel(&my_string);
```

```
    embExit();
    return 0;
}
```

AjPStr my_string=NULL; declares the object pointer and initialises it to NULL. Pointers should always be set to NULL when they are declared because EMBOSS functions presume that non-NULL pointers have had memory allocated to them. If you do not set the pointer to NULL then it may receive some junk value when the program runs and any function that uses it might mistakenly assume memory had been allocated for it. That might lead to a segmentation fault or bus error!

ajStrNew() is the constructor function. This conceptually allocates a block of memory for the object and returns the memory address of the allocated block. The memory address is held in the variable my_string. Disregard the calls to ajStrAssignC and ajFmtPrint for the time being.

ajStrDel() is the destructor function. This must not only free the memory but also set the pointer back to NULL so that it's ready for reuse. You see the address of my_string is passed. You may be wondering, as my_string is a pointer anyway, why do you need to pass the address of it? The answer is simple if you remember that in C the function arguments are passed 'by value'. A temporary copy of each argument is created and passed to the function rather than the originals. Although a copy of the pointer would be enough to free the memory that is pointed to, you need a handle on (the address of) the original if you want to set the original pointer to NULL. Hence the requirement for passing the address of my_string (&my_string).

5.4.3.2 Object construction

Constructor functions (constructors) return a pointer to a new object in memory. There are four basic types of constructor to consider:

- ACD data constructor functions
- Default constructor functions
- Alternative constructor functions
- Functions whose primary purpose is not object construction but which will construct an object if necessary as a failsafe measure, e.g. if a NULL pointer is passed for an output parameter of the function.

When managing memory for the objects a knowledge of the behaviour of any called functions is required. There are three cases to discern:

- A function requires a pre-existing object
- A function can use but does not require a pre-existing object and will allocate one if necessary
- A function always allocates an object and either returns a pointer to it or allocates an object pointer, the address of which has been passed as an argument.

In most but not all cases it is obvious from the function name whether a function is a constructor or merely uses an object.

5.4.3.2.1 ACD data construction

The ACD data constructor functions are used to return objects that are defined in the application ACD file. They are all defined in `ajacd.h/c` and have the general name:

```
ajAcdGetDatatype
```

where `Datatype` is one of the supported ACD datatypes (Section A.2, 'Datatypes').

Strictly speaking they are not constructor functions but instead return a pointer to an appropriate AJAX object that has been allocated by a call to the `embInit` function, a call which all EMBOSS applications must use (see Section 6.3, 'Handling ACD files'). For example `ajAcdGetString` returns a pointer to an AJAX string object (`AjPStr`) produced by parsing an ACD string (`string`) data definition:

```
AjPStr ajAcdGetString (const char *token);
```

The `token` parameter is the name of the ACD data definition to read. Attributes in the data definition and/or user input gathered during ACD file processing are used to initialise the object. Memory for any new objects must be freed later on in the `main()` function. The use of these functions is explained in detail elsewhere (see Section 6.3, 'Handling ACD files').

5.4.3.2.2 Default object construction

The default object constructor functions are the usual way to create new objects in your source code that are not defined in the ACD file. They normally have the suffix `New` in their name and have no parameters. For example:

```
AjPStr ajStrNew (void); /* Create a string object.   */
```

The use of such functions for individual datatypes is described in the library programming guides (see Section 6.2, 'Programming guides').

5.4.3.2.3 Alternative object construction

Alternative constructor functions provide different ways to create new objects and often have parameters used for initialising elements in the object. They have `New` in their name to make their behaviour clear. For example:

```
AjPStr ajStrNewC (const char *txt);                          /* Construct from
C-type string */
AjPStr  ajStrNewResC (const char *txt, ajuint size);         /* Construct from
C-type string with reserved size */
```

The use of such functions for individual datatypes is described in the library programming guides (see Section 6.2, 'Programming guides').

5.4.3.2.4 Failsafe construction

Most functions that write to an object passed by argument will allocate memory for the object if necessary. This safety measure prevents failures and errors in cases where an unallocated (`NULL`) pointer is passed. In practice it's recommended that, where appropriate, all object pointers are explicitly allocated in your code before they are used. Consider the string assignment function `ajStrAssignS` which copies one string value (`str`) to another (`Pstr`):

```
AjBool ajStrAssignS(AjPStr* Pstr, const AjPStr str);
```

It's not at all obvious from the name that this function will allocate a string object for `Pstr` if `NULL` is passed. Therefore, if you rely on the failsafe construction behaviour it will obfuscate your code. You should therefore code this behaviour into any new functions you write, but not rely on this behaviour in the functions that you call.

5.4.3.3 Object destruction

Destructor functions (destructors) free the memory pointed to by an object pointer and reset it to `NULL` so that it is ready for reuse. For most objects there is a default destructor function which is the typical method for deleting objects in your source code. These have a single parameter which is the address of the object pointer being freed. In a few cases there are alternative destructors with non-standard behaviour, for example with parameters to provide a handle on some elements of the object which are not freed. Usually all destructor functions for an object are listed under a single section in the C source (and documentation) for the library file and have `Del` in their name, most often as a suffix. For example:

```
AjPStr ajStrDel (AjPStr *Pstr);   /* Delete a string object.   */
```

5.5 Programming with objects

To cement your knowledge of pointers, objects and memory management we'll use three cases to illustrate how they are used when programming EMBOSS code. First a look at the `AjPPdbtosp` object which holds sequence codes. It is not widely used but does illustrate a lot of the key points nicely. Then the AJAX string object `AjPStr` and the special memory handling required to ensure failsafe memory allocation of the object functions. Finally the AJAX array objects to see how other dynamic objects are implemented.

5.5.1 An example object: `AjPPdbtosp`

5.5.1.1 Object definition

Consider the following object which holds SwissProt codes and accession numbers for a Protein Databank (PDB) code:

```
/* @data AjPPdbtosp *******************************************************
**
** Ajax Pdbtosp object.
**
** Holds swissprot codes and accession numbers for a PDB code.
**
** AjPPdbtosp is implemented as a pointer to a C data structure.
**
** @alias AjSPdbtosp
** @alias AjOPdbtosp
**
** @attr  Pdb     [AjPStr]    PDB code
** @attr  Acc     [AjPStr*]   Accession numbers
** @attr  Spr     [AjPStr*]   Swissprot codes
** @attr  n       [ajint]     No. entries for this pdb code
** @attr  Padding [char[4]]   Padding to alignment boundary
**
**
** @@
**************************************************************************/

typedef struct AjSPdbtosp
{
    AjPStr   Pdb;
    AjPStr  *Acc;
    AjPStr  *Spr;
    ajint    n;
    char Padding[4];
} AjOPdbtosp;
#define AjPPdbtosp AjOPdbtosp*
```

There is nothing new here other than `Acc` and `Spr` which are both pointers to `AjPStr` objects. These would have better been declared by using `AjPPStr` but that is not done here. As an `AjPStr` is itself a pointer (to a string object proper) you can see that we're dealing with pointers to pointers. In this case `Acc` and `Spr` are used to create two arrays of strings as you can see in the constructor function (below).

5.5.1.2 Object construction

Here is the constructor function:

```
/* @func ajPdbtospNew *****************************************************
 **
 ** Pdbtosp object constructor. Fore-knowledge of the number of entries is
 ** required. This is normally called by the ajPdbtospReadCNew /
 ** ajPdbtospReadNew functions.
 **
 ** @param [r] n [ajint] Number of entries
 **
 ** @return [AjPPdbtosp] Pointer to a Pdbtosp object
 ** @@
 *************************************************************************/
AjPPdbtosp ajPdbtospNew(ajint n)          /* 1. */
{

   AjPPdbtosp ret = NULL;                 /* 2. */
   ajint i=0;

   AJNEW0(ret);                           /* 3. */
```

```
    ret->Pdb = ajStrNew();                  /* 5. */

    if(n)
    }
        AJCNEW0(ret->Acc,n);                /* 4. */
        AJCNEW0(ret->Spr,n);                /* 4. */
        for(i=0; i< n; i++)
        {
            ret->Acc[i]=ajStrNew();         /* 6. */
            ret->Spr[i]=ajStrNew();         /* 6. */
        }
    {

    ret->n = n;                             /* 7. */

    return ret;
}
```

The commented lines are as follows:

1. The first line declares that the function returns an object pointer of type `AjPPdbtosp`. The parameter `ajint n` is the size the `Acc` and `Spr` arrays should be, i.e. the number of pairs of `Acc` / `Spr` values that the object will hold.
2. The next line declares a variable called `ret`. This is the object pointer that is going to have memory allocated to it and will be returned to the calling function.
3. `AJNEW0(ret);` is the line that allocates an object proper to the pointer `ret`. This will now point to an instance of an `AjPPdbtosp` memory. By the time `AJNEW0(ret);` returns, memory space for an `AjPPdbtosp` object is reserved. This means enough space for an `AjPStr`, an `ajint` and two pointers (`AjPStr *`). Note that the two arrays or any string objects proper have not yet been allocated!

   `AJNEW0` sets all the structure elements to 0, this means the element `n` is set to 0 and the three pointers are set to `NULL`. `AJNEW0` is a macro: it will allocate a single object of the correct type to any pointer that is passed to it – it can be used with *any* object.
4. Compare `AJNEW0` to the two `AJCNEW0` lines. `AJCNEW0` will allocate an array of objects of any type and initialises the new variables to 0 or `NULL` as required. In this case, arrays of `n` objects each will be created. It is important to bear in mind here that `ret->Acc` and `ret-Spr` are passed to the macro. These are of the type `AjPStr *`, which means that the 'object' which they point to is in fact another pointer variable. Therefore these macro calls will allocate arrays of `n` pointers, *not* arrays of instances of `AjPStr` objects as one might (incorrectly) first imagine. They create an array of `n AjPStr` object pointers allocated to each of `Acc` and `Spr`. In other words, `ret->Acc` and `ret->Spr` will point to blocks of memory each holding `n` pointer variables which are as yet `NULL` (unallocated).
5. The arrays are created but still no strings yet. `ret->Pdb = ajStrNew();` allocates memory for a string object to the pointer `Pdb` in the new object. Notice that `->` is used to dereference the pointer `ret`; and get to the `Pdb` element. This is the standard way in C of accessing elements in a data structure when you have a pointer to that data structure.

6. The lines `ret->Acc[i]=ajStrNew();` and `ret->Spr[i]=ajStrNew();` allocate memory for the `n` string objects for each array. It also illustrates how pointer and array notation can be used together. In this case, the `i`th element of the arrays that `ret->Acc` and `ret->Spr` point to are accessed. The elements in these arrays are `AjPStr` (object pointers) and a string object is allocated to each of them.
7. The rest is obvious. The integer in the object is set to the size of the arrays and the pointer to the new object, complete with an allocated string and two arrays of strings, is returned to the calling function by `return ret;`. Note that the constructor should be coded to deal with negative arguments in a safe way, but that is not done here.

5.5.1.3 Object destruction

It is the job of the destructor function to free the object itself and any memory that its member elements might point to. The destructor *safely* clears up *all* of the memory that was allocated by the constructor. This is achieved by calling other destructor functions as appropriate and by using `AJFREE`. The code is shown below:

```
/* @func ajPdbtospDel ************************************************************
**
** Destructor for Pdbtosp object.
**
** @param [w] thys [AjPPdbtosp*] Pdbtosp object pointer
**
** @return [void]
** @@
******************************************************************************/


void ajPdbtospDel(AjPPdbtosp *thys)                 /* 1. */
{
    AjPPdbtosp pthis = NULL;                        /* 2. */
    ajint i;

    if(!thys)
       return;

    pthis = *thys;

    if(!pthis)
       return;

    ajStrDel(&pthis->Pdb);                  /* 3. */

    if(pthis->n)
    {
       for(i=0; i< pthis->n; i++)
       {
           ajStrDel(&pthis->Acc[i]); /* 4. */
           ajStrDel(&pthis->Spr[i]); /* 4. */
       }

       AJFREE(pthis->Acc);                          /* 5. */
       AJFREE(pthis->Spr);                          /* 5. */
    }
```

```
        AJFREE(pthis);                              /* 6. */
        (*thys)=NULL;                               /* 6. */

        return;
    }
```

Again, this is broken down comment by comment:

1. The function, like all destructors, takes the address of the object pointer being deleted (`AjPPdbtosp *thys`).
2. For convenience a second pointer is declared and is used in the following lines to dereference `thys`. This is purely for reasons of clarity. The function returns if either an empty address was passed or if the pointer stored there is `NULL`.
3. The string object in `AjPPdbtosp` is deleted first by calling the default destructor function with the address of the string.
4. The string objects proper, referenced through the arrays, are deleted by calling `ajStrDel` in a loop for every array element in both arrays.
5. `AJFREE` is then called to delete the arrays themselves, referenced by `pthis->Acc` and `pthis->Spr`.
6. Once the loop terminates `AJFREE` is again called, this time freeing memory for the `AjPPdbtosp` object itself. The pointer is set to `NULL` so that it's ready for reuse by the calling function.

It should be clear that although `AJFREE` will free the memory pointed to by its argument, as used here it frees the arrays but not the string objects proper that are pointed to; that is the job of the `ajStrDel` calls in the preceding code.

5.5.1.4 Usage example

Here is a code snippet illustrating how the object constructor and destructor could be used. You'll notice they're used in exactly the same way as any other object:

```
int main()
{
    AjPPdbtosp ptr=NULL;

    ptr = ajPdbtospNew(10);
    ajPdbtospDel(&ptr);

    /* ptr will have been reset to NULL now, and is ready for reuse */
    ptr = ajPdbtospNew(10);
    ajPdbtospDel(&ptr);

}
```

5.5.2 AJAX dynamic string object

5.5.2.1 Introduction

The string object (AjPStr) is one of the simplest of all the AJAX objects. AJAX strings have more functions than any other datatype and are used by many other objects. Two features distinguish its use from standard C-type (char *) strings. First, AJAX strings are dynamic objects, meaning that memory is dynamically reallocated as needed so that you never run out of space when using the object functions; a string will grow automatically as required. Second, AJAX strings are reference counted. This means that the object itself keeps track of how many references (pointers) to the string there are in the code that have been requested by calling library functions. It is not until all references to a string are deleted that the string itself is freed. This ensures that broken references to a string do not occur and that you always have a handle on objects in memory.

5.5.2.2 String definition

The string definition is shown below:

```
typedef struct AjSStr
[
    ajuint Res;
    ajuint Len;
    char *Ptr;
    ajuint Use;
    ajint Padding;
] AjOStr;
#define AjPStr AjOStr*
typedef AjPStr* AjPPStr;
```

A structure is defined called AjSStr of four elements (Len, Res, Use and Ptr) and with three new datatype names, AjOStr for the object itself, AjPStr for the object pointer and AjPPStr for a pointer to an AjPStr.

The Ptr pointer is just a standard C one which holds a character string and Len is its length. In contrast to C-type strings, the character string may or may not be NULL terminated; the library functions for printing AjPStr objects use the length field (Len) for how many characters to print and won't stop at the first NULL if there is one.

The Res element indicates how much reserved dynamic memory is associated with the object. This is always at least equal to Len but is often more. Res is and should be outside your direct control. If you use a library call to add anything to the string then, if it'll fit within the memory given by Res then the operation is performed immediately. If the memory required is larger than Res then more memory is allocated and the Res item is updated. More memory than required is usually allocated.

Use is the string usage counter. Sometimes you'll want two or more references to a single string rather than making a genuine copy. EMBOSS functions that do this increment the string's usage counter. The usage counter is decremented when a call to destroy either the string itself, or a reference to it, is made. When the usage counter reaches zero the object will be deleted. All of this is function internals, so you don't need to worry about it as long as you don't change the object elements directly. If you intend altering the contents of an object then safety is guaranteed if you use the available library functions.

Finally, the `Padding` element indicates the number of characters used to pad the string to its alignment boundary and is used only to keep pedantic compilers happy.

5.5.2.3 String construction

Code for the default constructor function is shown below:

```
/* @func ajStrNew *************************************************************
**
** Default string constructor which allocates memory for a string.
**
** The null string usage pointer is incremented.
**
** @return [AjPStr] Pointer to an empty string
** @@
******************************************************************************/
AjPStr ajStrNew(void)
[
    return ajStrNewRef(strPNULL);
]
```

You can see that the `ajStrNew` is a wrapper function to `ajStrNewRef`. It passes to `ajStrNewRef` an interal string object called `strPNULL` which is defined in `ajstr.c`:

```
#define NULL_USE 1
char charNULL[1] = "";

AjOStr strONULL = [ 1,0,charNULL,NULL_USE,0];
AjPStr strPNULL = &strONULL;
```

`strPNULL` is a global variable for an empty object called the 'AJAX NULL string'. This has a single character of reserved memory, length of zero, a C-type string which is set to `NULL`, a reference count of 1 and zero padding.

If you look at `ajStrNewRef`:

```
/* @func ajStrNewRef **********************************************************
**
** String referencing function which returns a pointer to the string passed in
** and increases its reference count.
**
** Not a true string copy, but if modified the new string will behave
** as a copy.
**
** @param [u] refstr [AjPStr] AJAX string object
** @return [AjPStr] Pointer to the string passed as an argument,
**          with its use count increased by 1.
** @@
******************************************************************************/
AjPStr ajStrNewRef(AjPStr refstr)
{
    if(!refstr)
        ajFatal("ajStrNewRef source string NULL");
```

```
    refstr->Use++;

    return refstr;
}
```

All this function does is increase the reference count of the object that was passed and return the same pointer. It raises a fatal error if NULL was passed.

In other words, a call to ajStrNew doesn't immediately instantiate an AjSStr object, it just returns the address of the 'global AJAX NULL string'. It's only when the char * string (Ptr) is given a non-NULL value (by whatever means) that memory for the string object proper will be allocated. AJAX is programmed in this way for maximum speed and efficiency of string handling. You can see this for yourself if you print the reference count of a string which you have just allocated using ajStrNew but not yet used. You might be surprised at the value of the usage count which is higher than you might expect. The reference count may well be in the hundreds owing to the call to embInit in the application code, which itself makes, indirectly, many calls to ajStrNew. If true objects had been allocated for all these strings the code would be less efficient.

Things are different if you call the alternative constructor function ajStrNewRes, which allocates memory for a string with an initial reserved size:

```
/* @func ajStrNewRes **********************************************************
**
** String constructor which allocates memory for a string of an initial
** reserved size (including a possible null).
**
** @param [r] size [ajuint] Reserved size (including a possible null).
** @return [AjPStr] Pointer to an empty string of specified size.
** @@
******************************************************************************/
AjPStr ajStrNewRes(ajuint size)
{

    AjPStr thys;

    thys = ajStrNewResLenC("", size, 0);

    return thys;
}
```

The function declares an AJAX string object (thys) then calls ajStrNewResLenC to allocate the string, passing to that function the size (size) of the required string:

```
/* @func ajStrNewResLenC ******************************************************
**
** String constructor which allocates memory for a string of a given length
** and of an initial reserved size (including a possible null) and initialises
** it with the text string provided.
**
** @param [r] txt [const char*] Null-terminated character string to initialise
**        the new string.
** @param [r] size [ajuint] Reserved size (including a possible null).
** @param [r] len [ajuint] Length of txt to save calculation time.
** @return [AjPStr] Pointer to a string of the specified size
```

```
**          containing the supplied text.
** @@
******************************************************************************/

AjPStr ajStrNewResLenC(const char* txt, ajuint size, ajuint len)
{
    AjPStr thys;
    ajuint minlen;

    minlen = size;

    if(size <= len)
        minlen = len+1;

    thys = strNew(minlen);
    thys->Len = len;
    if(txt)
        memmove(thys->Ptr, txt, len+1);

    thys->Ptr[len] = '\0';

    return thys;
}
```

This function is an alternative constructor function. It sets a minimum string length (`minlen`) to the requested reserved size (`size`), or to the current length (`len`) plus 1 (for a terminating `NULL` character) if the requested size is not greater than the current length. It then calls `strNew` which is a static function in `ajstr.c` to allocate a string object (this function is shown below). The string length (`thys->Len`) is set, and the specified text (`txt`, which is an empty string when called by `ajStrNewRes` as in this example) is copied (using `memmove`) to the C-type string pointer (`thys-Ptr`) in the AJAX string object. A terminating `NULL` character is added.

Let's look at the static constructor function which actually allocates the string object:

```
/* @funcstatic strNew *******************************************************
**
** Internal constructor for modifiable AJAX strings. Used by all the string
** parameterized constructors to allocate the space for the text string.
** The exception is ajStrNew which returns a clone of the null string.
**
** @param [rE] size [ajuint] size of the reserved space, including the
**        terminating NULL character. Zero uses a default string size STRSIZE.
** @return [AjPStr] A pointer to an empty string
** @@
******************************************************************************/

static AjPStr strNew(ajuint size)
{
    AjPStr ret;

    if(!size)
        size = STRSIZE;

    AJNEW0(ret);
    ret->Res = size;
    ret->Ptr = AJALLOC0(size);
    ret->Len =0;
```

```
    ret->Use =1;
    ret->Ptr[0] = '\0';

    strAlloc += size;
    strCount++;
    strTotal++;

    return ret;
}
```

The function first checks that a reserved size has been specified, and sets this to the default length `STRSIZE` if not. `STRSIZE` is defined in `ajstr.c`:

```
#define STRSIZE 32
```

The macro `AJNEW0` is called. You'll recall that this is the equivalent of a `calloc` and allocates memory to an object pointer (`ret`) for a single object of a given type, in this case, a string. The memory is initialised to zero.

The rest of that block of code assigns correct values to the other elements in the string object. You can see that the reserved size is set to `size` and the first character of the string is set to a `NULL` character, meaning you have a new, empty string with the specified reserved size, a pointer to which is returned to the calling function.

The code also sets some global variables (`strAlloc`, `strCount` and `strTotal`) used internally for statistics and for debugging strings. You needn't worry about those.

5.5.2.4 String destruction

The internals of string destruction are simpler than construction. The default constructor `ajStrDel` is shown below:

```
/* @func ajStrDel ************************************************************
**
** Default string destructor which frees memory for a string.
**
** Decrements the use count. When it reaches zero, the string is removed from
** memory.  If the given string is NULL, or a NULL pointer, simply returns.
**
** @param  [d] Pstr [AjPStr*] Pointer to the string to be deleted.
**         The pointer is always deleted.
** @return [void]
** @cre    The default null string must not be deleted. Calling this
**         routine for copied pointers could cause this. An error message
**         is issued and the null string use count is restored.
** @@
******************************************************************************/

void ajStrDel(AjPStr* Pstr)
{
   AjPStr thys;

   if(!Pstr)
       return;

   if(!*Pstr)
```

```
        return;

    thys = *Pstr;

    --thys->Use;
    if(!thys->Use)
    {   /* any other references? */
        if(thys == strPNULL)
        {
            ajErr("Error - trying to delete the null string constant\n");
            thys->Use = NULL_USE;      /* restore the original value */
        }
        else
        {
            AJFREE(thys->Ptr);         /* free the string */

            strFree += thys->Res;
            strFreeCount++;
            strCount--;
            thys->Res =0;              /* in case of copied pointers */
            thys->Len =0;

            AJFREE(*Pstr);                /* free the object */
        }
    }

    *Pstr = NULL;

    return;
}
```

It is clear from `AjPStr* Pstr` that the function takes the address of a string object pointer. The function first checks that NULL is not passed (`if(!Pstr)`) and that the pointer itself is not NULL (`if(!*Pstr)`). In other words, it ensures that the `AjPStr` passed in by reference does actually point to something. The function must assume that it points to a string and this will be the case if there are no bugs in the code. This is why pointers when declared should be set to NULL. If they are not and receive some junk value on startup then this function (and many others like it) will mistakenly assume that it references valid memory and will, at best, head for a segmentation fault or bus error when it tries to address that memory.

The line `--thys->Use;` reduces the reference count of the string by 1. If this becomes zero then AJFREE is used to free the object. It is called twice, once to free the C-type string (`AJFREE(thys->Ptr);`) and again to free the object proper (`AJFREE(*Pstr);`). Some global variables (`strFree`, `strFreeCount` and `strCount`) used internally for debugging and statistics are also set.

Finally, the string object pointer that was passed is set to NULL (`*Pstr = NULL;`) so that it's ready for reuse by the program.

5.5.2.5 String functions

Now we'll look at two string functions to see how pointers and memory are handled. `ajStrMatchS` is a simple function for matching two AJAX strings:

```
/* @func ajStrMatchS *************************************************************
**
** Simple test for matching two strings.
```

```
**
** @param [r] str [const AjPStr] String
** @param [r] str2 [const AjPStr] Second String
** @return [AjBool] ajTrue if two complete strings are the same
** @@
******************************************************************************/

AjBool ajStrMatchS(const AjPStr str, const AjPStr str2)
{
    if(!str || !str2)
        return ajFalse;

    if(!strcmp(str->Ptr, str2->Ptr))
        return ajTrue;

    return ajFalse;
}
```

It is passed two AJAX string objects and uses the C function `strcmp` to compare the C-type strings in the object, returning `ajTrue` if they are the same or `ajFalse` otherwise. The function merely reads the value of the strings passed so will never need to allocate memory.

`ajStrAssignS` is different. This function assigns the value of one string to another. The string is copied rather than just setting a reference (pointer) to the original:

```
/* @func ajStrAssignS ********************************************************
**
** Copy a string to a string.
**
** This is a genuine copy (reference count isn't used). Useful where both
** strings will be separately overwritten later so that they can both remain
** modifiable.
**
** @param [w] Pstr [AjPStr*] Target string.
** @param [rN] str [const AjPStr] Source string.
** @return [AjBool] ajTrue if string was reallocated
** @@
******************************************************************************/

AjBool ajStrAssignS(AjPStr* Pstr, const AjPStr str)
{
    AjBool ret = ajFalse;
    AjPStr thys;

    if(!str)
    {
       return ajStrAssignC(Pstr, "");
    }

    ret = ajStrSetRes(Pstr, str->Len+1); /* minimum reserved size, OR more */
    thys = *Pstr;

    thys->Len = str->Len;
    memmove(thys->Ptr, str->Ptr, str->Len+1);

    return ret;
}
```

The function takes the source string that is being copied (`str`) and a destination string (`Pstr`). The destination string (`Pstr`) will be modified, therefore the address must be passed (`AjPStr* Pstr`).

You can see that if a `NULL` pointer is passed for the source string then an empty string is written to the destination string by calling `ajStrAssignC`. This is the failsafe mechanism that was mentioned before. The function should also check, as a safety measure, whether `NULL` is passed for the address of the destination string, but it currently doesn't do this.

`ajStrSetRes` is called (see below) to ensure that the destination string is a new string, not referenced by other string objects, and is big enough for its intended purpose. The length of the destination string (`thys->Len`) is set and the C-type string in the source string (`str->Ptr`) is copied (using `memmove`) to the destination string (`thys->Ptr`). `ajTrue` is returned if the string was reallocated or `ajFalse` otherwise.

`ajStrSetRes` ensures a string is big enough for its intended purpose:

```
/* @func ajStrSetRes *****************************************************
**
** Ensures a string is modifiable and big enough for its intended purpose.
**
** The target string is guaranteed to have a reference count of 1 and a
** minimum reserved size.
**
** @param [u] Pstr [AjPStr*] String
** @param [r] size [ajuint] Minimum reserved size.
** @return [AjBool] ajTrue if the string was reallocated
** @@
*************************************************************************/

AjBool ajStrSetRes(AjPStr* Pstr, ajuint size)
{
    AjPStr thys;
    ajuint savesize;

    savesize = size; /* often part of *Pstr, about to vanish */

    if(!*Pstr)
    {
        *Pstr = ajStrNewRes(savesize);
        return ajTrue;
    }

    thys = *Pstr;

    if((thys->Use > 1) || (thys->Res < savesize))
    {
        strCloneL(Pstr, savesize);
        return ajTrue;
    }

    return ajFalse;
}
```

It takes the address of a target string and a minimum size (`size`). If the target string is `NULL` then a string with a reserved size is allocated using `ajStrNewRes`. That function has already been explained. Otherwise, if the usage count is greater than 1 or if the current reserved size is less than that requested, the static function strCloneL is called (see below) to make a copy of the string but with a usage count of 1 and a minimum reserved size. ajTrue is returned if the string was reallocated or ajFalse otherwise.

strCloneL takes the address of a target string (Pstr) and a reserved size (size):

```
/* @funcstatic strCloneL ***************************************************
**
** Makes a new clone of a string with a usage count of one and a minimum
** reserved size.
**
** @param [w] Pstr [AjPStr*] String
** @param [r] size [ajuint] Minimum reserved size.
** @return [void]
** @@
******************************************************************************/

static void strCloneL(AjPStr* Pstr, ajuint size)
{
    AjPStr thys;
    AjPStr ret;

    thys = *Pstr;
    ret = ajStrNewResLenC(thys->Ptr, size, thys->Len);
    ajStrDel(Pstr);
    *Pstr = ret;

    return;
}
```

It calls ajStrNewResLenC to allocate a string with a reserved size, as has already been explained. The original target string that was passed is deleted by calling the destructor ajStrDel.

It should be said that strings are a special case and that the internals of memory management for most other objects are considerably simpler. It is only for strings, that are so widely used by the other libraries, that special handling is needed for reasons of safety and efficiency.

5.5.3 AJAX dynamic array objects

The memory management macros are nicely illustrated by the array handling functions in ajarr.c. Here we'll consider the constructor and destructor functions for the AjPInt and AjPInt2d objects. These are dynamic one-dimensional (1D) (AjPInt) and two-dimensional (2D) (AjPInt2d) arrays of integers.

5.5.3.1 Object definition

Here are the object definitions:

```
typedef struct AjSInt
{
  ajuint Res;
  ajuint Len;
  ajint *Ptr;
} AjOInt;
```

```
#define AjPInt AjOInt*

typedef struct AjSInt2d
{
  ajuint Res;
  ajuint Len;
  AjPInt *Ptr;
} AjOInt2d;
#define AjPInt2d AjOInt2d*
```

Both objects include variables for the current length of the array (`Len`) and the reserved size (`Res`). `AjPInt` includes a pointer (`Ptr`) to `ajint` which, when allocated, will point to an array of `ajint` values. In contrast, `AjPInt2d` includes a pointer (`Ptr`) to `AjPInt` which will eventually point to an array of `AjPInt` object pointers.

5.5.3.2 `AjPInt` construction and destruction

`ajIntNewRes` is a constructor for `AjPInt` objects, allocating an array with an initial reserved size. The code is shown below:

```
/* @func ajIntNewRes ***********************************************************
**
** Constructor given an initial reserved size.
**
** @param [r] size [ajuint] Reserved size
** @return [AjPInt] Pointer to an empty integer array struct of specified size.
** @category new [AjPInt] Constructor with reserved size
** @@
******************************************************************************/

AjPInt ajIntNewRes(ajuint size)
{
    AjPInt thys;

    size = ajRound(size,RESERVED_SIZE);

    AJNEW0(thys);
    thys->Ptr = AJALLOC0(size*sizeof(ajint));
    thys->Len =0;
    thys->Res = size;

    arrTotal++;
    arrAlloc += size*sizeof(ajint);

    ajDebug("ajIntNewRes size %d*%d %d\n",
            size, sizeof(ajint), size*sizeof(ajint));

    return thys;
}
```

`AJNEW0` is used to allocate memory for a single `AjPInt` object. `AJALLOC0` is called to create an array of AJAX integers (`ajint`) of size `size`. `arrTotal` and `arrAlloc` are also set which are global variables used for debugging arrays.

`ajIntDel` is the `AjPInt` destructor function:

```
/* @func ajIntDel ***********************************************************
**
** Default destructor for AJAX integer arrays.
**
** If the given array is a NULL pointer, simply returns.
**
** @param  [d] thys [AjPInt*] Pointer to the ajint array to be deleted.
**         The pointer is always deleted.
** @return [void]
** @category delete [AjPInt] Default destructor
** @@
****************************************************************************/

void ajIntDel(AjPInt *thys)
{
    if(!thys || !*thys)
        return;
    ajDebug("ajIntDel Len %u Res %u\n",
            (*thys)->Len, (*thys)->Res);

    AJFREE((*thys)->Ptr);
    AJFREE(*thys);

    *thys = NULL;

    arrFreeCount++;

    return;
}
```

AJFREE is called twice. The first call (`AJFREE((*thys)->Ptr);`) frees the array of integers. The second call (`AJFREE(*thys);`) frees the object itself. You can see that the pointer (`thys`) that is passed to the function is set to NULL using the code `*thys = NULL;`.

5.5.3.3 AjPInt2d construction and destruction

`ajInt2dNewRes` is a constructor for `AjPInt2d` objects, allocating a 2D array with an initial reserved size for the first dimension. The code is shown below:

```
/* @func ajInt2dNewRes ******************************************************
**
** Constructor given an initial reserved size.
**
** @param [r] size [ajuint] Reserved size 1st dim
** @return [AjPInt2d] Pointer to an empty integer 2d array struct of
**                    specified size.
** @category new [AjPInt2d] Constructor with reserved size
** @@
****************************************************************************/

AjPInt2d ajInt2dNewRes(ajuint size)
{
    AjPInt2d thys;
    ajuint i;

    size = ajRound(size,RESERVED_SIZE);

    AJNEW0(thys);
```

```
    thys->Ptr = AJALLOC0(size*sizeof(AjPInt));
    thys->Len = 0;
    thys->Res = size;

    for(i=0;i<size;++i)
        thys->Ptr[i] = NULL;

    arr2dAlloc++;

    return thys;
}
```

AJNEW0 is again called to create the basic object, an instance of an AjPInt2d in this case. AJALLOC0 is called to create an array of pointers to 1D integer array objects (AjPInt) of size size. You can deduce from the code that the second dimensions of the array (the arrays of integers themselves) is not created until it is needed. This is for reasons of efficiency.

ajInt2dDel is the AjPInt2d destructor function:

```
/* @func ajInt2dDel ***********************************************************
**
** Default destructor for AJAX integer arrays.
**
** If the given array is a NULL pointer, simply returns.
**
** @param  [d] thys [AjPInt2d*] Pointer to the ajint array to be deleted.
**         The pointer is always deleted.
** @return [void]
** @category delete [AjPInt2d] Default destructor
** @@
******************************************************************************/

void ajInt2dDel(AjPInt2d *thys)
{
    ajint i;

    if(!thys || !*thys)
        return;

    ajDebug("ajInt2dDel Len %u Res %u\n",
            (*thys)->Len, (*thys)->Res);

    for(i=(*thys)->Res-1;i>-1;--i)
        if((*thys)->Ptr[i])
            ajIntDel(&((*thys)->Ptr[i]));

    AJFREE((*thys)->Ptr);
    AJFREE(*thys);

    *thys = NULL;

    arr2dFreeCount++;

    return;
}
```

The function takes the address of the AjPInt2d object (thys) that is to be freed. To get to the object proper you must dereference thys, i.e. everywhere in the function body where you see *thys. You will recall that Ptr references an array of AjPInt object pointers, each of which points to array of integers.

The integer arrays are freed by calling the destructor function ajIntDel in a loop. This destructor takes the address of a AjPInt. Array notation is used to index the ith element of the AjPInt array, having first dereferenced thys ((*thys)->Ptr[i]). This retrieves an individual AjPInt object, the address of which is needed by the destructor which is why you have ajIntDel(&((*thys)->Ptr[i]));.

AJFREE is then called twice. The first call (AJFREE((*thys)->Ptr);) frees the array of AjPInt. The second call (AJFREE(*thys);) frees the AjPInt2d object itself. You can see that the pointer (thys) that is passed to the function is set to NULL using the code *thys = NULL;.

5.5.3.4 AjPInt2d putting and getting array elements

The function ajInt2dGet is used to retrieve a value from a 2D integer array. The source code is below:

```
/* @func ajInt2dGet *****************************************************
**
** Retrieve an element from an AJAX 2d integer array.
**
** If the given array is a NULL pointer, simply returns.
**
** @param [r] thys [const AjPInt2d] Pointer to the ajint array.
** @param [r] elem1 [ajuint] array element.
** @param [r] elem2 [ajuint] array element.
**
** @return [ajint] contents of array element
** @category cast [AjPInt2d] Retrieve an integer from an array
** @@
*************************************************************************/

ajint ajInt2dGet(const AjPInt2d thys, ajuint elem1, ajuint elem2)
{
    AjPInt t;

    if(!thys || elem1>=thys->Len)
        ajErr("Attempt to access bad ajint array index [%d][%d]\n",elem1,
              elem2);

    t = thys->Ptr[elem1];
    if(!t)
        ajErr("Attempt to access bad 1st dimension [%d][]\n",elem1);

    return ajIntGet(t,elem2);
}
```

The element in column elem1 and row elem2 will be retrieved from the array thys. An error is raised if you try to inspect an element that has not been allocated. Otherwise the value of the element is returned.

The function `ajInt2dPut` is used to load a 2D integer array element with a value. If the array is of insufficient size then the memory is extended as required. The source code is below:

```
/* @func ajInt2dPut ***********************************************************
**
** Load an integer 2d array element.
**
** If the given array is a NULL pointer an error is generated.
** If the array is of insufficient size then the array is extended.
** Negative indices generate an error.
**
** @param  [w] thys [AjPInt2d*] Pointer to the ajint array.
** @param  [r] elem1 [ajuint] array element.
** @param  [r] elem2 [ajuint] array element.
** @param  [r] v [ajint] value to load.
**
** @return [AjBool] true if any array was extended.
** @category modify [AjPInt2d] Load an integer array element
** @@
******************************************************************************/

AjBool ajInt2dPut(AjPInt2d *thys, ajuint elem1, ajuint elem2, ajint v)
{
    if(!thys || !*thys)
        ajErr("Attempt to write to illegal array value [%d][%d]\n",elem1,
              elem2);

    if(elem1 < (*thys)->Res)
    {
        ajDebug("ajInt2dPut [%u][%u] %d ([%u] %x)\n",
                elem1, elem2, v, (*thys)->Len, (*thys)->Ptr[elem1]);

        if(elem1>=(*thys)->Len)
            (*thys)->Len = elem1+1;

        if(!(*thys)->Ptr[elem1])
            (*thys)->Ptr[elem1] = ajIntNew();

        return ajIntPut(&(*thys)->Ptr[elem1],elem2,v);
    }

    arrInt2dResize(thys, elem1);

    if(!(*thys)->Ptr[elem1])
        (*thys)->Ptr[elem1] = ajIntNew();

    ajIntPut(&(*thys)->Ptr[elem1],elem2,v);

    return ajTrue;
}
```

There is nothing new here that you haven't seen already. Work through it to test your understanding of pointers, particularly how pointer and array notation may be mixed.

Programming with AJAX 6

6.1 Demonstration applications

EMBOSS includes, for certain AJAX and NUCLEUS library files, applications which illustrate the correct usage of each function in that library. Currently these 'demonstration applications' are kept in the **myembossdemo** EMBASSY package and have the prefix 'demo'. There is, of course, an ACD file for each application. For example the following files illustrate the use of the string library:

```
/home/auser/emboss/emboss/embassy/myembossdemo/src/demostring.c
/home/auser/emboss/emboss/embassy/myembossdemo/emboss_acd/demostring.acd
```

Caution

Demonstration applications are a work in progress and only available for a few of the library files and the coverage of the available functions is by no means comprehensive in all cases. The file **demostring.c** is the most complete and gives an idea of what is to come.

6.1.1 Compiling the demonstration applications

If you open the two Makefile.am files (in the src and acd directories) and find the entries for demostring you'll notice it occurs in the check_PROGRAMS section of the 'src' Makefile.am file. This section is reserved for programs that are included for testing, are undergoing active development or have yet to be fully tested, or which should be considered incomplete for some other reason, e.g. they are undocumented or quality assurance test data are not yet available for them. The demo* applications are there, reflecting their 'demonstration' status.

To compile the "check_PROGRAMS" applications, type:

```
make check
```

from the applications directory. This will produce all the "`check_PROGRAMS`" applications, including the demonstration applications.

You should now have a working `demostring` application. The source code and ACD file should contain sufficient comments to be comprehensible. To get a feel for the scope of the string library run `demostring` and inspect the output and the corresponding source code for each function call in turn. If it's not obvious what each function is doing, or to test your understanding, edit the source code, recompile and run `demostring` again.

6.2 Programming guides

Programming guides are available for most files in AJAX. These summarise the available C data structures and functions and methods for their use. For some files a programming guide is *'Not available'*, others say *'See'* a guide only, without any deep coverage. In most cases, you are unlikely to need such files and should see the online library documentation for a description of the datatypes and functions included.

AJAX library file	**Programming guide**
acd/ajacd.c/h	*Handling ACD Files*
core/ajalign.c/c	*Handling Alignments*
core/ajalign.h/c	*Handling Alignments*
core/ajarch.h	*C Basic Datatypes*
core/ajarr.c/h	*Handling Arrays*. Also see *Handling Simple*
core/ajassert.h/c	Not available
core/ajax.h	*C Basic Datatypes, Your First EMBOSS Application*
core/ajbase.h/c	*Handling Sequences*
core/ajcall.h/c	See *Handling Graphics*
core/ajcod.h/c	*Handling Codon*
core/ajdan.h/c	*Handling Sequences*
core/ajdmx.h/c	Not available
core/ajdefine.h	*C Basic Datatypes*
core/ajdomain.h/c	Not available
core/ajdom.h/c	See *Handling Graphics*
core/ajexcept.h/c	*Handling Exceptions*
core/ajfeat.h/c	*Handling Features*
core/ajfeatdata.h	*Handling Features*
core/ajfile.h/c	*Handling Files*
core/ajfiledata.h/c	Not available
core/ajfileio.h/c	Not available
core/ajfmt.h/c	*Handling Strings*

`graphics/ajgraph.h/c`	*Handling Graphics*
`graphics/ajgraphstruct.h`	*Handling Graphics*
`graphics/ajgraphxml.h/c`	See *Handling Graphics*
`graphics/ajhist.h/c`	*Handling Graphics*
`core/ajindex.h/c`	Not available
`core/ajjava.h/c`	Not available
`core/ajlist.h/c`	*Handling Lists*
`core/ajmath.h/c`	*Handling Maths*
`core/ajmatrices.h/c`	*Handling Matrices*
`core/ajmem.h/c`	*Objects (C Data Structures)*
`core/ajmess.h/c`	*Handling Exceptions*
`core/ajnam.h/c`	Not available
`core/ajnexus.h/c`	See *Handling Phylip*
`core/ajpat.h/c`	*Handling Patterns*. Also see *Handling Simple*
`core/ajpdb.h/c`	Not available
`core/ajpdbio.h/c`	Not available
`core/ajphylo.h/c`	*Handling Phylip*
`core/ajrange.h/c`	*Handling Simple*
`core/ajreg.h/c`	*Handling Patterns*. Also see *Handling Simple*
`core/ajreport.h/c`	*Handling Reports*
`core/ajseqabi.h/c`	See *Handling Sequences*
`core/ajseq.h/c`	*Handling Sequences*
`core/ajseqdata.h`	*Handling Sequences*
`ajaxdb/ajseqdb.h/c`	See *Handling Sequences*
`core/ajseqread.h/c`	*Handling Sequences*
`core/ajseqtype.h/c`	*Handling Sequences*
`core/ajseqwrite.h/c`	*Handling Sequences*
`core/ajsort.h/c`	*Handling Arrays*
`core/ajsql.h/c`	Not available
`core/ajstr.h/c`	*Handling Strings*. Also see *Handling Simple*
`core/ajsys.h/c`	*Handling System Code*
`core/ajtable.h/c`	*Handling Tables*
`core/ajtime.h/c`	*Not available*
`core/ajtranslate.h/c`	*Handling Sequences*
`core/ajtree.h/c`	*Not available*
`core/ajutil.h/c`	*Handling System Code*
`core/ajvector.h/c`	See *Handling Maths*

pcre/pcre.c	See *Handling Patterns*
pcre/pcre_chartables.c	See *Handling Patterns*
pcre/pcre_config.h	See *Handling Patterns*
pcre/pcre_get.c	See *Handling Patterns*
pcre/pcre.h	See *Handling Patterns*
pcre/pcre_internal.h	See *Handling Patterns*
pcre/pcreposix.c	See *Handling Patterns*
pcre/pcreposix.h	See *Handling Patterns*
pcre/pcre_printint.c	See *Handling Patterns*
pcre/pcre_study.c	See *Handling Patterns*

Note

Programming guides are not currently available for the NUCLEUS library. In some cases algorithms and data structures that you might expect to find in NUCLEUS are in fact kept in AJAX. This is usually necessary for purposes of compilation. For example any functions that are called during ACD file processing, which is handled by a part of AJAX, must also live in AJAX.

6.3 Handling ACD files

6.3.1 Introduction

Basic housekeeping code is required by all EMBOSS and EMBASSY applications. This includes code to process the command line and application ACD file, handle user inputs, retrieve AJAX objects corresponding to data definitions in the ACD file, and to exit cleanly.

6.3.2 Program initialisation

Every application must process the ACD file and user input. It must:

- Read in local database definitions
- Find the right ACD file to use and parse it
- Parse the command line
- Prompt the user for required values not specified on the command line
- Validate user input and reprompt for any incorrect values
- Allocate memory for an AJAX object for each ACD data definition
- Open input and output files
- Read input files (the first sequence from any input sequence stream is read)
- Initialise the AJAX objects (from reading the input files)

A single call is made to handle all of the above.
For EMBOSS applications:

```
embInit("ApplicationName", argc, argv);
```

For EMBASSY applications:

```
embInitP("ApplicationName", argc, argv, "PackageName");
```

All applications must call one of the above right at the start of the application. **ApplicationName** is the name of the ACD file to parse (**ApplicationName.acd**). **PackageName** is the name of the EMBASSY package, for example "myemboss". The command line is required which is why argc and argv from main are passed. Once these functions return then no further interaction with the user occurs. All input is read and held in memory before the application proper begins. An AJAX object for each ACD data definition is allocated.

For a simple program with no ACD data definitions, the first few lines of the program would look like this:

```
int main(int argc, char **argv)
[
      embInit("helloworld", argc, argv);
```

6.3.3 Retrieving ACD values

The ajAcdGet* family of functions return AJAX objects for data definitions in the application ACD file. They are defined in ajacd.h/c and have the general name:

```
ajAcdGetDatatype
```

where **Datatype** is one of the supported ACD datatypes (Section A.2, 'Datatypes').

A prefix ajAcdGet function is provided for each ACD datatype. They are not constructor functions as such, but instead return a pointer to an appropriate AJAX object that has been allocated by a call to embInit or embInitP. For example, when retrieving an ACD string, ajAcdGetString returns a pointer to the string (an AjPStr object) created by embInit. Attributes in the data definition and user input gathered at the command line are used to initialise the object. Memory for any new objects must be freed later on in main().

ACD datatype	AJAX datatype (return value)	AJAX function
align	AjPAlign	ajAcdGetAlign
array	AjPFloat	ajAcdGetArray

boolean	AjBool	ajAcdGetBoolean
codon	AjPCod	ajAcdGetCodon
cpdb	AjPFile	ajAcdGetCpdb
datafile	AjPFile	ajAcdGetDatafile
directory	AjPDir	ajAcdGetDirectory
	AjPStr	ajAcdGetDirectoryName
dirlist	AjPList	ajAcdGetDirlist
discretestates	AjPPhyloState*	ajAcdGetDiscretestates
	AjPPhyloState	ajAcdGetDiscretestatesSingle
distances	AjPPhyloDist*	ajAcdGetDistances
	AjPPhyloDist	ajAcdGetDistancesSingle
featout	AjPFeattabOut	ajAcdGetFeatout
features	AjPFeattable	ajAcdGetFeatures
filelist	AjPList	ajAcdGetFilelist
float	float	ajAcdGetFloat
	double	ajAcdGetFloatDouble
frequencies	AjPPhyloFreq	ajAcdGetFrequencies
graph	AjPGraph	ajAcdGetGraph
graphxy	AjPGraph	ajAcdGetGraphxy
infile	AjPFile	ajAcdGetInfile
int	ajint	ajAcdGetInt
	ajlong	ajAcdGetIntLong
list	AjPStr*	ajAcdGetList
	AjPStr	ajAcdGetListSingle
matrix	AjPMatrix	ajAcdGetMatrix
matrixf	AjPMatrixf	ajAcdGetMatrixf
outcodon	AjPOutfile	ajAcdGetOutcodon
outcpdb	AjPOutfile	ajAcdGetOutcpdb
outdata	AjPOutfile	ajAcdGetOutdata
outdir	AjPDirout	ajAcdGetOutdir
	AjPStr	ajAcdGetOutdirName
outdiscrete	AjPOutfile	ajAcdGetOutdiscrete
outdistance	AjPOutfile	ajAcdGetOutdistance
outfile	AjPFile	ajAcdGetOutfile
outfileall	AjPFile	ajAcdGetOutfileall
outfreq	AjPOutfile	ajAcdGetOutfreq

outmatrix	AjPOutfile	ajAcdGetOutmatrix
outmatrixf	AjPOutfile	ajAcdGetOutmatrixf
outproperties	AjPOutfile	ajAcdGetOutproperties
outscop	AjPOutfile	ajAcdGetOutscop
outtree	AjPOutfile	ajAcdGetOuttree
pattern	AjPPatlistSeq	ajAcdGetPattern
properties	AjPPhyloProp	ajAcdGetProperties
range	AjPRange	ajAcdGetRange
regexp	AjPPatlistRegex	ajAcdGetRegexp
	AjPRegexp	ajAcdGetRegexpSingle
report	AjPReport	ajAcdGetReport
scop	AjPFile	ajAcdGetScop
select	AjPStr*	ajAcdGetSelect
	AjPStr	ajAcdGetSelectSingle
seq	AjPSeq	ajAcdGetSeq
seqall	AjPSeqall	ajAcdGetSeqall
seqout	AjPSeqout	ajAcdGetSeqout
seqoutall	AjPSeqout	ajAcdGetSeqoutall
seqoutset	AjPSeqout	ajAcdGetSeqoutset
seqset	AjPSeqset	ajAcdGetSeqset
seqsetall	AjPSeqset*	ajAcdGetSeqsetall
	AjPSeqset	ajAcdGetSeqsetallSingle
string	AjPStr	ajAcdGetString
toggle	AjBool	ajAcdGetToggle
tree	AjPPhyloTree*	ajAcdGetTree
	AjPPhyloTree	ajAcdGetTreeSingle

It's recommended that variables for handling ACD datatypes should have the same name as the parameter or qualifier in question, i.e. the name given in the ACD data definition. This is not strictly required but it makes the code much easier to understand. For the same reason all calls to `ajAcdGet*` functions should be given in a single block of code for ease of reading.

Example

Consider the following ACD file:

```
application: example
[
    documentation: "Example application."
]
```

```
string: astring
[
    default: "String to be printed to screen."
]
```

Here is the C source code to print `astring` to the screen:

```
int main(int argc, char **argv)
[
    AjPStr astring = NULL;

    embInit("example", argc, argv);

    astring = ajAcdGetString("astring")
    ajFmtPrint("%S\n", astring);
    ajStrDel(&astring);

    embExit();

    return 0;
]
```

The code declares an AJAX string object (`AjPStr`) and calls `embInit` to invoke ACD file processing. `embInit` allocates memory for the string object which is why the above code does not call a string constructor function explicitly. Nonetheless a string object was created by `embInit` and should be freed once you are done with it. That is why `ajStrDel` is called.

6.3.4 Alternative ACD retrieval functions

There are several alternative ACD retrieval functions. In all cases `token` is the name of the ACD data definition (the name of the parameter or qualifier):

```
AjPStr          ajAcdGetDirectoryName (const char *token);
AjPPhyloState   ajAcdGetDiscretestatesSingle (const char *token);
AjPStr          ajAcdGetListSingle(const char *token);
AjPStr          ajAcdGetOutdirName (const char *token);
AjPRegexp       ajAcdGetRegexpSingle (const char *token);
AjPStr          ajAcdGetSelectSingle (const char *token);
AjPSeqset       ajAcdGetSeqsetallSingle (const char *token);
AjPPhyloTree    ajAcdGetTreeSingle (const char *token);
```

In contrast to the standard retrieval functions these return a value *derived* from the ACD datatype, such as the first sequence from a set of sequences or the name of a directory. For example `ajAcdGetOutdirName` returns an `AjPStr` holding the name of an output directory whereas the standard retrieval function `ajAcdGetOutdir` returns an `AjPDir`, i.e. the directory itself. Their use is explained in the appropriate programming guide.

The alternative functions are provided for convenience where the full object is not required. ACD takes care of the memory management for any objects that have not been passed to the main program. Alternative functions with the suffix `Single` return an element of the object that would normally be returned by the standard retrieval function. This saves the calling program from stepping through a list of values when only one value can be selected form the list (i.e. the ACD file defines a minimum and maximum of one value to be returned). You only need to free the single string that was returned. Similarly, alternative functions with the suffix `Name` return an entirely new string. You need only free this string later, ACD takes care of freeing the full object when the program exits.

Consider the following ACD file:

```
application: example
[
    documentation: "Example application."
]
directory: dir
[
    help: "Directory for reading."
]
```

The program below would print the name of the directory:

```
int main(int argc, char **argv)
[
    embInit("example", argc, argv);
    AjPStr name = NULL;
    name = ajAcdGetOutdirName("dir");
    ajFmtPrint("Directory name is %S\n", name);
    ajStrDel(&name);
    embExit();
    return 0;
```

6.3.5 Exiting cleanly

Your application must exit cleanly. In other words all memory that has been allocated must be freed and an appropriate code returned to the operating system.

Memory management is covered in detail elsewhere (Section 5.5, 'Programming with objects'). In brief, memory is allocated by:

- `embInit` or `embInitP` allocate memory for an AJAX object for each ACD data definition, a pointer to which is returned by the `ajAcdGet*` functions
- `embInit` or `embInitP` also allocate some memory for housekeeping purposes
- Explicit calls to memory allocation macros
- Explicit calls to constructor functions
- Implicit calls to constructor functions, which are made by some functions as a failsafe mechanism where an object is required but an unallocated object pointer was passed.

All allocation macros must be matched to a corresponding freeing macro. All constructor calls, explicit or implicit, including calls to `ajAcdGet*` functions, must be matched to a corresponding destructor function. To free the memory allocated by EMBOSS for housekeeping you must call one of:

```
void embExit (void);
void embExitBad (void);
```

These functions are defined in embExit.h/c. embExit returns the success code (0) whereas embExit returns the failure code (0).

The last two lines of most EMBOSS applications are therefore:

```
embExit();
return 0;
```

6.4 Handling simple ACD datatypes

6.4.1 Introduction

This section covers ACD file processing for the 'simple' ACD datatypes (Section A.2.1, 'Description of simple ACD datatypes'):

- integer
- float
- boolean
- toggle
- string
- array
- range

Values are retrieved by calls to ajAcdGet* functions which return an AJAX datatype of the appropriate type: a fundamental type is returned for integer, float, boolean and toggle and an EMBOSS object for the other types. Functions for handling ranges are covered and include:

- Get and set elements of the range object
- Query the properties of a range object
- Process a string (AjPStr) and sequence (AjPSeq) according to the specification in a range object.

For regular expressions (regexp ACD datatype) and sequence patterns (pattern ACD datatype) see Section 6.6, 'Handling sequence patterns'. Array handling (Section 6.17, 'Handling arrays') and string handling (Section 6.5, 'Handling strings') are described in depth elsewhere.

6.4.2 AJAX library files

AJAX library files for handling simple ACD datatypes are listed in the table below. Library file documentation, including a complete description of datatypes and functions, is available at: http://emboss.open-bio.org/rel/dev/libs/

Library file documentation	Description
ajrange	Handling of AJAX range expressions
ajstr	String handling
ajarr	Array handling

ajrange.h/c. Defines the range specification object (AjPRange) and contain functions for handling of range specifications (see Section A.2.1.5, 'range').

ajstr.h/c. Defines the string object (AjPStr) used for handling strings from the ACD file. They contain most of the functions you will ever need for general string handling (Section 6.5, 'Handling strings').

ajarr.h/c. Most of the functions you will ever need for general array handling (Section 6.17, 'Handling arrays'). They define the AjPFloat object used for handling arrays from an ACD file. They contain static data structures and functions for handling arrays at a low level.

6.4.3 ACD datatypes

The simple ACD datatypes are used for application input:

integer	Simple integer number
float	Simple floating point number
boolean	Simple boolean value
toggle	Simple boolean value
string	Simple string
array	List of either integer or floating point numbers
range	Range of sequence positions

6.4.4 ACD data definition

Typical ACD definitions are shown below.

6.4.4.1 integer

For integer number input:

```
integer: wordsize
[
    default: "4"
    minimum: "2"
    maximum: "20"
    information: "Word size"
]
```

6.4.4.2 float

For floating point number input:

```
float: minscore
[
    default: "0.0"
    minimum: "0.0"
    information: "Minimum score of feature to display"
]
```

6.4.4.3 boolean

For boolean input:

```
boolean: feature
[
    default: "N"
    information: "Use feature information"
]
```

6.4.4.4 toggle

For toggle input:

```
toggle: tolower
[
    default: "N"
    information: "Change masked region to lower case"
]
```

6.4.4.5 string

For string input:

```
string: delimiter
[
    default: "|"
    information: "Delimiter of records in text output file"
    knowntype: "output delimiter"
  ]
```

6.4.4.6 array

For array input:

```
array: thresholds
[
    information: "Values to represent 'identical', 'similar' and 'related'"
    default: "-1.5,0.0,1.5"
    minimum: "0.0"
    size: "3"
    sum: "0"
    sumtest: "Y"
]
```

6.4.4.7 range

For range input:

```
range: regions
[
    information: "Regions to put in uppercase (eg: 4-57,78-94)"
    default: ""
    help: "Regions to put in uppercase. If this is left blank, the sequence case is
left alone. A set of regions is specified by a set of pairs of integer positions
separated by any non-digit, non-alpha character. For example: \
           24-45, 56-78 \
           1:45, 67=99;765..888 \
           1,5,8,10,23,45,57,99"
]
```

6.4.4.8 Parameter name

A standard parameter name might be available depending on the specific use-case of the data definition; for example `gap penalty` for any `float` input that defines a gap penalty. See Appendix A, *ACD syntax reference*.

6.4.4.9 Common attributes

Attributes that are typically specified are summarised below. They are datatype-specific (Section A.5, 'Datatype-specific attributes') unless they are indicated as being global attributes (Section A.4, 'Global attributes').

`default:` A global attribute and specifies a default value.

`minimum:` Specifies the minimum permitted value.

`maximum:` Specifies the maximum permitted value.

`information:` A global attribute that specifies the user prompt and is also used in the application documentation.

`knowntype:` This global attribute should always be specified for string inputs. If the output is not of any of the standard EMBOSS known types then *`ApplicationName`* `output` is the recommended value.

`size:` Specifies the permissible number of elements in an `array` data definition.

sum: Specifies the total of all values in an `array` data definition and is tested for unless the `sumtest:` attribute is false.

sumtest: A boolean attribute which, if set to false, turns off testing for the `sum:` attribute for an array data definition.

6.4.5 AJAX datatypes

For handling simple ACD datatypes defined in the ACD file use:

`ajint`	Simple integer number (for `integer` ACD datatype).
`float`	Simple floating point number (for `float` ACD datatype).
`AjBool`	Simple boolean value (for `boolean` and `toggle` ACD datatypes).

`AjPStr`	String (for `string` ACD datatype).
`AjPFloat`	Array of floating point numbers (for `array` ACD datatype).
`AjPRange`	AJAX sequence range specification (for `range` ACD datatype). See Section A.2.1.5, '`range`'.

6.4.6 ACD file handling

Datatypes and functions for handling simple ACD datatypes via the ACD file are shown below.

ACD datatype	AJAX datatype	To retrieve from ACD
`integer`	`ajint`	`ajAcdGetInt`
`float`	`float`	`ajAcdGetFloat`
`boolean`	`AjBool`	`ajAcdGetBoolean`
`toggle`	`AjBool`	`ajAcdGetToggle`
`string`	`AjPStr`	`ajAcdGetString`
`array`	`AjPFloat`	`ajAcdGetArray`
`range`	`AjPRange`	`ajAcdGetRange`

Your application code will call `embInit` to process the ACD file and command line (see Section 6.3, 'Handling ACD files'). All values from the ACD file are read into memory. You have a handle on memory through the `ajAcdGet*` family of functions which return pointers to appropriate objects.

> **Note**
>
> You wouldn't normally retrieve a **`toggle`** from ACD as they're intended for use within the ACD file only, usually to control the prompting for another parameter (see Section 4.5, 'Controlling the prompt').

6.4.6.1 Retrieval of simple ACD datatypes

To retrieve data from the ACD file a simple variable or object pointer is declared as required, and then initialised using the appropriate `ajAcdGet*` function.

6.4.6.1.1 `integer`

```
ajint wordsize =0;

wordsize = ajAcdGetInt("wordsize");
```

6.4.6.1.2 `float`

```
float minscore =0.;

minscore = ajAcdGetFloat("minscore");
```

6.4.6.1.3 `boolean`

```
AjBool showall = ajFalse;

showall = ajAcdGetBoolean("showall");
```

6.4.6.1.4 `toggle`

```
AjBool tolower = ajFalse;

tolower = ajAcdGetToggle("tolower");
```

6.4.6.1.5 `string`

```
AjPStr delimiter = NULL;

delimiter = ajAcdGetString("delimiter");
```

6.4.6.1.6 `array`

```
AjPFloat thresholds = NULL;

    thresholds = ajAcdGetArray("thresholds");
```

6.4.6.1.7 range

```
AjPRange regions = NULL;

regions = ajAcdGetRange("regions");
```

6.4.6.2 Processing command line options and ACD attribute

6.4.6.2.1 Setting range object properties

Functions to set range object properties are:

```
/* Set the start and end values of a range element. */
AjBool ajRangeElementSet (AjPRange thys, ajuint element, ajuint start, ajuint end);

/* Sets range values offset relative to a sequence -sbegin value */
AjBool ajRangeSetOffset (AjPRange thys, ajuint begin);
```

`ajRangeSetOffset` will set the range values relative to a specified position (`begin`), usually the start position of a range of positions in a sequence as specified on the command line with **`-sbegin`** or in the USA of a sequence (see the *EMBOSS User's Guide*). If, for example, `begin` is `11` and the range is `11-12` the new range is changed to `1-2`.

6.4.6.3 Memory management

It is your responsibility to free memory at the end of the program. You must call the default destructor function (see below) on any `AjPStr`, `AjPFloat` or `AjPRange` objects returned by calls to `ajAcdGet*`. This is not necessary, of course, for the primitive datatypes.

6.4.7 Object memory management

6.4.7.1 Default object construction

To use a string, array or range object that is not defined in the ACD file you must first instantiate the appropriate object pointer. The default constructor functions are:

```
AjPStr ajStrNew (void);            /* String object. */
AjPFloat ajFloatNew (void);        /* Float array */
AjPRange ajRangeNewI (ajuint n);   /* Range object   */
```

`ajRangeNewI` takes an integer (`n`) which is the number of ranges the object can hold.

All constructors return the address of a new object. The pointers do not need to be initialised to `NULL` but it is good practice to do so:

```
AjPStr    delimiter  = NULL;
AjPFloat  thresholds = NULL;
AjPRange  regions    = NULL;

delimiter    = ajStrNew();
```

```
thresholds  = ajFloatNew();

/* Object with a single range */
regions     = ajRangeNewI(1);

/* Do something with objects */

ajStrDel(&delimiter);
ajFloatDel(&thresholds);
ajRangeDel(&regions);
```

6.4.7.2 Default object destruction

You must free the memory for objects once you are finished with them. The destructor functions are:

```
AjPStr   ajStrDel  (AjPStr *Pstr);      /* String object. */
void   ajFloatDel (AjPFloat* Parr);    /* Float array    */
void   ajRangeDel (AjPRange *Prange); /* Range object   */
```

They are used as follows:

```
AjPStr delimiter    = NULL;
AjPFloat thresholds = NULL;
AjPRange regions    = NULL;

delimiter  = ajAcdGetString("delimiter");
thresholds = ajAcdGetArray("thresholds");
regions    = ajAcdGetRange("regions");

/* Do something with objects */

ajStrDel(&delimiter);
ajFloatDel(&thresholds);
ajRangeDel(&regions);
```

6.4.7.3 Alternative object construction and loading

There are a variety of alternative constructor functions for the `AjPRange` object:

```
/* Copy a  range object. */
AjPRange   ajRangeNewRange (const AjPRange src);

/* Construct from a string. */
AjPRange ajRangeNewString (const AjPStr str);

/* Construct from a string with explicit specification. */
AjPRange ajRangeNewStringLimits (const AjPStr str, ajuint imin, ajuint imax,
                                 ajuint minsize, ajuint size);

/* Construct from a file. */
AjPRange ajRangeNewFilename (const AjPStr name);
```

```
/* Construct from a file with explicit specification. */
AjPRange ajRangeNewFilenameLimits  (const AjPStr name,
                                    ajuint imin, ajuint imax,
                                    ajuint minsize, ajuint size);
```

ajRangeNewStringLimits and ajRangeNewFilenameLimits both construct a range object with specified limits: minimum value (imin), maximum value (imax), minimum number of ranges (minsize) and the required number of ranges (size). A value of zero for size indicates that there is no required number.

ajRangeNewFilename and ajRangeNewFilenameLimits construct an object from a 'range file' (see Section A.2.1.5, 'range').

For alternative constructor functions for the AjPStr object see Section 6.5, 'Handling strings'.

6.4.8 Getting range object elements

The following functions retrieve elements from a range object:

```
/* Get the number of ranges  */
ajuint ajRangeGetSize (const AjPRange thys);

/* Get text value of a range */
AjBool ajRangeElementGetText  (const AjPRange thys, ajuint element,
                               AjPStr *text);

/* Get start and end values  */
AjBool ajRangeElementGetValues  (const AjPRange thys, ajuint element,
                                 ajuint *start, ajuint *end);
```

ajRangeElementGetText will retrieve text from the specified range element (element). The text is defined as any non-digit characters after the pair of range numbers. For example, for the pair of ranges 10-20 potential exon 50-60 repeat the text values are: "potential exon" and "repeat". The address of the string object (text) to hold the text is passed.

6.4.9 Querying range object properties

Functions for querying the properties of a range object include:

```
/* Tests if the set of ranges are in ascending non-overlapping order */
AjBool  ajRangeIsOrdered (const AjPRange thys);

/* Tests if any range elements overlap to a region (of a sequence). */
ajuint  ajRangeCountOverlaps (const AjPRange thys, ajuint pos, ajuint  length);

/* Tests for a single range from the start to end of a sequence.     */
AjBool  ajRangeIsWhole (const AjPRange thys, const AjPSeq seq);
```

ajRangeCountOverlaps returns the number of ranges in a range object which overlap with a sequence region defined by a start position (pos) and a length (length).

ajRangeIsWhole tests whether the range object contains a single range from the start to end of the given sequence (seq).

6.4.10 Sequence manipulation functions

These functions process a sequence object (AjPSeq) according to the specification in a range object:

```
/* Remove all subsequences not corresponding to ranges */
AjBool  ajRangeSeqExtract (const AjPRange thys, AjPSeq seq);

/* Store retained text as a list of strings. */
AjBool  ajRangeSeqExtractList (const AjPRange thys,const AjPSeq seq, AjPList
outliststr);

/* Insert spaces into sequence to pad out to the ranges.   */
AjBool  ajRangeSeqStuff (const AjPRange thys, AjPSeq seq);

/* Mask ranges of positions in a sequence. */
AjBool  ajRangeSeqMask (const AjPRange thys, AjPSeq seq, const AjPStr maskchar);

/* Convert the ranges of characters in a sequence to lower-case. */
AjBool  ajRangeSeqToLower (const AjPRange thys, AjPSeq seq);
```

ajRangeSeqExtract retains regions in a sequence corresponding to the ranges: regions not in a range are removed. A sequence processed by ajRangeSeqExtract will comprise regions from the original sequence concatenated in the order specified in the set of ranges. If these are not in ascending order then the resulting sequence won't be in position order either.

ajRangeSeqExtractList is the same as ajRangeSeqExtract except that the retained subsequences are written to a list. The order of the list is the same as that specified in the set of ranges. If these are not in ascending order then the resulting list of strings won't be either.

ajRangeSeqStuff takes a string and an ordered, non-overlapping set of ranges and writes a string padded with whitespace such that a space is given for all positions not within a range. For example, for the string "abcde" and ranges 3-5,7-8 the string generated will be " abc de".

ajRangeSeqMask will mask the ranges of positions in a sequence, replacing all characters within range with the mask character (maskchar).

6.4.11 String manipulation functions

A set of functions equivalent to the sequence manipulation functions are provided for strings:

```
AjBool ajRangeStrExtract (const AjPRange thys, const AjPStr instr,
AjPStr *outstr);
AjBool ajRangeStrExtractList (const AjPRange thys, const AjPStr instr, AjPList
outliststr);
AjBool ajRangeStrStuff (const AjPRange thys, const AjPStr instr, AjPStr *outstr);
AjBool ajRangeStrMask (const AjPRange thys, AjPStr *str, const AjPStr maskchar);
AjBool ajRangeStrToLower (const AjPRange thys, AjPStr *str);
```

Their functions are identical to their sequence counterparts except that a string (`AjPStr`) rather than a sequence (`AjPSeq`) is taken. Also, `ajRangeStrStuff` has individual arguments for the input and output strings.

6.5 Handling strings

6.5.1 Introduction

Efficient and flexible string handling is fundamental to molecular sequence manipulation. Accordingly, string handling is the best-developed area in the AJAX library. The functionality, which is spread over several library files, is comprehensive and includes:

- String construction
- String referencing and dereferencing where a handle on, but not a copy of, a string is required
- Assignment functions to assign a value to a string
- Functions to combine two strings or parts of a string. The types of operation include appending, insertion and pasting (overwriting character positions)
- Cut functions to remove substrings, regions or characters from a target string
- Substitutions of characters or substrings of a string with other characters/substrings
- Query functions to test the properties of a string
- Retrieval of characters and string properties (such as length)
- Conversion functions to convert a string to some other datatype
- String formatting
- String comparison functions
- Search functions to find substrings or characters in strings
- String parsing functions to parse text tokens from strings
- String iteration, which allows you to step through a string a single character at a time
- String tokenisation
- Formatting and printing. Conversion characters are defined for all the EMBOSS fundamental datatypes (Section 5.1, 'Basic datatypes') and are an extension of the basic C conversion codes.

For convenience, most functions implemented for an AJAX string parameter have a corresponding function with a C-type (`char *`) string parameter. A string may be defined in the ACD file and retrieved from the C source code by a call to `ajAcdGetString`. More typically though, strings are created directly in the code.

In contrast to standard C-type (`char *`) strings, the AJAX string object (`AjPStr`) is dynamic; memory is (re)allocated as needed so that you never run out of space when calling string functions. AJAX strings are reference counted. The object keeps track of how many references (pointers to) the string there are in the code. It is not until all references to a string have been deleted that the string itself is freed.

The string object definition is shown below:

```
typedef struct AjSStr
[
    ajuint Res;
    ajuint Len;
    char *Ptr;
    ajuint Use;
    ajint Padding;
] AjOStr;
#define AjPStr AjOStr*
typedef AjPStr* AjPPStr;
```

`Ptr` holds the character string and `Len` is its length. In contrast to C-type strings the character string may or may not be `NULL` terminated. The library functions for printing `AjPStr` objects uses the length field (`Len`) for how many characters to print and won't stop at the first `NULL` if there is one.

`Res` is the reserved dynamic memory associated with the object and is always at least equal to `Len` but is often more. It is used for handling dynamic reallocation of string memory. `Use` is the string reference counter mentioned above. Finally, the `Padding` element pads the string to an alignment boundary (to mollify strict compilers).

The string object and the internals of string memory management are described in greater detail elsewhere (Section 5.5, 'Programming with objects').

6.5.2 AJAX library files

AJAX library files for handling strings are listed in the table below. Library file documentation, including a complete description of datatypes and functions, is available at: http://emboss.open-bio.org/rel/dev/libs/

Library file documentation	Description
`ajstr`	General string handling
`ajfmt`	String formatting functions

`ajstr.h/c`. Most of the functions you will ever need for general string handling. They define the basic string object `AjPStr`, string iteration object (`AjIStr`) and string token parser object (`AjPStrTok`) for use with the functions.

`ajfmt.h/c`. Functions for string formatting. The functions are similar to the C functions `printf`, `fprintf`, etc., but the set of conversion specifiers and other functionality is extended. They also contain a static data structure and functions for handling formatting at a low level (Section 6.5.23, 'Handling string formatting').

You are unlikely to need the static data structures and functions unless you plan to extend the string library.

6.5.3 ACD datatypes

The ACD datatype for handling string input is:

`AjPStr` String

6.5.4 ACD data definition

A typical ACD definition for string input:

```
string: delimiter
[
    default: "|"
    information: "Delimiter of records in text output file"
    knowntype: "output delimiter"
]
```

6.5.4.1 Parameter name

A standard parameter name (Section A.1.3, 'Parameter naming conventions') might be used. This depending on the specific use-case of the data definition.

6.5.4.2 Common attributes

Attributes that are typically specified are summarised below. They are all global attributes (Section A.4, 'Global attributes').

`default:` Default value (global attribute)

`information:` User-prompt used in the application documentation (global attribute)

`knowntype:` Should always be specified for string inputs. If the output is not of any of the standard EMBOSS known types (Section 4.3.5.3.1, 'Application data known types file (`knowntypes.standard`)') then *`ApplicationName`* `output` is the recommended value (global attribute).

6.5.5 AJAX datatypes

For handling strings, including those defined in the ACD file (`string` ACD datatype), use:

`AjPStr` String

Two datatypes are for string-related operations:

`AjIStr`	String iteration object
`AjPStrTok`	String token parser object

6.5.6 ACD file handling

Datatypes and functions for handling string input via the ACD file are shown below.

Title	To read a string
ACD datatype	`string`
AJAX datatype	`AjPStr`
To retrieve from ACD	`ajAcdGetString`

6.5.6.1 Input string retrieval

To retrieve an input string an object pointer is declared and then initialised using `ajAcdGetString`:

```
AjPStr delimiter = NULL;

delimiter = ajAcdGetString("delimiter");
```

6.5.6.2 Processing command line options and ACD attribute

Currently there are no functions for this.

6.5.7 String object memory management

6.5.7.1 Default object construction

To use a string object that is not defined in the ACD file you must first instantiate the appropriate object pointer. The default string construction function is:

```
/* Create a string object.    */
AjPStr ajStrNew (void);
```

All constructors return the address of a new object. The pointers do not need to be initialised to `NULL` but it is good practice to do so:

```
AjPStr        str = NULL;

str    =  ajStrNew();

/* The object is instantiated and ready for use */
```

6.5.7.2 Default object destruction

You must free the memory for an object once you are finished with it. The default string destructor function is:

```
/* Delete a string object. */
AjPStr ajStrDel (AjPStr *Pstr);
```

It is the responsibility of the calling function to destroy any objects

```
AjPStr str = NULL;

str = ajStrNew();

/* Do something with the instantiated object */

ajStrDel(&str);

/* The memory is freed and the pointer reset to NULL, ready for re-use. */

str = ajStrNew();

/* Do something else with the new object. The pointer variable is reallocated. */

ajStrDel(&str);

/* Done with the object so the memory is freed. */
```

6.5.7.3 Alternative object construction and loading

A variety of alternative string constructor functions are available. A string can be constructed from an existing string object (`AjPStr`) or C-type (`char *`) string, with an optional reserved size:

```
/* Construct from C-type string */
AjPStr  ajStrNewC (const char *txt);

/* Construct from C-type string with reserved size */
AjPStr  ajStrNewResC (const char *txt, ajuint size);

/* Construct from C-type string with explicit reserved size */
AjPStr  ajStrNewResLenC (const char *txt, ajuint size, ajuint len);

/* Construct with reserved size */
AjPStr  ajStrNewRes(ajuint size);

/* Construct from string object */
AjPStr  ajStrNewS (const AjPStr str);

/* Construct from string object with reserved size */
AjPStr  ajStrNewResS (const AjPStr str, ajuint size);
```

`ajStrNewResLenC` is identical to `ajStrNewResC` except that the string length is passed to `ajStrNewResLenC` for speed.

They are all used in same way as the default constructor, i.e. they return a pointer to the new object.

6.5.8 String referencing and dereferencing functions

There is a string referencing function:

```
/* Reference an existing string */
AjPStr  ajStrNewRef (AjPStr str);
```

In contrast to the other constructor functions ajStrNewRef does not create a new object but instead returns a pointer to the string passed in and increases its reference count.

There is a string dereferencing function:

```
/* Dereference an existing string */
AjBool ajStrDelStatic (AjPStr* Pstr);
```

ajStrDelStatic will set the string pointer to NULL and decrement the use count of the string to which it refers. In contrast to the default destructor function, strings with a use count of 1 are not freed to avoid freeing and reallocating memory when they are reused. Memory reserved for the string is never deleted by this function and can be reused.

6.5.9 String assignment functions

The string assignment functions will assign a value to a string. A string can be assigned from a character, an existing string object (AjPStr) or C-type (char *) string, or a substring of an appropriate datatype. Some function variants allow optional reserved sizes to be specified:

```
/* Assign from character */
AjBool  ajStrAssignK (AjPStr* Pstr, char chr);

/* Assign from C-type string */
AjBool  ajStrAssignC (AjPStr* Pstr, const char* txt);

/* Assign from string object */
AjBool  ajStrAssignS (AjPStr* Pstr, const AjPStr str);

/* Assign from C-type string up to a given length */
AjBool  ajStrAssignLenC (AjPStr* Pstr, const char* txt, ajuint ilen);

/* Assign using a pointer only. The reference count is incremented */
AjBool  ajStrAssignRef (AjPStr* Pstr, AjPStr refstr);

/* Assign from C-type string with reserved size */
AjBool  ajStrAssignResC (AjPStr* Pstr, ajuint size, const char* txt);

/* Assign from string object with reserved size */
AjBool  ajStrAssignResS (AjPStr* Pstr, ajuint i, const AjPStr str);

/* Assign from substring of C-type string */
AjBool  ajStrAssignSubC (AjPStr* Pstr, const char* txt,  ajint pos1, ajint pos2);

/* Assign from substring of string object */
AjBool  ajStrAssignSubS (AjPStr* Pstr, const AjPStr str, ajint pos1, ajint pos2);
```

ajStrAssignLenC is identical to ajStrAssignC except that the source string is only copied up to a specified length.

Memory for the string is allocated to NULL target pointers if necessary, although to keep the calling code intuitive we strongly recommend that a string object is first instantiated by calling ajStrNew before any of these functions are used.

For example, in the following code it is clear you are dealing with two separate strings:

```
AjPStr str      = NULL;
AjPStr strcopy  = NULL;

str     = ajStrNewC("A string");
strcopy = ajStrNew();

if(!ajStrAssignC(&strcopy, str))
    ajFatal("String not assigned");

ajStrDel(&str);
ajStrDel(&strcopy);
```

Whereas the following code is perfectly valid but is less clear:

```
AjPStr str      = NULL;
AjPStr strcopy  = NULL;

str = ajStrNewC("A string");

if(!ajStrAssignC(&strcopy, str))
    ajFatal("String not assigned");

ajStrDel(&str);
ajStrDel(&strcopy);
```

6.5.10 String combination functions

The string combination functions will combine two strings together. They fall into a variety of classes described below.

6.5.10.1 String append functions

The string append functions will append a source string to a target string. An individual character or multiple characters, an existing string object (AjPStr) or C-type (char *) string, or a substring of either of the latter can be appended:

```
/* Append a C-type string */
AjBool  ajStrAppendC (AjPStr* Pstr, const char* txt);

/* Append a single character */
AjBool  ajStrAppendK (AjPStr* Pstr, char chr);

/* Append a string object */
AjBool  ajStrAppendS (AjPStr* Pstr, const AjPStr str);

/* Append multiples of a single character */
AjBool  ajStrAppendCountK (AjPStr* Pstr, char chr, ajuint num);

/* Append a C-type string up to a given length */
AjBool  ajStrAppendLenC (AjPStr* Pstr, const char* txt, ajuint len);

/* Append a substring of a string object */
AjBool  ajStrAppendSubS (AjPStr* Pstr, const AjPStr str, ajint pos1, ajint pos2);
```

ajStrAppendLenC is identical to ajStrAppendC except that a region from the source string up to a specified length is appended.

6.5.10.2 String join functions

The string join functions are similar to the append functions except that they cut the source and target strings at specified positions before appending:

```
/* Cut down string at pos1 and add string2 from position pos2. */
AjBool  ajStrJoinC (AjPStr* Pstr, ajint pos1, const char* txt, ajint pos2);
AjBool  ajStrJoinS (AjPStr* Pstr, ajint pos1,  const AjPStr str, ajint pos2);
```

6.5.10.3 String insert functions

The string insert functions will insert a character, an existing string object (AjPStr) or C-type (char *) string into a string:

```
/* Insert a C-type string */
AjBool  ajStrInsertC (AjPStr* pthis, ajint pos, const char* str);
/* Insert a character     */
AjBool  ajStrInsertK (AjPStr* pthis, ajint begin, char insert);
/* Insert a string        */
AjBool  ajStrInsertS  (AjPStr* pthis, ajint pos, const AjPStr str);
```

6.5.10.4 String paste functions

The string paste functions will overwrite the target string with the source string (or character) at a specified point (pos), using (optionally) up to a specified number of characters from the source string:

```
/* Paste string */
AjBool  ajStrPasteS( AjPStr* Pstr, ajint pos, const AjPStr str);
/* Paste specified number of characters */
AjBool  ajStrPasteCountK(AjPStr* Pstr, ajint pos, char chr, ajuint num);
/* Paste portion of C-type string */
AjBool  ajStrPasteMaxC (AjPStr* Pstr, ajint pos, const char* txt, ajuint n);
/* Paste portion of string object */
AjBool  ajStrPasteMaxS( AjPStr* Pstr, ajint pos, const AjPStr str, ajuint n);
```

In addition there is a string masking function which will replace all characters in the target string with a mask character over a specified range:

```
/* Replace all characters in a region with mask characters */
AjBool  ajStrMaskRange(AjPStr* str, ajint begin, ajint end, char maskchar);
```

6.5.11 String cut functions

The string cut functions will remove regions or individual characters from a target string. A selection of the available functions in various functional categories are described below. All the functions return ajTrue if the operation was performed successfully or ajFalse otherwise.

6.5.11.1 Simple cut functions

A number of characters can be removed from the start, end or interior of a string using:

```
/* Removes a number of characters from the start of a string. */
AjBool  ajStrCutStart(AjPStr* Pstr, ajuint len);

/* Removes a number of characters from the end of a string. */
AjBool  ajStrCutEnd(AjPStr* Pstr, ajuint len);

/* Removes a region from a string. */
AjBool  ajStrCutRange(AjPStr* Pstr, ajint pos1, ajint pos2);
```

6.5.11.2 Removing characters from a string

Functions to remove characters from a string include:

```
/* Removes non-sequence characters (all but alphabetic characters and asterisk) */
AjBool  ajStrRemoveGap(AjPStr* thys);

/* Removes HTML mark-up from a string. */
AjBool  ajStrRemoveHtml(AjPStr* pthis);

/* Removes last character from a string if it is a newline character. */
AjBool  ajStrRemoveLastNewline(AjPStr* Pstr);

/* Removes all of a given set of characters from a string. */
AjBool  ajStrRemoveSetC(AjPStr* Pstr, const char *txt);

/* Removes all whitespace characters from a string. */
AjBool  ajStrRemoveWhite(AjPStr* Pstr);

/* Removes excess whitespace characters from a string. */
AjBool  ajStrRemoveWhiteExcess(AjPStr* Pstr);

/* Removes excess space characters from a string. */
AjBool  ajStrRemoveWhiteSpaces(AjPStr* Pstr);

/* Removes all characters after the first wildcard character (if found). */
AjBool  ajStrRemoveWild(AjPStr* Pstr);
```

ajStrRemoveWhiteExcess and ajStrRemoveWhiteSpaces both remove the leading/trailing whitespace from a string and replace multiple spaces with a single space. Additionally, ajStrRemoveWhiteSpaces converts tabs to spaces but leaves newline characters unchanged.

6.5.11.3 Retaining characters in a string

Functions are available to remove a region from a string or all characters in a string other than those in a defined set. The character sets can be provided either as a string object (AjPStr) or C-type (char *) string:

```
/* Trim sequence down to a defined range */
AjBool  ajStrKeepRange(AjPStr* Pstr, ajint pos1, ajint pos2);

/* Removes all characters that are not in a given set. */
AjBool  ajStrKeepSetC(AjPStr* Pstr, const char* txt);

/* Removes all characters that are not in a given set. */
AjBool  ajStrKeepSetS(AjPStr* Pstr, const AjPStr str);

/* Removes all characters that are not alphabetic.
AjBool  ajStrKeepSetAlpha(AjPStr* Pstr);

/* Removes all characters that are not alphabetic and are not in a given set. */
AjBool  ajStrKeepSetAlphaC(AjPStr* Pstr, const char* txt);
```

6.5.11.4 String trimming functions

The string trim functions below will remove region(s) of a given character composition (provided in the string `txt`) from the start and/or end of a string:

```
/* Remove from start of a string */
AjBool  ajStrTrimStartC (AjPStr* Pstr, const char* txt);

/* Remove from end of a string */
AjBool  ajStrTrimEndC (AjPStr* Pstr, const char* txt);

/* Remove from start and end of a string */
AjBool  ajStrTrimC (AjPStr* pthis, const char* txt);
```

All characters will be removed from the start and/or end up to the first character that is not in the set provided.

Similar functions are provided to remove regions composed of whitespace characters only from the start and end of a string.

```
/* Remove from start and end of a string. */
AjBool  ajStrTrimWhite (AjPStr* Pstr);

/* Remove from start of a string. */
AjBool  ajStrTrimWhiteStart (AjPStr* Pstr);

/* Remove from end of a string. */
AjBool  ajStrTrimWhiteEnd (AjPStr* Pstr);
```

There are also two truncate functions which remove characters from the end of a string reducing it to a defined length (`ajStrTruncateLen`) or cut the end off a string at a defined position (`ajStrTruncatePos`):

```
/* Remove characters from end, reduce to defined length */
AjBool ajStrTruncateLen (AjPStr* Pstr, ajuint len);

/* Cut end of defined position */
AjBool ajStrTruncatePos (AjPStr* Pstr, ajint pos);
```

6.5.12 String substitution functions

The string substitution functions will perform substitutions of characters or substrings of a string with other characters/substrings.

Functions with the prefix `ajStrExchange` will replace all occurrences in a string of one substring (or character) with another string (or character). Variants of the function support string objects (`AjPStr`) and C-type (`char *`) strings for the target and replacement substrings:

```
/* C-type string target and replacement.                 */
AjBool  ajStrExchangeCC (AjPStr* Pstr, const char* txt, const char* txtnew);

/* C-type string target, string replacement  */
AjBool  ajStrExchangeCS (AjPStr* Pstr, const char* txt, const AjPStr strnew);

/* Character target and replacement           */
AjBool  ajStrExchangeKK (AjPStr* Pstr, char chr, char chrnew);

/* String target, C-type string replacement  */
AjBool  ajStrExchangeSC (AjPStr* Pstr, const AjPStr str, const char* txtnew);

/* String target and replacement              */
AjBool  ajStrExchangeSS (AjPStr* Pstr, const AjPStr str, const AjPStr strnew);
```

Functions with the prefix `ajStrExchangeSet` are similar except that they replace all occurrences in a string of one set of characters with another character or set of characters. Variants of the function use string objects (`AjPStr`) and C-type (`char *`) strings to define the sets:

```
/* C-type string target and replacement sets         */
AjBool  ajStrExchangeSetCC (AjPStr* Pstr, const char* txt,const char* newc);

/* String target and replacement sets         */
AjBool  ajStrExchangeSetSS (AjPStr* Pstr, const AjPStr str,const AjPStr strnew);

/* Replace C-type target with single character */
AjBool  ajStrExchangeSetRestCK (AjPStr* Pstr, const char* txt, char chr);

/* Replace string target with single character */
AjBool  ajStrExchangeSetRestSK (AjPStr* Pstr, const AjPStr str, char chr);
```

6.5.13 String query functions

The string query functions test the properties of a string. All functions with the prefix `ajStrIs` return `ajTrue` if some basic test of a string is satisfied. The following functions illustrate the scope of the query tests that can be performed but you should see the online documentation for a full list:

```
/* Alphanumeric characters only. */
AjBool  ajStrIsAlnum (const AjPStr str);
```

```
/* Alphabetic characters only. */
AjBool  ajStrIsAlpha (const AjPStr str);

/* Represents Boolean value. */
AjBool  ajStrIsBool (const AjPStr str);

/* Represents integer value. */
AjBool  ajStrIsInt (const AjPStr str);

/* Represents float value. */
AjBool  ajStrIsFloat (const AjPStr str);

/* No uppercase alphabetic characters. */
AjBool  ajStrIsLower (const AjPStr str);

/* Decimal digits only. */
AjBool  ajStrIsNum (const AjPStr str);

/* Uppercase alphabetic characters only. */
AjBool  ajStrIsUpper (const AjPStr str);
```

6.5.14 String properties and character retrieval functions

For convenience, macros are provided to retrieve the properties of a string including its length, the C-type (`char *`) string, the usage count and the current reserved size. These functions all return an element of the string C-data structure:

```
#define MAJSTRGETLEN(str) str->Len /* String length          */
#define MAJSTRGETPTR(str) str->Ptr /* String char * pointer  */
#define MAJSTRGETRES(str) str->Res /* Reserved length        */
#define MAJSTRGETUSE(str) str->Use /* Usage count            */
```

Functions are available to return individual characters from a string.

```
/* Get first character */
char  ajStrGetCharFirst (const AjPStr str);

/* Get last character */
char  ajStrGetCharLast (const AjPStr str);

/* Get character from specified position */
char  ajStrGetCharPos (const AjPStr str, ajint pos);
```

6.5.15 String conversion functions

A string may be converted to some other datatype using one of the following functions:

```
AjBool    ajStrToBool (const AjPStr str, AjBool* Pval);     /* To boolean
*/
AjBool    ajStrToDouble (const AjPStr str, double* Pval); /* To double
*/
AjBool    ajStrToFloat (const AjPStr str, float* Pval);     /* To float
```

```
*/
AjBool    ajStrToHex (const AjPStr str, ajint* Pval);       /* To hexadecimal
*/
AjBool    ajStrToInt (const AjPStr str, ajint* Pval);       /* To integer
*/
AjBool    ajStrToLong (const AjPStr thys, ajlong* result); /* To long
*/
AjBool    ajStrToUint (const AjPStr str, ajuint* Pval);     /* To unsigned integer
*/
```

In all cases, the functions return `ajTrue` if the conversion was performed successfully. They take the address of a variable of the appropriate type. For example, to convert a string to an integer value:

```
ajint val = 0;
AjPStr str = NULL;

str = ajStrNewC("10");

if(!ajStrToInt(str, &val))
    ajFatal("This error message will not be printed.");

ajStrDel(&str);
```

Conversely, the C datatypes can be converted to an EMBOSS string using the following:

```
AjBool  ajStrFromBool (AjPStr* Pstr, AjBool val);                      /*
From double                         */
AjBool  ajStrFromDouble (AjPStr* Pstr, double val, ajint precision);     /*
From double                         */
AjBool  ajStrFromDoubleExp (AjPStr* Pstr, double val, ajint precision); /*
From double in exponential form. */
AjBool  ajStrFromFloat (AjPStr* Pstr, float val, ajint precision);       /*
From float                          */
AjBool  ajStrFromInt (AjPStr* Pstr, ajint val);                         /*
From integer                        */
AjBool  ajStrFromLong (AjPStr* Pstr, ajlong val);                       /*
From long                           */
AjBool  ajStrFromUint (AjPStr* Pstr, ajuint val);                       /*
From unsigned integer               */
```

Again, these functions return `ajTrue` if the conversion was performed successfully, and take the address of a string. For example, to convert an integer to a string:

```
ajint val = 0;
AjPStr str = NULL;

str = ajStrNew();
val = 100;

if(!ajStrFromInt(&str, val))
    ajFatal("This error message will not be printed.");

ajStrDel(&str);
```

6.5.16 String formatting functions

Functions to reformat a string have the prefix `ajStrFmt`. For example, a string or region of a string can be converted to upper- or lower-case by using:

```
/* Convert to lower-case         */
AjBool  ajStrFmtLower (AjPStr* Pstr);

/* Convert region to lower-case */
AjBool  ajStrFmtLowerSub (AjPStr* Pstr, ajint pos1, ajint pos2);

/* Convert to upper-case         */
AjBool  ajStrFmtUpper (AjPStr* Pstr);

/* Convert region to upper-case */
AjBool  ajStrFmtUpperSub (AjPStr* Pstr, ajint pos1, ajint pos2);
```

The address of the string to be reformatted is passed and `ajTrue` is returned if the reformatting was successful. You should see the online documentation for other formatting functions.

6.5.17 String comparison functions

EMBOSS provides comprehensive string comparison functions.

Functions with the prefix `ajStrMatch` compare one string with another. The functions perform case-sensitive and case-insensitive comparisons with or without wildcard characters. Variants that take a C-type (`char *`) string as the second argument are available but not shown:

```
/* Simple string to C-type string comparison */
AjBool  ajStrMatchC (const AjPStr thys, const char* txt);

/* Simple string to string comparison */
AjBool  ajStrMatchS (const AjPStr thys, const AjPStr str);

/* Case-insensitive string to string comparison */
AjBool  ajStrMatchCaseS (const AjPStr thys, const AjPStr str);

/* String to string comparison with wildcards */
AjBool  ajStrMatchWildS (const AjPStr thys, const AjPStr wild);

/* Case-insensitive string to string comparison with wildcards */
AjBool  ajStrMatchWildCaseS (const AjPStr thys, const AjPStr wild);
```

The following functions will compare the first two words in a string:

```
/* String to C-type string comparison with wildcards. */
AjBool  ajStrMatchWildWordC (const AjPStr str, const char* text);

/* String to string comparison with wildcards.*/
AjBool  ajStrMatchWildWordS (const AjPStr str, const AjPStr text);

/* Case-insensitive string to C-type string comparison with wildcards.*/
```

```
AjBool  ajStrMatchWildWordCaseC (const AjPStr str, const char* text);

/* Case-insensitive string to string comparison with wildcards.*/
AjBool  ajStrMatchWildWordCaseS (const AjPStr str, const AjPStr text);
```

Functions with the prefix `ajStrPrefix` or the prefix `ajStrSuffix` will compare the start or end of a string to the given prefix or suffix respectively. Variants that take a C-type (`char *`) string as the second argument are available but not shown:

```
/* Prefix comparison */
AjBool  ajStrPrefixS(const AjPStr str, const AjPStr str2);

/* Case-insensitive prefix comparison */
AjBool  ajStrPrefixCaseS (const AjPStr str, const AjPStr pref);

/* Suffix comparison */
AjBool  ajStrSuffixS (const AjPStr thys, const AjPStr suff);

/* Case-insensitive suffix comparison */
AjBool  ajStrSuffixCaseS (const AjPStr str, const AjPStr pref);
```

6.5.18 String search functions

String search functions have the prefix `ajStrFind` and are used to find substrings or characters within strings:

```
/* Find a string */
ajint  ajStrFindS (const AjPStr str, const AjPStr str2);

/* Find a character */
ajint  ajStrFindAnyK(const AjPStr str, char chr);

/* Find any character in a set */
ajint  ajStrFindAnyS (const AjPStr str, const AjPStr str2);

/* Find a string (case-insensitive) */
ajint  ajStrFindCaseS (const AjPStr str, const AjPStr str2);

/* Find any character not in a set */
ajint  ajStrFindRestS (const AjPStr str, const AjPStr str2);

/* Find any character not in a set (case-insensitive) */
ajint  ajStrFindRestCaseS (const AjPStr str, const AjPStr str2);

/* Find last occurence of a string */
ajint  ajStrFindlastS (const AjPStr str, const AjPStr str2);
```

These functions return the position of the start of the search text in the sequence, or `-1` if the text was not found.

`ajStrFindAnyS`, `ajStrFindRestS` and `ajStrFindRestCaseS` use a set of characters provided as a string (`str2`).

6.5.19 String parsing functions

Functions for parsing text tokens from strings have the prefix `ajStrExtract` or the prefix `ajStrParse`.

To extract the first word (`Pword`) and the remainder of the string (`Prest`) from an input string (`str`) use either of:

```
/* Remove first word (with no leading spaces) from a string *
AjBool ajStrExtractFirst (const AjPStr str, AjPStr* Prest, AjPStr* Pword);

/* Remove first word from a string, skipping spaces */
AjBool ajStrExtractWord (const AjPStr str, AjPStr* Prest, AjPStr* Pword);
```

`ajStrExtractWord` will skip any leading whitespace whereas `ajStrExtractFirst` will return `ajFalse` if the input string starts with a space. Like most of the string functions they will allocate memory for the strings if necessary although it is cleaner to allocate the strings manually. In the example below, `ajStrExtractFirst` will return `ajFalse` and the printed strings will be empty, whereas `ajStrExtractFirst` will print the first word and the rest of the string:

```
AjPStr inputstring = NULL;
AjPStr word        = NULL;
AjPStr rest        = NULL;

inputstring = ajStrNewC(" First word in this string is 'First'");
word        = ajStrNew();
rest        = ajStrNew();

ajStrExtractFirst(inputstring, &rest, &word);
ajFmtPrint("word: %S\n", word);    /* Empty */
ajFmtPrint("rest: %S\n", rest);    /* Empty */

ajStrExtractWord(inputstring, &rest, &word);
ajFmtPrint("word: %S\n", word);    /* First */
ajFmtPrint("rest: %S\n", rest);    /*  word in this string is 'First' */

ajStrDel(&inputstring);
ajStrDel(&word);
ajStrDel(&rest);
```

There is a function to split a newline-separated multi-line string into an array of strings:

```
ajuint ajStrParseSplit(const AjPStr str, AjPStr **PPstr);
```

The function allocates memory for an array of strings (which must be freed later) and returns the number of array elements created:

```
AjPStr  inputstring = NULL;
AjPStr *array       = NULL;
ajint   dim;
ajint   x;
```

```
inputstring = ajStrNewC("First line\nSecond line\nThird line\n");

dim = ajStrParseSplit(inputstring, &array);

for(x=0; x<dim; x++)
    ajFmtPrint("array[%d]: %S\n", x, array[x]);

ajStrDel(&inputstring);

for(x=0; x<dim; x++)
    ajStrDel(&array[x]);

AJFREE(array);
```

6.5.20 String iteration

String iteration allows you to step through a string a single character at a time. The AJAX datatype for this is:

`AjIStr` String iteration object

To iterate through a string you must first instantiate the string iteration object. Two constructors are provided for forward (start to end) or reverse (end to start) iteration:

```
/* Constructor for forward iteration */
AjIStr ajStrIterNew (const AjPStr thys);

/* Constructor for reverse iteration */
AjIStr ajStrIterNewBack (const AjPStr thys);
```

To iterate through a string use either of the following functions. They return `NULL` if iteration cannot continue:

```
/* Forward iteration */
AjIStr ajStrIterNext (AjIStr iter);

/* Reverse iteration */
AjIStr ajStrIterNextBack (AjIStr iter);
```

To retrieve the character or the remainder of the string at the current position use:

```
/* Retrieve the character */
char ajStrIterGetK (const AjIStr iter);

/* Retrieve the remainder of the string */
const char* ajStrIterGetC (const AjIStr iter);
```

To change the character at the current position use:

```
/* Change character */
void ajStrIterPutK (AjIStr iter, char chr);
```

The following functions return `ajTrue` if iteration can continue and are also used to control iteration:

```
/* Test if iteration can continue */
AjBool ajStrIterDone (const AjIStr iter);

/* Test if reverse iteration can continue */
AjBool ajStrIterDoneBack (const AjIStr iter);
```

A string iterator can be reset so that it points to the start or end of the string:

```
/* Reset forward iteration to start of string */
void ajStrIterBegin (AjIStr iter);

/* Reset reverse iteration to end of string */
void ajStrIterEnd(AjIStr iter);
```

Once you are done, you must free the string iteration object:

```
/* Destructor for iteration object */
void ajStrIterDel (AjIStr *iter);
```

The example code below uses the iteration functions to iterate through a string, replacing all dash characters with full stops and writing a new string:

```
AjPStr   str  = NULL;
AjPStr   seq  = NULL;
AjIStr   iter = NULL;
char     chr;

str = ajStrNewC("--AALIY---TIWLASL--");
seq = ajStrNew();

iter = ajStrIterNew(str);

while(ajStrIterNext(iter))
[
     if((chr = ajStrIterGetK(iter)) == '-')
         ajStrIterPutK(iter, '.');
     else
         ajStrAppendK(&seq, chr);
]

ajFmtPrint("str: %S\n", str);
ajFmtPrint("seq: %S\n", seq);

ajStrDel(&str);
ajStrDel(&seq);
ajStrIterDel(&iter);
```

6.5.21 String tokenisation

There is a dedicated AJAX datatype for string tokenisation:

`AjPStrTok`	String tokenisation object

String tokenisation functions have the prefix `ajStrToken` and are used to delimit a string into text tokens and extract them.

To tokenise a string you must first instantiate the string tokenisation object. The constructors provided take the delimiter characters either as a C-type (`char *`) or as an EMBOSS string (`AjPStr`) and return a pointer to a string tokenisation object:

```
/* Create string tokenisation object */
AjPStrTok ajStrTokenNewC (const AjPStr str, const char* txtdelim);
AjPStrTok ajStrTokenNewS (const AjPStr str, const AjPStr strdelim);
```

The sequence to be tokenised and a delimiter string can also be set for an existing string tokenisation object:

```
/* Specify string to be tokenised only */
AjBool ajStrTokenAssign (AjPStrTok* Ptoken, const AjPStr str);

/* Specify string and delimiters */
AjBool ajStrTokenAssignC (AjPStrTok* Ptoken, const AjPStr str, const char* txtdelim);

/* Specify string and delimiters */
AjBool ajStrTokenAssignS (AjPStrTok* Ptoken, const AjPStr str, const AjPStr strdelim);
```

These functions will allocate the string tokenisation object if necessary and can therefore be used as an alternative to the constructor function. It is however much clearer if they are only used to update an object that has been created using the standard constructors.

To parse individual tokens from a string call:

```
/* Parse tokens from a string */
AjBool ajStrTokenNextFind (AjPStrTok* Ptoken, AjPStr* Pstr);
AjBool ajStrTokenNextFindC (AjPStrTok* Ptoken, const char* strdelim, AjPStr* Pstr);
```

`ajStrTokenNextFindC` will update the string tokenisation object with the string of delimiters (`strdelim`).

To return the remainder of a string that's been partially parsed call:

```
/* Get a string remainder */
AjBool ajStrTokenRestParse (AjPStrTok* Ptoken, AjPStr* Pstr);
```

If you want the delimiter to be treated as a string rather than individual characters, in other words to tokenise the string using another string, use the following to parse individual tokens from the string:

```
/* Tokenise a string using another string */
AjBool ajStrTokenNextParse (AjPStrTok* Ptoken, AjPStr* Pstr);
```

```
AjBool ajStrTokenNextParseC (AjPStrTok* Ptoken, const char* txtdelim, AjPStr* Pstr);
AjBool ajStrTokenNextParseS (AjPStrTok* Ptoken, const AjPStr strdelim, AjPStr* Pstr);
```

These functions return `ajTrue` if the token was parsed successfully or `ajFalse` otherwise. `ajStrTokenNextParseC` and `ajStrTokenNextParseS` will update the string tokenisation object with the string of delimiters (`strdelim` or `txtdelim`) provided. Note that these functions can return `ajTrue` but write an empty token (`Pstr`) in cases where the delimeter has been changed since the previous call.

The string tokenisation object can be reset (all strings cleared) so that it is ready for reuse by calling:

```
void ajStrTokenReset (AjPStrTok* Ptoken);
```

Once you are done you must free the string tokenisation object:

```
/* String tokenisation object destructor */
void ajStrTokenDel  (AjPStrTok* Ptoken);
```

The example code below uses the string tokenisation object and its functions to retrieve individual lines from a string that contains multiple newline characters:

```
AjPStr inputstring    = NULL;
AjPStr token          = NULL;
AjPStrTok tokens      = NULL;

inputstring = ajStrNewC("First line\nSecond line\nThird line\n");
tokens = ajStrTokenNewC(inputstring, "\n");

while(ajStrTokenNextFind(&tokens, &token))
     ajFmtPrint("token: %S\n", token);

ajStrTokenDel(&tokens);
ajStrDel(&inputstring);
```

A string (`str` or `txt`) may be tokenised with either whitespace characters or a specified set of delimiters given as a text string (`txtdelim`):

```
/* Tokenise C-type string by set of delimiters */
AjPStr ajCharParseC (const char* txt, const char* delim);

/* Tokenise by set of delimiters */
const AjPStr ajStrParseC (const AjPStr str, const char* txtdelim);

/* Tokenise by whitespace */
const AjPStr ajStrParseWhite (const AjPStr str);
```

These functions usc the C `strtok` function and return tokens from the string. The first time the function is called it is passed the string to be parsed. For subsequent calls on the same string it is passed `NULL` as the first argument. A pointer to the token is returned or `NULL` when all tokens have been parsed:

```
AjPStr inputstring = NULL;
AjPStr token       = NULL;

inputstring = ajStrNewC(" First word in this string is 'First'");

token = ajStrParseWhite(inputstring)

/* Prints 'First' to the screen */
ajFmtPrint("token: %S\n", token);

/* Prints the rest of the words, one word at a time */
while(token = ajStrParseWhite(NULL))
    ajFmtPrint("token: %S\n", token);

ajStrDel(&inputstring);
```

To count the tokens in a string in which the tokens are delimited by either whitespace characters or a specified set of delimiters (`strdelim`) use:

```
/* Count tokens delimited by whitespace */
ajuint ajStrParseCount (const AjPStr line);

/* Count tokens delimited by set of delimiters */
ajuint ajStrParseCountS (const AjPStr line, const AjPStr
strdelim);
```

6.5.22 Handling C-type (`char *`) strings

For convenience, several groups of functions are provided for handling C-type (`char *`) strings. They all have the prefix `ajChar` to distinguish them from the other functions.

6.5.22.1 C-type string constructor and destructor functions

In the same way as for EMBOSS strings, a C-type string can be created from a starting C-type (`char *`) or string object, with or without a reserved size:

```
/* Create a string from a C-type string      */
char*   ajCharNewC (const char* txt);

/* Create a string from a string object      */
char*   ajCharNewS (const AjPStr thys);

/* Create an empty string of reserved size. */
char*   ajCharNewRes(ajuint size);

/* Create a string with reserved size from a C-type string */
char*   ajCharNewResC(const char* txt, ajuint size);

/* Create a string with reserved size from a string object */
char*   ajCharNewResS(const AjPStr str, ajuint size);

/* Create a string from a C-type string with specified length */
char*   ajCharNewResLenC(const char* txt, ajuint size, ajuint len);
```

In all cases a pointer to the allocated memory is returned, which must be freed once you are done with it. To delete a C-type string call:

```
void ajCharDel (char** Ptxt);
```

For example:

```
char *string = NULL;

string = ajCharNewC("This is a text string");

ajCharDel(&string);
```

6.5.22.2 C-type string comparison functions

Most of the string comparison functions (Section 6.5.17, ‘String comparison functions’) are available for C-type strings too:

```
AjBool  ajCharMatchC (const char* txt1, const char* txt2);
AjBool ajCharMatchCaseC (const char* txt1, const char* txt2);
AjBool ajCharMatchWildC (const char* txt1, const char* txt2);
AjBool ajCharMatchWildCaseC (const char* txt1, const char* txt2);
AjBool ajCharMatchWildNextC (const char* txt1, const char* txt2);
AjBool ajCharMatchWildWordC (const char* str, const char* txt);
AjBool ajCharMatchWildNextCaseC (const char* txt1, const char* txt2);
AjBool ajCharMatchWildWordCaseC (const char* str, const char* txt);
AjBool ajCharPrefixC (const char* txt, const char* pref);
AjBool ajCharPrefixCaseC (const char* txt, const char* pref);
AjBool ajCharSuffixC (const char* txt, const char* suff);
AjBool ajCharSuffixCaseC (const char* txt, const char* suff);
```

Variants of these functions that take a string object as the second argument are available (not shown) and have the suffix `S` rather than `C`.

6.5.23 Handling string formatting

Functions for formatting and printing a string are defined in the library file `ajfmt.h/c`. Conversion characters are defined for all the EMBOSS fundamental datatypes (Section 5.1, ‘Basic datatypes’) and are an extension of the basic C conversion codes. They are:

- An AJAX string object `AjPStr` is printed by `%S`
- A C-type (`char*`) string is printed by `%s`. If the pointer is `NULL` then `<null>` is printed
- The filename of an AJAX file object `AjPFile` is printed by `%F`
- A boolean variable `AjBool` is printed by `%b` (for output as “`Y/N`”) or `%B` (“`Yes/No`”)
- The date and time (`AjPDate`) is printed by `%D`.

Functions with the prefix `ajFmtScan` are equivalent to the C `*scan*` functions:

```
ajint ajFmtScanS  (const AjPStr str, const char* fmt, ...);
ajint ajFmtScanC (const char* txt, const char* fmt, ...);
ajint ajFmtScanF (AjPFile thys, const char* fmt, ...);
```

Functions with the prefix `ajFmtPrint` are equivalent to the C `*print*` functions:

```
/* format and emit the "..." arguments according to fmt; writes to stdout */
void  ajFmtPrint (const char *fmt, ...);

/* format and emit the "..." arguments according to fmt; writes to a file object */
void  ajFmtPrintF (AjPFile file,
                   const char *fmt, ...);

/* format and emit the "..." arguments according to fmt; writes to C FILE stream */
void  ajFmtPrintFp (FILE *stream,
                    const char *fmt, ...);

/* formats the "..." arguments into a buffer with a maximum size according to fmt */
ajint  ajFmtPrintCL (char *buf, ajint size,
                     const char *fmt, ...);

/* Block and print a string. String is split at given delimiters */
void  ajFmtPrintSplit(AjPFile outf, const AjPStr str,
                      const char *prefix, ajint len,
         const char *delim);

/* Formats the "..." arguments into an AjPStr according to fmt */
AjPStr  ajFmtPrintS (AjPStr *pthis, const char *fmt, ...) ;

/* Formats the "..." arguments and appends to an AjPStr according to fmt */
AjPStr  ajFmtPrintAppS (AjPStr *pthis, const char *fmt, ...) ;
```

6.6 Handling sequence patterns

6.6.1 Introduction

Regular expression and other patterns are used to identify motifs in molecular sequences.

6.6.2 AJAX library files

AJAX library files for handling patterns are listed in the table below. Library file documentation, including a complete description of datatypes and functions with usage notes is available at: http://emboss.open-bio.org/rel/dev/libs/

Library file documentation	Description
`ajreg`	Regular expression handling
`ajpat`	Pattern handling

`ajreg.h/c`. Defines the regular expression object (`AjPRegexp`) and functions for handling regular expressions.

ajpat.h/c. Defines the sequence pattern list object (`AjPPatlistSeq`) and general pattern list object (`AjPPatlistRegex`) and functions for handling lists of regular expression patterns. They contain static data structures and functions for handling sequence patterns at a low level.

You are unlikely to need the static data structures and functions unless you plan to implement code to extend the functionality of the libraries themselves.

In addition to the above library files, EMBOSS includes a library of functions to support regular expressions whose syntax and semantics are as close as possible to those of the Perl 5 language. The library files are:

```
pcre_config.h
pcre_internal.h
pcre.h
pcre.c
pcreposix.h
pcreposix.c
pcre_printint.c
pcre_chartables.c
pcre_get.c
pcre_study.c
```

They are not described here and you should see the online library documentation for further information.

6.6.3 ACD datatypes

There are two datatypes for handling pattern input:

`regexp`	A regular expression pattern
`pattern`	A sequence pattern

6.6.4 ACD data definition

Typical ACD definitions are shown below.

6.6.4.1 `regexp`

Regular expression input:

```
regexp: patterns
[
    information: "Regular expression patterns"
    upper: "Y"
    minlength: "3"
    minlength: "10"
]
```

6.6.4.2 `pattern`

Pattern input:

```
pattern: seqpatterns
[
    information: "Sequence patterns"
    type: "nucleotide"
    pmismatch: "0"
]
```

6.6.4.3 Parameter name

All data definitions for pattern or regular expression input should have the standard parameter name `pattern` (see).

6.6.4.4 Common attributes

Attributes that are typically specified are summarised below. These include various datatype-specific attributes (Section A.5, 'Datatype-specific attributes').

`minlength:` Specifies the minimum length of a regular expression. This is used to ensure an expression has been defined.

`maxlength:` Specifies the maximum length of a regular expression.

`maxsize:` Specifies the maximum number of patterns that will be read from a file.

`upper:` Sets the case of a regular expression to upper-case. `lower:` is also available (only one should be given).

`type:` Defines a regular expression or patern to be `nucleotide` or `protein` which alows pattern matching to use sequence ambuguity codes. By default the type is `string` which uses exact character matching.

`pmismatch:` Sets the number of mismatches allowed in a pattern. This is not allowed by regular expression algorithms so is only available for sequence patterns.

`pname:` Sets a pattern name to be used in output. Multiple input patterns have numbers appended to this name.

`pformat:` Defines the pattern or regular expression input format. The default is to use the string as the pattern, or (as for sequence input) to read a string in the form `@filename` as a file of patterns. This file, if the format is defined as 'fasta', has a sequence FASTA-style identifier line with an optional `mismatch=nn` term to set the number of mismatches for each pattern (mismatches are not applicable to regular expressions). The default format is simply one sequence per line with a name automatically defined using the `-pname` qualifier.

6.6.5 AJAX datatypes

The basic AJAX datatype for handling patterns is:

`AjPRegexp`	Regular expression

Two AJAX datatypes are provided for handling lists of patterns including input patterns defined in the ACD file. These are lists of individual patterns and regular expressions using the datatypes below.:

`AjPPatlistSeq`	Sequence pattern list object holding a list of sequence patterns and associated information (for `pattern` ACD datatype)
`AjPPatlistRegex`	General pattern list object holding a list of general patterns and associated information (for `regexp` ACD datatype)

The following AJAX datatypes are provided for handling regular expressions at a low level beyond that provided by the static data structures and functions. You are unlikely to need to use these directly:

`AjPPatComp`	All required data for compiling and searching
`AjPPatternSeq`	Definition of feature pattern
`AjPPatternRegex`	Holds definition of feature pattern

6.6.6 ACD file handling

Datatypes and functions for handling pattern input via the ACD file are shown below.

Title	To read a regular expression	To read a sequence pattern
ACD datatype	`regexp`	`pattern`
AJAX datatype	`AjPPatlistRegex`	`AjPPatlistSeq`
To retrieve from ACD	`ajAcdGetRegexp`	`ajAcdGetPattern`

Your application code will call `embInit` to process the ACD file and command line (see Section 6.3, 'Handling ACD files'). All values from the ACD file are read into memory. You have a handle on memory through the `ajAcdGet*` family of functions which return pointers to appropriate objects.

6.6.6.1 Input pattern retrieval

To retrieve an input pattern an object pointer is declared and then initialised using the appropriate `ajAcdGet*` function.

6.6.6.1.1 `regexp`

```
AjPPatlistRegex patterns=NULL;
patterns = ajAcdGetRegexp("patterns");
```

6.6.6.1.2 `pattern`

```
AjPPatlistSeq seqpatterns=NULL;
seqpatterns = ajAcdGetPattern("seqpatterns");
```

6.6.6.2 Alternative ACD retrieval functions

In cases where just a single regular expression is required `ajAcdGetRegexpSingle` is used to return a single `AjPRegexp` object:

```
AjPRegexp ajAcdGetRegexpSingle (const char *token);
```

It is used as follows:

```
AjPRegexp regexp=NULL;
regexp = ajAcdGetRegexpSingle("patterns");
```

The memory for any other patterns or regular expressions is automatically cleared by a final call to `embExit`.

6.6.6.3 Processing command line options and ACD attributes

Currently there are no functions for this.

6.6.6.4 Memory management

It is your responsibility to close any files and free up memory at the end of the program. You must call the default destructor function (see below) on any objects returned by calls to `ajAcdGetRegexp` or `ajAcdGetPattern`.

6.6.7 Pattern object memory management

6.6.7.1 Default object construction

To use a pattern object that is not defined in the ACD file you must first instantiate the appropriate object pointer. The default constructor functions are:

```
/* Create a new regular expression. */
AjPPatlistRegex ajPatlistRegexNew (void);

/* Create a new sequence pattern    */
AjPPatlistSeq   ajPatlistSeqNew (void);
```

```
/* Create a new prosite pattern object */
AjPPatComp ajPatCompNew (void);
```

All constructors return the address of a new object. The pointers do not need to be initialised to NULL but it is good practice to do so:

```
AjPPatlistRegex    expression = NULL;
AjPPatlistSeq      pattern = NULL;

expression = ajPatlistRegexNew();
pattern    = ajPatlistSeqNew();
/* The objects are instantiated and ready for use */
```

6.6.7.2 Default object destruction

You must free the memory for an object before the pointer is reused and also once you are finished with it. The default destructor functions are:

```
/* Delete a regular expression.  */
void ajPatlistRegexDel (AjPPatlistRegex* pthys);

/* Delete a pattern list object */
void ajPatlistSeqDel (AjPPatlistSeq* pthys);
```

```
/* Delete a pattern object */
void ajPatternSeqDel (AjPPatternSeq* pthys);

/* Delete a prosite pattern object */
void ajPatCompDel (AjPPatComp* pthys);
```

They are used as follows:

```
AjPPatlistSeq plist = NULL;
 AjPPatlistRegex rlist = NULL;
 AjPRegexp patexp = NULL;
 AjPSeqall seqall = NULL;
 AjPSeq seq = NULL;

 plist = ajAcdGetPattern("pattern");
 rlist = ajAcdGetRegexp("regexp");
 seqall = ajAcdGetSeqall("sequence");
 while(ajSeqallNext(seqall, &seq))
 /* Do something */

 ajPatlistSeqDel(&plist);
 ajPatlistRegexDel(&rlist);
```

6.6.7.3 Alternative object construction and loading

These functions are used internally in processing lists of patterns and regular expressions. They are available for developers to use when working with lists:

```
/* Constructor for a sequence pattern list object
 (adds a pattern to a supplied list). */
AjPPatternSeq ajPatternSeqNewList (AjPPatlistSeq plist, const AjPStr name,

                                   const AjPStr pat, ajuint mismatch);
AjPPatternRegex ajPatternRegexNewList (AjPPatlistRegex plist,
                                       const AjPStr name,
                                       const AjPStr pat);
/* Constructor for a sequence pattern list object (parameter is true for a protein
pattern) */
AjPPatlistSeq ajPatlistSeqNewType (AjBool Protein);

/* Constructor for a pattern list object with a specified type */
AjPPatlistRegex ajPatlistRegexNewType (ajuint type);

/* Removes current pattern from pattern list. */
void ajPatlistSeqRemoveCurrent (AjPPatlistSeq thys);
void ajPatlistRegexRemoveCurrent (AjPPatlistRegex thys);

/* Adds pattern into patternlist */
void ajPatlistAddRegex (AjPPatlistRegex thys, AjPPatternRegex pat);
void ajPatlistAddSeq (AjPPatlistSeq thys, AjPPatternSeq pat);
```

6.6.8 Read functions

These functions are used in ACD processing to convert a pattern qualifier value (a pattern or a filename reference) into a pattern or regular expression list. They can be used to define patterns from strings or files:

```
/* Parses a file of regular expressions into a pattern list object. */
AjPPatlistRegex ajPatlistRegexRead(const AjPStr patspec,
                                   const AjPStr patname,
                                   const AjPStr fmt,
                                   ajuint type,
                                   AjBool upper, AjBool lower);

/* Parses a file into pattern list object. */
AjPPatlistSeq ajPatlistSeqRead(const AjPStr patspec,
                               const AjPStr patname,
                               const AjPStr fmt,
                               AjBool protein,
                               ajuint mismatches);

/* Resets the pattern list iteration. */
void ajPatlistSeqRewind (AjPPatlistSeq thys);
void ajPatlistRegexRewind (AjPPatlistRegex thys);
```

6.6.9 Getting elements of objects

Applications work at the level of lists of patterns and regular expressions. For these, functions are available to return the number of items in the list, and to return the next item.

```
/* Gets next available pattern from list. */
AjBool ajPatlistSeqGetNext (AjPPatlistSeq thys,
                            AjPPatternSeq* pattern);

/* Gets number of patterns from list. */
ajuint ajPatlistSeqGetSize (const AjPPatlistSeq plist);

/* Gets number of patterns from list. */
ajuint ajPatlistRegexGetSize (const AjPPatlistRegex plist);

/* Gets next available pattern from list. */
AjBool ajPatlistRegexGetNext (AjPPatlistRegex thys,
                              AjPPatternRegex* pattern);
```

There are further functions to return elements of each item retrieved from a list:

```
/* Returns the name of the pattern. */
const AjPStr ajPatternSeqGetName (const AjPPatternSeq thys);

/* Returns pattern in string format. */
const AjPStr ajPatternSeqGetPattern (const AjPPatternSeq thys);

/* Returns void pointer to compiled pattern. */
AjPPatComp ajPatternSeqGetCompiled (const AjPPatternSeq thys);

/* Returns true if the pattern is for a protein sequence. */
AjBool ajPatternSeqGetProtein (const AjPPatternSeq thys);

/* Returns the mismatch of the pattern. */
ajuint ajPatternSeqGetMismatch (const AjPPatternSeq thys);

/* Returns the name of the pattern */
const AjPStr ajPatternRegexGetName (const AjPPatternRegex thys);

/* Returns pattern in string format. */
const AjPStr ajPatternRegexGetPattern (const AjPPatternRegex thys);

/* Returns void pointer to compiled pattern. */
AjPRegexp ajPatternRegexGetCompiled (const AjPPatternRegex thys);

/* Returns the type of the pattern. */
ajuint ajPatternRegexGetType (const AjPPatternRegex thys);
```

6.6.10 Setting elements of objects

There are no elements that should be set within the pattern list objects.

6.6.11 Debugging functions

These functions report the internals of pattern objects to the debug fiule when **-debug** is set on the command line.

```
void ajPatternSeqDebug (const AjPPatternSeq pat);
void ajPatternRegexDebug (const AjPPatternRegex pat);
```

6.6.12 Miscellaneous functions

These functions copy a list of patterns to a string for use in debugging:

```
ajuint ajPatlistRegexDoc (AjPPatlistRegex thys, AjPStr* pdoc);
ajuint ajPatlistSeqDoc (AjPPatlistSeq thys, AjPStr* pdoc);
```

There is also a function to return the type associated with a named type of regular expression:

```
ajuint ajPatternRegexType (const AjPStr type);
```

6.7 Handling sequences

6.7.1 Introduction

There are a wide variety of sequence formats currently in use. They define the permitted layout and content of text in a file, including required text tokens that are used to define the fields in the databank. These fields include:

- The sequence identifier name
- Accession number
- Bibliographic information
- Sequence annotation including a description of any features
- The sequence itself
- Molecular features associated with the sequence, such as protein secondary structure or molecular recognition sites, are kept in a feature table.

The bibliographic information includes data such as:

- Literature references
- Experimental details
- Author contact information
- Cross-links to other databases

- Descriptions, annotations and comments around the sequence
- Much more besides!

An entry in a sequence databank will usually include a code to uniquely identify an entry in the database. The code is often an *accession number* and/or *identifier name (ID)*. An accession number is a unique alphanumeric identifier guaranteed to stay with that sequence for the life of the database.

Note

If two sequences are merged into one then the new sequence will get a new accession number; the accession numbers of the merged sequences will be retained as 'secondary' accession numbers.

EMBL, GenBank and SwissProt share an accession numbering scheme, so a given accession number uniquely identifies a sequence within these three databases. ID names, in contrast, are not guaranteed to remain the same between different versions of a database, although in practice they usually do.

EMBOSS recognises nearly all of the known formats for input, and most can be used for output too. Sequences are referred to on the EMBOSS command line by their Uniform Sequence Address (USA) (see the *EMBOSS User's Guide*). This is a standard sequence naming scheme used by all EMBOSS applications. A USA specifies one or more sequences that might be read from or written to a file or a larger databank. Other sequence sources such as applications or a web server can also be specified.

There is also a set of command line qualifiers that are used for sequence input and output. They allow you to set such things as file format, sequence regions, database and entry names, etc. Such qualifiers (Section A.5, 'Datatype-specific attributes') may also be hard-coded as attributes in the ACD file.

As you would expect, AJAX includes very well-developed functionality for handling sequences. This includes:

- Recovering or setting elements in the sequence objects.
- Testing sequence properties, for instance whether a sequence is protein or nucleotide.
- Calculating sequence properties, for example base composition of a nucleotide sequence or the molecular weight of a protein sequence.
- Interconversion of sequences and strings.
- Processing sequences, for instance trimming it to defined regions or removing gap characters.
- Sequence type validation; whether a sequence is protein or nucleic acid, contains gaps and so on. This is usually handled at the level of the ACD file but can be done directly.
- Conversion of sequence to numerical form for convenient processing, for example as an index for a scoring matrix.
- Translation of a genetic sequence into a protein.

- Interconversion of nucleotide bases and amino acid codes.
- Calculating the properties of nucleotide sequences, for example DNA melting and annealing temperatures and energies.

Application input and output sequences are typically defined in the ACD file and retrieved by appropriate calls to `ajAcdGet*` functions. There are ACD datatypes for input of single sequences, sequential access to multiple sequences or multiple sequences in one or more sets. There are corresponding datatypes for sequence output. Sequence objects can also be created directly if that's required. Sequences are held internally as strings and required memory is dynamically reallocated as needed. This means that there are no arbitrary limits when using a function with a sequence object.

6.7.2 AJAX library files

AJAX library files for handling sequences are listed in the table below. Library file documentation, including a complete description of datatypes and functions, is available at: http://emboss.open-bio.org/rel/dev/libs/

Library file documentation	Description
`ajseq`	General sequence handling
`ajseqdata`	Sequence datatypes
`ajseqread`	Sequence reading
`ajseqwrite`	Sequence writing
`ajseqdb`	Sequence database access methods
`ajseqabi`	Sequence parsing from ABI trace file
`ajseqtype`	Validation and processing for sequences types
`ajtranslate`	Sequence translation
`ajbase`	Base conversion functions
`ajdan`	Nucleic acid functions

`ajseq.h/c`. Most of the functions you will ever need for general sequence handling. They define the `AjPSeqCvt` object used for sequence conversion (Section 6.7.14, 'Sequence conversion').

`ajseqdata.h/c`. Define the basic sequence objects (`AjPSeq`, `AjPSeqset` and `AjPSeqall`) for general use, e.g. retrieving sequences via ACD file processing. They also define a sequence input object (`AjPSeqin`) and a sequence query object (`AjPSeqQuery`) used for low-level sequence retrieval.

`ajseqread.h/c`. Functions for handling the sequence query and sequence input objects and a few general sequence handling functions. They also define static data structures and functions for handling sequence input from the supported formats at a low level.

ajseqwrite.h/c. Defines the sequence output object (`AjPSeqout`) and include functions for writing sequences to file. They define static data structures and functions for handling sequence output, in the supported formats, at a low level.

ajseqdb.h/c. Defines a sequence database reading object (`SeqSAccess`) used in the definition of `AjPSeqin` and static data structures and functions for handling all aspects of sequence database access.

ajseqabi.h/c. Provides functions for parsing sequences (and other data) from application binary interface (ABI) trace files.

ajseqtype.h/c. Functions for handling of sequence types. They also contain static data structures and functions for validation and processing of sequence types at a low level.

ajtranslate.h/c. Defines a sequence translation object (`AjPTrn`) and include functions for handling sequence translation (Section 6.8, 'Handling sequence translation').

ajbase.h/c. Contains functions for character conversion of nucleotide bases and other character conversion functions for use with molecular sequences. For further information see Section 6.7.16, 'Handling IUB base codes'.

ajdan.h/c. Contains functions for calculating properties of nucleotide sequences, mostly related to DNA melting/annealing. For further information see Section 6.7.17, 'DNA calculations'.

Note

You are unlikely to need the static data structures and functions in the above files unless you plan to implement code to support new sequence formats for EMBOSS. For advice on how to do this ask the EMBOSS developers. Detailed usage notes for **ajseqdata.h**, **ajseqread.h/c** and **ajseqwrite.h/c** are not given here. For advice on how to use the datatypes and functions ask the EMBOSS developers.

6.7.3 ACD datatypes

There are four ACD datatypes and corresponding methods for handling sequence input. Often just a single sequence is required. Where multiple sequences are needed they may be read either sequentially, for example from a database, or read and held in memory as a set, for example from a sequence alignment:

`sequence`	Read a single sequence
`seqall`	Read multiple sequences sequentially
`seqset`	Read multiple sequences as a single set
`seqsetall`	Read multiple sequences as multiple sets

There are three datatypes for handling sequence output to match their input equivalents. They have identical behaviour:

`seqout`	Write a single sequence
`seqoutset`	Write multiple sequences as a single set
`seqoutall`	Write multiple sequences sequentially, one at a time

6.7.4 ACD data definition

Typical ACD definitions for sequence input:

```
# single input sequence
sequence: sequence
[
    parameter: "Y"
    type:      "gapany"
]

# multiple sequence input read sequentially
seqall: sequence
[
    parameter: "Y"
    type:      "gapany"
]

# multiple sequence input read as a single set
seqset: sequence
[
    parameter: "Y"
    type:      "gapany"
    aligned:   "N"
]

# multiple sequence input read as multiple sets
seqsetall: sequence
[
    parameter: "Y"
    type:      "gapany"
    aligned:   "N"
]
```

Typical ACD definitions for sequence output:

```
# single sequence
seqout: outseq
[
    parameter: "Y"
    type:      "gapany"
]

# multiple sequence output (sequential writing)
seqoutall: outseq
[
    parameter: "Y"
    type:      "gapany"
```

```
]

# multiple sequence output (written as a set)
seqoutset: outseq
[
    parameter: "Y"
    type:      "gapany"
]
```

6.7.4.1 Parameter name

All data definitions for sequence input and output should have standard parameter names:

- `sequence` for any sequence inputs
- `outseq` for any outputs
- Alternatives and variations (e.g. `asequence` and `bsequence` for multiple inputs are allowed).

See Appendix A, *ACD syntax reference*.

6.7.4.2 Common attributes

Attributes that are typically specified are summarised below. They are datatype-specific (Section A.5, 'Datatype-specific attributes') unless they are indicated as being global attributes (Section A.4, 'Global attributes').

`parameter:` Sequences are typically the primary input or output of an EMBOSS application and, as such, should be defined as parameters by using the global attribute `parameter: "Y"`.

`type:` Specifies the type of the input or output sequence(s) for validation purposes. See Section A.7, 'Sequence types'.

`aligned:` Any `seqset` or `seqsetall` datatype *must* have the `aligned:` attribute set to indicate whether the sequences are aligned or not.

`features:` Sequence features (Section 6.9, 'Handling features') will (if available) be read or written if the `features:` ACD attribute is set for the input or output sequence(s).

`osformat:` FASTA format is used by default for the output sequence(s). The format is normally set at the command line but a default may be hard-coded with `osformat:`. All common sequence formats are supported (see the *EMBOSS User's Guide*).

6.7.5 AJAX datatypes

There are nine AJAX datatypes for handling sequence. Four of these are for general use including handling sequences defined in the ACD file:

`AjPSeq`	A single sequence (for `sequence` ACD datatype)
`AjPSeqall`	Sequence input stream for sequential access to multiple sequences (for `seqall` ACD datatype)
`AjPSeqset`	Multiple sequences in one or more sets (for `seqset` and `seqsetall` ACD datatypes)
`AjPSeqout`	Sequence output stream (for `seqout` ACD datatype)

Two are for sequence-related operations:

`AjPSeqCvt`	Used for sequence conversion into numerical form
`AjPTrn`	Holds the genetic code specification and information needed to translate a nucleotide sequence and find initiation sites

The rest are for lower-level sequence handling beyond that provided by the static datatypes in the various library files:

`AjPSeqin`	Holds the sequence USA specification and information needed to read the sequence and possible further sequences
`SeqSAccess`	Holds information needed to read a sequence from a database: access methods are defined for each known database type and sequences are read from the database using the defined database access function. It is used in the definition of `AjPSeqin`
`AjPSeqQuery`	Used for interpreting the entry specification part of a USA (see the *EMBOSS User's Guide*).

You are unlikely to need these specialised objects unless you plan to implement code to support new sequence formats for EMBOSS. For advice on how to do this ask the EMBOSS developers.

6.7.6 ACD file handling

Datatypes and functions for handling sequences via the ACD file are shown below.

Title	**To read a single sequence**	**To read a set of sequences (e.g. an alignment)**	**To read sequences sequentially from a database**
Sequence input			
ACD datatype	`sequence`	`seqall`	`seqset`
AJAX datatype	`AjPSeq`	`AjPSeqall`	`AjPSeqset`
To retrieve from ACD	`ajAcdGetSeq`	`ajAcdGetSeqall`	`ajAcdGetSeqset`

To get individual sequences	(ajAcdGetSeq)	ajSeqallNext	ajSeqsetGetSize, ajSeqsetGetseqSeq

Sequence output

	To write a single sequence	To write sequences to a database	To write a set of sequences (e.g. an alignment)
ACD datatype	seqout	seqoutall	seqoutset
AJAX datatype	AjPSeqout	AjPSeqout	AjPSeqout
To open the output stream	ajAcdGetSeqout	ajAcdGetSeqoutall	ajAcdGetSeqoutset
To write sequences to file	ajSeqoutWriteSeq	ajSeqoutWriteSeq	ajSeqoutWriteSet

Your application code will call `embInit` to process the ACD file and command line (see Section 6.3, 'Handling ACD files'). All values from the ACD file are read into memory and files are opened as necessary. You have a handle on the files and memory through the `ajAcdGet*` family of functions which return pointers to appropriate objects.

6.7.6.1 Input sequence retrieval

To retrieve an input sequence an object pointer is declared and initialised using the appropriate `ajAcdGet*` function.

6.7.6.1.1 Single input sequence

```
AjPSeq seq = NULL;
seq = ajAcdGetSeq("sequence");
```

6.7.6.1.2 Multiple sequential input sequences

A `seqall` object is really a means to an end of returning individual sequences in a loop. `ajSeqallNext` is used to iterate through the sequences:

```
AjPSeqall seqall = NULL;
 AjPSeq    seq    = NULL;

 seqall = ajAcdGetSeqall("sequence");

 while(ajSeqallNext(seqall, &seq))
 {
      /* Do something with 'seq' */
 }
```

6.7.6.1.3 Multiple input sequences in a set

```
AjPSeqset seqset = NULL;
seqset = ajAcdGetSeqset("sequence");
```

6.7.6.1.4 Multiple input sequence in multiple sets

ajAcdGetSeqsetall is used which returns a pointer for an array of seqset objects:

```
AjPSeqset *seqset = NULL;
seqset = ajAcdGetSeqsetall("sequence");
```

Of course, to access an individual set the array notation is used. Code to achieve the same effect as ajAcdGetSeqsetallSingle is as follows:

```
AjPSeqset *seqset = NULL;
AjPSeqset seqsetFirst=NULL;

seqset = ajAcdGetSeqsetall("sequence");
seqsetFirst = seqset[0]; /* Picks up the pointer to the first sequence set. */
```

6.7.6.2 Output sequence stream retrieval

To retrieve an output sequence stream an object pointer is declared and initialised using the appropriate ajAcdGet* function. Note that multiple sequence sets are written with using ajAcdGetSeqoutset.

6.7.6.2.1 Single output sequence

```
AjPSeqout seqout = NULL;
seqout = ajAcdGetSeqout("outseq");
```

6.7.6.2.2 Multiple output sequences

```
AjPSeqout seqout = NULL;
seqout = ajAcdGetSeqoutall("outseq");
```

6.7.6.2.3 Output sequence set

```
AjPSeqout seqout = NULL;
seqout = ajAcdGetSeqoutset("outseq");
```

6.7.6.2.4 Multiple output sequence sets

```
AjPSeqout seqout = NULL;
seqout = ajAcdGetSeqoutset("outseq");
```

6.7.6.3 Alternative ACD retrieval functions

In cases where just the first of multiple sets is required `ajAcdGetSeqsetallSingle` is used to return a single `seqset` object:

```
AjPSeqset ajAcdGetSeqsetallSingle (const char *token);
```

It is used as follows:

```
AjPSeqset seqset = NULL;
seqset = ajAcdGetSeqsetallSingle("sequence");
```

The memory for any additional sequence sets is automatically cleaned when `embExit` is called.

6.7.6.4 Processing command line options and ACD attributes

6.7.6.4.1 Setting the sequence range

All sequence input datatypes have various inbuilt command line qualifiers (see the *EMBOSS User's Guide*) including **`-sbegin`** and **`-send`** which specify a start and end position for a sequence.

Whenever a sequence is read these values are held in the appropriate sequence object. Regardless of their value you still get the entire sequence loaded into memory. The function `ajSeqTrim` (or `ajSeqsetTrim` for a `AjPSeqset` object) is used to trim the sequence to the region defined by **`-sbegin`** and **`-send`**:

```
AjPSeq seq = NULL;
seq = ajAcdGetSeq("sequence");
ajSeqTrim(seq);
```

6.7.6.4.2 Setting the output format

There is a function for setting the output format although this is typically specified by the user and handled during ACD file processing:

```
AjBool ajSeqoutSetFormatS (AjPSeqout thys, const AjPStr format);
```

6.7.6.5 Writing sequences

There are two AJAX functions for writing out sequence information.

```
void ajSeqoutWriteSeq (AjPSeqout outseq, const AjPSeq seq);
void ajSeqoutWriteSet (AjPSeqout seqout, const AjPSeqset seq);
```

Note

An **AjPSeq** object is used to write sequences from a **seqall** object.

6.7.6.5.1 Writing a sequence

```
AjPSeq seq = NULL;
AjPSeqout seqout = NULL;
seq    = ajAcdGetSeq("sequence");
seqout = ajAcdGetSeqout("outseq");
ajSeqoutWriteSeq(seqout, seq);
```

6.7.6.5.2 Writing sequences sequentially

Typically `ajSeqallNext` is used to iterate through the sequences in loop and pick up a pointer (`seqPtr` in this case) to the next sequence for writing:

```
AjPSeqall seqall = NULL;
AjPSeqout seqout = NULL;
AjPSeq seqPtr = NULL;
seqall = ajAcdGetSeqall("sequence");
seqout = ajAcdGetSeqout("outseq");
while(ajSeqallNext(seqall, &seqPtr))
{
    ajSeqoutWriteSeq(seqout, seqPtr);
}
```

6.7.6.5.3 Writing a sequence set

Here the entire sequence set is written to file:

```
AjPSeqset seqset = NULL;
AjPSeqout seqout = NULL;
seqset = ajAcdGetSeqset("sequence");
seqout = ajAcdGetSeqout("outseq");
ajSeqoutWriteSet(seqout, seqset);
```

6.7.6.5.4 Writing multiple sequence sets

Multiple sequence sets must be written one at a time. Array notation is used to access individual sets:

```
AjPSeqset *seqset = NULL;
AjPSeqout seqout = NULL;
ajint     i = 0;
seqset = ajAcdGetSeqsetall("sequence");
seqout = ajAcdGetSeqout("outseq");
```

```
for(i=0; seqset[i] != NULL; i++)
    ajSeqoutWriteSet(seqout, seqset[i]);
```

6.7.6.6 Memory and file management

It is your responsibility to close any files and free up memory at the end of the program.

6.7.6.6.1 Closing output sequence files

To close an output sequence stream use `ajSeqoutClose`:

```
AjPSeqout seqout = NULL;
seqout = ajAcdGetSeqout("outseq");
...
ajSeqoutClose(seqout);
```

6.7.6.6.2 Freeing memory

You must call the default destructor function (see below) on any sequence objects returned by calls to `ajAcdGet*`.

6.7.7 Sequence object memory management

6.7.7.1 Default object construction

To use a sequence object that is not defined in the ACD file you must first instantiate the appropriate object pointer. Three default construction functions are provided, one for each access method:

```
AjPSeq     ajSeqNew(void);     /* Sequence         */
AjPSeqset ajSeqsetNew (void); /* Sequence set     */
AjPSeqall ajSeqallNew (void); /* Sequence stream */
```

All constructors return the address of a new object. The pointers do not need to be initialised to `NULL` but it is good practice to do so:

```
AjPSeq       seq    =  NULL;
AjPSeqset seqset  =  NULL;
AjPSeqall seqall  =  NULL;

seq    = ajSeqNew();
seqset = ajSeqsetNew();
seqall = ajSeqallNew();

/* The objects are instantiated and ready for use */
```

Sequence output objects are typically loaded from ACD file processing (see above). In the unlikely event you do need to create a sequence output object manually you can use the default `seqout` object constructor:

```
AjPSeqout    seqout = NULL;

 seqout = ajSeqoutNew();

 /* The object is instantiated and ready for use */
```

6.7.7.2 Default object destruction

There is a sequence object destruction function for each access method, and one for arrays of sequence sets:

```
void ajSeqDel (AjPSeq* Pseq);
void ajSeqallDel (AjPSeqall *Pseq);
void ajSeqsetDel (AjPSeqset *Pseq);
void ajSeqsetDelarray (AjPSeqset **PPseq);
```

They are used as follows:

```
 AjPSeq seq = NULL;
 AjPSeqall seqall = NULL;
 AjPSeqset seqset = NULL;
 AjPSeqset *seqsets = NULL;

 seq     = ajAcdGetSeq("asequence");
 seqall  = ajAcdGetSeqall("bsequence");
 seqset  = ajAcdGetSeqset("csequence");
 seqsets = ajAcdGetSeqsetall("dsequence");

 ...
 ajSeqDel(&seq);
 ajSeqallDel(&seqall);
 ajSeqsetDel(&seqset);
 ajSeqsetDelarray(&seqsets);
```

You must free the memory for an object once you are finished with it. Destructor functions are provided:

```
void  ajSeqDel (AjPSeq* Pseq);        /* Sequence        */
void  ajSeqallDel (AjPSeqall *Pseq); /* Sequence stream */
void  ajSeqsetDel (AjPSeqset *Pseq); /* Sequence set    */
```

For convenience, there is also a destructor for arrays of `seqset` objects:

```
/* Destroy an array of sequence set objects and the array. */
void ajSeqsetDelarray (AjPSeqset **PPseq);
```

```
AjPSeq       seq = NULL;
AjPSeqset seqset = NULL;
AjPSeqall seqall = NULL;

seq    = ajSeqNew();
seqset = ajSeqsetNew();
seqall = ajSeqallNew();

/* Do something with the instantiated objects */

ajSeqDel(&seq);
ajSeqsetDel(&seqset);
ajSeqallDel(&seqall);

/* The memory is freed and the pointers reset to NULL, ready for re-use. */

seq    = ajSeqNew();
seqset = ajSeqsetNew();
seqall = ajSeqallNew();

* Do something else with the new objects. The pointer variable is reallocated.*/

ajSeqDel(&seq);
ajSeqsetDel(&seqset);
ajSeqallDel(&seqall);

/* Done with the objects so the memory is freed. */
```

Similarly, for arrays of sequence sets:

```
AjPSeqset *seqset = NULL;
seqset = ajAcdGetSeqsetall("sequence");

/* Do something with the sequence set array */

ajSeqsetDelarray(&seqset);
```

6.7.7.3 Alternative object construction and loading

When managing memory for the sequence objects, knowledge of a function's behaviour is required. There are three cases to discern:

- A function requires a pre-existing object.
- A function can use but does not require a pre-existing object, and will allocate one if necessary.
- A function always allocates an object, and either returns a pointer to it or allocates an object pointer to it, the address of which is passed as an argument.

6.7.7.3.1 Single sequence

There are a variety of alternative ways to create a sequence object. To create a sequence with a reserved size use:

```
AjPSeq ajSeqNewRes (size_t size);
```

A sequence can be created from a string where the name of the sequence is also provided. Other strings are used in order to create the sequence so three objects need to be freed later:

```
AjPSeq seq     = NULL;
AjPStr str     = NULL;
AjPStr strname = NULL;

str     = ajStrNewC("ASEQUENCE");
strname = ajStrNewC("Sequence Name");

seq = ajSeqNewNameS(str, strname);

/* Do something with sequence */

ajStrDel(&str);
ajStrDel(&strname);
ajSeqDel(&seq);
```

Or from a string where the start and end region (given as offsets) are defined. The string is copied into the sequence object and the sequence is trimmed to the region defined:

```
AjPSeq seq = NULL;
AjPStr str   = NULL;
ajint offstart  = 1;
ajint offend    = 8;

str = ajStrNewC("ASEQUENCE");

seq = ajSeqNewRangeS(str, offstart, offend, ajFalse);

ajStrDel(&str);
ajSeqDel(&seq);
```

The `ajSeqNewRangeS` call above will create the subsequence "`SEQUENCE`". The last function argument is a boolean which, if set true, means that the specified offset values are reversed and the sequence is set to be reversed when printed. Had you called the function like this:

```
seq = ajSeqNewRangeS(str, offstart, offend, ajTrue);
```

then the start of the sequence is set to `offend` (8), the end set to `offstart` (1) and the printed (reverse) sequence would be ECNEUQES:

```
seq = ajSeqNewRangeS(str, offstart, offend, 1);
```

A sequence can also be created and initialised from an existing sequence by calling `ajSeqNewSeq`, which copies the original sequence, so two objects need destroying later:

```
AjPSeq seq     = NULL;
AjPSeq seqCopy = NULL;
```

```
seq = ajAcdGetSeq("sequence");

seqCopy = ajSeqNewSeq(seq);
/* Do something with the sequences */

ajSeqDel(&seq);
ajSeqDel(&seqCopy);
```

An array of sequences can be created from an existing sequence set using `ajSeqsetGet-Seqarray`. The array is created and copies of the sequences are taken. This function will be renamed in the future to reflect the fact that it is a constructor function. To ensure that all memory is freed, `ajSeqsetGetSize` is called to get the size of the array, and the array itself is freed by calling `AJFREE`:

```
AjPSeqset seqset  = NULL;
AjPSeq *   seqs   = NULL;
ajint      nseq;   /*Number of sequences in set */
ajint      x;

seqset=ajAcdGetSeqset("sequences");

seqs = ajSeqsetGetSeqarray(seqset);

/* Do something with the array of sequences */

ajSeqsetDel(&seqset);
nseq = ajSeqsetGetSize(seqset);

for(x=0; x<nseq; x++)
    ajSeqDel(&seqs[x]);

AJFREE(seqs);
```

6.7.7.3.2 Sequence set

Three functions are provided for loading a pre-existing `AjPSeqset` object with sequences.

A sequence set can be built from a pair of sequences (in this case from ACD file processing) by using `ajSeqsetFromPair`. This copies the sequences:

```
AjPSeq       seqa = NULL;
AjPSeq       seqb = NULL;
AjPSeqset seqset = NULL;
ajint        nset = 0; /* Number of sequences in set */

seqa = ajAcdGetSeq("asequence");
seqb = ajAcdGetSeq("bsequence");

seqset = ajSeqsetNew();
nset   = ajSeqsetFromPair(seqset, seqa, seqb);

/* Do something with set */

ajSeqDel(&seqa);
ajSeqDel(&seqb);
ajSeqsetDel(&seqset);
```

Or from a list of sequences by using `ajSeqsetFromList`, which in contrast to `ajSeqsetFromPair` loads the set with the actual sequences (not copies). Therefore you have to be careful not to try to free the same objects twice. `ajListFree` is called to delete the list but not the nodes (sequences). The latter were deleted by `ajSeqsetDel`:

```
AjPSeq      seqa = NULL;
AjPSeq      seqb = NULL;
AjPList     list = NULL;
AjPSeqset seqset = NULL;
ajint       nset = 0; /* Number of sequences in set */

seq1 = ajAcdGetSeq("asequence");
seq2 = ajAcdGetSeq("bsequence");

/* Build list of sequences */
list = ajListNew();
ajListPush(list, (void *)seq1);
ajListPush(list, (void *)seq2);

seqset = ajSeqsetNew();
nset   = ajSeqsetFromList(seqset, list);

/* Do something with set */

ajSeqsetDel(&seqset);
ajListFree(&list);
```

You can append to a sequence set using `ajSeqsetApp`, which copies the sequences:

```
AjPSeq      seqa = NULL;
AjPSeq      seqb = NULL;
AjPSeqset seqset = NULL;
ajint       nset = 0; /* Number of sequences in set */

seq1 = ajAcdGetSeq("asequence");
seq2 = ajAcdGetSeq("bsequence");

seqset = ajSeqsetNew();
ajSeqsetApp(seqset, seqa);
ajSeqsetApp(seqset, seqb);

/* Do something with set */

ajSeqDel(&seqa);
ajSeqDel(&seqb);
ajSeqsetDel(&seqset);
```

6.7.7.3.3 Sequence output

An output sequence object can be created from an open file. In this case an output file isn't created during ACD file processing. The file is appropriated by the function so only one destructor needs to be called:

```
AjPFile      file = NULL;
AjPSeqout seqout = NULL;

/* Open a file called "InputFileName" */
file = ajFileNewInNameC ("InputFileName");

/* Create an output sequence object that uses this file */
seqout = ajSeqoutNewFile(file);

/* Do something with the output sequence object */
ajSeqoutDel(&seqout);
```

An existing output sequence object can be loaded (and a file opened for writing) by using `ajSeqoutOpenFilename` which, given a valid filename, returns a boolean which is `ajTrue` if the file was opened OK:

```
AjPSeqout seqout = NULL;
AjPStr name = NULL;

/* Create an empty seqout object */
seqout = ajSeqoutNew();

ajStrAssignC(&name, "InputFileName");

if(!ajSeqoutOpenFilename(seqout, name))
    /* Error handling */;

/* Do something with the output sequence object */

ajSeqoutDel(&seqout);
```

6.7.7.3.4 Loading sequence objects using a USA

Functions are provided to load the sequence input objects with one or more sequences, or initialise the output sequence object, given a USA (see the *EMBOSS User's Guide*) of an input or output sequence(s) or sequence stream.

`Single sequence`

A sequence can be loaded into a sequence object using `ajSeqGetFromUsa`. The second argument is true if the sequence is protein, and the function returns `ajTrue` if the sequence was loaded successfully:

```
AjPSeq seq = NULL;
AjPStr usa = NULL;

usa = ajStrNewC("asis::ALVLLYNITEF"); /* "asis::ALVLLYNITEF" is a valid USA */
seq = ajSeqNew(); /* Create sequence object */

if(!ajSeqGetFromUsa(usa, 1, &seq))
    /* Do something to handle the failure */;

else
   /* Sequence was loaded successfully .. do something with it */;

   ajSeqDel(&seq);
```

Sequence set

A sequence set can be created using `ajSeqsetGetFromUsa` in the same way as `ajSeqGetFromUsa`:

```
AjPSeq seqset = NULL;
AjPStr usa = NULL;
seqset = ajSeqsetNew(); /* Create sequence object */

usa = ajStrNewC("swiss:hba_*"); /*  ""swiss:hba_*"" is a valid USA */

if(!ajSeqsetGetFromUsa(usa, 1, &seqset))
    /* Do something to handle the failure */;
else
   /* Sequence set was loaded successfully .. do something with it */;

ajSeqsetDel(&seqset);
```

Sequence stream

A sequence input stream is created from a USA using `ajSeqallFile`, which in contrast to the previous two functions returns a pointer to the new object.

```
AjPSeq eqall = NULL;
AjPStr usa = NULL;

usa = ajStrNewC("swiss:hba_*"); /* ""swiss:hba_*"" is a valid USA */

if(!(seqall=ajSeqallFile(usa)))
    /* Do something to handle the failure */;
else
    /* Sequence input stream was opened successfully .. do something with it */;

ajSeqallDel(&seqall);
```

Sequence output

`ajSeqoutClearUsa` creates (if necessary) or resets a sequence output object using a new USA:

```
AjPSeqout seqout = NULL;
AjPStr      usa  = NULL;

usa = ajStrNewC("asis::ALVLLYNITEF"); /* "asis::ALVLLYNITEF" is a valid USA */

seqout = ajSeqoutNew();
ajSeqoutClearUsa (&seqout, usa);

ajSeqoutDel(&seqout);
```

The code would work equally well if the call to the constructor function was omitted.

6.7.8 Getting and setting elements

AJAX functions are provided to recover or to set elements (data fields) in the basic sequence objects.

6.7.8.1 Setting single values

Functions to assign a value to one element of an existing sequence object have the prefix `ajSeqAssign`.

In the following example, `ajSeqAssignDescC` and `ajSeqAssignNameS` are used to assign a name and description to a region (user-defined by **-sbegin** and **-send** values) of the input sequence:

```
AjPSeq    seq = NULL;
AjPSeq seqsub = NULL;

seq    = ajAcdGetSeq("sequence");
seqsub = ajSeqNewSeq(seq);

ajSeqTrim(seqsub);
ajSeqAssignNameC(seqsub, "Subsequence");
ajSeqAssignDescC(seqsub, "Subsequence defined by sbegin and send values.");
ajSeqAssignAccC(seqsub, "YZ456789");
```

There are assignment functions for other sequence object elements:

```
ajSeqAssignEntryC  /* entry name */
ajSeqAssignFileC   /* file name */
ajSeqAssignFullC   /* full sequece name */
ajSeqAssignGiC     /* NCBI GI number */
ajSeqAssignSeqC    /* sequence as a string */
ajSeqAssignSvC     /* EMBL sequence version */
ajSeqAssignUfoC    /* uniform feature object string for feature input */
ajSeqAssignUsaC    /* sequence input uniform sequence address */
```

6.7.8.2 Other assignment functions

The `ajSeqSet*` family are functions are used to set other sequence properties (not single elements). For example `ajSeqSetRange` sets the range of a sequence object, in this case to the first 10 residues:

```
AjPSeq seq=NULL;
ajint start=0;
ajint end=9;

seq = ajSeqNew();

...

ajSeqSetRange(seq, start, end);
```

There are set functions for other sequence object elements: `ajSeqSetOffsets` (sets the offset from the start, and the full length to define the offset at the end), `ajSeqSetRangeRev` (sets the range as above and sets the sequence to be reversed), `ajSeqSetunique` (makes a sequence unique so that the sequence characters can be modified without side effects)

'Set' functions are available for the other datatypes too:

```
AjPSeqset seqset=NULL;
AjPSeqset seqall=NULL;
ajint start = 0;
ajint end = 9;

seqset = ajSeqNew();
seqall = ajSeqNew();

...

ajSeqsetSetRange(seq, start, end);
ajSeqallSetRange(seq, start, end);
```

These are the only useful functions at the sequence stream and sequence set level. Other properties are set for each individual sequence in the stream ort set using the sequence 'Set' or 'Assign' functions.

6.7.8.3. Getting single values

The `ajSeqGet*` family of functions are used to return the contents of a sequence object. The functions return pointers to the object's actual contents so the caller must take care not to unintentionally change the returned value in any way. Here, two flavours of function are used. `ajSeqGetDescC` returns the sequence description whereas `ajSeqGetNameS` returns the sequence name. The suffix `S` is used when an EMBOSS string is returned whereas `C` indicates a C-type (`char *`) string is returned:

```
AjPSeq seq          = NULL;
AjPStr str          = NULL;
AjPStr strcopy      = NULL;
char * strc         = NULL;

seq = ajAcdGetSeq("sequence");

strc = ajSeqGetDescC(seq);
str  = ajSeqGetNameS(seq);

/* Do something (but don't change) the returned strings */

ajStrAssignS(&strcopy, str);

/* Copy of string is now safe to change ... do something with it */

ajSeqDel(&seq);
ajStrDel(&str);
```

All the common sequence attributes can be retrieved like this, for example the accession and GI numbers using:

```
const AjPStr ajSeqGetGiS(const AjPSeq seq)
const AjPStr ajSeqGetAccS (const AjPSeq seq);
```

The sequence itself is retrieved using one of:

```
const char* ajSeqGetSeqC (const AjPSeq seq);
const AjPStr ajSeqGetSeqS (const AjPSeq seq);
```

The function `ajSeqGetOffset` will return the **`-sbegin`** value the user specified, i.e. the offset to the start of the sequence. `ajSeqGetOffend` will return the number of positions to be removed from the end as defined by the sequence length and the **`-send`** value:

```
 AjPSeq seq = NULL;
 ajint offset;
 ajint offend;

 seq = ajAcdGetSeq("sequence");

 offset = ajSeqGetOffset(seq);
 offend = ajSeqGetOffend(seq);

 /* Do something here using offset and offend, perhaps code to
retain the trimmed regions. */

 ajSeqTrim(seq);
```

'Get' functions are available for the other datatypes. For instance to get the number of sequences and range of a sequence set, or the filename and USA of a sequence input stream:

```
AjPSeq      seqset  = NULL;
AjPSeqall seqall   = NULL;
ajint      start;
ajint      end;
ajint      nseq;
AjPStr     name = NULL;
AjPStr     usa = NULL;

seqset = ajAcdGetSeqset("sequence");
seqall = ajAcdGetSeqall("sequences");

nseq = ajSeqsetGetSize(seqset);
ajSeqsetGetRange(seqset, &start, &end);

name = ajSeqallGetFilename(seqall);
usa  = ajSeqallGetUsa(seqall);
```

There are 'Get' functions that return a property of a sequence in a set. These have the prefix `ajSeqsetGetseq`. The number of the required sequence is specified. For instance three functions are provided to return a given sequence in a set, either as a sequence object pointer, an EMBOSS string or as a C-type (`char *`) string. Again, care should be taken not to unwittingly change the object data in any way:

```
AjPSeq seqset = NULL;

AjPSeq seqPtr
AjPStr seqS;
char * seqC;

seqset = ajAcdGetSeqset("sequence");

/* First sequence in call cases */
seqPtr = ajSeqsetGetseqSeq (seqset, 1);
seqS   = ajSeqsetGetseqSeqS (seqset, 1);
seqC   = ajSeqsetGetseqSeqC (seqset, 1);
```

Functions to return a property of a sequence in a stream have the prefix `ajSeqallGetseq` and return properties of the *current* sequence in the stream. For instance, to get the name and length of the current sequence:

```
AjPSeq seqall = NULL;
AjPStr name=NULL;
ajint  len;

seqall = ajAcdGetSeqall("sequences");

name = ajSeqallGetseqName (seqall);
len  = ajSeqallGetseqLen(seqall)
```

6.7.9 Testing sequence properties

Sequence properties can be tested, e.g. for whether a sequence is protein or nucleotide. Such functions return `AjBool`:

```
AjPSeq        seq = NULL;
AjPSeqset  seqset = NULL;

seq    = ajAcdGetSeq("sequence");
seqset = ajAcdGetSeq("sequences");

if(ajSeqIsProt(seq))
   /* Protein sequence */;
else if(ajSeqIsNuc(seq))
   /* Nucleotide sequence */;
else
   /* Error handling */;

if(ajSeqsetIsNuc(seqset)
```

```
{
   if(ajSeqsetIsRna (seqset))
   /* Do something. */;
}
else if(ajSeqsetIsNuc(seqset)
{
   /* Do something else */
}
```

Such tests are seldom required however as the sequence type can be enforced at the level of the ACD file.

6.7.10 Calculating sequence properties

Various properties can be calculated for a sequence. For example, `ajSeqCalcCount` writes an array with the number of `A`, `C`, `G` and `T` nucleotides in a nucleotide sequence, whereas `ajSeqCalcMolwt` calculates the molecular weight of a protein sequence:

```
AjPSeq     seq = NULL;
ajint nbases[5];
float molwt;

seq = ajAcdGetSeq("sequence");

if(ajSeqIsNuc(seq))
    ajSeqCalcCount(seq, *nbases);
else if(ajSeqIsProt(seq))
    molwt = ajSeqCalcMolwt(const AjPSeq seq);
```

Note

Note that various calculated attributes of sequences (Section A.6, 'Calculated attributes'), such as length and type, are available at the level of the ACD file.

6.7.11 Sequence string functions

There are two functions to copy a sequence into a string, using as a target either an EMBOSS string or a C-type (`char *`) string:

```
char* ajSeqGetSeqCopyC (const AjPSeq seq);
AjPStr ajSeqGetSeqCopyS (const AjPSeq seq);
```

There are a few functions for processing sequences that are held as strings using an `AjPStr` object. These have the prefix `ajSeqstr`. Here `ajSeqGetSeqCopyS` is first used to copy the sequence into a string. The molecular weight of a protein sequence is calculated, or the reverse and complement of a nucleotide sequence:

```
AjPSeq seq = NULL;
AjPStr str = NULL;
float molwt;

seq = ajAcdGetSeq("sequence");

str = ajSeqGetSeqCopyS(seq);

if(ajSeqIsNuc(seq))
    ajSeqstrReverse(&str);
else if(ajSeqIsProt(seq))
     molwt = ajSeqCalcMolwt(str);
```

6.7.12 Processing sequences

Various functions for processing a sequence are provided. You've already come across `ajSeqTrim` and `ajSeqsetTrim` for trimming sequences to their start and end regions but other operations are possible. For example, gap characters can be processed using the following functions:

```
void ajSeqGapFill (AjPSeq seq, ajuint len);
void ajSeqGapStandard (AjPSeq seq, char gapchar);
void ajSeqGapLen (AjPSeq thys, char gapc, char padc, ajint ilen);
```

`ajSeqGapFill` appends gap characters to a sequence up to a specified maximum sequence length. `ajSeqGapStandard` makes all the gaps in a string use the standard gap character provided. Any dash, tilde, full stop, question mark or blank space characters (`- ~ . ?`) in the string are converted. `ajSeqGapLen` sets any non-sequence characters to valid gap characters, and pads with extra gaps if necessary to a specified length.

To count the gaps in a sequence (counting any possible gap character) call `ajSeqCountGaps`:

```
ajuint ajSeqCountGaps (const AjPSeq seq);
```

Functions are available to complement, or reverse and complement a nucleotide sequence, if it hasn't already been done from ACD or command line processing:

```
void ajSeqComplement (AjPSeq seq);
void ajSeqReverseDo (AjPSeq seq)
```

There are a few functions to convert a sequence, or all sequences in a set, to lower- or upper-case:

```
void ajSeqsetFmtLower (AjPSeqset seq);
void ajSeqsetFmtUpper (AjPSeqset seq);
void ajSeqFmtLower (AjPSeq thys);
void ajSeqFmtUpper (AjPSeq thys);
```

`ajSeqsetFill` can be used to fill a sequence set by putting gaps at the ends of any short sequences:

```
ajint ajSeqsetFill(AjPSeqset seq)
```

6.7.13 Sequence type validation

In most cases the validation of sequence type (see Section A.7, 'Sequence types'), e.g. whether a sequence is for a protein or nucleic acid, contains gaps and so on, is handled at the level of the ACD file. This means that the application always receives (via a call to `ajAcdGet*`) a sequence of the exact type specified in the ACD file.

Several functions are provided in AJAX for handling sequence type directly. The type of a sequence can be set automatically, i.e. determined from the sequence itself, by calling `ajSeqType`. Otherwise it can be set directly for a sequence, or a sequence set, to one of `nucleotide` or `protein` using the following functions:

```
void ajSeqType (AjPSeq thys);
void ajSeqSetNuc (AjPSeq thys);
void ajSeqSetProt (AjPSeq thys);
void ajSeqsetSetNuc (AjPSeqset thys);
void ajSeqsetSetProt (AjPSeqset thys);
```

6.7.14 Sequence conversion

It is sometimes necessary to convert a sequence into numeric form for convenient processing. The AJAX datatype for this is:

`AjPSeqCvt`	Used for sequence conversion into numeric form

The common use-case is where a sequence character conversion table is generated for a scoring matrix, such as an amino acid substitution scoring matrix (see Section 6.10, 'Handling comparison matrices').

6.7.15 Handling sequence translation

Translation of a genetic sequence into a protein is a common task and is covered elsewhere (Section 6.8, 'Handling sequence translation').

6.7.16 Handling IUB base codes

EMBOSS supports the official IUB base codes as shown below and also `X` to indicate an unknown base.

6.7.16.1 Unambiguous codes

Unambiguous base codes are:

- A (Adenine)
- C (Cytosine)
- G (Guanine)
- T (Thymine) or U (Uracil)

6.7.16.2 Ambiguity codes

Either of two bases possible:

- M (A or C)
- R (A or G)
- W (A or T/U)
- S (C or G)
- Y (C or T/U)
- K (G or T/U)

Any of three bases possible:

- B (C G or T/U)
- D (A G or T/U)
- H (A C or T/U)
- V (A C or G)

Any of the four bases possible:

- N (A C G or T/U)

An unknown (or non-standard) base:

- X (not A C G or T/U)

6.7.16.3 Processing an IUB base code

The standard IUB base codes have four possibilities which can be represented in binary as 4 bits for rapid comparison including testing of ambiguity codes. Functions are provided for the character and binary representations of base codes.

The ambiguity codes and their numeric representations are read from local data file Ebases.iub.

```
/* Sets up binary OR representation of all IUB bases in
 a table aj_base_table where A=1,
```

```
C=2, G=4 and T=8. Also sets up a match probability array aj_base_prob holding the
probability of one IUB base matching any other. Uses the Ebases.iub file. Is
initialised if necessary from other AJAX functions. */

/* Complements a nucleotide base. */
char   ajBaseAlphacharComp(char base);

/* Returns an element of the base match probability array */
float   ajBaseAlphaCompare(ajint base1, ajint base2);
```

To clean up internal memory used for housekeeping of sequence base processing call `ajBaseExit`:

```
void ajBaseExit (void);
```

This is called automatically by `embExit`.

6.7.16.4 Conversion of base and amino acid codes

Functions for converting an integer representation of a base code into something else:

```
/* Convert to string of matching base code(s). */
const AjPStr ajBaseGetCodes (ajint ibase);

/* Convert to base code: 0 to 'A', 1 to 'B' … 25 to 'Z'. Returns an integer. */
ajint ajBasecodeFromInt (ajint n);
```

To convert base code into something else:

```
/*Convert to integer: 'A' to 0, 'B' to 1, … 'Z' to 25. Returns an integer.*/
ajint ajBasecodeToInt (ajint c);

/* Convert to binary OR representation of an IUB base. */
ajint ajBaseAlphaToBin (ajint c);

/* Convert to binary OR representation of an IUB base. */
char ajBaseAlphacharToBin (char c);
```

`ajBaseAlphaToBin` and `ajBaseAlphacharToBin` return a binary OR representation of an IUB base where A = 1, C = 2, G = 4 and T = 8. These functions use an internal base table.

To convert a binary OR representation of an IUB base into something else:

```
/* Converts a binary OR representation of an IUB base to an ambiguous DNA base code. */
char ajBaseBinToAlpha (ajint c);
```

`ajBaseBinToAlpha` converts a binary OR representation of an IUB base where A = 1, C = 2, G = 4 and T = 8 into an ambiguous DNA base code (uses T rather than U). It uses the internal base table.

To interconvert single-letter and three-letter amino acid codes:

```
/* Convert single letter to three letter amino acid code. */
AjBool ajResidueToTriplet (char c, AjPStr *Paa3);

/* Convert three letter to single letter amino acid code. */
AjBool ajResidueFromTriplet (char *aa4, char *Pc);
```

6.7.17 DNA calculations

Various functions for calculating the properties of nucleotide sequences are provided in `ajdan.h/c`. These functions use a DNA melting object (`AjPMelt`) but this is transparent to the application programmer, you will never need to instantiate or use the object directly:

`AjPMelt` Used for DNA melting

6.7.17.1 General functions for DNA melting

The `AjPMeltInit` and `AjPMeltExit` functions are used internally to initialise and clean up memory. There is no need to call them directly. The `AjPMeltGC` function is a utility to calculate G+C content:

```
/* Initialises melt entropies, enthalpies and energies. Different data files are
read for DNA or RNA heteroduplex. Also sets optional flag for array saving of the
above. */
void ajMeltInit (AjBool isdna, ajint savesize);

/* Cleans up DNA melting processing internal memory */
void ajMeltExit (void);

/* Calculates GC fraction of a sequence allowing for ambiguity */
float ajMeltGC (const AjPStr strand, ajint len);
```

6.7.17.2 Calculating DNA melting and annealing temperatures and energies

`AjPMeltEnergy` calculates melt energy and enthalpy/entropy for a sequence string. `AjPMeltEnergy2` is an alternative interface that initialises a set of 'save' arrays to make later calls more efficient by inspecting the sequence only once. `AjPMeltTemp` calculates a melting temperature for an oligonucleotide, with a `AjPMeltTempSave` version that uses 'save' arrays for efficiency. `AjPMeltTempProd` calculated the melting temperature of an oligonucleotide product under a given salt concentration. `AjPMeltTemp` caclulates the annealing temperature of a product and primer, given their respective melting temperatures.

```
/* Calculate melt energy and enthalpy/entropy for a sequence */
float ajMeltEnergy(const AjPStr strand, ajint len,
                   ajint shift, AjBool isDNA,
                   AjBool maySave, float *enthalpy, float *entropy);

/* Calculate melt energy for programs like prima */
float ajMeltEnergy2(const char *strand, ajint pos, ajint len,
```

```
                    AjBool isDNA,
                    float *enthalpy, float *entropy,
                     float **saveentr, float **saveenth, float **saveener);

/* Calculate melt temperature of DNA or RNA */
float ajMeltTemp (const AjPStr strand, ajint len, ajint shift, float saltconc,
                 float DNAconc, AjBool isDNA);

/* Calculate melt temperature of DNA or RNA */
float ajMeltTempSave(const char *strand, ajint pos,
                     ajint len, float saltconc,
                     float DNAconc, AjBool isDNA,
                     float **saveentr, float **saveenth, float **saveener);

/* Calculate product melt temperature of DNA */
float ajMeltTempProd (float gc, float saltconc, ajint len);

/* Calculate annealing temperature of product and primer */
float ajAnneal (float tmprimer, float tmproduct);
```

6.8 Handling sequence translation

6.8.1 Introduction

Translation of a nucleotide sequence into a protein sequence is a common task. AJAX provides all the basic functionality you would expect. The nucleic sequence can be in a variety of forms (an AJAX string (`AjPStr`), C-type string (`char *`) or AJAX sequence object (`AjPSeq`) and can be translated in all reading frames. The reverse complement of a sequence can also be translated.

6.8.2 AJAX library files

The AJAX library file for handling sequence translation is listed in the table below. Library file documentation, including a complete description of datatypes and functions, is available at: http://emboss.open-bio.org/rel/dev/libs/

Library file documentation	Description
`ajtranslate`	Sequence translation

`ajtranslate.h/c.` Defines a sequence translation object (`AjPTrn`) and include functions for handling sequence translation.

6.8.3 ACD datatypes

There is no dedicated ACD datatype for handling translation. Such operations are performed on a nucleotide sequence and so require a sequence input of the appropriate type. A genetic code is also required, a choice of which is provided to the user (usually) via a menu implemented by a `list` ACD datatype. The ACD datatypes you'll require are therefore:

`list` A list of options (text descriptions) with text labels. The user is presented with a limited list of options they can choose from. The choices can be labelled by any arbitrary text label.

`sequence` A single input sequence.

Note

For general information on menu and sequence handling see: Section 6.19, ('Handling menus') and Section 6.7 'Handling sequences').

6.8.4 ACD data definition

A typical ACD definition for single sequence input:

```
sequence: sequence
[
    parameter: "Y"
    type: "nucleotide"
]
```

The available genetic codes must be defined in the ACD file and the `list` datatype may be used for this. EMBOSS supports a standard set of genetic codes which are given as follows:

```
list: table
[
    additional: "Y"
    default: "0"
    minimum: "1"
    maximum: "1"
    header: "Genetic codes"
    values: "0:Standard; 1:Standard (with alternative initiation
             codons); 2:Vertebrate Mitochondrial; 3:Yeast Mitochondrial;
             4:Mold, Protozoan, Coelenterate Mitochondrial and
             Mycoplasma/Spiroplasma; 5:Invertebrate Mitochondrial; 6:Ciliate
             Macronuclear and Dasycladacean; 9:Echinoderm Mitochondrial;
             10:Euplotid Nuclear; 11:Bacterial; 12:Alternative Yeast Nuclear;
             13:Ascidian Mitochondrial; 14:Flatworm Mitochondrial;
             15:Blepharisma Macronuclear; 16:Chlorophycean Mitochondrial;
             21:Trematode Mitochondrial; 22:Scenedesmus obliquus;
             23:Thraustochytrium Mitochondrial"
    delimiter: ";"
    codedelimiter: ":"
    information: "Code to use"
]
```

The order of the codes is currently important; the list must be given in the *exact* order shown above.

6.8.5 AJAX datatypes

For handling sequence translation, which requires sequence and menu input, use:

`AjPTrn`	Used for sequence translation
`AjPSeq`	Single input sequence (for `sequence` ACD datatype)
`AjPStr`	Single selection from a menu (for `list` ACD datatype)

6.8.6 ACD file handling

Datatypes and functions for handling translation via the ACD file are shown below. Here a single selection from the list is retrieved but other types of menu, input sequence or access methods could be used.

Title	To read a sequence	To read a single selection from a list
ACD datatype	`sequence`	`list`
Object	`AjPSeq`	`AjPStr`
To retrieve from ACD	`ajAcdGetSeq`	`ajAcdGetListSingle`

Your application code will call `embInit` to process the ACD file and command line (see Section 6.3, 'Handling ACD files'). All values from the ACD file are read into memory. You have a handle on the memory through the `ajAcdGet*` family of functions which return pointers to appropriate objects.

6.8.6.1 Sequence or menu selection retrieval

To retrieve the sequence or menu selection object pointers are declared then initialised using the appropriate `ajAcdGet*` function.

6.8.6.1.1 Input sequence

To retrieve an input sequence:

```
AjPSeq seq=NULL;
seq = ajAcdGetSeq("sequence");
```

6.8.6.1.2 Menu selection

The option selected from the list of genetic codes is required as an integer. `ajAcdGetListSingle` returns the selection as a string, the initial part of which is converted to an integer using `ajStrToInt`. This integer is passed to `ajTrnNewI` for the creation of the translation table object. This is why the list order in the ACD file is important! You must also declare a translation object pointer:

```
AjPTrn trnTable = NULL; /* Translation object pointer */
AjPStr gcode = NULL; /* Genetic code (selection from list) */
```

```
ajint n = 0; /* Selection */

gcode = ajAcdGetListSingle("table");
ajStrToInt(gcode,&n);

trnTable = ajTrnNewI(n);
```

6.8.6.2 Memory management

It is your responsibility to free up memory at the end of the program. You must call the default destructor function for the translation, sequence and string objects used for the ACD data definitions:

```
/* Deletes a translation table object */
void ajTrnDel(AjPTrn* pthis);
/* Delete a string object. */
void ajStrDel (AjPStr *Pstr);

/* Delete a sequence object. */
void ajSeqDel (AjPSeq* Pseq);
```

Function `ajTrnExit` is automatically called on exit to clean up internal memory used for housekeeping of translation processing:

```
void ajTrnExit(void);
```

6.8.7 Translation object memory management

6.8.7.1 Default object construction

To use a translation object you must first instantiate the appropriate object pointer. Default construction functions are provided. They will read a translation data file from the EMBOSS data search directory (see the *EMBOSS User's Guide*) called `EGC.n`, where `n` is the number of the genetic code to use. This number can be provided explicitly to `ajTrnNewI`. Alternatively a file can be opened by filename by calling `ajTrnNew` or `ajTrnNewC`:

```
/* Reads EGC.trnFileNameInt where trnFileNameInt is supplied as a parameter. */
AjPTrn ajTrnNewI (ajint trnFileNameInt);

/* Reads trnFileName. */
AjPTrn ajTrnNew (const AjPStr trnFileName);

/* Reads trnFileName. */
AjPTrn ajTrnNewC (const char *trnFileName);
```

All constructors return the address of a new object. The pointers do not need to be initialised to `NULL` but it is good practice to do so:

```
AjPStr gcode =NULL;
AjPTrn trnTable = NULL;
ajint n = 0;

gcode = ajAcdGetListSingle("table");
ajStrToInt(gcode,&n);

trnTable = ajTrnNewI(n);
/* The object is instantiated and ready for use */
```

Alternatively:

```
AjPStr  name   =  NULL;
AjPStr gcode = NULL;
AjPTrn trnTable = NULL;
ajint n = 0;

gcode = ajAcdGetListSingle("table");
ajStrToInt(gcode,&n);

name = ajStrNew();
ajFmtPrintS(&name, "EGC.%d", n); /* Create the string EGC.n */
trnTable = ajTrnNew(name);
/* The object is instantiated and ready for use */
```

6.8.7.2 Default object destruction

For the examples above you must free a single string and sequence:

```
AjPSeq seq    = NULL;
AjPStr gcode = NULL;

seq   = ajAcdGetSeq("sequence");
gcode = ajAcdGetListSingle("table");
ajStrToInt(gcode,&n);

/* Do something */

ajSeqDel(&seq);
ajStrDel(&str);
```

You must free the memory for the translation object before the pointer is reused and also once you are finished with it. A default destructor function is provided:

```
/* Deletes a translation table object */
void ajTrnDel(AjPTrn* pthis);
```

It is used as follows:

```
AjPStr  gcode =NULL;
AjPTrn trnTable = NULL;
```

```
ajint n = 0;

gcode = ajAcdGetListSingle("table");
ajStrToInt(gcode,&n);

trnTable = ajTrnNewI(n);
/* The object is instantiated and ready for use */

ajTrnDel(&trnTable);

/* The memory is freed and the pointer reset to NULL, ready for re-use. */
```

6.8.8 Translation

`ajTrnSeqSeqOrig` creates a peptide sequence containing the full translation of a nucleotide sequence, including any trailing partial codon (one or two base) which translate to X unless the first two bases can only define one amino acid:

```
AjPSeq ajTrnSeqSeqOrig (const AjPTrn trnObj, const AjPSeq seq, ajint frame);
```

A nucleotide sequence held in an AJAX string (`AjPStr`), C-type string (`char *`) or AJAX sequence object (`AjPSeq`) can be translated into protein using:

```
void ajTrnSeqS (const AjPTrn trnObj, const AjPStr str, AjPStr *pep);
void ajTrnSeqC (const AjPTrn trnObj, const char *str, ajint len, AjPStr *pep);
void ajTrnSeqSeq (const AjPTrn trnObj, const AjPSeq seq, AjPStr *pep);
```

These functions translate in frame 1 (from the first base) to the last full triplet codon. If there are one or two bases extra at the end then they are ignored.

To translate the reverse complement of a sequence call:

```
void ajTrnSeqRevC (const AjPTrn trnObj, const char *str, ajint len, AjPStr *pep);
void ajTrnSeqRevS (const AjPTrn trnObj, const AjPStr str, AjPStr *pep);
void ajTrnSeqRevSeq (const AjPTrn trnObj, const AjPSeq seq, AjPStr *pep);
```

These functions translate in frame −1 (from the last base) to the first full triplet codon. If there are one or two bases extra at the start then they are ignored. All functions will *append* the translation to the input peptide.

Alternative translation is available for people who define frame −1 as being the frame starting from the first base of a reverse-complemented sequence. To translate the reverse complement of a sequence call:

```
void ajTrnSeqAltRevC (const AjPTrn trnObj, const char *str, ajint len, AjPStr *pep);
void ajTrnSeqAltRevS (const AjPTrn trnObj, const AjPStr str, AjPStr *pep);
void ajTrnSeqAltRevSeq (const AjPTrn trnObj, const AjPSeq seq, AjPStr *pep);
```

These functions translate in frame −4 (from the last base) to the first full triplet codon, i.e. if there are one or two bases extra at the start then they are ignored. All functions will *append* the translation to the input peptide.

The frame of translation may be specified:

```
void ajTrnSeqFrameC (const AjPTrn trnObj, const char *seq, ajint len, ajint frame,
AjPStr *pep);
void ajTrnSeqFrameS (const AjPTrn trnObj, const AjPStr seq, ajint frame, AjPStr
*pep);
void ajTrnSeqFrameSeq (const AjPTrn trnObj, const AjPSeq seq, ajint frame, AjPStr
*pep);
AjPSeq ajTrnSeqFramePep (const AjPTrn trnObj, const AjPSeq seq, ajint frame);
```

All functions will *append* the translation to the input peptide. In contrast, `ajTrnSeqFramePep` returns an `AjPSeq` object with the new peptide.

These functions translate in the specified frame (which must be one of 1,2,3,−1,−2,−3,4, 5,6,−4,−5,−6) to the last full triplet codon, i.e. if there are one or two bases extra at the end, they are ignored. Frames −6 to −1 give translations in the reverse sense, frames 1 to 3 give normal forward translations. Frames 4 to 6 reverse complement the DNA sequence then reverse the peptide sequence. Frames 4 to 6 are therefore reversed protein sequences useful mainly for displaying beneath the original DNA sequence.

Frame −1 is defined as the translation of the reverse complemented sequence which matches the codons used in frame 1. For example, in the sequence ACGT the first codon of frame 1 is ACG and the last codon of frame −1 is the reverse complement of ACG, i.e. CGT.

Frame −4 is defined as the translation of the reverse complement, starting the translation in the first codon of the reversed sequence. In the sequence ACGT, the last codon is CGT and so frame −4 translates from the reverse complement of CGT (i.e. ACG) – this is for those people who define frame −1 as using the first codon when the sequence is reverse-complemented. This is also known as the 'alternative frame −1'.

Frame −5 starts on the penultimate base (alternative frame −2). Frame −6 starts on the antepenultimate base (alternative frame −3). Frame 4 is the same as frame −1, 5 is −2, 6 is −3.

To complete a translation by attempting to translate the last one or two bases of a frame call:

```
ajint ajTrnSeqDangleC (const AjPTrn trnObj, const char *seq, ajint frame, AjPStr *pep);
ajint ajTrnSeqDangleS (const AjPTrn trnObj, const AjPStr seq, ajint frame, AjPStr *pep);
```

In both cases, the translation is *appended* to the input peptide.

There are functions to translate a single codon into one-letter or three-letter amino acid codes. Alternative functions that take a C-type (`char *`) string are available but not shown:

```
/* Translates a codon into a 3-letter code. */
char ajTrnCodonS (const AjPTrn trnObj, const AjPStr codon);

/* Translates the reverse complement of a codon into a 3-letter code. */
char ajTrnCodonRevS (const AjPTrn trnObj, const AjPStr codon);

/* Translates a codon into a 1-letter code. */
char ajTrnCodonC (const AjPTrn trnObj, const char *codon);

/* Translates the reverse complement of a codon into a 1-letter code. */
char ajTrnCodonRevC (const AjPTrn trnObj, const char *codon);
```

6.8.9 Miscellaneous functions

There are a couple of functions for retrieving the elements of a translation object:

```
AjPStr ajTrnGetTitle (const AjPTrn thys);
AjPStr ajTrnGetFilename (const AjPTrn thys);
```

To check whether the input codon is a start codon, a stop codon or something else, call

```
ajint ajTrnCodonstrTypeC (const AjPTrn trnObj, const char *codon, char *aa);
ajint ajTrnCodonstrTypeS (const AjPTrn trnObj, const AjPStr codon, char *aa);
```

To return the genetic code description as a string for a given translation table file name number call:

```
const AjPStr ajTrnName(ajint trnFileNameInt);
```

To create a suitably named sequence object to hold a peptide translation call:

```
AjPSeq ajTrnNewPep(const AjPSeq nucleicSeq, ajint frame);
```

To read a translaton data file (used internally when translation is initialised):

```
void ajTrnReadFile(AjPTrn trnObj, AjPFile trnFile);
```

6.9 Handling features

6.9.1 Introduction

A feature is a region of interest in a molecular sequence such as a restriction enzyme cut site, protein secondary structure prediction state, exon position, regions of motif match and so on. EMBOSS supports, for input and output, most of the formats (see the *EMBOSS User's Guide*) that were developed for input of features into the major genome sequence databases.

Many applications do not read and write features directly. Features are also used to store the results of sequence analysis, and can be written out as 'reports' where a report format defined through ACD is used to write out a feature table and (for some formats) the original sequence (see Section 6.15, 'Handling application reports').

Features are annotations of simple ranges in a sequence (start and end) or of a numbered group of features which have a 'join' (to combine exons in a coding sequence) or some other combination ('group' or 'one-of' in the EMBL/Genbank feature table). These complex features are stored as a parent feature with a set of simpler features for each component. Currently these are stored in the same feature table. In a future release these may become sub-features to simplify sorting operations.

A feature object is modelled on the General Feature Format 3 (GFF3) feature data format, where a feature is described by:

- Start and end position
- Name describing the feature
- The strand direction (in a nucleic sequence)
- A score.

A feature object also holds data on:

- A second start and end position for features where the start or end is wider than a single base or residue.
- 'Source' that records the names of the program or database from which the features were derived.
- The feature type (feature key) using an internal name derived from the Sequence Ontology (SO) and defined in the `Efeatures.emboss` (nucleotide) or `Efeatures.protein` (protein) data files and include all *EMBL/GenBank* and *UniProt* feature types.
- List of tag names and values which are defined in the `Etags.emboss` (nucleotide) or `Etags.protein` (protein) data files and include all *EMBL/GenBank* or *UniProt* feature qualifiers.
- Frame 1...3, −1...−3 for coding nucleotide features or 0 for non-coding or protein features.
- Flag bit mask for *EMBL* location to record features between bases (11^12), types of join/order/one-of, and other attributes.
- Group number for the individual exons and the parent of join/order/one-of feature locations in the *EMBL/GenBank* feature table.
- Remote ID where the feature location (e.g. for an exon used by a join) is in another entry in the same database.
- A label for the location of the feature in another entry.
- Exon number.

The feature types need to be standardised to allow interconversion of formats. EMBOSS uses a set of data files installed in the `share/EMBOSS/data/` directory to define types and tag names for each input/output format, and for internal use. The master internal naming files are `Efeatures.emboss` and `Etags.emboss` for nucleotid efeatures, and `Efeatures.protein` and `Etags.protein` for protein features. These include the files for the major feature format definitions so that most feature types and tags (where there is no clash between formats) will be stored and returned unchanged. For any type or tag that does not appear in these files, the first name defined is used as a default ('misc_feature' for nucleotide type, 'polypeptide region' for protein type, 'note' for a tag name).

A feature table is simply a group of features and is stored in one of three contexts:

- As part of a sequence file
- As part of a database entry
- As a *raw feature table* (a file that does not contain the sequence the features refer to).

Note

Feature table definitions have a controlled vocabulary, i.e. there is a specified list of feature types and tag names that can be used. This means that you cannot edit feature tables to add in features with new keys. If you edit a feature table you must stick to the allowed set of feature keys.

'Named note' tags are a way to store feature tag names that are not in the allowed set. The default (note) feature tag is stored with a value that begins with '`*name`' followed by the value. This preserves the annotation in a readable form when features are written out using a standard format such as EMBL or GFF3.

Features to be read or written by an application are defined in the application ACD file, although it is possible to create feature tables directly if this is required.

The name of a features file and the format of the features in the file are specified on the command line using a Uniform Feature Object (UFO) (see the *EMBOSS User's Guide*). A set of command line qualifiers are available for features. These allow you to set such things as filename and format and the region of the sequence containing the features of interest. These qualifiers may be hard-coded as attributes in the ACD file (see Section A.5.2.8, '`features`' and Section A.5.3.2, '`featout`').

AJAX provides comprehensive functionality for handling features including:

- Features may be read and written directly as an alternative to ACD processing
- Elements of the objects for handling features may be retrieved or set directly
- Handling of feature tags
- Querying the properties of features and feature tables
- Processing of features and feature tables.

6.9.2 AJAX library files

AJAX library files for handling features are listed in the table below. Library file documentation, including a complete description of datatypes and functions, is available at: http://emboss.open-bio.org/rel/dev/libs/

Library file documentation	Description
`ajfeat`	General sequence feature handling
`ajfeatdata`	Feature datatypes

`ajfeat.h/c`. Most of the functions you will ever need for general feature handling. They also contain static functions for handling features at a low level. You are unlikely to need these unless you plan to implement code to extend the core functionality of the library.

ajfeatdata.h/c. Basic feature objects (`AjPFeattable`, `AjPFeature` and `AjPFeattabOut`) for general use, e.g. retrieving features via ACD file processing. It also defines a feature input object (`AjPFeattabIn`) used for low-level feature input handling.

6.9.3 ACD datatypes

A feature table (not individual features) may be specified for input or output in an ACD file.

The datatype for handling feature table input is:

`features`	Feature table input

The datatype for handling feature table output is:

`featout`	Feature table output

Features can also be read from an input sequence and written alongside an output sequence if the `features:` ACD attribute is set. If set then the sequence output will include feature information either in the same file (if the sequence format supports it) or in a separate file (by default in GFF format).

ACD datatypes for sequence input include:

`sequence`	Read a single sequence
`seqall`	Read multiple sequences sequentially, one at a time
`seqset`	Read multiple sequences as a single set
`seqsetall`	Read multiple sequences as multiple sets

ACD datatypes for sequence output include:

`seqout`	Write a single sequence
`seqoutset`	Write multiple sequences as a single set
`seqoutall`	Write multiple sequences sequentially, one at a time

6.9.4 ACD data definition

A typical ACD definition for feature input:

```
features: features
[
    parameter: "Y"
    type: "protein"
]
```

A typical ACD definition for feature output:

```
featout: outfeat
[
    parameter: "Y"
    type: "protein"
    multiple: "N"
]
```

A typical ACD definition for sequence input with features:

```
# single input sequence
sequence: sequence
[
    parameter: "Y"
    type: "gapany"
    features: "Y"
]
```

A typical ACD definition for sequence output with features:

```
# single sequence
seqout: outseq
[
    parameter: "Y"
    type: "gapany"
    features: "Y"
]
```

The use of the `sequence` datatype is for illustrative purposes; the other sequence input and output types could also have been given.

6.9.4.1 Parameter name

All data definitions for feature input and output should have standard parameter names. These include:

- `features` for any feature inputs
- `outfeat` for any outputs
- Alternatives and variations (e.g. `afeatures`, `bfeatures` for multiple inputs) are allowed.

For more information see Appendix A, *ACD syntax reference*.

6.9.4.2 Common attributes

Attributes that are typically specified are summarised below. They are datatype-specific (Section A.5, 'Datatype-specific attributes') unless they are indicated as being global attributes (Section A.4, 'Global attributes').

parameter: Features are typically the primary input or output of an EMBOSS application and, as such, should be defined as parameters by using the global attribute `parameter: "Y"`.

type: Specifies the type of the sequence (`protein` or `nucleotide`) to which the features pertain and is used for validation purposes.

multiple: A boolean attribute that can be set for a `featout` data definition to specify the feature annotation is for multiple sequences.

offormat: GFF format is used by default for the output feature(s). The format is normally set at the command line but a default may be hardcoded with `offormat:`. All common feature formats are supported (see the *EMBOSS User's Guide*).

All sequence objects can include a feature table. On input through an ACD datatype features will be read if the `features:` attribute is true in the ACD definition.

6.9.5 AJAX datatypes

For handling feature tables, including input feature tables defined in the ACD file, use:

`AjPFeattable`	Feature table which includes a list of `AjPFeature` objects (for the `features` ACD datatype)

For handling feature table outputs defined in the ACD file use:

`AjPFeattabOut`	Feature table output (for the `featout` ACD datatype)

There is also a basic object for handling individual features:

`AjPFeature`	Biological feature

There is a datatype for low-level feature input beyond that provided by the static datatypes in the various library files:

`AjPFeattabIn`	Low-level feature table input

You are unlikely to need `AjSFeattabIn` unless you plan to implement code to support new feature formats for EMBOSS. For advice on how to do this ask the EMBOSS developers.

In developing applications, feature tables are most likely to be used as part of a report output. A sequence is read, analysis results are generated as features, and both are output as a report format (see Section 6.15, 'Handling application reports').

6.9.6 ACD file handling

Datatypes and functions for handling features via the ACD file are shown below.

Title	To read features	To write features
ACD datatype	`features`	`featout`

AJAX datatype	AjPFeattable	AjPFeattabOut
To retrieve from ACD	ajAcdGetFeatures	ajAcdGetFeatout

Your application code will call `embInit` to process the ACD file and command line (see Section 6.3, 'Handling ACD files'). All values from the ACD file are read into memory and files are opened as necessary. You have a handle on the files and memory through the `ajAcdGet*` family of functions which return pointers to appropriate objects.

6.9.6.1 Input feature retrieval

To retrieve input features an object pointer is declared and then initialised using `ajAcdGetFeatures`:

```
AjPFeattable features=NULL;
features = ajAcdGetFeatures("features");
```

6.9.6.2 Output feature retrieval

To retrieve an output feature stream an object pointer is declared and initialised using `ajAcdGetFeatout`:

```
AjPFeattabOut outfeat=NULL;
outfeat = ajAcdGetFeatout("outfeat");
```

6.9.6.3 Processing command line options and ACD attributes

6.9.6.3.1 Setting the features range

The `features` input datatype has various inbuilt command line qualifiers including **-fbegin** and **-fend** which specify a start and end position for the features, and **-freverse** to reverse the orientation of nucleotide features.

When a feature table is read the feature values are held in the appropriate feature table object. Regardless of the range values you still get the entire table loaded into memory. The functions `ajFeattableTrim` or `ajFeattableTrimOff` are used to trim the features to the region defined by **-fbegin** and **-fend**:

```
AjFeattable ftable=NULL;

ftable = ajAcdGetFeatures("features");

ajFeattableTrim(ftable);
/* ajFeattableTrimOff(ftable,0,ajFeattableGetLen(ftable));*/
```

When a sequence is read the feature values are held in a feature table in the appropriate sequence object. Regardless of the sequence range values you still get the entire sequence loaded into memory. The function `ajSeqTrim` (or `ajSeqsetTrim` for a `AjPSeqset` object) is used to trim the sequence and features to the region defined by **`-sbegin`** and **`-send`**:

```
AjPSeq seq=NULL;

seq = ajAcdGetSeq("sequence");

ajSeqTrim(seq);
```

6.9.6.4 Memory management

It is your responsibility to free up memory at the end of the program. You must call the default destructor function (see below) on any objects returned by calls to `ajAcdGet*`.

Additionally you must call `embExit` to clean up internal memory including that allocated for the housekeeping of feature tables:

```
embExit();
```

6.9.7 Memory management

6.9.7.1 Default object construction

To use a feature table object that is not defined in the ACD file you must first instantiate the appropriate object pointer. The default construction function is `ajFeattableNew`. `ajFeattableNew` leaves the type of feature table uninitialised. It is set when a feature is added to the table. Type-specific functions `ajFeattableNewDna` and `ajFeattableNewProt` create feature tables for nucleotide and protein features respectively. Function `ajFeattableNewSeq` creates a feature table with the type and length matching a sequence object.

Feature table output objects are typically loaded from ACD file processing (see above). In the unlikely event you need to create one manually you can use the default constructor `ajFeattabOutNew` or functions `ajFeattabOutNewCSF` or `ajFeattabOutNewSSF` which use an existing output file and a specified type and sequence name.

To create a feature object (usually to be stored within a feature table), a similar set of functions is available. `ajFeatNew` creates a feature with all attributes including the type. `ajFeatNewProt` creates a feature with all attributes required by protein features (no strand or frame). `ajFeatNewII` and `ajFeatNewIIRev` are generic constructors requiring only the start and end values. The feature type will default to a 'miscellaneous feature' or 'polypeptide region' value. `ajFeatNewFeat` is a copy constructor making a feature object from an existing feature.

```
/* Constructor for feature which is added to table */
AjPFeature ajFeatNew (AjPFeattable thys,
                      const AjPStr source,
                      const AjPStr type,
                      ajint Start,
                      ajint End,
```

```
                        float score,
                        char strand,
                        ajint frame);
/* Constructor for protein feature */
AjPFeature ajFeatNewProt(AjPFeattable thys,
                         const AjPStr source,
                         const AjPStr type,
                         ajint Start, ajint End,
                         float score);

/* Simple constructor with only start and end positions */
AjPFeature ajFeatNewII (AjPFeattable thys,
                        ajint Start, ajint End);

/* Simple constructor, sets features to be on reverse strand */
AjPFeature ajFeatNewIIRev (AjPFeattable thys,
                           ajint Start, ajint End);

/* Copies a feature */
AjPFeature ajFeatNewFeat (const AjPFeature orig);

/*  Feature Table Object Constructors */
AjPFeattable ajFeattableNew (const AjPStr name);
AjPFeattable ajFeattableNewDna (const AjPStr name);
AjPFeattable ajFeattableNewProt (const AjPStr name);
AjPFeattable ajFeattableNewSeq (const AjPSeq seq);

/* Output Feature Table Object constructors */
AjPFeattabOut ajFeattabOutNew (void);
AjPFeattabOut ajFeattabOutNewCSF (const char* fmt, const AjPStr name,
       const char* type, AjPFile file);
AjPFeattabOut ajFeattabOutNewSSF (const AjPStr fmt, const AjPStr name,
       const char* type, AjPFile file);
```

The parameters to `ajFeatNew` are as follows:

`thys`	Pointer to the feature table to which the new feature is added
`source`	Analysis basis for feature
`type`	Type of feature (e.g. exon)
`Start`	Start position of the feature
`End`	End position of the feature
`score`	Analysis score for the feature
`strand`	Strand of the feature
`frame`	Frame of the feature

All constructors return the address of a new object.

6.9.7.2 Default object destruction

You must free the memory for an object once you are finished with it. The functions are:

```
void ajFeatDel (AjPFeature *pthis);
void ajFeattableDel (AjPFeattable *pthis) ;
void ajFeattabOutDel (AjPFeattabOut* pthis);
```

Example

In the example below, a feature table and individual features are built manually using the default constructor functions. The features are written out to a feature table retrieved from ACD processing:

```
AjPFeattable feattable;
AjPStr name   = NULL;
AjPStr source = NULL;
AjPStr type   = NULL;
char strand   = '+';
ajint frame   = 0;
AjPFeature feature = NULL;
AjPFeattabOut output = NULL;
ajint i;
float score = 0.0;
embInit("demofeatures", argc, argv);

output = ajAcdGetFeatout("outfeat");

ajStrAssignC(&name,"seq1");

feattable = ajFeattableNew(name);

ajStrAssignC(&source,"demofeature");
score = 1.0;

for(i=1;i<11;i++)
{
    if(i & 1)
        ajStrAssignC(&type,"CDS");
    else
        ajStrAssignC(&type,"misc_feature");
    feature = ajFeatNew(feattable, source, type, i, i+10, score, strand,
                        frame);
}

ajFeattableWrite(output, feattable);

ajStrDel(&source);
ajStrDel(&name);
ajStrDel(&type);
ajFeattableDel(&feattable);
ajFeattabOutDel(&output);
```

6.9.7.4 Alternative object construction and loading

6.9.7.4.1 Single feature

There are a variety of alternative ways to create a feature object. The start and end position of the features may be specified:

```
AjPFeature ajFeatNewII (AjPFeattable thys, ajint Start, ajint End);
AjPFeature ajFeatNewIIRev (AjPFeattable thys, ajint Start, ajint End);
```

```
AjPFeature ajFeatNewProt (AjPFeattable thys,
                          const AjPStr source, const AjPStr type,
                          ajint Start, ajint End, float score);
```

`ajFeatNewIIRev` sets features to be on the reverse strand whereas `ajFeatNewProt` is for protein features.

For cases where a copy of a feature is required that can be safely changed and/or deleted you can use `ajFeatNewFeat`:

```
AjPFeature ajFeatNewFeat (const AjPFeature orig);
```

6.9.7.4.2 Feature table

There are a variety of alternative ways to create a feature table object, either by name or from an existing sequence object:

```
/* DNA feature table       */
AjPFeattable ajFeattableNewDna (const AjPStr name);

/* Protein feature table */
AjPFeattable ajFeattableNewProt (const AjPStr name);

/* From existing sequence; type is determined by the sequence type. */
AjPFeattable ajFeattableNewSeq (const AjPSeq seq);
```

For cases where a copy of a feature table is required that can be safely changed and/or deleted you can use:

```
/* Copy whole feature table */
AjPFeattable ajFeattableNewFtable (const AjPFeattable orig);

/* Copy limited number of features */
AjPFeattable ajFeattableNewFtableLimit (const AjPFeattable orig, ajint limit);
```

A feature table may be retrieved from a sequence object using these functions (defined in `ajseq.c`):

```
AjPFeattable ajSeqGetFeatCopy (const AjPSeq thys);
const AjPFeattable ajSeqGetFeat (const AjPSeq thys);
```

6.9.7.4.3 Editing a feature table

To add a new feature (`AjPFeature`) to a feature table (`AjPFeattable`) call:

```
void ajFeattableAdd (AjPFeattable thys, AjPFeature feature);
```

To clear a feature table of all features call:

```
void ajFeattableClear (AjPFeattable thys);
```

To clear an output feature table of all features call:

```
void ajFeattabOutClear (AjPFeattabOut *thys);
```

6.9.8 Reading features

Features may be read directly, using feature table input objects (these are the functions used by ACD processing):

```
AjPFeattable ajFeattableNewRead (AjPFeattabIn ftin);
AjPFeattable ajFeattableNewReadUfo (AjPFeattabIn tabin,
                                    const AjPStr Ufo);
```

`ajFeattableNewReadUfo` will parse a UFO, open an input file and read a feature table. `ajFeattableNewRead` is a generic interface function for reading in features from a feature table input object.

6.9.9 Writing features

Features may be written directly, i.e. without using ACD processing (which uses either GFF3 format by default or the format defined by the environment variable `EMBOSS_OUTFEATFORMAT`). Features may be written in any format defined in the data structure `FeatOOutFormat` defined in `ajfeat.c`:

```
AjBool ajFeatOutFormatDefault (AjPStr* pformat);

AjBool ajFeattableWriteUfo (AjPFeattabOut tabout, const AjPFeattable thys,
                            const AjPStr Ufo);

AjBool ajFeattableWrite (const AjPFeattable ft, AjPFeattabOut ftout);
```

`ajFeatOutFormatDefault` sets the default output format which is “gff” unless the `EMBOSS_OUTFEATFORMAT` variable is defined or a format is passed in the `pformat` parameter.

`ajFeattableWriteUfo` and `ajFeattableWrite` are equivalent to `ajFeattableNewRead` and `ajFeattableNewReadUfo` but for writing. `ajFeattableWriteUfo` will parse a UFO, open an output file and write a feature table to it. `ajFeattableWrite` is a generic interface function for writing features to a file given the file handle, class of map, data format of output and possibly other associated data.

The following functions write the feature table in the indicated format:

```
/* DDBJ format      */
AjBool ajFeattableWriteDdbj (const AjPFeattable features, AjPFile file);

/* EMBL format      */
AjBool ajFeattableWriteEmbl (const AjPFeattable features, AjPFile file);

/* Genbank format   */
AjBool ajFeattableWriteGenbank (const AjPFeattable features, AjPFile file);

/* GFF format       */
AjBool ajFeattableWriteGff2 (const AjPFeattable features, AjPFile file);

/* GFF format       */
AjBool ajFeattableWriteGff3 (const AjPFeattable features, AjPFile file);

/* PIR format       */
AjBool ajFeattableWritePir (const AjPFeattable features, AjPFile file);

/* SwissProt format */
AjBool ajFeattableWriteSwiss (const AjPFeattable features, AjPFile file);
```

Feature tables may be written to an application report using the following functions defined in `ajreport.h/c`:

```
void ajReportSetType (AjPReport thys, const AjPFeattable ftable, const AjPSeq seq);
AjBool ajReportWrite (AjPReport thys, const AjPFeattable ftable, const AjPSeq seq);
void ajReportWriteHeader (AjPReport thys, const AjPFeattable ftable, const
AjPSeq seq);
void ajReportWriteTail (AjPReport thys, const AjPFeattable ftable, const AjPSeq seq);
```

For more information on reports, see Section 6.15, 'Handling application reports'.

6.9.10 Output feature table functions

Functions described here are for manipulating an output feature table object.

To open the output file call:

```
AjBool ajFeattabOutOpen (AjPFeattabOut thys, const AjPStr ufo);
```

Elements of the output feature table object may be retrieved and queried using:

```
AjPFile ajFeattabOutFile (const AjPFeattabOut thys);
AjPStr ajFeattabOutFilename (const AjPFeattabOut thys);

/* These functions are used internally to test whether the output file
   has been opened and used */
AjBool ajFeattabOutIsLocal (const AjPFeattabOut thys);
AjBool ajFeattabOutIsOpen (const AjPFeattabOut thys);
```

Elements of an output feature table are set with:

```
/* sets the UFO (format and filename) for feature output */
AjBool ajFeattabOutSet (AjPFeattabOut thys, const AjPStr ufo);

/* sets the base file name (.format) for feature output */
void ajFeattabOutSetBasename (AjPFeattabOut thys, const AjPStr basename);

/* sets the feature table type 'any', 'N' 'nucleotide' or 'P' 'protein' */
AjBool ajFeattabOutSetType (AjPFeattabOut thys, const AjPStr type);
AjBool ajFeattabOutSetTypeC (AjPFeattabOut thys, const char* type);
```

6.9.11 Retrieving elements of a feature object

The elements of a feature object may be retrieved using the following:

```
/* End position */
ajuint ajFeatGetEnd (const AjPFeature thys);

/* Direction (ajTrue for a forward direction, ajFalse for reverse) */
AjBool ajFeatGetForward (const AjPFeature thys);

/* Reading frame */
ajint ajFeatGetFrame (const AjPFeature thys);

/* Sequence length */
ajuint ajFeatGetLength(const AjPFeature thys);

/* Finds a named note tag (a general tag value with a *name prefix) */
AjBool ajFeatGetNoteS (const AjPFeature thys, AjPStr* val, const AjPStr name);
AjBool ajFeatGetNoteSI (const AjPFeature thys, AjPStr* val, const AjPStr name,
ajint count);
AjBool ajFeatGetNoteC (const AjPFeature thys, AjPStr* val, const char* name);
AjBool ajFeatGetNoteCI (const AjPFeature thys,
AjPStr* val, const char* name, ajint count);

/* Returns the nth value of a named
 feature tag. If not found as a tag, also searches for a named note tag*/
AjBool ajFeatGetTagC (const AjPFeature thys, const char* tname, ajint num,
        AjPStr* Pval)
AjBool ajFeatGetTagS (const AjPFeature thys, const AjPStr name, ajint num,
                     AjPStr* val);

/* Score */
float ajFeatGetScore (const AjPFeature thys);

/* Source name */
const AjPStr ajFeatGetSource (const AjPFeature thys);

/* Start position */
ajuint ajFeatGetStart (const AjPFeature thys);

/* Strand */
char ajFeatGetStrand (const AjPFeature thys);

/* Returns the type (key) of a feature object. */
const AjPStr ajFeatGetType (const AjPFeature thys);
```

Caution

Note that **ajFeatGetType** returns a copy of the pointer to the type (key) of the specified feature object. The key is still owned by the feature and should not to be destroyed.

6.9.12 Retrieving elements of a feature table object

The elements of a feature table object may be retrieved be using the following:

```
/* Returns the feature table start position, or 1 if no start has been set.*/
ajint ajFeattableGetBegin (const AjPFeattable thys);

/* Returns the features table end position, or the
feature table length if no end has been set.*/
ajint ajFeattableGetEnd (const AjPFeattable thys);

/* Returns the name of a feature table object. */
const AjPStr ajFeattableGetName (const AjPFeattable thys);
const char* ajFeattableGetTypeC (const AjPFeattable thys);
const AjPStr ajFeattableGetTypeS (const AjPFeattable thys);

/* Returns the sequence length of a feature table */
ajint ajFeattableGetLen (const AjPFeattable thys);

/* Returns the number of features */
ajint ajFeattableGetSize (const AjPFeattable thys);
```

Caution

ajFeattableGetName, **ajFeattableGetTypeC** and **ajFeattableGetTypeS** return a copy of the pointer to the name or type (key). This is still owned by the feature table and so should not to be destroyed.

6.9.13 Setting elements of a feature object

The elements (indicated in comments) of a feature object may be set using the following:

```
/* Description */
void ajFeatSetDesc (AjPFeature thys, const AjPStr desc);

/* Append to description */
void ajFeatSetDescApp (AjPFeature thys, const AjPStr desc);

/* Score */
void ajFeatSetScore (AjPFeature thys, float score);

/* Strand */
void ajFeatSetStrand (AjPFeature thys, AjBool rev);
```

6.9.14 Setting elements of a feature table object

The elements of a feature table object may be set using the following:

```
/* Name */
void ajFeattableSetDefname (AjPFeattable thys, const AjPStr setname);

/* Sequence length */
void ajFeattableSetLength (AjPFeattable thys, ajuint len)

/* Type to nucleotide */
void ajFeattableSetNuc (AjPFeattable thys);

/* Type to protein */
void ajFeattableSetProt (AjPFeattable thys);

/* Begin and end range */
void ajFeattableSetRange (AjPFeattable thys, ajint fbegin, ajint fend) ;
*/
```

`ajFeattableSetDefname` will provides a unique name for the current program run for a feature table.

6.9.15 Functions for handling feature tags

Feature tags (names and values) are stored as pairs in a list. Tags can be returned as arrays or iterated over. When values are added (annotating a feature) they are usually simply defined as a name and value pair which replaces any existing value with the same tag name. Some tag names allow multiple values (for example the *EMBL/Genbank* feature 'note' used for general annotation and for 'named note tags' with a '*name' prefix to the value). These can be added as extra values using the `ajFeatTagAdd` functions. The tag values are validated against the most recent *EMBL/GenBank* feature table documentation. Warning messages are generated if variable `EMBOSS_FEATWARN` is set true, but turned off by default to avoid excessive warnings on data from other sources. Functions for handling feature tags include:

```
/* Sets a feature tag value, creating a new
feature tag even if one already exists. */
AjBool ajFeatTagAdd (AjPFeature thys, const AjPStr tag, const AjPStr value);
AjBool ajFeatTagAddC (AjPFeature thys, const char* tag, const AjPStr value);
AjBool ajFeatTagAddCC (AjPFeature thys, const char* tag, const char* value);

/* Sets a feature tag value */
AjBool ajFeatTagSet (AjPFeature thys, const AjPStr tag, const AjPStr value);
AjBool ajFeatTagSetC (AjPFeature thys, const char* tag, const AjPStr value);

/* Returns an iterator over all feature tag-value pairs */
AjIList ajFeatTagIter (const AjPFeature thys);

/* Returns the tag-value pairs of a feature object */
AjBool ajFeatTagval (AjIList iter, AjPStr* tagnam, AjPStr* tagval);

/* Traces (to the debug file) the tag-value pairs of a feature object */
void ajFeatTagTrace (const AjPFeature thys);
```

Functions for handling the `FeattabIn` object are available but not covered here as you will not normally need to use this object.

6.9.16 Querying properties of features

Functions are available to examine complex feature locations to process joins and their child (exon) features. Functions are provided to test whether a feature has a remote ID defined (the feature refers to another sequence) and to test the base range is within the range required for processing or for output. The properties of features may be queried using the following:

```
/* Tests whether the feature is a child member of a join */
AjBool ajFeatIsChild (const AjPFeature gf);

/* Tests whether the feature is a member of a complement around a multiple
(join, etc.) */
AjBool ajFeatIsCompMult (const AjPFeature gf);

/* Tests whether the feature is a member of a join, group order or one_of */
AjBool ajFeatIsMultiple (const AjPFeature gf);

/* Checks whether the feature is in another (remote id) sequence */
AjBool ajFeatIsLocal (const AjPFeature gf);

/* ... and tests the location is within a specified range */
AjBool ajFeatIsLocalRange (const AjPFeature gf, ajuint start, ajuint end);
```

6.9.17 Querying properties of feature tables

The type (nucleotide or protein) of a feature table may be queried using:

```
/* Returns ajTrue if nucleotide */
AjBool ajFeattableIsNuc (const AjPFeattable thys);

/* Returns ajTrue if protein */
AjBool ajFeattableIsProt (const AjPFeattable thys);
```

6.9.18 Processing features

There are a couple of functions for processing features:

```
void ajFeatReverse (AjPFeature thys, ajint ilen) ;
AjBool ajFeatTrimOffRange (AjPFeature ft, ajuint ioffset,
                           ajuint begin, ajuint end,
                           AjBool dobegin, AjBool doend);
```

`ajFeatReverse` will reverse a feature by reversing all positions and strand data.

`ajFeatTrimOffRange` trims a feature table using the specified `begin` and `end` values. `ajFeatTrimOffRange` is called where a sequence has been trimmed, so it is necessary to specify any missing sequence positions at the start (`ioffset`).

6.9.19 Processing feature tables

All features in a feature table may be reversed or trimmed using:

```
/* Reverse the features in a feature table by iterating through and reversing all
positions and strands. */
void ajFeattableReverse (AjPFeattable thys) ;

/* Trim a feature table using the Begin and Ends. */
AjBool ajFeattableTrimOff (AjPFeattable thys, ajuint ioffset, ajuint ilen);
```

To convert a position (start or end value) in a feature table into a true position in the source sequence, using any offset information from trimming the feature table within a set range:

```
ajuint ajFeattablePos (const AjPFeattable thys, ajint ipos);
ajuint ajFeattablePosI (const AjPFeattable thys, ajuint imin, ajint ipos);
ajuint ajFeattablePosII (ajuint ilen, ajuint imin, ajint ipos);
```

If `ipos` is negative, it is counted from the end of the string rather than the beginning. For strings the result can go off the end to the terminating `NULL`. For sequences the maximum is the last base.

Finally, features in a feature table may be sorted using:

```
/* End position */
void ajFeatSortByEnd (AjPFeattable Feattab);

/* Start position */
void ajFeatSortByStart (AjPFeattable Feattab);

/* Type */
void ajFeatSortByType (AjPFeattable Feattab);
```

6.9.20 Miscellaneous functions

There are a few miscellaneous functions for handling features:

```
/* Returns a sequence from a feature. */
AjBool ajFeatGetSeq(const AjPFeature thys, const AjPFeattable table,
                    const AjPSeq seq, AjPStr* Pseqstr);
```

6.10 Handling comparison matrices

6.10.1 Introduction

Matrices are commonly used in molecular sequence analysis to compare sequence characters at the same position in two or more aligned sequences.

Matrix objects are created by reading a local matrix data file, either through ACD (where the user has a choice of files), or by directly reading a named file where the filename is fixed.

EMBOSS includes sets of comparison matrix files in the data directory which can be used as examples when creating new files.

Matrix objects are in two very similar forms, using integers (`AjPMatrix`) for speed, and floating point numbers (`AjPMatrixf`) for flexibility. Both types include a set of column labels (usually sequence characters), and a set of row labels which usually matches the column labels (asymmetric matrices are used in some applications). For rows and columns there is also a matrix size value. The numbers in the data file become a two-dimensional table of comparison values.

Applications that analyse sequence alignments, for example **prettyplot**, can directly use the conversion table and character codes in a matrix to look up comparison scores using an `AjPSeqcvt` object.

6.10.2 AJAX library files

AJAX library files for handling matrices are listed in the table below. Library file documentation, including a complete description of datatypes and functions, is available at: http://emboss.open-bio.org/rel/dev/libs/

Library file documentation	Description
`ajmatrices`	Comparison matrix handling functions

`ajmatrices.h/c`. Defines the `AjPMatrix` and `AjPMatrixf` objects and functions for handling comparison matrices.

6.10.3 ACD datatypes

There are two datatypes for handling comparison matrix input:

`matrix`	Integer comparison matrix
`matrixf`	Floating point comparison matrix

There are two datatypes for handling comparison matrix output:

`outmatrix`	Output integer comparison matrix
`outmatrixf`	Output floating point comparison matrix

6.10.4 ACD data definition

Typical ACD definitions for comparison matrix input:

```
#Integer matrix (input)
matrix: matrix
[
    information: "Matrix file"
    protein: "$(acdprotein)"
]
# Floating point matrix (input)
matrixf: matrixf
[
    information: "Matrix file"
    protein: "$(acdprotein)"
]
```

Typical ACD definitions for comparison matrix output:

```
# Integer matrix (output)
outmatrix: outmatrix
[
    information: "Matrix file"
    protein: "$(acdprotein)"
]
# Floating point matrix (output):
outmatrixf: outmatrixf
[
    information: "Matrix file"
    protein: "$(acdprotein)"
]
```

6.10.4.1 Parameter name

All data definitions for comparison matrix input and output should have a standard parameter name, which is `matrix`. For further information see Appendix A, *ACD syntax reference*.

6.10.4.2 Common attributes

Attributes that are typically specified are summarised below. They are datatype-specific (Section A.5, 'Datatype-specific attributes') unless they are indicated as being global attributes (Section A.4, 'Global attributes').

`information:` A global attribute.
It specifies the user prompt and is used in the application documentation.

`protein:` A boolean attribute which if set specifies that the matrix is for proteins.
If not set the matrix is presumed to be for nucleic acids.

6.10.5 AJAX datatypes

For handling comparison matrices, including input matrices defined in the ACD file, use:

`AjPMatrix`	Integer comparison matrix (for `matrix` ACD datatype)
`AjPMatrixf`	Floating point comparison matrix (for `matrixf` ACD datatype)

For handling comparison matrix output use:

`AjPOutfile`	General output file (for `outmatrix` and `outmatrixf` ACD datatypes)

It is sometimes necessary to convert a sequence into numerical form for convenient processing. The AJAX datatype for this is:

`AjOSeqCvt`	Used for sequence conversion into numerical form

6.10.6 ACD file handling

Datatypes and functions for handling comparison matrices via the ACD file are shown below.

ACD datatype	AJAX datatype	To retrieve from ACD
Comparison matrix input		
`matrix`	`AjPMatrix`	`ajAcdGetMatrix`
`matrixf`	`AjPMatrixf`	`ajAcdGetMatrixf`
Comparison matrix output		
`outmatrix`	`AjPMatrix`	`ajAcdGetOutmatrix`
`outmatrixf`	`AjPMatrixf`	`ajAcdGetOutmatrixf`

Your application code will call `embInit` to process the ACD file and command line (see Section 6.3, 'Handling ACD files'). All values from the ACD file are read into memory and files are opened as necessary. You have a handle on the files and memory through the `ajAcdGet*` family of functions which return pointers to appropriate objects.

6.10.6.1 Input comparison matrix retrieval

To retrieve a comparison matrix, an object pointer is declared and then initialised using the appropriate `ajAcdGet*` function.

6.10.6.1.1 Integer comparison matrix

```
AjPMatrix matrix = NULL;
matrix = ajAcdGetMatrix("matrix");
```

6.10.6.1.2 Floating point comparison matrix

```
AjPMatrixf matrix = NULL;
matrix = ajAcdGetMatrixf("matrix");
```

6.10.6.2 Output comparison matrix retrieval

To retrieve an output comparison matrix an object pointer is declared and initialised using the appropriate `ajAcdGet*` function.

6.10.6.2.1 Integer comparison matrix

```
AjPOutfile outmatrix=NULL;
outmatrix = ajAcdGetOutmatrix("outmatrix");
```

6.10.6.2.2 Floating point comparison matrix

```
AjPOutfile outmatrixf=NULL;
outmatrixf = ajAcdGetOutmatrixf("outmatrixf");
```

6.10.6.3 Processing command line options and ACD attributes

Currently there are no functions for this.

6.10.6.4 Memory and file management

It is your responsibility to close any files and free up memory at the end of the program.

6.10.6.4.1 Closing files

You must close the output file for any `outmatrix` or `outmatrixf` definitions in the ACD file by calling `ajOutfileClose` with the address of the output file object:

```
AjPOutfile outmatrix  = NULL;
AjPOutfile outmatrixf = NULL;

outmatrix = ajAcdGetOutmatrix("outmatrix");
outmatrixf = ajAcdGetOutmatrixf("outmatrixf");

/* Do something with matrices */

ajOutfileClose(&outmatrix);
ajOutfileClose(&outmatrixf);
```

6.10.6.4.2 Freeing memory

You must call the default destructor function (see below) on any comparison matrix objects returned by calls to `ajAcdGet*`.

6.10.7 Matrix object memory management

6.10.7.1 Default object construction

Matrix objects are usually created through an ACD definition, reading a named EMBOSS local data file. To create a matrix object directly (perhaps where there is no choice of matrix filename) the object must be constructed from a given local data file by calling:

```
AjBool ajMatrixNewFile  (AjPMatrix* pthis, const AjPStr filename);
AjBool ajMatrixfNewFile (AjPMatrixf* pthis, const AjPStr filename);
```

The functions take the name of the data file to open. The file must be found in the EMBOSS data path, including the current directory and the installed data files.

All constructors return the address of a new object. The pointers do not need to be initialised to `NULL` but it is good practice to do so:

```
AjPMatrix intmatrix = NULL;
AjPMatrixf floatmatrix = NULL;
AjPStr filename = NULL;

filename = ajStrNewC("EBLOSUM62");

intmatrix = ajMatrixNewFile(filename);
floatmatrix = ajMatrixfNewFile(filename);
```

All constructors return the address of a new object. The pointers do not need to be initialised to `NULL` but it is good practice to do so:

6.10.7.2 Default object destruction

You must close any output files and free the memory for your objects once you are finished with them.

To close an output file (`AjPOutfile`) call `ajOutfileClose`, or call `ajFileClose` for general file objects (`AjPFile`):

```
void ajFileClose    (AjPFile* Pfile);
void ajOutfileClose (AjPOutfile* Pfile);
```

The objects are freed by calling the destructor functions:

```
void ajMatrixDel (AjPMatrix *thys);
void ajMatrixfDel (AjPMatrixf *thys);
```

They are used as follows:

```
AjPMatrix matrix  = NULL;
AjPMatrix matrixf = NULL;
```

```
matrix = ajAcdGetMatrix("matrix");
matrixf = ajAcdGetMatrixf("matrixf");

/* Do something with matrices */

ajMatrixDel(&matrix);
ajMatrixfDel(&matrixf);
```

6.10.7.3 Alternative object construction and loading

Internally, the matrix object constructor functions are:

```
AjPMatrix ajMatrixNew (const AjPPStr codes, ajint n, const AjPStr filename);
AjPMatrixf ajMatrixfNew (const AjPPStr codes, ajint n, const AjPStr filename);
```

These will create a new matrix with values initialised to zero. The functions take the matrix name (`filename`), a string (`codes`) containing characters for the matrix labels, and an integer (`n`) that is the number of labels. If the matrix is a residue substitution matrix then the string should contain defined sequence characters.

The matrices that are created by `ajMatrixNew` and `ajMatrixfNew` are square, having the same number of rows and columns. To create a matrix with an unequal number of rows and columns call:

```
AjPMatrix ajMatrixNewAsym (const AjPPStr codes, ajint n, const AjPPStr rcodes,
ajint rn, const AjPStr filename);
AjPMatrixf ajMatrixfNewAsym (const AjPPStr codes, ajint n, const AjPPStr rcodes,
ajint rn, const AjPStr filename);
```

These will create a new matrix with values initialised to zero. The functions take the matrix name (`filename`), and two strings (`codes` and `rcodes`) containing characters for the matrix column and row labels respectively, and two integers (`n` and `rn`) that are the number of column and row labels.

EMBOSS requires all matrix objects to be loaded from data files. No functions are provided to add or change the matrix object values. The following functions read from an EMBOSS data file. For more information on EMBOSS data files, see the *EMBOSS User's Guide*.

A matrix can be constructed from a given local data file by calling:

```
AjBool ajMatrixNewFile (AjPMatrix* pthis, const AjPStr filename);
AjBool ajMatrixfNewFile (AjPMatrixf* pthis, const AjPStr filename);
```

The functions take the name of the data file to open.

6.10.8 Functions for retrieving the properties of a matrix

Most elements of a matrix object can be retrieved by calling one of the `ajMatrixGet*` or `ajMatrixfGet*` functions:

To return the comparison matrix as an array of integer or floating point arrays call:

```
AjIntArray* ajMatrixGetMatrix (const AjPMatrix thys);
AjFloatArray* ajMatrixfGetMatrix (const AjPMatrixf thys);
```

Sequence characters are indexed in this array using the internal sequence conversion table in the matrix. `AjIntArray` and `AjFloatArray` are defined `ajdefine.h` as arrays of C-type integer and floating point numbers:

```
typedef float* AjFloatArray;
typedef int* AjIntArray;
```

To return the label (sequence character or the column name for an asymmetric matrix) for a matrix row or column in position `i` call:

```
AjPStr ajMatrixGetLabelNum (const AjPMatrix thys, ajint i);
AjPStr ajMatrixfGetLabelNum (const AjPMatrixf thys, ajint i);
```

To return the sequence character conversion table for a matrix call:

```
AjPSeqCvt ajMatrixGetCvt (const AjPMatrix thys);
AjPSeqCvt ajMatrixfGetCvt (const AjPMatrixf thys);
```

This table converts any character defined in the matrix to a positive integer, and any other character is converted to zero.

To return the character codes for each offset in the matrix call:

```
AjPStr ajMatrixGetCodes (const AjPMatrix thys);
AjPStr ajMatrixfGetCodes (const AjPMatrixf thys);
```

To return the name of a matrix object (which typically is the filename from which it was read), call:

```
const AjPStr ajMatrixGetName (const AjPMatrix thys);
const AjPStr ajMatrixfGetName (const AjPMatrixf thys);
```

To return the comparison matrix size (or the number of columns for an asymmetric matrix) call:

```
ajuint ajMatrixGetSize (const AjPMatrix thys);
ajuint ajMatrixfGetSize (const AjPMatrixf thys);
```

For an asymmetric matrix the number of rows can be returned by calling:

```
ajuint ajMatrixGetRows (const AjPMatrix thys);
ajuint ajMatrixfGetRows (const AjPMatrixf thys);
```

6.10.9 Functions for indexing a matrix

To convert a sequence to index numbers using the matrix's internal conversion table call:

```
AjBool ajMatrixSeqIndex (const AjPMatrix thys, const AjPSeq seq, AjPStr* numseq);
AjBool ajMatrixfSeqIndex (const AjPMatrixf thys, const AjPSeq seq, AjPStr* numseq);
```

Sequence characters not defined in the matrix are converted to zero.

6.10.10 Sequence conversion

These functions handle sequence conversion objects. The basic constructor `ajSeqcvtNewStr` uses an array of strings as the column labels. For sequence comparison matrices these strings will be one character each. The other constructors renumbers the codes for the specific expectations of some older legacy code and are not recommended for general use. The most useful functions are those which return the numeric code for a base or residue, and are frequently used to look up a sequence character in a conversion table.

```
/* constructors with base codes as a string*/
AjPSeqCvt ajSeqcvtNewStr (const AjPPStr bases, ajint n);
AjPSeqCvt ajSeqcvtNewC (const char* bases);
AjPSeqCvt ajSeqcvtNewNumberC (const char* bases);
AjPSeqCvt ajSeqcvtNewEndC (const char* bases);

/* asymmetric conversion table constructor */

AjPSeqCvt ajSeqcvtNewStrAsym (const AjPPStr bases, ajint n, const AjPPStr rbases,
ajint rn);

/* destructor */
void        ajSeqcvtDel (AjPSeqCvt* thys);

/* return conversion table length */
ajuint      ajSeqcvtGetLen (const AjPSeqCvt thys);

/* return numeric code for a residue code (matrix column label) */
ajint       ajSeqcvtGetCodeK (const AjPSeqCvt thys, char ch);
ajint       ajSeqcvtGetCodeS (const AjPSeqCvt thys, const AjPStr ch);

/* return numeric code for a column or row in an asymnmetric matrix */
ajint       ajSeqcvtGetCodeAsymS (const AjPSeqCvt cvt, const AjPStr str);
ajint       ajSeqcvtGetCodeAsymrowS (const AjPSeqCvt cvt, const AjPStr str);
```

6.11 Handling alignments

6.11.1 Introduction

EMBOSS supports most of the common sequence alignment formats for input and output (see the *EMBOSS User's Guide* for a complete list). Alignments to be read or written are defined in the application ACD file although it is possible to create alignment objects directly if this is required.

Most of the alignment formats can include a standard header (given at the start of the alignment file) and in some cases a tail (given at the end) which gives information including the program, date, output filename, ID names of the sequences and some of the parameters and statistics of the alignment. There is also a subheader and subtail used for additional comments and annotation.

The tail section is used by some applications (e.g. **merger** in EMBOSS) to report special features of the alignment.

All alignments have certain basic properties:

- Type of sequences (protein or nucleotide)
- Alignment format
- Number of sequences in the alignment
- Minimum permissible number of sequences
- Maximum permissible number of sequences
- Output width.

In addition there is often associated data including:

- An integer or floating point matrix
- Name of matrix
- Gap insertion penalty
- Gap extension penalty.

For global alignments the full sequence or just the matching regions can be displayed. Optionally the alignment can include the accession number, sequence description and full USA of the aligned sequences.

The options above are usually set in the application ACD file (via attributes of the data definition) or on the command line (via qualifiers that are specific to alignments). Functions are provided to set these directly in case this is required. For a description of these attributes and qualifiers see Section A.5, 'Datatype-specific attributes'.

Functions for manipulating alignments are organised into four groups:

- Writing alignments
- Retrieving elements of an alignment object
- Setting elements of an alignment object
- Miscellaneous functions.

6.11.2 AJAX library files

AJAX library files for handling alignments are listed in the table below. Library file documentation, including a complete description of datatypes and functions, is available at: http://emboss.open-bio.org/rel/dev/libs/

Library File Documentation	Description
ajalign	Functions for handling sequence alignments
ajseq	General sequence handling

ajalign.h/c. Defines the main alignment object (AjPAlign). It can be used for retrieving an input sequence alignment via ACD file processing. The header file contains most of the functions you will ever need for general handling of sequence alignments. It includes static datatypes and functions for handling alignments at a low level. You are unlikely to need the latter unless you plan to implement code to support new alignment formats. For advice on how to do this ask the EMBOSS developers.

ajseq.h/c. Defines the AjPSeqset object. This is a set of sequences for general use and is used for handling input alignments from ACD files. ajseq.h/c contain extensive functions for handling sequence sets and thereby for general manipulations of sequence alignments.

6.11.3 ACD datatypes

Alignment input is handled as a special case of general sequence set input. The seqset ACD datatype is used:

seqset Read multiple sequences as a single set. The ACD data definition must include the aligned: attribute (see below).

There is a dedicated datatype for output alignments:

align Alignment output

6.11.4 ACD data definition

A typical ACD definition for an input alignment:

```
# multiple sequence input read as a single aligned set
seqset: sequence
[
    parameter: "Y"
    type: "protein"
```

```
    aligned: "Y"
]
```

A typical ACD definition for an output alignment:

```
align: outfile
[
    parameter: "Y"
    aformat: "srspair"
    type: "protein"
    minseqs: "2"
    maxseqs: "2"
    aglobal: "Y"
]
```

6.11.4.1 Parameter name

All data definitions for alignment input and output should have standard parameter names. These include:

- `sequence` for any aligned input sequences
- `outfile` for alignment output
- Alternatives and variations (e.g. `afile`, `bfile` for multiple outputs) are allowed.

For more information see Appendix A, *ACD syntax reference*.

6.11.4.2 Common attributes

Attributes that are typically specified are summarised below. They are datatype-specific (Section A.5, 'Datatype-specific attributes') unless they are indicated as being global attributes (Section A.4, 'Global attributes').

`parameter:` Alignments are typically the primary input or output of an EMBOSS application and, as such, should be defined as parameters by using the global attribute `parameter: "Y"`.

`type:` Specifies the type of the sequences in the input or output alignment and is used for validation purposes.

`aligned:` Any `seqset` or `seqsetall` datatype *must* have the `aligned:` attribute set to indicate whether the sequences are aligned or not.

`aformat:` The output format is normally set at the command line but a default may be hard-coded with `aformat:`. All common alignment formats are supported (see the *EMBOSS User's Guide*).

`minseqs:` Specifies the minimum number of expected sequences and is used for validation of output.

maxseqs: Specifies the maximum number of expected sequences and is used for validation of output.

aglobal: A boolean attribute which is set to "Y" if the output can contain more than one alignment from the same input.

6.11.5 AJAX datatypes

For handling alignments, including input alignments defined in the ACD file, use:

`AjPSeqset` Input alignment

For handling output alignments defined in the ACD file use:

`AjPAlign` Output alignment

6.11.6 ACD file handling

Datatypes and functions for handling alignments via the ACD file are shown below.

Title	To read an alignment	To write an alignment
ACD datatype	seqset	align
AJAX datatype	AjPSeqset	AjPAlign
To retrieve from ACD	ajAcdGetSeqset	ajAcdGetAlign

Your application code will call `embInit` to process the ACD file and command line (see Section 6.3, 'Handling ACD files'). All values from the ACD file are read into memory and files are opened as necessary. You have a handle on the files and memory through the `ajAcdGet*` family of functions which return pointers to appropriate objects.

6.11.6.1 Input alignment retrieval

To retrieve an input alignment an object pointer is declared and then initialised using `ajAcdGetSeqset`:

```
AjPSeqset seqset=NULL;
seqset = ajAcdGetSeqset("sequence");
```

6.11.6.2 Output alignment retrieval

To retrieve an output alignment stream an object pointer is declared and initialised using `ajAcdGetAlign`:

```
AjPAlign outfile=NULL;
outfile = ajAcdGetAlign("outfile");
```

6.11.6.3 Memory and file management

It is your responsibility to close any files and free up memory at the end of the program.

6.11.6.3.1 Closing output alignment files
To close an output alignment stream the AJAX function `ajAlignClose` is used:

```
ajAlignClose (outfile);
```

6.11.6.3.2 Freeing memory
You must call the default destructor function (see below) on any objects returned by calls to `ajAcdGet*`.

Additionally you must call `ajAlignExit` to clean up internal memory allocated for housekeeping of alignment processing:

```
ajAlignExit();
```

6.11.7 Alignment object memory management

6.11.7.1 Default object construction

Alignment output objects are typically loaded from ACD file processing (see above). In the unlikely event that you need to create one manually you can use the default alignment object constructor `ajAlignNew`. All constructors return the address of a new object. In the following code the pointer does not need to be initialised to `NULL` but it is good practice to do so:

```
AjPAlign align = NULL;

align = ajAlignNew();
/* The object is instantiated and ready for use */
```

6.11.7.2 Default object destruction

You must free the memory for an object, once you are finished with it. The default destructor function is:

```
void ajAlignDel (AjPAlign* pthys); /* Destructor for Alignment objects */
```

It is used as follows:

```
AjPAlign align=NULL;

align = ajAcdGetAlign("align");
```

```
/* Do something with alignment */

ajAlignDel(&align);
```

6.11.7.3 Alternative object construction and loading

Applications that create alignment outputs usually generate aligned sequences which are then used to populate the alignment object. The following functions are available:

```
/* Defines a sequence set as an alignment. The sequences
 are stored internally and may
be edited by alignment processing. */
AjBool ajAlignDefine (AjPAlign pthys, AjPSeqset seqset);

/* Defines a sequence pair as an alignment.
The sequences are stored internally and may
be edited by alignment processing. */
AjBool ajAlignDefineSS (AjPAlign pthys,
                        AjPSeq seqa, AjPSeq seqb);

/* Defines a pair of char* strings as an alignment (names of sequences are also
required) */
AjBool ajAlignDefineCC (AjPAlign pthys,
                        const char* seqa, const char* seqb,
                        const char* namea,const char* nameb);
```

6.11.8 Writing alignments

There are several AJAX functions for writing out alignment information. Applications will usually create an alignment object through ACD processing, populate it with aligned sequences (see above) and call `ajAlignWrite`.

```
/* Writes an alignment file */
void ajAlignWrite (AjPAlign thys);

/* Reset to allow resue of Alignment objects */
void ajAlignReset (AjPAlign thys);

/* Opens a new align file. Called bvy ACD processing*/
AjBool ajAlignOpen (AjPAlign thys, const AjPStr name);

/* Writes an alignment header. Called by ajAlignWrite */
void ajAlignWriteHeader (AjPAlign thys);

/* Writes an alignment tail Called by ajAlignWrite */
void ajAlignWriteTail (AjPAlign thys);

/* Sets the default format for an alignment to 'gff' if not already defined */
AjBool ajAlignFormatDefault (AjPStr* pformat);
```

6.11.9 Retrieving elements of an alignment object

Alignment object elements rarely need to be examined by the programmer. Functions are available to retrieve internal values.

```
/*Returns the filename for an alignment. If the alignment has more than
one subalignment, returns the total.*/
ajint ajAlignGetLen (const AjPAlign thys);

/* Returns the filename */
const char* ajAlignGetFilename (const AjPAlign thys);

/* Returns the sequence format */
const AjPStr ajAlignGetFormat (const AjPAlign thys);
```

6.11.10 Setting elements of an alignment object

Alignment objects have elements that are used to populate the header and tail sections of the output (where the output format can include such extra detail).

Elements that can be set in the alignment header include:

- Gap penalties
- Matrix name
- Alignment score.

Functions for this are below. Note the matrix name can be set directly or from a matrix object:

```
/* Setting elements of the alignment header */
void ajAlignSetGapI (AjPAlign thys, ajint gappen, ajint extpen);
void ajAlignSetGapR (AjPAlign thys, float gappen, float extpen);
void ajAlignSetMatrixName (AjPAlign thys, const AjPStr matrix);
void ajAlignSetMatrixNameC (AjPAlign thys, const char* matrix);
void ajAlignSetMatrixInt (AjPAlign thys, AjPMatrix matrix);
void ajAlignSetMatrixFloat (AjPAlign thys, AjPMatrixf matrix);
void ajAlignSetScoreI (AjPAlign thys, ajint score);
void ajAlignSetScoreL (AjPAlign thys, ajlong score);
void ajAlignSetScoreR (AjPAlign thys, float score);
```

The standard properties in alignment subheader are:

- Length
- Identity
- Gaps
- Similarity
- Score.

The function to set these is:

```
void ajAlignSetSubStandard (AjPAlign thys, ajint iali);
```

Alternatively, you may set these manually using:

```
void ajAlignSetStats (AjPAlign thys, ajint iali, ajint len,
                      ajint ident, ajint sim, ajint gaps,
                      const AjPStr score);
```

The header section can include an optional comment. Similarly, the tail section is free text available to report any special notes on the alignment. Comments can be set in the (sub) header and (sub)tail, or appended or prepended too, using the following functions:

```
/* Setting comments of the alignment (sub)header and (sub)tail */
void   ajAlignSetHeader (AjPAlign thys, const AjPStr header);
void   ajAlignSetHeaderApp (AjPAlign thys, const AjPStr header);
void   ajAlignSetHeaderC (AjPAlign thys, const char* header);
void   ajAlignSetSubHeader (AjPAlign thys, const AjPStr subheader);
void   ajAlignSetSubHeaderApp (AjPAlign thys, const AjPStr subheader);
void   ajAlignSetSubHeaderC (AjPAlign thys, const char* subheader);
void   ajAlignSetSubHeaderPre (AjPAlign thys, const AjPStr subheader);
void   ajAlignSetSubTail(AjPAlign thys, const AjPStr tail);
void   ajAlignSetSubTailC(AjPAlign thys, const char* tail);
void   ajAlignSetSubTailApp(AjPAlign thys, const AjPStr tail);
void   ajAlignSetTail (AjPAlign thys, const AjPStr tail);
void   ajAlignSetTailApp (AjPAlign thys, const AjPStr tail);
void   ajAlignSetTailC (AjPAlign thys, const char* tail);
```

The alignment type (protein or nucleic) may be set directly (if it is not set already):

```
void ajAlignSetType (AjPAlign thys);
```

A range (or sub-ranges) of sequences to output can be set using:

```
/* Setting sequence range to output */
AjBool  ajAlignSetRange (AjPAlign thys,
                         ajint start1, ajint end1,
                         ajint len1, ajint off1,
                         ajint start2, ajint end2,
                         ajint len2, ajint off2);

AjBool  ajAlignSetSubRange (AjPAlign thys,
                            ajint substart1, ajint start1,
                            ajint end1, AjBool rev1, ajint len1,
                            ajint substart2, ajint start2,
                            ajint end2, AjBool rev2, ajint len2);
```

Finally, there is a function to set the alignment object to use external sequence references, which are references (copied pointers) rather than clones of the actual sequences:

```
/* Setting properties of alignment object */
void  ajAlignSetExternal (AjPAlign thys, AjBool external);
```

This is intended for alignments of large sequences where it is undesirable to keep many copies.

6.11.11 Miscellaneous functions

The `ajAlignConsStats` function calculates a consensus sequence and statistics (percent identity and similarity and alignment length) for a multiple alignment.
`ajAlignFindFormat`is used in ACD processing to match the specified alignment format to the internal list of known formats.

```
/* Calculates alignment statistics (and consensus) */
AjBool ajAlignConsStats (const AjPSeqset thys, AjPMatrix mymatrix,
                         AjPStr *cons, ajint* retident,
                         ajint* retsim, ajint* retgap,
                         ajint* retlen);

/* Finds alignment format in internal definitions, returns index */
AjBool ajAlignFindFormat (const AjPStr format, ajint* iformat);
```

6.12 Handling phylogenetic data

6.12.1 Introduction

The EMBASSY **phylipnew** package includes various applications for phylogenetic analysis.

A set of phylogenenetic data types is available to replicate the available input data types for **phylip** with automatic detection of the data formats (for example, distance matrix files):

AJAX datatype	ACD datatype (for reading)	ACD datatype (for writing)
`AjPPhyloDist` distance matrix data	`distances`	`outdistance`
`AjPPhyloFreq` frequency data	`frequencies`	`outfreq`
`AjPPhyloProp` properties data	`properties`	`outproperties`
`AjPPhyloState` state data	`discretestates`	`outdiscrete`
`AjPPhyloTree` phylogenetic tree data	`tree`	`outtree`

6.12.2 AJAX library files

AJAX library files for handling phylogenetic data are listed in the table below. Library file documentation, including a complete description of datatypes and functions, is available at: http://emboss.open-bio.org/rel/dev/libs/

Library file documentation	Description
ajphylo	Data structures and functions for handling the **phylipnew** applications
ajnexus	Data structures and functions for parsing the NEXUS file format

ajphylo.h/c. Defines the objects and functions for handling phylogenetic data. These include:

- Phylogeny distance matrix object (AjPPhyloDist)
- Phylogeny frequencies object (AjPPhyloFreq)
- Phylogeny properties object (AjPPhyloProp)
- Phylogeny discrete state data object (AjPPhyloState)
- Phylogeny tree object (AjPPhyloTree).

They also include static functions for handling phylogenetic data at a low level. You are unlikely to need these unless you plan to extend the phylogeny handling code.

ajnexus.h/c. Functions and objects (including static data structures and functions) for parsing the NEXUS file format. You are unlikely to need this library file. See the online library documentation for further information.

6.12.3 AJAX datatypes

For handling phylogenetic data input files defined in the ACD file use:

AjPPhyloState*	Phylogeny discrete state data object (for discretestates ACD datatypes)
AjPPhyloDist	Phylogeny distance matrix object (for distances ACD datatype)
AjPPhyloFreq	Phylogeny frequencies object (for frequencies ACD datatype)
AjPPhyloProp	Phylogeny properties object (for properties ACD datatype)
AjPPhyloTree*	Phylogeny tree object (for tree ACD datatype)

For handling phylogenetic data output files defined in the ACD file use:

AjPOutfile   Output file (for all phylogenetic output ACD datatypes).

6.12.4 ACD datatypes

The datatypes for handling phylogenetic data input are:

`discretestates`	Discrete states file
`distances`	Distance matrix
`frequencies`	Frequency value(s)
`properties`	Property value(s)
`tree`	Phylogenetic tree

The datatypes for handling phylogenetic data output are:

`outdiscrete`	Output file for phylogenetics discrete characteristics data
`outdistances`	Output file for phylogenetics distance matrix data
`outfreq`	Output file for phylogenetics character frequency data
`outproperties`	Output file for phylogenetics property data
`outtree`	Output file for phylogenetic tree data

6.12.5 ACD data definition

Typical ACD definitions for phylogenetic data input and output are shown below.

6.12.5.1 discretestates

Input of discrete states data:

```
discretestates: discretestatesfile
[
    parameter: "Y"
    characters: "01PB?"
    knowntype: "discrete states"
    information: "Phylip discrete states file"
]
```

6.12.5.2 distances

Input of distances data:

```
distances: distancesfile
[
    parameter: "Y"
    knowntype: "distance matrix"
    information: "Phylip distance matrix file"
]
```

6.12.5.3 frequencies

Input of frequencies data:

```
frequencies: frequenciesfile
[
    parameter: "Y"
]
```

6.12.5.4 properties

Input of properties data:

```
properties: propertiesfile
[
    characters: "01"
    length: "$(infile.discretelength)"
    knowntype: "ancestral states"
    information: "Phylip ancestral states file"
]
```

6.12.5.5 tree

Input of tree data:

```
tree: treefile
[
    parameter: "Y"
    knowntype: "newick"
    information: "Phylip tree file (optional)"
]
```

6.12.5.6 outdiscrete

Output of discrete states data:

```
outdiscrete: outdiscretefile
[
    parameter: "Y"
]
```

6.12.5.7 outdistance

Output of distances data:

```
outdistance: outdistancefile
[
    parameter: "Y"
]
```

6.12.5.8 `outfreq`

Output of frequencies data:

```
outfreq: outfreqfile
[
    parameter: "Y"
]
```

6.12.5.9 `outproperties`

Output of properties data:

```
outproperties: outpropertiesfile
[
    parameter: "Y"
]
```

6.12.5.10 `outtree`

Output of tree data:

```
outtree: outtreefile
[
    parameter: "Y"
]
```

6.12.5.11 Parameter name

All data definitions for phylogenetic data input and output should have intuitive names. There are some general guidelines but currently no specific naming rules are enforced. See Appendix A, *ACD syntax reference*.

6.12.5.12 Common attributes

Attributes that are typically specified are summarised below. They are datatype-specific (Section A.5, 'Datatype-specific attributes') unless they are indicated as being global attributes (Section A.4, 'Global attributes').

`parameter:` If the phylogenetic data is the primary input or output of an EMBOSS application then it should be defined as a parameter by using the global attribute `parameter: "Y"`.

`characters:` Specifies the allowed discrete state or property characters for a `discretestates` or `properties` object respectively.

`knowntype:` This global attribute is typically specified for all the phylogenetic input and output types.

`information:` A global attribute used for the user prompt and in the application documentation.

length: Specifies the number of property values per set (`properties` datatype) or the number of frequency loci / values per set (`frequencies` datatype).

size: Specifies the number of discrete state sets (`discretestates` datatype), the number of frequency sets (`frequencies` datatype) or the number of trees (`tree` datatype).

> **Note**
>
> Various calculated attributes (Section A.6, 'Calculated attributes') of the datatypes are available at the level of the ACD file.

6.12.6 ACD file handling

Datatypes and functions for handling phylogenetic data via the ACD file are shown below.

ACD datatype	AJAX datatype	To retrieve from ACD
Phylogenetic data input		
discretestates	AjPPhyloState*	ajAcdGetDiscretestates
distances	AjPPhyloDist	ajAcdGetDistances
frequencies	AjPPhyloFreq	ajAcdGetFrequencies
properties	AjPPhyloProp	ajAcdGetProperties
tree	AjPPhyloTree*	ajAcdGetTree
Phylogenetic data output		
outdiscrete	AjPOutfile	ajAcdGetOutdiscrete
outdistance	AjPOutfile	ajAcdGetOutdistance
Outfreq	AjPOutfile	ajAcdGetOutfreq
outproperties	AjPOutfile	ajAcdGetOutproperties
Outtree	AjPOutfile	ajAcdGetOuttree

Your application code will call `embInit` to process the ACD file and command line (see Section 6.3, 'Handling ACD files'). All values from the ACD file are read into memory and files are opened as necessary. You have a handle on the files and memory through the `ajAcdGet*` family of functions which return pointers to appropriate objects.

6.12.6.1 Phylogenetic data retrieval

6.12.6.1.1 Input phylogenetic data

To retrieve input phylogenetic data an object pointer is declared and then initialised using the appropriate `ajAcdGet*` function.

discretestates

```
AjPPhyloState *data=NULL;
```

```
data = ajAcdGetDiscretestates("discretestatesfile");
```

distances

```
AjPPhyloDist data=NULL;

data = ajAcdGetDistances("distancesfile");
```

frequencies

```
AjPPhyloFreq data=NULL;

data = ajAcdGetFrequencies("frequenciesfile");
```

properties

```
AjPPhyloProp data=NULL;

data = ajAcdGetProperties("propertiesfile");
```

tree

```
AjPPhyloTree* data=NULL;

data = ajAcdGetTree("treefile");
```

6.12.6.1.2 Output phylogenetic data

To retrieve an output phylogenetic data stream an object pointer is declared and initialised using the appropriate `ajAcdGet*` function.

outdiscrete

```
AjPOutfile outfile=NULL;

outfile = ajAcdGetOutdiscrete("outdiscretefile");
```

outdistance

```
AjPOutfile outfile=NULL;

outfile = ajAcdGetOutdistance("outdistancefile");
```

outfreq

```
AjPOutfile outfile=NULL;

outfile = ajAcdGetOutfreq("outfreqfile");
```

`outproperties`

```
AjPOutfile outfile=NULL;

outfile = ajAcdGetOutproperties("outpropertiesfile");
```

`outtree`

```
AjPOutfile outfile=NULL;

outfile = ajAcdGetOuttree("outtreefile");
```

6.12.6.1.3 Alternative ACD retrieval functions

There are functions to retrieve a single (the first) state or tree object from file:

```
AjPPhyloState ajAcdGetDiscretestatesSingle (const char *token);
AjPPhyloTree  ajAcdGetTreeSingle (const char *token);
```

Where these are used, it is still necessary to call the appropriate destructor function (see below) to ensure that the array of state or tree objects allocated during ACD file processing is freed.

6.12.6.2 Processing command line options and ACD attributes

Currently there are no functions for this.

6.12.6.3 Memory and file management

It is your responsibility to close any files and free up memory at the end of the program.

6.12.6.3.1 Closing output phylogenetic data files

To close an output phylogenetic data stream call `ajOutfileClose` with the address of the output file:

```
ajOutfileClose(&outfile);
```

6.12.6.3.2 Freeing memory

You must call the appropriate destructor function (see below) on any phylogenetic data objects returned by calls to `ajAcdGet*`.

Additionally, you must call `ajPhyloExit` to free up any internal memory allocated internally for housekeeping:

```
void ajPhyloExit(void);
```

6.12.7 Phylogenetic object memory management

6.12.7.1 Default object construction

To use a phylogenetic data object that is not defined in the ACD file you must first instantiate the appropriate object pointer. The default constructor functions are:

```
AjPPhyloDist    ajPhyloDistNew (void);
AjPPhyloFreq    ajPhyloFreqNew (void);
AjPPhyloProp    ajPhyloPropNew (void);
AjPPhyloState   ajPhyloStateNew (void);
AjPPhyloTree    ajPhyloTreeNew (void);
```

6.12.7.2 Default object destruction

You must free the memory for an object once you are finished with it. The default destructor functions are:

```
void   ajPhyloDistDel (AjPPhyloDist* pthis);
void   ajPhyloFreqDel (AjPPhyloFreq* pthis);
void   ajPhyloPropDel (AjPPhyloProp* pthis);
void   ajPhyloStateDel (AjPPhyloState* pthis);
void   ajPhyloTreeDel (AjPPhyloTree* pthis);
```

The default constructor and destructor functions are used as follows:

```
AjPPhyloDist   dist  = NULL;
AjPPhyloFreq   freq  = NULL;
AjPPhyloProp   prop  = NULL;
AjPPhyloState  state = NULL;
AjPPhyloTree   tree  = NULL;

/* Call constructor functions */
dist  = ajPhyloDistNew();
freq  = ajPhyloFreqNew();
prop  = ajPhyloPropNew();
state = ajPhyloStateNew();
tree  = ajPhyloTreeNew();

/* Do something with instantiated objects */
...

/* Call destructor functions */
ajPhyloDistDel (&dist);
ajPhyloFreqDel (&freq);
ajPhyloPropDel (&prop);
ajPhyloStateDel (&state);
ajPhyloTreeDel (&tree);
```

There are two alternative destructor functions used to free arrays of state and tree objects:

```
void   ajPhyloStateDelarray(AjPPhyloState** pthis);
void   ajPhyloTreeDelarray(AjPPhyloTree** pthis);
```

They are used for state and tree objects instead of the default destructor to free memory from ACD file processing:

```
AjPPhyloState* states   = NULL;
AjPPhyloTree* trees     = NULL;

states   = ajAcdGetDiscretestates(discretestatesfile);
trees    = ajAcdGetTree(treefile);

/* Do something with objects */

ajPhyloStateDelarray(&states);
ajPhyloTreeDelarray(&trees);
```

6.12.7.3 Alternative object construction and loading

6.12.7.3.1 Phylogenetic data input

Currently there are no functions for this.

6.12.7.3.2 Phylogenetic data output

Currently there are no functions for this.

6.12.8 Reading phylogenetic data from file

The functions for this are:

```
AjPPhyloDist*    ajPhyloDistRead (const AjPStr filename, ajint size, AjBool missing);
AjPPhyloFreq     ajPhyloFreqRead (const AjPStr filename, AjBool contchar, AjBool
genedata, AjBool indiv);
AjPPhyloProp     ajPhyloPropRead (const AjPStr filename, const AjPStr propchars, ajint
len, ajint size);
AjPPhyloState*   ajPhyloStateRead (const AjPStr filename, const AjPStr statechars);
AjPPhyloTree*    ajPhyloTreeRead (const AjPStr filename, ajint size);
```

They are provided in case phylogenetic data needs to be processed outside the context of ACD file processing. See the online documentation for further information.

6.12.9 Getting elements of phylogenetic objects

Currently there is a single function for this. It returns the size of a properties object:

```
ajint   ajPhyloPropGetSize (const AjPPhyloProp thys);
```

6.12.10 Debug functions

These report the elements of each object to the debug file. The functions are:

```
void   ajPhyloDistTrace (const AjPPhyloDist thys);
void   ajPhyloFreqTrace (const AjPPhyloFreq thys);
void   ajPhyloPropTrace (const AjPPhyloProp thys);
void   ajPhyloStateTrace (const AjPPhyloState thys);
void   ajPhyloTreeTrace (const AjPPhyloTree thys);
```

6.13 Handling codon usage tables

6.13.1 Introduction

EMBOSS provides a codon usage table object that holds arrays describing codon usage in a nucleotide sequence.

Codon usage tables for input and output are usually defined in the ACD file, but they can be created directly if this is required. Functions are provided for reading and writing the data, getting and setting elements of the object, calculating properties of the codon usage table and for back translation. There are also some miscellaneous functions, mostly used for interconverting triplet base codes to codon indices.

Code that creates a codon usage table needs to first create an output object, then populate it with codon usage data from coding sequences, calculate statistics on codon frequencies and write to a file which uses the user-selected format.

6.13.2 AJAX library files

AJAX library files for handling codon usage tables are listed in the table below. Library file documentation, including a complete description of datatypes and functions, is available at: http://emboss.open-bio.org/rel/dev/libs/

Library file documentation	Description
ajcod	Codon usage table handling

ajcod.h/c. Defines the AjPCod object and functions for handling codon usage tables. It also contains static data structures and functions for handling them at a low level. You are unlikely to need these unless you plan to extend the core functionality of the library.

6.13.3 ACD datatypes

The ACD datatype for handling codon usage tables is:

codon	Codon usage table file

6.13.4 ACD data definition

A typical ACD definition for codon usage table input:

```
codon: infile
[
    parameter: "Y"
]
```

A typical ACD definition for codon usage table output:

```
outcodon: outfile
[
```

```
    parameter: "Y"
]
```

6.13.4.1 Parameter name

All data definitions for codon usage table input and output should have an intuitive name: no standard names are currently defined.

6.13.4.2 Common attributes

Attributes that are typically specified are summarised below.

`parameter:` If a codon usage table is the primary input or output of an EMBOSS application it should be defined as a parameter by using `parameter: "Y"` (see Section A.4, 'Global attributes').

6.13.5 AJAX datatypes

For handling codon usage tables, including input codon usage tables defined in the ACD file, use:

`AjPCod` Codon usage table object (for `codon` ACD datatype)

For handling output codon usage tables defined in the ACD file use:

`AjPOutfile` General output file (for `outcodon` ACD datatype)

6.13.6 ACD file handling

Datatypes and functions for handling codon usage tables via the ACD file are shown below.

	To read a codon usage table	To write a codon usage table
ACD datatype	`codon`	`outcodon`
AJAX datatype	`AjPCod`	`AjPOutfile`
To retrieve from ACD	`ajAcdGetCodon`	`ajAcdGetOutcodon`

Your application code will call `embInit` to process the ACD file and command line (see Section 6.3, 'Handling ACD files'). All values from the ACD file are read into memory and files are opened as necessary. You have a handle on the files and memory through the `ajAcdGet*` family of functions which return pointers to appropriate objects.

6.13.6.1 Input codon usage table retrieval

To retrieve an input codon usage table an object pointer is declared and then initialised using `ajAcdGetCodon`:

```
AjPCod   cod    = NULL;

cod    = ajAcdGetCodon("infile");
```

6.13.6.2 Output codon usage table retrieval

To retrieve an output codon usage table stream an object pointer is declared and initialised using ajAcdGetOutcodon:

```
AjPOutfile codout = NULL;

codout = ajAcdGetOutcodon("outfile");
```

6.13.6.3 Processing command line options and ACD attributes

Currently there are no functions for this.

6.13.6.4 Memory and file management

It is your responsibility to close any files and free up memory at the end of the program.

6.13.6.4.1 Closing output codon usage table files

To close an output codon usage table use ajOutfileClose:

```
ajOutfileClose(&codout);
```

6.13.6.4.2 Freeing memory

You must call the default destructor function (see below) on any objects returned by calls to ajAcdGetCodon or ajAcdGetOutcodon.

Additionally, you must call ajCodExit to clean up internal memory allocated for housekeeping of codon usage processing:

```
ajCodExit();
```

6.13.7 Codon usage table object memory management

6.13.7.1 Default object construction

To use a codon usage table object that is not defined in the ACD file you must first instantiate the appropriate object pointer. The default construction function is:

```
AjPCod ajCodNew (void);
```

All constructors return the address of a new object. In the following code the pointer does not need to be initialised to NULL but it is good practice to do so:

```
AjPCod   cod   = NULL;

cod = ajCodNew();

/* The object is instantiated and ready for use */
```

6.13.7.2 Default object destruction

You must free the memory for an object when you are finished with it. The default destructor function is:

```
void ajCodDel (AjPCod *pthys);
```

It is used as follows:

```
AjPCod    cod       = NULL;
AjPOutfile codout   = NULL;

cod      = ajAcdGetCodon("infile");
codout   = ajAcdGetOutcodon("outfile");

...

ajCodDel(&cod);
ajOutfileClose(&codout);
```

6.13.7.3 Alternative object construction and loading

There are two alternative constructor functions. `ajCodNewCodenum` creates a codon usage object with the amino acid assignments taken from a standard genetic code. In contrast, `ajCodNewCod` will duplicate an existing codon object and return a pointer to the new object:

```
AjPCod ajCodNewCodenum (ajint code);
AjPCod ajCodNewCod (const AjPCod thys);
```

6.13.8 Reading and writing codon usage tables

6.13.8.1 Reading a codon usage table

To use a codon usage table created directly (i.e. not one defined in the ACD file) it's necessary to assign a codon index to it. `ajCodRead` will read codon usage data from a file (fn). The file format can be specified explicitly, given as a prefix (format::) to the filename, or be given as NULL (in which case all known formats are tried):

```
AjBool  ajCodRead (AjPCod thys, const AjPStr fn, const AjPStr format);
```

Codon input and output formats are printed by the **entrails** utility using the ajCodPrintFormat function.

6.13.8.2 Writing a codon usage table

There are two analogous functions for writing codon usage table information to file which differ in the type of file object (AjPFile and AjPOutfile) passed:

```
void  ajCodWrite (AjPCod thys, AjPFile outf);
void  ajCodWriteOut (const AjPCod thys, AjPOutfile outf);
```

Usually codon usage table output files (AjPOutfile) are loaded from ACD file processing (see above). Therefore ajCodWriteOut is called for such an AjPOutfile corresponding to an outcodon: ACD data definition:

```
AjPCod      cod    = NULL;
AjPOutfile  codout = NULL;

cod    = ajAcdGetCodon("infile");
codout = ajAcdGetOutcodon("outfile");

ajCodWriteOut(cod, codout);
```

Where ajCodWrite is used it's necessary to handle the creation of the output file manually. Use ajFileNewOutNameC (or other functions) for doing so:

```
AjPCod  cod    = NULL;
AjPFile  codout = NULL;

/* Open a file called "OutputFileName" */
codout = ajFileNewOutNameC ("OutputFileName");

ajCodWrite(cod, codout);
```

6.13.9 Getting and setting elements

The following elements of a codon usage table may be retrieved or set:

- Name
- Release
- Description
- Division

- Species
- Genetic code
- Number of CDSs
- Number of codons.

The functions to get elements have `Get` in their name. Variants of these functions for returning C-type (`char *`) strings instead of `AjPStr` are available but not shown:

```
const AjPStr  ajCodGetName (const AjPCod thys);          /* Name */
const AjPStr  ajCodGetRelease (const AjPCod thys);       /* Release */
const AjPStr  ajCodGetDesc (const AjPCod thys);          /* Description */
const AjPStr  ajCodGetDivision (const AjPCod thys);      /* Division */
const AjPStr  ajCodGetSpecies (const AjPCod thys);       /* Species */
ajint         ajCodGetCode (const AjPCod thys);          /* Genetic code */
ajint         ajCodGetNumcds (const AjPCod thys);        /* Number of CDSs */
ajint         ajCodGetNumcodon (const AjPCod thys);      /* Number of codons */
```

Additionally, `ajCodGetCodonlist` will write the codon triplets to a list of strings:

```
void          ajCodGetCodonlist(const AjPCod cod, AjPList list);
```

The functions to set elements have `Set` in their name. Again, variants for C-type strings are available:

```
void  ajCodSetDescS (AjPCod thys, const AjPStr desc);             /* Description */
void  ajCodSetDivisionS (AjPCod thys, const AjPStr division);     /* Division */
void  ajCodSetNameS (AjPCod thys, const AjPStr name);             /* Name */
void  ajCodSetReleaseS (AjPCod thys, const AjPStr release);       /* Release */
void  ajCodSetSpeciesS (AjPCod thys, const AjPStr species);       /* Species */
void  ajCodSetCodenum (AjPCod thys, ajint geneticcode);           /* Genetic code */
void  ajCodSetNumcds (AjPCod thys, ajint numcds);                 /* Number of CDSs
*/
void  ajCodSetNumcodons (AjPCod thys, ajint numcodon);            /* Number of
codons */
void  ajCodSetBacktranslate (AjPCod thys);                        /* Amino acid
index used for back-translation. */
void  ajCodSetTripletsS (AjPCod thys, const AjPStr s, ajint *c); /* Number of
codons */
```

6.13.9.1 Clearing a codon usage table

In case you want to reuse a codon usage table there are two functions for clearing the data:

```
/* Zero all entries including the genetic codes (the amino acids for each codon) */
void ajCodClear(AjPCod thys);

/* Zero the name, number count and fraction codon elements. */
void ajCodClearData(AjPCod thys);
```

6.13.10 Calculated properties

Various properties of codon usage tables may be calculated including:

- Codon adaptive index (see *Nucleic Acids Research* **15**:1281–1295)
- Codon adaptive index W values (see *Nucleic Acids Research* **15**:1281–1295)
- Gribskov statistic (count per thousand)
- Effective number of codons (*Gene* **87**:23–29)
- Fractional count and codons per thousand
- Sequence composition.

The functions are as follows:

```
/* Codon adaptive index from a codon usage table */
double ajCodCalcCaiCod (const AjPCod thys);

/* Codon adaptive index for a coding sequence */
double ajCodCalcCaiSeq (const AjPCod cod, const AjPStr str);

/* Gribskov statistic (count per thousand) */
void ajCodCalcGribskov (AjPCod thys, const AjPStr s);

/* Effective number of codons */
double ajCodCalcNc (const AjPCod thys);

/* Fractional count and codons per thousand */
void ajCodCalcUsage (AjPCod thys, ajint c);
```

6.13.11 Back translation

There are functions to backtranslate a string representing an amino acid sequence. `ajCodBacktranslate` will perform a simple back translation whereas `ajCodBacktranslateAmbig` will backtranslate a string to a fully ambiguous nucleotide sequence:

```
void  ajCodBacktranslate (AjPStr *b, const AjPStr a, const AjPCod thys);
void  ajCodBacktranslateAmbig (AjPStr *b, const AjPStr a, const AjPCod thys);
```

Before you call these functions you should first call `ajCodSetBacktranslate`. This initialises the codon usage object with the most commonly used triplet index for the amino acids.

6.13.12 Miscellaneous functions

These include:

```
/* Return one codon value given a possibly ambiguous base */
ajint ajCodBase (ajint c);
```

```
/* Convert triplet index to triple */
char* ajCodTriplet (ajint idx);
/* Return a codon index given a three character codon */
ajint ajCodIndex (const AjPStr s);

/* Return a codon index given a three character codon */
ajint ajCodIndexC (const char *codon);

/* Tests the output format for an outcodon ACD type */
ajint ajCodOutFormat (const AjPStr name);
```

6.14 Handling files

6.14.1 Introduction

EMBOSS provides comprehensive and flexible functionality for file handling, including:

- Input and output general files
- Input and output data files
- Input and output directories
- Lists of files
- Lists of files from directories.

ACD datatypes are provided for handling all of the above. In addition, functions are provided to read and write files and handle directories directly if this is required.

6.14.2 AJAX library files

AJAX library files for handling files are listed in the table below. Library file documentation, including a complete description of datatypes and functions, is available at: http://emboss.open-bio.org/rel/dev/libs/

Library file documentation	Description
`ajfile`	General file handling

`ajfile.h/c`. Objects and functions for handling of data files, general files and directories and static functions for manipulating them at a low level. You are unlikely to need the static functions unless you plan to implement code to extend the core functionality of the library.

6.14.3 ACD datatypes

There are five datatypes for handling file input:

`infile`	General input file
`datafile`	Input data file

directory	A directory with an optional specified file extension and filename prefix
dirlist	A list of file names that are read from a directory with an optional specified file extension and filename prefix
filelist	A list of input files

There are four datatypes for handling file output:

outfile	General output file
outdata	Output data file
outdir	Output directory for writing of multiple output files
directory	A directory that can be used for input or output

6.14.4 ACD data definition

Typical ACD definitions for file input and output are given below.

6.14.4.1 infile

Input file:

```
infile: pepstatsfile
[
   parameter: "Y"
   information: "Pepstats program output file (optional)"
   knowntype: "pepstats output"
]
```

6.14.4.2 filelist

File list input:

```
filelist: files
[
   parameter: "Y"
   information: "Molecular weights file list"
   knowntype: "molwt"
]
```

6.14.4.3 datafile

Input data file:

```
datafile: aaproperties
[
   name: "Eaa_properties.dat"
```

```
   information: "Amino acid chemical classes data file"
   knowntype: "amino acid classification"
]
```

6.14.4.4 dirlist

Input directory list:

```
dirlist: directory
[
   parameter: "Y"
   information: "Codon usage directories"
   knowntype: "codon usage"
]
```

6.14.4.5 directory

Input directory:

```
directory: directory
[
   parameter: "Y"
   information: "Database directory"
   knowntype: "emboss database directory"
]
```

6.14.4.6 outfile

Output file:

```
outfile: outfile
[
   parameter: "Y"
   information: "Pepstats program output file"
   knowntype: "pepstats output"
]
```

6.14.4.7 outdata

Output data file:

```
outdata: outfile
[
   default: "Eaa_properties.dat"
   information: "Amino acid chemical classes data file"
   knowntype: "amino acid classification"
]
```

6.14.4.8 `outdir`

Output directory:

```
outdir: indexoutdir
[
   parameter: "Y"
   information: "Index file output directory"
   knowntype: "emboss database index"
]
```

6.14.4.9 `directory`

Output directory:

```
directory: directory
[
   parameter: "Y"
   information: "Database directory"
   knowntype: "emboss database directory"
]
```

6.14.4.10 Parameter names

All data definitions for file input and output should have standard parameter names. These include:

- "`infile`" for `infile`
- "`*files`" for `filelist`
- "`directory`" for `dirlist`
- "`directory`" for `directory`
- "`outfile`" for `outfile`
- "`outdir`" for `outdir`.

Alternatives and variations are allowed. For more information, see Appendix A, *ACD syntax reference*.

6.14.4.11 Common attributes

Attributes that are typically specified are summarised below. They are datatype-specific (Section A.5, 'Datatype-specific attributes') unless they are indicated as being global attributes (see Section 4, 'Global attributes').

`parameter:` File datatypes are typically the primary input or output of an EMBOSS application and, as such, should be defined as parameters by using the global attribute `parameter: "Y"`.

`information:` A global attribute that specifies the user prompt. It is used in the application documentation.

`knowntype:` A global attribute typically specified for all file input and output types and is required for `outfile` and `outdir`. If the output is not any of the standard

EMBOSS known types (Section 4.3.5.3.1, 'Application data known types file (`knowntypes.standard`)') then *`ApplicationName`* `output` is the recommended value.

`name:` Data files often have a hard-coded filename. You are free to define this using the `name:` attribute.

`default:` Data files often have a hard-coded filename. A default filename is defined for `outdata:` using the `default:` global attribute (no `name:` attribute is available as you might have expected).

6.14.5 AJAX datatypes

For handling input files and directories defined in the ACD file use:

`AjPFile`	An unbuffered file that can be used for input or output (for `infile` and `datafile` ACD datatypes)
`AjPList`	General list (for lists of files from `filelist` and `dirlist` ACD datatypes)
`AjPDir`	A directory that can be used for input or output (for `directory` ACD datatype)

For handling output files and directories defined in the ACD file use:

`AjPFile`	General file (for `outfile` ACD datatype)
`AjPDir`	Directory (for `outdir` and `directory` ACD datatypes)

For handling general output files including output data files defined in the ACD file use:

`AjPOutfile`	An unbuffered output file

For handling buffered files use:

`AjPFilebuff`	Buffered file object holding information for a buffered input file
`AjPFilebufflist`	File buffer holding a simple linked list of buffered lines. This is a substructure of the `AjPFilebuff` object

6.14.6 ACD file handling

Datatypes and functions for handling files via the ACD file are shown below.

ACD datatype	AJAX datatype	To retrieve from ACD
File input		
`infile`	`AjPFile`	`ajAcdGetInfile`
`filelist`	`AjPList`	`ajAcdGetFilelist`

datafile	AjPFile	ajAcdGetDatafile
dirlist	AjPList	ajAcdGetDirlist
directory	AjPDir	ajAcdGetDirectory
File output		
outfile	AjPFile	ajAcdGetOutfile
outdata	AjPOutfile	ajAcdGetOutdata
outdir	AjPDir	ajAcdGetOutdir
directory	AjPDir	ajAcdGetDirectory

Your application code will call `embInit` to process the ACD file and command line (see Section 6.3, 'Handling ACD files'). All values from the ACD file are read into memory and files are opened as necessary. You have a handle on the files and memory through the `ajAcdGet*` family of functions which return pointers to appropriate objects.

6.14.6.1 Input file retrieval

To retrieve an input file, list of files, etc. an object pointer is declared and then initialised using the appropriate `ajAcdGet*` function.

6.14.6.1.1 Input file

```
AjPFile pepstatsfile = NULL;

pepstatsfile = ajAcdGetInfile("pepstatsfile");
```

6.14.6.1.2 Input list of files

```
AjPList files = NULL;

files = ajAcdGetFilelist("files");
```

6.14.6.1.3 Input data file

```
AjPFile aaproperties = NULL;

aaproperties = ajAcdGetDatafile("aaproperties");
```

6.14.6.1.4 Input list of files from a directory

```
AjPList filelist = NULL;

filelist = ajAcdGetDirlist("directory");
```

6.14.6.1.5 Directory (input or output)

```
AjPDir directory = NULL;

directory = ajAcdGetDirectory("directory");
```

6.14.6.2 Output file retrieval

To retrieve an output file stream, directory etc. an object pointer is declared and initialised using the appropriate `ajAcdGet*` function.

6.14.6.2.1 Output file

```
AjPFile outfile = NULL;

outfile = ajAcdGetOutfile("outfile");
```

6.14.6.2.2 Output data file

```
AjPOutfile outfile = NULL;

outfile = ajAcdGetOutdata("outfile");
```

6.14.6.2.3 Output directory

```
AjPDir indexoutdir = NULL;

indexoutdir = ajAcdGetOutdir("indexoutdir");
```

6.14.6.2.4 Directory (input or output)

```
AjPDir directory = NULL;

directory = ajAcdGetDirectory("directory");
```

6.14.6.3 Alternative ACD retrieval functions

To retrieve a directory name rather than the directory itself call either `ajAcdGetDirectoryName` (`directory` ACD object) or `ajAcdGetOutdirName` (`outdir` ACD object). Both functions return the directory name as a string object. The directory and name strings are owned by the caller and should be freed when the application exits. For example:

```
    AjPDir directory   = NULL;
    AjPDir indexoutdir = NULL;
    AjPStr indir       = NULL;
    AjPStr outdir      = NULL;

    directory   = ajAcdGetDirectory("directory");
    indexoutdir = ajAcdGetOutdir("indexoutdir");

    indir  = ajAcdGetDirectoryName("directory");
    outdir = ajAcdGetOutdirName("indexoutdir");

/* … */
```

```
ajStrDel(&indir);
ajStrDel(&outdir);
ajDirDel(&indexoutdir);
```

6.14.6.4 Processing command line options and ACD attributes

Currently there are no functions for this.

6.14.6.5 Memory and file management

It is your responsibility to close any files and free up memory at the end of the program.

6.14.6.5.1 Closing files

To close an input file or input data file call `ajFileClose` with the address of the input file. To close an output file call `ajOutfileClose` with the address of the output file.

Following from the examples above:

```
AjPFile pepstatsfile = NULL;
AjPFile outfile      = NULL;

pepstatsfile = ajAcdGetInfile("pepstatsfile");
outfile      = ajAcdGetOutfile("outfile");

/* Do something with files */

ajFileClose (&pepstatsfile);
ajOutfileClose(&outfile);
ajOutfileClose(&outfiles);
```

In both cases the file is closed and memory for the file object is freed.

6.14.6.5.2 Freeing memory

You must call the destructor functions (see below) to free the memory used for any objects returned by calls to `ajAcdGet*`.

6.14.7 File and directory object memory management

6.14.7.1 Default object construction

To use a file or directory object that is not defined in the ACD file you must first instantiate the appropriate object pointer. The default constructor functions described below take the name of the file or directory to open.

6.14.7.1.1 General input and output files

The basic constructors for general input and output files are:

```
/* Input file */
AjPFile ajFileNewInNameS (const AjPStr name);

/* Input file */
```

```
AjPFile ajFileNewInNameC (const char* name);

/* Output file */
AjPFile ajFileNewOutNameS (const AjPStr name);

/* Output file */
AjPFile ajFileNewOutNameC (const char* name);

/* Perl-style Pipe. */
AjPFile ajFileNewInPipe (const AjPStr str);

/* Output file */
AjPOutfile ajOutfileNewNameS (const AjPStr name);
```

You will typically use an `AjPFile` for input and an `AjPOutfile` for output and therefore use `ajOutfileNewNameS` in preference to `ajFileNewOutNameS` or `ajFileNewOutNameC`.

6.14.7.1.2 Data files

The default constructor functions for input data files are:

```
/* Input data file */
AjPFile ajDatafileNewInNameS (const AjPStr name);

/* Input data file */
AjPFile ajDatafileNewInNameC (const char* name);
```

The functions take the file name as a C-type (`char *`) or string object (`AjPStr`) but are otherwise identical. They will open the named file if it exists in one of the following paths which are searched in the order shown:

1. . (current working directory)
2. `./.embossdata` (EMBOSS data directory)
3. ~/ (users home directory)
4. `~/.embossdata` (EMBOSS data directory)
5. `$EMBOSS_DATA` (EMBOSS data directory)

The default constructor function for output data files is:

```
AjPFile ajDatafileNewOutNameS (const AjPStr name);
```

`ajDatafileNewOutNameS` will return an pointer to a file object for a file in the EMBOSS data directory (`$EMBOSS_DATA`).

6.14.7.1.3 Buffered input files

There is no ACD datatype for buffered files although there is an AJAX datatype (`AjPFilebuff`). To use a buffered file you must first instantiate the object pointer using a filename, an open file, or buffered lines of data. The basic constructor functions are:

```
/* Named buffered input file */
AjPFilebuff ajFilebuffNewNameS (const AjPStr name);

/* Named buffered input file and directory */
AjPFilebuff ajFilebuffNewNamePathS (const AjPStr name, const AjPStr path)

/* Named directory and wildcard filename to read all matching files */
AjPFilebuff ajFilebuffNewPathWild (const AjPStr path,
                                   const AjPStr wildname);

/* Named directory and wildcard filename to read all matching files with exceptions */
AjPFilebuff ajFilebuffNewPathWildExclude(const AjPStr path,
                                         const AjPStr wildname,
                                         const AjPStr exclude)

/* no file - to be loaded with data for re-reading */
AjPFilebuff ajFilebuffNewNofile (void);
```

`ajFilebuffNewNofile` is a primitive constructor creating the object but no file stream used in several places in EMBOSS where the buffer will be loaded with data records. Examples can be found in processing sequence databases. The usual constructor is `ajFilebuffNewNameS` and will open the named file.

6.14.7.1.4 Directories

Default constructors are provided for input and output:

```
/* Directory for input */
AjPDir ajDirNewPath (const AjPStr path);

/* Directory for output */
AjPDir ajDiroutNewPath (const AjPStr path);
```

The constructors return the address of a new object. The pointers do not need to be initialised to `NULL` but it is good practice to do so:

```
AjPFile       filein   = NULL;
AjPFilebuff   buffin   = NULL;
AjPFile       datain   = NULL;

AjPOutfile    fileout  = NULL;

AjPDir        dirin    = NULL;
AjPDir        dirout   = NULL;

AjPStr        fileout_name = NULL;
AjPStr        buffin_name  = NULL;
AjPStr        dirin_name   = NULL;
AjPStr        dirout_name  = NULL;
```

```
fileout_name = ajStrNewC("outfile");
buffin_name  = ajStrNewC("bufffile");
dirin_name   = ajDirNewPath("./dirin");
dirout_name  = ajDirNewPath("./dirout");

filein  = ajFileNewInNameC("infile");
fileout = ajOutfileNewNameS(fileoutname);
datain  = ajDatafileNewInNameC("datafile");
buffin  = ajFilebuffNewNameS(buffname);
dirin   = ajDirNewPath(dirin_name);
dirout  = ajDiroutNewPath(dirout_name);

/* The objects are instantiated and ready for use */
```

6.14.7.2 Default object destruction

You must free the memory for any object once you are finished with them. All the default destructor functions take the address of the object being freed:

```
/* File object destructor */
void ajFileClose (AjPFile* Pfile);

/* Output file object destructor */
void ajOutfileClose (AjPOutfile* Pfile);

/* Buffered file object constructor */
void ajFilebuffDel (AjPFilebuff* Pfile);

/* Directory destructor */
void ajDirDel (AjPDir* Pdir);
```

The destructor functions close the file (if appropriate) and free memory for the file or directory object. Following from the example above:

```
ajFileClose(&filein);
ajOutfileClose(&fileout);
ajFilebuffDel(&buffin);
ajDirDel(&dirin);
ajDirDel(&dirout);
```

For a list of files (`filelist` or `dirlist`) it is necessary to free the list of filenames (which are stored as strings), once you are done with them, by calling `ajListstrFreeData`:

```
AjPList files    = NULL;
AjPList filelist = NULL;
AjIList iter     = NULL;
AjPFile infile   = NULL;
AjPStr filename  = NULL;

files    = ajAcdGetFilelist("files");
filelist = ajAcdGetDirlist("directory");

/* both lists are filenames as strings */
```

```
iter = ajListIterNewread(filelist);
while(!ajListIterDone(iter))
{
    filename = (AjPStr) ajListIterGet(iter);
    infile = ajFileNewInNameS(filename);

    if(infile)
    {

         /* Do something with the file */
         ajFileClose(&infile);
    }
}
ajListIterDel(&iter);

ajListstrFreeData(&files);
ajListstrFreeData(&directory);
```

Where a directory is used, either for input (`directory`) or output (`outdir`) directory, you must call the directory destructor function `ajDirDel` once you are done with it:

```
AjPDir indexoutdir = NULL;
AjPDir directory = NULL;
AjPList filelist = NULL;

indexoutdir = ajAcdGetOutdir("indexoutdir");
directory = ajAcdGetDirectory("directory");

filelist = ajListstrNew();
ajFilelistAddDirectory(filelist, directory);

/* Do something with directory or list of files */

ajDirDel(&indexoutdir);
ajDirDel(&directory);
ajListstrFreeData(&filelist);
```

6.14.7.3 Alternative object construction and loading

The numerous alternative constructor functions described below provide maximum flexibility in file object construction and loading. All constructors return the address of a new object.

6.14.7.3.1 General input file

For `ajFileNewListinPathWild` and `ajFileNewListinPathWildExclude` the filename may be wildcarded and all files matching the name will be opened for reading. `ajFileNewListinPathWildExclude` also takes a second string (`exclude`) which is a wildcard string specifying which file names to exclude. `ajFileNewListinNameDirS` and `ajFileNewListinDirPre` will also open a named file from a directory but in contrast to the other functions mentioned adds a default extension defined in a directory object (`AjPDir`):

```
/* List of input file names as strings (AjPStr) */
AjPFile ajFileNewListinList (AjPList list);

/* Specify a directory object. */
AjPFile ajFileNewListinNameDirS (const AjPStr name, const AjPDir dir);

/* Specify a directory object and filename prefix. The directory object supplies the
filename extension. */
AjPFile ajFileNewListinDirPre (const AjPDir dir, const AjPStr prefix);

/* Read all files matching wild card. */
AjPFile ajFileNewListinPathWild (const AjPStr name, const AjPStr path);

/* Read all files matching / excluding wild card. */
AjPFile ajFileNewListinPathWildExclude (const AjPStr name,const AjPStr path, const
AjPStr exclude);

/* Read using blocked fread calls */
AjPFile ajFileNewInBlocksS (const AjPStr name, ajuint blocksize);
```

Functions with the prefix `ajFileNewFrom` construct the file object from some other entity; either a UNIX pipe, a list of input files or from a C-type (`FILE *`) file pointer:

```
/* C-type FILE pointer. */
AjPFile ajFileNewFromCfile(FILE* file);
```

6.14.7.3.2 **`infile`** (data file)

A named input data file may be read from a specific data sub-directory by calling:

```
AjPFile  ajDatafileNewInNamePathS (const AjPStr name, const AjPStr path);
```

The file will be searched for in the usual EMBOSS data directories (see above) *and* in the sub-directory `path` of the active EMBOSS data directory.

6.14.7.3.3 Buffered input files

Buffered input files may be read from a named directory using the following functions, which are otherwise identical to the general file handling functions with similar names (see above:)

```
/* Specify a path. */
AjPFilebuff ajFilebuffNewNamePathS (const AjPStr name, const AjPStr path);

/* Specify a path. */
AjPFilebuff ajFilebuffNewNamePathC (const char* name, const AjPStr path);

/* Read all files matching wild card. */
AjPFilebuff ajFilebuffNewPathWild (const AjPStr name, const AjPStr path);

/* Construct from all files matching / excluding wild card. */
AjPFilebuff ajFilebuffNewPathWildExclude (const AjPStr name, const AjPStr path,
const AjPStr exclude);
```

Buffered input files may be constructed from some other entity; either a file object, a C-type (`FILE *`) file pointer, another buffered input file (in which case the file is copied), a string (rather than a file) or a list of filenames:

```
/* from an already open file */
AjPFilebuff ajFilebuffNewFromFile (AjPFile file);

/* from an already open C file */
AjPFilebuff ajFilebuffNewFromCfile (FILE* file);

/* no file, but with one line of buffered data */
AjPFilebuff ajFilebuffNewLine (const AjPStr line);

/* list of filenames to be read in turn */
AjPFilebuff ajFilebuffNewListinList(AjPList);
```

The string passed to `ajFilebuffNewLine` should contain one line of buffered data for reading.

`ajFilebuffNewFromFile` and `ajFilebuffNewFromCfile` both appropriate the file pointer that is passed, so you must be careful to avoid calling the destructor twice on the same object when managing the memory for your objects:

```
AjPFile     file  = NULL;
AjPFilebuff buff  = NULL;

file  = ajFileNewInNameC("infile");
buff  = ajFilebuffNewFromCfile(file);

/* Do something with buffered file */

/* file is appropriated by buff so we only need to call one destructor (close) function
*/
ajFilebuffDel(&buff);
```

6.14.7.3.4 **`directory`**

A directory object can be constructed to exclusively read or write files with a specified file extension (`ext`) or a specified prefix (`pre`) and file extension:

```
/* Input directory */
AjPDir ajDirNewPathExt (const AjPStr path, const AjPStr ext);
AjPDir ajDirNewPathPreExt (const AjPStr path, const AjPStr pre, const AjPStr ext);

/* Output directory */
AjPDirout ajDiroutNewPath (const AjPStr path);
AjPDirout ajDiroutNewPathExt (const AjPStr path, const AjPStr ext);
```

6.14.7.3.5 General output file

A named general output file may be constructed in a given directory which may be specified as a directory object (`AjPDir`) or as a file path:

```
/* Directory object. */
AjPFile ajFileNewOutNameDirS (const AjPStr name, const AjPDir dir);

/* Path. */
AjPFile ajFileNewOutNamePathS (const AjPStr name, const AjPStr path);
```

To open a named general output file for appending you can use:

```
AjPFile  ajFileNewOutappendNameS (const AjPStr name);
```

6.14.8 Reading from file

Functions for reading lines from a general file have the prefix `ajReadline`. The file position before the read can be retained, newline characters can be trimmed from the line, and the line can be appended to a provided buffer:

```
/* Read a line from a file. */
AjBool ajReadline (AjPFile file, AjPStr* Pstr);

/* Read and retain file position before read. */
AjBool ajReadlinePos (AjPFile file, AjPStr* Pstr, ajlong* fpos);

/* Read and trim any trailing newline character. */
AjBool ajReadlineTrim (AjPFile file, AjPStr* Pstr);

/* Read, trim newline characters and retain file position before read. */
AjBool ajReadlineTrimPos (AjPFile file, AjPStr* Pstr, ajlong* fpos);

/* Read and append to buffer. */
AjBool ajReadlineAppend (AjPFile file, AjPStr* Pbuff);
```

To read a binary file call:

```
/* Binary read from a file using the C 'fread' function. */
AjBool ajReadbinBinary (AjPFile file, size_t count, size_t element_size,
                        void* buffer);

/* Read unsigned int */
AjBool ajReadbinUint (AjPFile file, ajuint* ret);

/* Read unsigned int bigendian and converts if needed. */
AjBool ajReadbinUintEndian (AjPFile file, ajuint* ret);
```

`ajReadbinBinary` will read `count` elements of size `element_size` from an input file. `ajReadbinUint` and `ajReadbinUintEndian` read an unsigned integer and convert from littleendian (the form used for binary data output, e.g. database index files) or bigendian respectively. Both functions use the C function `fread`.

6.14.8.1 Buffered files

Functions with the prefix ajBuffRead read a line from a buffered file. The line that's read can be appended to a string and the file position before the read can be retained:

```
/* Read a line from a buffered file. */
AjBool ajBuffreadLine (AjPFilebuff file, AjPStr* Pstr);

/* Read a line and note the file position. */
AjBool ajBuffreadLinePos (AjPFilebuff file, AjPStr* Pstr, ajlong* fpos);

/* Read a line and optionally append to a Pstore buffer */
AjBool ajBuffreadLinePosStore (AjPFilebuff buff, AjPStr* Pdest, ajlong* Ppos,
                               AjBool dostore, AjPStr *Pstore);
```

6.14.9 Writing to files

6.14.9.1 General files

There are two functions from ajfmt.c/h for writing to files. The most widely used is ajFmtPrintF which writes to an AJAX file object. Alternative functions write to a C FILE or to standard output:

```
/* Format and emit the "..." arguments according to fmt; writes to a file object */
void ajFmtPrintF (AjPFile file,
                  const char *fmt, ...);

/* Format and emit the "..." arguments according to fmt; writes to C FILE stream */
void ajFmtPrintFp (FILE *stream,
                   const char *fmt, ...);

/* Format and emit the "..." arguments according to fmt; writes to stdout */
void ajFmtPrint (const char *fmt, ...);
```

6.14.9.2 Binary files

Functions for writing to a binary file have the prefix ajWritebin. A single byte, a C-type string (char *) or a string object can be written to a field of defined size (len), or a numeric values can be written with the correct byte orientation:

```
/* Writes a single byte in a defined field. */
ajint ajWritebinByte (AjPFile file, char chr);
ajint ajWritebinChar (AjPFile file, const char* txt, ajint len);
ajint ajWritebinStr (AjPFile file, const AjPStr str, ajint len);

/* Writes a 2-byte integer . */
ajint ajWritebinInt2 (AjPFile file, ajlong i);

/* Writes a 4-byte integer. */
ajint ajWritebinInt4 (AjPFile file, ajlong i);
```

```
/* Writes a 8-byte integer. */
ajint ajWritebinInt8 (AjPFile file, ajlong i);
```

6.14.9.3 Buffered files

In some cases it is necessary to write to a buffered file that is otherwise being used for input. Three functions are provided to write lines to a buffered file for input. These are used internally, for example when reading binary sequence formats.

```
/* Writes a C-type string to the buffer */
void ajFilebuffLoadC (AjPFilebuff file, const char* txt);

/* Writes a string object to the buffer */
void ajFilebuffLoadS (AjPFilebuff file, const AjPStr str);

/* Writes all lines from a file into the buffer. */
void ajFilebuffLoadAll (AjPFilebuff file);
```

6.14.10 Manipulating files

Three functions are provided to set the properties of a file object:

```
/* Close current file and open next one. */
AjBool ajFileReopenNext (AjPFile* Pfile);

/* Reopens a file with a new name. */
FILE* ajFileReopenName (AjPFile* Pfile, const AjPStr name);

/* Sets the current position in an open file. */
ajint ajFileSeek (AjPFile* Pfile, ajlong offset, ajint from);
```

For a file object that includes a list of input files, `ajFileReopenNext` will close the current input file and open the next one. `ajTrue` is returned on success or `ajFalse` otherwise. `ajFileReopenName` calls the C function `freopen` to close the current file and open a file with a new name. `ajFileSeek` is, essentially, a wrapper to the C function `fseek` and sets the current position in an open file. In cases where end-of-file was reached and this function is used to move back somewhere in the file, then the end-of-file flag in the file object is reset.

Functions for setting the properties of buffered file objects have the prefix `ajFilebuffSet` and include:

```
/* Sets file to be buffered. */
AjBool ajFilebuffSetBuffered (AjPFilebuff* Pfile);

/* Sets file to be unbuffered. */
AjBool ajFilebuffSetUnbuffered (AjPFilebuff* Pfile);

/* Sets next read to start at first buffered line. */
AjBool ajFilebuffReset (AjPFilebuff* Pfile);
```

```
/* Sets next read position and resets file position. */
AjBool ajFilebuffResetPos (AjPFilebuff* Pfile);

/* Resets the pointer and current record and optionally clears Pstore buffer */
void ajFilebuffResetStore (AjPFilebuff buff, AjBool dostore, AjPStr *Pstore)

/* Deletes processed lines from a file buffer. */
AjBool ajFilebuffClear (AjPFilebuff* Pfile, ajint n);

/* Clears processed records and removes extra record(s) from the Pstore buffer */
AjBool ajFilebuffClearStore (AjPFilebuff buff, ajint lines, const AjPStr lastline,
                             AjBool dostore, AjPStr *Pstore);
```

Buffering of file input can be turned on and off by calling `ajFilebuffSetBuffered` and `ajFilebuffSetUnbuffered` respectively (buffering for an `AjPFilebuff` object is on by default). Functions with the prefix `ajFilebuffReset` set the next read position to the start of the first buffered line. `ajFilebuffResetPos` also resets the current file position to the last known read as a precaution in cases where this might have been changed by some other function. `ajFilebuffResetStore` also clears the caller's record of processed lines, using the same control parameter as `ajBuffreadLinePosStore`. `ajFilebuffClear` and `ajFilebuffClear` will delete the buffered lines processed so far. `ajFilebuffClearStore` will also clear the caller's record of processed lines, using the same control parameter as `ajBuffreadLinePosStore`.

6.14.11 Querying properties of files

6.14.11.1 Basic properties

The file name, C-type file pointer (`FILE *`) and file length of a file object are available by calling:

```
/* File name. */
AjPStr ajFileGetNameS (const AjPFile file);

/* File name. */
const char* ajFileGetNameC (const AjPFile file);

/* C file pointer. */
FILE* ajFileGetFileptr (const AjPFile file);
```

These functions (and other `*Get*` functions below) return elements (or values calculated from them) of the object itself, therefore care must be taken not to inadvertently change the object in cases where a pointer is returned.

Functions with the prefix `ajFileIs` check whether a file is a certain type or if the object satisfies some other property:

```
/* File is stderr. */
AjBool ajFileIsStderr (const AjPFile file);

/* File is stdin. */
AjBool ajFileIsStdin (const AjPFile file);
```

```
/* File is stdout. */
AjBool ajFileIsStdout (const AjPFile file);

/* File is set for appending. */
AjBool ajFileIsAppend (const AjPFile file);

/* File position is at end-of-file. */
AjBool ajFileIsEof (const AjPFile file);
```

These functions (and `*Is*` functions for other datatype below) return `ajTrue` if the condition is met.

6.14.11.2 Output files

The file object, C-type file pointer (`FILE *`) from the file object and format are available for retrieval from an output file object:

```
/* File object. */
AjPFile ajOutfileGetFile (const AjPOutfile file);

/* Output format. */
AjPStr ajOutfileGetFormat (const AjPOutfile file);

/* C file pointer. */
FILE* ajOutfileGetFileptr (const AjPOutfile file);
```

6.14.11.3 Buffered files

The file object, C-type file pointer (`FILE *`) from the file object and standard record buffer size are available for retrieval from a buffered file object:

```
/* File object */
AjPFile ajFilebuffGetFile (const AjPFilebuff file);

/* C file pointer from file object /
FILE* ajFilebuffGetFileptr (const AjPFilebuff file);
```

A buffered file can be tested for whether it is empty, buffered, whether end-of-file has been reached or whether the file is exhausted (end-of-file is reached and the buffer is empty):

```
/*File buffer is empty. */
AjBool ajFilebuffIsEmpty (const AjPFilebuff file);

/* Input file is buffered. */
AjBool ajFilebuffIsBuffered (const AjPFilebuff file);

/* File is exhausted (EOF is reached and buffer is empty). */
AjBool ajFilebuffIsEmpty (const AjPFilebuff file);

/* EOF is reached. */
AjBool ajFilebuffIsEof (const AjPFilebuff file);
```

6.14.11.4 Directories

The directory name and file extension of files for reading are available for a directory object:

```
AjPStr ajDirGetPath (const AjPDir Pdir);
AjPStr ajDirGetExt (const AjPDir Pdir);
AjPStr ajDirGetPrefix (const AjPDir Pdir);
```

The directory object can be used to generate a list of full path names of files in the directory and matching any specified prefix and extension:

```
ajint   ajFilelistAddDirectory (AjPList list, const AjPDir dir);
```

6.14.12 Querying and manipulating file and directory names

It's possible to check a string representing a filename includes a path by calling `ajFilenameHasPath`, which returns return `ajTrue` if the property is satisfied. Functions with the prefix `ajFilenameTest` test a filename:

```
/* Checks for path. */
AjBool ajFilenameHasPath (const AjPStr name);
```

There are various functions to test a filename (with or without a path specification) against wildcards defining filenames to include or exclude. Filenames may be included or excluded by default. The inclusion wildcard is used to select files, and the exclusion wildcard is then used to exclude selected files again:

```
/* Tests a filename. Default is to exclude */
AjBool ajFilenameTestExclude(const AjPStr filename,
                             const AjPStr exclude,
                             const AjPStr include);

/* Tests a full path file name. Default is to exclude. */
AjBool ajFilenameTestExcludePath(const AjPStr filename,
                                 const AjPStr exclude,
                                 const AjPStr include);

/* Tests a file name. Default is to include. */
AjBool ajFilenameTestExclude(const AjPStr filename,
                             const AjPStr exclude,
                             const AjPStr include);

/* Tests a full path file name. Default is to include). */
AjBool ajFilenameTestExcludePath(const AjPStr filename,
                                 const AjPStr exclude,
                                 const AjPStr include);
```

There are functions to check whether a named file exists and is in a specified mode (read, write or executable):

```
/* File exists */
AjBool ajFilenameExists (const AjPStr filename);

/* Exists and is a directory */
AjBool ajFilenameExistsDir (const AjPStr filename);

/* Exists and is executable */
AjBool ajFilenameExistsExec (const AjPStr filename);

/* Exists and is readable */
AjBool ajFilenameExistsRead (const AjPStr filename);

/* Exists and is writeable */
AjBool ajFilenameExistsWrite (const AjPStr filename);
```

To return the file size call `ajFilenameGetSize`:

```
ajlong   ajFilenameGetSize (const AjPStr name);
```

Functions for setting a string representing a filename have the prefix `ajFilenameTrim` if they remove the path and/or file extension from the name. The trim functions include:

```
/* Truncates a filename to a basic file name. */
AjBool ajFilenameTrimAll (AjPStr* Pname);

/* Remove directory path and file extension from a file name. */
AjBool ajFilenameTrimPathExt (AjPStr* Pname);

/* Remove directory path (if any) only. */
AjBool ajFilenameTrimPath (AjPStr* Pname);

/* Remove extension (if any) only. */
AjBool ajFilenameTrimExt (AjPStr* Pname);
```

The prefix `ajFilenamereplace` functions allow the file extension and path to be replaced with some other value. They have the prefix `ajFilenameSet` if they set the filename in some other way. For example, the filename can also be set to an available temporary filename (with optional path) by calling `ajFilenameSetTempname`:

```
/* Sets the file extension.*/
AjBool ajFilenameReplaceExtS (AjPStr* Pname, const AjPStr extn);
AjBool ajFilenameReplaceExtC (AjPStr* Pname, const char* extn);

/* Sets the directory path. */
AjBool ajFilenameReplacePathS (AjPStr* Pname, const AjPStr path);
AjBool ajFilenameReplacePathC (AjPStr* Pname, const char* path);

/* Sets name to an available temporary file name with a defined path. */
AjBool ajFilenameSetTempnamePathS (AjPStr* Pfilename, const AjPStr str);

/* Sets name to an available temporary file name. */
AjBool ajFilenameSetTempname (const char** Pname);
```

There are a few functions for manipulating strings representing file paths:

```
/* Changes directory path to one level up. */
AjBool ajDirnameUp (AjPStr* Pname);

/* Add trailing '/' if missing. */
AjBool ajDirnameFix (AjPStr* Path);

/* Checks for valid path, and appends trailing '/' if missing. */
AjBool ajDirnameFixExists (AjPStr* Ppath);

/* Checks for valid path and ensures full path definition is included. */
AjBool ajDirnameFillPath (AjPStr* Ppath);
```

`ajDirnameFixExists` and `ajDirnameFillPath` check that the specified path (`Ppath`) is a valid directory. A trailing / is appended to the path if missing. `ajDirnameFillPath` will navigate up the directory structure and modify the path if necessary to ensure that the full path definition is included.

6.15 Handling application reports

6.15.1 Introduction

Applications should use the standard set of report formats (see the *EMBOSS User's Guide*) for output where possible. Each report defines a content and layout of text in a file. Most of the reports are for describing features, usually short sequence motifs, in a sequence. The sequence motifs, ID codes, sequence position numbers and the corresponding features can be output in a variety of formats. A set of command line qualifiers (Section A.5.3.17, '`report`') are used to control the data written to the report and set such things as the name and format of the file. Non-printable control characters are not used therefore and so are suitable for viewing on-screen or for printing.

Reports have a consistent look and feel, which helps the end-user and improves interoperation of the applications. In principle an application can use different report formats for different purposes, depending on what the output file will be used for. A human-readable report is required for the end-user whereas a more compact format might be suitable if the output is to be processed by another application. All of the standard sequence feature table formats (see the *EMBOSS User's Guide*) are also report formats. Other report formats cater for non-sequence information. For a complete list, descriptions and examples of the supported report formats see the *EMBOSS User's Guide*.

All report formats, excluding the `TAB-delimited (excel)` one and those using the standard feature table formats, have a block of reference information at the start of the report. This gives:

- The program name
- Date the report was generated
- The output file name
- The ID name of the sequence

- The region of the sequence for which features are being reported
- Some of the parameters and statistics of the report.

The reported data are taken from the report feature table. Different report formats report different values or different tags from the feature table. The tags that might be reported are as follows:

`start`	Start position of match in sequence
`end`	End position of match in sequence
`length`	Length of match in sequence
`name`	Name of sequence
`sequence`	Sequence of match
`strand`	DNA strand in which feature occurs
`tagname`	Printable name of feature
`tagvalue`	Feature in sequence
`type`	Type of feature

Finally there is a block of information at the end of the report with summary information.

Internally all reports are held in a very close analogue of the General Feature Format (GFF). This flexible file format holds sequence feature information in a feature table as tag/value pairs. Features can easily be converted to any of the supported report formats (see the *EMBOSS User's Guide*), and can be selected by using the **`-rformat`** built-in qualifier.

GFF has some fields that must exist including:

- A feature type that can be any string but is best specified using the Sequence Ontology (SO) (for example, the EMBOSS **antigenic** uses 'SO:0001018' as the predicted features are of type 'epitope'. Using SO terms allows some report formats, for example GFF3 or DAS (distributed annotation system), to correctly label features with biological properties. Any unrecognised string is reported as a 'miscellaneous feature' or 'region' by these formats, with the original string retained and reported in the feature tags.
- The sequence itself which may or may not be output depending on the output report format used.
- The start and end positions of the features. GFF requires that the start is before the end, with the strand identifying nucleotide features in the reverse direction.
- A score (in case the program reports a score for the feature).

The score and most other standard GFF tags are set to zero or `NULL` automatically by the library. You have to explicitly use them (assign values to them) if you need them, otherwise they'll be ignored. To cope with non-standard feature values the AJAX library uses the default `'/note'` tag. This is transparent to the programmer, unless you select EMBL or another output format in which the `'/note'` tag is visible.

Any report output files should be defined in the application ACD file and retrieved from within the C source code by a call to `ajAcdGetReport`. Some programming is required to load the report with data and write it to file as described below.

6.15.2 AJAX library files

AJAX library files for handling reports are listed in the table below. Library file documentation, including a complete description of datatypes and functions, is available at: http://emboss.open-bio.org/rel/dev/libs/

Library file documentation	Description
`ajreport`	Application reports

`ajreport.h/c`. Defines the `AjPReport` object and functions for handling of reports. They also contain static data structures and functions for handling reports at a low level. You are unlikely to need the static data structures and functions unless you plan to implement code to support new report formats for EMBOSS. For advice on how to do this ask the EMBOSS developers.

6.15.3 AJAX datatypes

For handling application report output files defined in the ACD file use:

`AjPReport` Application report (for `report` ACD datatypes)

6.15.4 ACD datatypes

The datatype for handling report output is:

`report` Output file for sequence annotation (report)

6.15.5 ACD data definition

A typical ACD definition for report output:

```
report: outfile
[
    parameter: "Y"
    rformat: "diffseq"
    taglist: "int:start int:end int:length str:name str:sequence
              str:first_feature str:second_feature"
]
```

6.15.5.1 Parameter name

All data definitions for report output should have standard parameter names (see Appendix A, *ACD syntax reference*):

- `outfile`
- `*file`

6.15.5.2 Common attributes

Attributes that are typically specified are summarised below. They are datatype-specific (Section A.5, 'Datatype-specific attributes') unless they are indicated as being global attributes (Section A.4, 'Global attributes').

`parameter:` Reports are typically the primary output of an EMBOSS application and, as such, should be defined as parameters by using the global attribute `parameter: "Y"`.

`rformat:` Specifies the report format to use, which must be one of the supported report formats (see the *EMBOSS User's Guide*). There is a default report format that will be used if the user doesn't change it with **`-rformat`** on the command line.

`multiple:` A boolean attribute which should be set to `"Y"` if the output can contain more than one report from the same input. This will generally have a value of `"N"` if you are using a single input sequence (the `sequence` ACD datatype) and `"Y"` if you're processing multiple sequences from an input stream (the `seqall` ACD datatype).

`type:` Specifies where the report format is one of the standard feature table formats (see the *EMBOSS User's Guide*). `type:` defines whether the report output is `"protein"` or `"nucleotide"`. There is a default based on the type of any input sequence, but a value should always be specified.

`taglist:` Defines the tag / value pairs from the internal feature table that are to be reported in the output. Each tag is in the general format **`type:tagname`** and can (optionally) be appended with `[=columnname]` to indicate a column name to use in the report:

```
type:tagname[=columnname]
```

For example:

```
int:length
string:gc=GC%
```

Typical taglist datatypes are:

```
float
int
str
```

The name can be anything within reason but tag names and types *must* match those used in the C source code of the application.

In the given example `int : length` indicates that one of the columns is called `length` and that it will contain integer values.

6.15.6 ACD file handling

Datatypes and functions for handling report output via the ACD file are shown below.

	To write a report
ACD datatype	`report`
AJAX datatype	`AjPReport`
To retrieve from ACD	`ajAcdGetReport`

Your application code will call `embInit` to process the ACD file and command line (see Section 6.3, 'Handling ACD files'). All values from the ACD file are read into memory and files are opened as necessary. You have a handle on the files and memory through the `ajAcdGet*` family of functions which return pointers to appropriate objects.

6.15.6.1 Report retrieval

To retrieve an output report stream an object pointer is declared and initialised using `ajAcdGetReport`:

```
AjPReport report = NULL;

report = ajAcdGetReport ("report");
```

6.15.6.2 Processing command line options and ACD attributes

Currently there are no functions for this.

6.15.6.3 Memory and file management

It is your responsibility to close any files and free up memory at the end of the program.

6.15.6.3.1 Closing report files

Any report file must be closed by calling `ajReportClose`

```
void ajReportClose (AjPReport pthys);
```

6.15.6.3.2 Freeing memory

You must call the default destructor function (see below) on any report objects returned by calls to `ajAcdGetReport`.

Additionally you must call `ajReportExit` to free up any internal memory allocated for report housekeeping:

```
void ajReportExit (void);
```

6.15.7 Report object memory management

6.15.7.1 Default object construction

Report objects are typically loaded from ACD file processing via a call to `ajAcdGetReport` (see above). You should therefore never need to instantiate the report object (`AjPReport`) manually. In the unlikely event you do then the default constructor is:

```
AjPReport ajReportNew (void);
```

The constructor returns the address of a new report object. In the following code the pointer does not need to be initialised to `NULL` but it is good practice to do so:

```
AjPReport        report = NULL;

report   = ajReportNew();

/* The object is instantiated and ready for use */
```

6.15.7.2 Default object destruction

You must free the memory for an object once you are finished with it. The default destructor function is:

```
void ajReportDel (AjPReport* Preport);
```

It is used as follows:

```
AjPReport      report =NULL;

report = ajAcdGetReport("report");

...

ajReportDel(&report);
```

Or if you manually instantiated the object:

```
AjPReport report = NULL;

report = ajReportNew();

/* Do something with the instantiated objects */

ajReportDel(&report);

/* Done with the objects so the memory is freed. */
```

6.15.7.3 Alternative object construction and loading

Currently there are none of these.

6.15.8 Preparing a report

Each report has a header, body and tail information. The process of preparing a report for output is:

1. Define a suitable report datatype in ACD.
2. Collect the ACD report information within the application in a report object (`AjPReport`).
3. Create a feature table object (`AjPFeattable`) using `ajFeattableNewSeq`.
4. For each 'line' of information in the body of the report create a feature object (`AjPFeature`) and load the feature object with your data values (by using `ajFeatNewII` and `ajFeatTagAdd`).
5. Set any data for the header and tail of the report.
6. Pass the report (`AjPReport`) and feature table (`AjPFeattable`) objects to the general report writing function (`ajReportWrite`).
7. Clean up memory as normal and exit.

6.15.9 Example report application

Consider a simple application to report the sequence length and molecular weight of a sequence.

6.15.9.1 ACD file

In the following ACD file, a `sequence` input and a `report` output are defined. The `datafile` datatype is just there so that a file of molecular weight data in the EMBOSS data area can be read.

```
application: wreport
[
   documentation: "Example report program"
]
section: input [ info: "input Section" type: page ]
sequence: sequence
[
   parameter: "Y"
   type: "Protein"
]

endsection: input

section: advanced [ info: "advanced Section" type: page ]

datafile: aadata
[
   information: "Amino acid data file"
   help: "Molecular weight data for amino acids"
```

```
    default: "Eamino.dat"
]
endsection: advanced

section: output [ info: "output Section" type: page ]

report: outfile
[
    parameter: "Y"
    rformat: "table"
    multiple: "N"
    precision: "1"
    taglist: "float:molwt int:len"
]
endsection: output
```

The `section:` and `endsection:` definitions provide a means by which GUIs can be instructed to organise the ACD information on the screen. Each `section` must always have a corresponding `endsection`. It is standard practice to have at least an input and an output section definition, adding others as appropriate.

6.15.9.2 C source code

The complete source code for the application is given below.

```
/* @prog wreport ***************************************************************
**
** Show sequence length and molwt as a report
**
*******************************************************************************/
int main(int argc, char **argv)

{
    AjPSeq          seq    = NULL;
    AjPReport       report = NULL;
    AjPFeattable    ftable = NULL;
    AjPFeature      feat   = NULL;
    AjPStr          tmpstr = NULL;

    double          molwt;
    int             len;

    AjPFile         mfptr  = NULL;

    embInit ("wreport", argc, argv);

    seq    = ajAcdGetSeq ("sequence");
    report = ajAcdGetReport("outfile");

    /* This bit just reads in an EMBOSS data table of molwt info */
    mfptr  = ajAcdGetDatafile("aadata");

    embPropEaminoRead(mfptr);
    /* End of data file reading */

    /* Calculate the values to output */
    len    = ajSeqGetLen(seq);
    molwt = embPropCalcMolwt(ajSeqGetSeqC(seq),0,len-1);
```

```
/* Create a feature table */
ftable = ajFeattableNewSeq(seq);

tmpstr = ajStrNew();

/* Fill head and tail information for the report */
ajFmtPrintS(&tmpstr,"This is some Header Text");
ajReportSetHeaderS(report, tmpstr);

ajFmtPrintS(&tmpstr,"This is some Tail Text");
ajReportSetTailS(report, tmpstr);

/* Create feature object and load with the output values */
feat = ajFeatNewII(ftable,1,len);

ajFmtPrintS(&tmpstr,"*molwt %.1f", (float)molwt);
ajFeatTagAdd(feat,NULL,tmpstr);

ajFmtPrintS(&tmpstr,"*len %d", len);
ajFeatTagAdd(feat,NULL,tmpstr);

/* Write report and clean up */
ajReportWrite(report,ftable,seq);

ajFeattableDel(&ftable);
ajStrDel(&tmpstr);
ajFileClose(&mfptr);

embExit();
return 0;}
```

First the variables are defined. Objects are needed for a sequence (AjPSeq) and a feature table (AjPFeattable) to hold the feature objects containing the column values. Also needed is an object for individual features (AjPFeature) to contain the values that will be printed in the report. Some housekeeping variables are required and, of course, the report itself (AjPReport):

```
AjPSeq        seq    = NULL;   /*  Input sequence      */
AjPReport     report = NULL;   /*  Output report       */
AjPFeattable  ftable = NULL;   /*  Feature table       */
AjPFeature    feat   = NULL;   /*  Individual feature */
AjPStr        tmpstr = NULL;

double        molwt;
int           len;

AjPFile       mfptr  = NULL;
```

The program can then be initialised and the ACD file processed:

```
embInit ("wreport", argc, argv);

seq    = ajAcdGetSeq ("sequence");
report = ajAcdGetReport("outfile");
```

Code is then needed to read in the EMBOSS data file containing amino acid molecular weight data and calculate the molecular weight and length values to be reported. Note that `len` and `molwt` are used for both the variable names and for the column names. This is not necessary but does make the code more clear:

```
mfptr  = ajAcdGetDatafile("aadata");

embPropEaminoRead(mfptr);

len   = ajSeqGetLen(seq);
molwt = embPropCalcMolwt(ajSeqGetSeqC(seq),0,len-1);
```

The report head and tail information is then set. This could have been done any time after the `ajAcdGetReport` but before the report is printed. They are optional but recommended. The temporary string object is used for this:

```
tmpstr = ajStrNew();

ajFmtPrintS(&tmpstr,"This is some Header Text");
ajReportSetHeaderS(report, tmpstr);

ajFmtPrintS(&tmpstr,"This is some Tail Text");
ajReportSetTailS(report, tmpstr);
```

The feature table object is then created. Only one feature table is required per report. The sequence object is passed as a parameter so that the sequence name can be automatically loaded into the internal GFF format. A feature object can now be created into which the column values can be loaded. A single feature object with a tag/value pair is used for each output value. The column values are printed into a string object (`tmpstr`) which is used to load the feature object.

`ajFeatNewII` requires three parameters. The first is the feature table object, the last two are the sequence start and end positions. Rather naughtily they are hard-coded in this example (as mentioned above). A feature object has to be created per line of output in the final report. `ajFeatTagAdd` is used to load the molecular weight and length values into the feature object:

```
ftable = ajFeattableNewSeq(seq);

feat = ajFeatNewII(ftable,1,len);

ajFmtPrintS(&tmpstr,"*molwt %.1f", (float)molwt);
ajFeatTagAdd(feat,NULL,tmpstr);

ajFmtPrintS(&tmpstr,"*len %d", len);
ajFeatTagAdd(feat,NULL,tmpstr);
```

This is the bit where the ACD column names (`molwt` and `len`) must match the ones in the `ajFmtPrintS` calls. Within the C program these names must be preceded by an asterisk (`*`). The datatypes specified in the ACD file (`float` and `int`) must also match what's given in the

C code. The NULL parameter just means that only a value is being added (the library will add the /note tag automatically).

The report can now be written. Three objects are passed. The reason for the sequence object being passed is in case you choose a report format that prints out the (sub)sequence used:

```
ajReportWrite(report,ftable,seq);
```

Finally the dynamic memory is recovered in a clean-up prior to exiting:

```
ajFeattableDel(&ftable);
ajStrDel(&tmpstr);
ajFileClose(&mfptr);

embExit ();

return 0;
```

6.15.10 Report file management

As an alternative to ACD processing, a report file can be opened directly by calling `ajReportOpen`. A report file is closed by calling `ajReportClose`:

```
AjBool   ajReportOpen (AjPReport thys, const AjPStr name);
void     ajReportClose (AjPReport thys);
```

To add an additional file name and description to the report call `ajReportAddFileF`:

```
void ajReportAddFileF (AjPReport report, AjPFile file, const AjPStr type);
```

The file name and type will be listed in the report header as "Additional files" to let the user know that there are associated output files available.

To write the report file call `ajReportWrite`:

```
AjBool ajReportWrite (AjPReport report, const AjPFeattable ftable, const AjPSeq seq);
```

The function writes a report file (`report`) from the feature table object (`ftable`) for the given sequence (`seq`). It returns `ajTrue` if data was written or `ajFalse` if maximum output has already been reached. The steps needed to prepare the report are described in detail below.

The report header and tail can be written separately by calling `ajReportWriteHeader` or `ajReportWriteTail`. These take as parameters the sequence object being reported on:

```
void ajReportWriteHeader (AjPReport report, const AjPFeattable ftable, const AjPSeq
seq);
void ajReportWriteTail (AjPReport report, const AjPFeattable ftable, const AjPSeq seq);
```

You will never normally need to call these functions as the header and tail are written automatically when you call `ajReportWrite`.

6.15.11 Setting elements of a report object

The functions in this section are used to set the elements in a report object directly.

To set some preliminary text (`header`) as the report header or subheader then call `ajReportSetHeaderS` or `ajReportSetSubheaderS`:

```
void  ajReportSetHeaderS (AjPReport report, const AjPStr header);
void  ajReportSetHeaderC (AjPReport report, const char* header);
void  ajReportSetSubheaderS (AjPReport report, const AjPStr header);
void  ajReportSetSubheaderC (AjPReport report, const char* header);
```

To append some text to the report header or subheader to the report then call `ajReportAppendHeaderS` or `ajReportAppendSubheaderS`:

```
void  ajReportAppendHeaderS (AjPReport report, const AjPStr header)
void  ajReportAppendHeaderC (AjPReport report, const char* header);
void  ajReportAppendSubheaderS (AjPReport report, const AjPStr header)
void  ajReportAppendSubheaderC (AjPReport report, const char* header);
```

To set some trailing text (`tail`) as the report tail or subtail then call `ajReportSetTailS` or `ajReportSetSubtailS`:

```
void  ajReportSetTailS (AjPReport report, const AjPStr tail);
void  ajReportSetTailC (AjPReport report, const char* tail);
void  ajReportSetSubtailS (AjPReport report, const AjPStr tail);
void  ajReportSetSubtailC (AjPReport report, const char* tail);
```

To append some text to the report tail or subtail to the report then call `ajReportAppendTailS` or `ajReportAppendSubtailS`:

```
void ajReportAppendTailS(AjPReport report, const AjPStr tail)
void ajReportAppendTailC (AjPReport report, const char* tail);
void ajReportAppendSubtailS (AjPReport report, const AjPStr tail)
void ajReportAppendSubtailC (AjPReport report, const char* tail);
```

To set the tag list for a report call `ajReportSetTagsS`:

```
AjBool ajReportSetTagsS (AjPReport report, const AjPStr taglist);
```

To set the report type if it is not set already call `ajReportSetType`:

```
void ajReportSetType (AjPReport report, const AjPFeattable ftable, const AjPSeq seq);
```

To set the default format for a feature report call `ajReportFormatDefault`:

```
AjBool ajReportFormatDefault (AjPStr* pformat);
```

6.15.12 Getting elements of a report object

It is sometimes necessary to query the properties of a report object. Although you will typically not need the functions in this section, they are useful if you have created a report object manually (using the constructor) rather than by picking up one from ACD processing.

To return the sequence name or USA depending on the setting in the report object (derived from the ACD and command line **-rusashow** option) call `ajReportGetSeqnameSeq`:

```
const AjPStr ajReportGetSeqnameSeq (const AjPReport report, const AjPSeq seq);
```

6.15.13 Debugging report objects

To test that is a report is valid call:

```
AjBool ajReportValid (AjPReport report);
```

This will check that the specified format works with the number of tags and with the type (protein or nucleotide) of the sequence. The sequence format is set if it is not already defined.

To print information about the current report format call `ajReportPrintFormat`. If the `full` argument is `ajTrue` then information on all reports formats is generated:

```
void ajReportPrintFormat (AjPFile outf, AjBool full);
```

6.16 Handling lists

6.16.1 Introduction

AJAX provides objects for implementing doubly linked lists and iterating over them. All the basic functionality that you would expect is provided. This includes:

- Node retrieval without removal (peeking)
- List iteration to move backwards and forwards through a list
- List querying to return nodes identified by a provided query function
- List editing to add (push) and remove (pop) nodes from a list, to push an entire list onto another and reverse the node order
- Garbage collection to remove nodes identified by a provided query function
- List sorting on up to three fields (C data structure elements) of the node data
- List to array conversion
- List processing invoking a provided function on the list nodes.

The list and list iterator objects are created and handled directly from the C source code. The `list` ACD datatype is not for generic lists; it is used to handle selections from a menu.

6.16.2 AJAX library files

AJAX library files for handling lists are listed in the table below. Library file documentation, including a complete description of datatypes and functions, is available at:

```
http://emboss.open-bio.org/rel/dev/libs/
```

Library file documentation	Description
`ajlist`	General list handling

ajlist.h/c. All the functions you are likely to ever need for the creation and control of linked lists. They define the basic list object (`AjPList`) which includes the list node object (`AjPListNode`) as a nested substructure, and a list iteration object (`AjIList`). It also includes static data structures and functions for handling lists at a low level. You are unlikely to need the low-level code unless you plan to implement code to extend the core functionality of the library.

6.16.3 ACD datatypes

There is no dedicated ACD datatype for handling general lists, the `list` ACD datatype being used to handle selections from a menu (see Section 6.19, 'Handling menus'). For numeric arrays the `array` datatype can be used (see Section 6.17, 'Handling arrays').

6.16.4 AJAX datatypes for handling lists

There are three AJAX datatypes for handling lists:

`AjPList`	Basic linked list
`AjPListNode`	Node in a linked list, included as a nested substructure in `AjPList`
`AjIList`	Used for list iteration.

6.16.5 Types of list

The functions handle lists of nodes which include a pointer variable to the node data. The generic list handling functions have `void *` parameters which must be cast to the correct type when an argument is passed. For convenience, a dedicated set of functions for handling string lists are provided that are cast to the appropriate type. Such functions all have the prefix `ajListstr` to distinguish them from the general functions (which have the prefix `ajList`).

6.16.6 List object memory management

6.16.6.1 Default object construction

To use a list you must first instantiate the list object pointer. The default constructor functions allocate an empty list (no nodes or node data):

```
/* Default generic list constructor */
AjPList ajListNew (void)

/* Default string list constructor */
AjPList ajListstrNew (void);
```

All constructors return the address of a new list. The pointers do not need to be initialised to NULL but it is good practice to do so:

```
AjPList   list    = NULL;
AjPList   liststr = NULL;

list      = ajListNew();
liststr   = ajListstrNew();

/* The objects are instantiated and ready for use */
```

6.16.6.2 Default object destruction

You must free the memory for a list once you are finished with it. The default destructor functions will free the list and the nodes but not the node data itself:

```
/* Default generic list destructor */
AjPList ajListFree (AjPList* Plist);

/* Default string list destructor */
AjPList ajListstrFree (AjPList* Plist);
```

Typical code for object destruction is as follows:

```
AjPList   list    = NULL;
AjPList   liststr = NULL;

list      = ajListNew();
liststr   = ajListstrNew();

/* The objects are instantiated and ready for use … do something with the list */

ajListFree(&list);
ajListFree(&liststr);

/* The lists are freed and the pointers reset to NULL, ready for reuse.
```

```
list    = ajListNew();
liststr = ajListstrNew();

/* The pointer variable is reallocated. Do something else with the new lists. */

ajListFree(&list);
ajListstrFree(&liststr);

/* Free the lists once you are done with them. */
```

6.16.6.3 Managing node data

In practice you will allocate data to the nodes in a list which must be freed once you are done with them. You must be careful not to lose your handle (C pointers) on the node data in your program, or to forgot to free the nodes, or to try and free the node data twice. These are all easy mistakes to make.

In the following example, two strings (`str1` and `str2`) are pushed onto both a generic list (`list`) and a string list (`liststr`) using `ajListPushAppend` and `ajListstrPushAppend` respectively. This is for illustrative purposes only. In practice you would always use the list string functions (`ajListstr*`) to handle lists of strings:

```
AjPList    list    = NULL;
AjPList    liststr = NULL;
AjPStr     str1    = NULL;
AjPStr     str2    = NULL;
AjPStr     strtmp  = NULL;
AjIList    iter    = NULL;

/* Create the strings and lists */
str1 = ajStrNewC("String 1");
str2 = ajStrNewC("String 2");

list    = ajListNew();
liststr = ajListstrNew();

/* Push the strings onto the generic list. The strings must be cast to void *. */
ajListPushAppend(list, (void *)str1);
ajListPushAppend(list, (void *)str2);

/* Push the same strings onto the string list. The cast is not needed. */
ajListstrPushAppend(liststr, str1);
ajListstrPushAppend(liststr, str2);
```

As the same two strings are pushed onto both lists, two string objects and both the lists need freeing. For string lists only, there is a destructor to free both the list *and* the node data (strings) referenced by the list:

```
/* String list and nodes destructor */
AjPList ajListstrFreeData (AjPList* Plist);
```

Therefore the memory allocated by the code above could be freed as follows:

```
/* Free the string list and strings using ajListstrFreeData. Free the generic list
using ajListFree */
ajListFree(&list);
ajListstrFreeData(&liststr);

/* All memory is freed now and the list pointers are set to NULL and ready for reuse. */
```

For generic lists, you must free the node data manually. The following code is typical and uses the list iterator object (`iter`) to iterate through a list `list`) using `ajListIterGet` calling the destructor function (in this case a string destructor) on each node in turn. Instead of using `ajListstrFreeData` the memory allocated by the code above could be freed as follows:

```
/* Iterate through the list deleting the node data */
iter = ajListIterNew(list);

while(strtmp = (AjPStr) ajListIterGet(iter))
      ajStrDel(&strtmp);

/* Free the list and list iterator */
ajListIterDel(&iter);
ajListFree(&list);
```

Putting all this together:

```
AjPList  list    = NULL;
AjPList  liststr = NULL;
AjPStr   str1    = NULL;
AjPStr   str2    = NULL;
AjPStr   strtmp  = NULL;
AjIList  iter    = NULL;

/* Create the strings and lists */
str1 = ajStrNewC("String 1");
str2 = ajStrNewC("String 2");

list    = ajListNew();
liststr = ajListstrNew();

/* Push the strings onto the generic list. The strings must be cast to void *. */
ajListPushAppend(list, (void *)str1);
ajListPushAppend(list, (void *)str2);

/* Push the same strings onto the string list. The cast is not needed. */
ajListstrPushAppend(liststr, str1);
ajListstrPushAppend(liststr, str2);

/* Free the string list and strings using ajListstrFreeData. Free the generic list
using ajListFree */
ajListFree(&list);
ajListstrFreeData(&liststr);

/* All memory is freed now and the list pointers are set to NULL and ready for
reuse. */

/* Re-allocate the strings and push them onto the generic list. */
str1 = ajStrNewC("String 1");
```

```
str2 = ajStrNewC("String 2");

list = ajListNew();
ajListPushAppend(list, (void *)str1);
ajListPushAppend(list, (void *)str2);

/* Iterate through the list deleting the node data */
iter = ajListIterNew(list);
while(strtmp = (AjPStr) ajListIterGet(iter))
    ajStrDel(&strtmp);

/* Free the list and list iterator */
ajListFree(&list);
ajListIterDel(&iter);

/* All memory is freed now */
```

6.16.6.4 Alternative object construction and loading

Various alternative constructor and destructor functions are provided. For the constructors you must know whether only the nodes are copied by a function (the node data being shared between two lists) or the nodes and node data are physically copied from one list to another. Similarly, for the destructors, you must distinguish whether just the list or the list and the node data are deleted.

6.16.6.4.1 List copying

Functions to copy a list of nodes but not the node data itself include:

```
/* Copy a generic list */
AjPList ajListNewListref (const AjPList list);

 /* Copy a string list */
ajListstrNewListref (const AjPList list);
```

In the following example, `ajListstrNewListref` is used to copy a list:

```
AjPList   list      = NULL;
AjPList   listcopy  = NULL;
AjPStr    str1      = NULL;
AjPStr    str2      = NULL;

/* Create a list of strings */
list     = ajListNew();
str1 = ajStrNewC("String 1");
str2 = ajStrNewC("String 2");
ajListPushAppend(list, (void *)str1);
ajListPushAppend(list, (void *)str2);

/* Copy the list */
listcopy = ajListstrNewListref(list);

/* Do something with the lists */

/* Free the original list and node data */
```

```
ajListstrFreeData(&list);

/* Free the copy of the list */
ajListstrFree(&listcopy);
```

6.16.6.4.2 List creation (from supplied node data)

Step through the data and use ajListPushAppend calls.

For example, to create a list from two sequences:

```
AjPList  list   = NULL;
AjPSeq   seq1   = NULL;
AjPSeq   seq2   = NULL;
AjPSeq   seqtmp = NULL;
AjIList  iter   = NULL;

/* Create the sequences */
list = ajListNew();
seq1 = ajSeqNewNameC("AAAAAAAAAA");
seq2 = ajSeqNewNameC("LLLLLLLLLL");

/* Create the list */
list = ajListNew;
ajListPushAppend(list, seq1);
ajListPushAppend(list, seq2);

/* Do something with the lists */

/* Free the list and node data */
iter = ajListIterNew(list);

while(seqtmp = (AjPSeq) ajListIterGet(iter))
     ajSeqDel(&seqtmp);

ajListFree(&list);
```

6.16.7 List node retrieval

Functions are provided to return a node from a list without removing it. The current node, first node, last node or a numbered node may be returned:

```
/* Retrieve the current node in a generic list */
AjBool ajListPeek (const AjPList list, void** Pptr);

/* Retrieve the current node in a string list  */
AjBool ajListstrPeek (const AjPList list, AjPStr* Pstr);

/* Retrieve the nth node in a generic list     */
AjBool ajListPeekNumber (const AjPList list, void** Pptr, ajint n);

/* Retrieve the first node in a generic list    */
AjBool ajListPeekFirst (const AjPList list, void** Pptr);
```

```
/* Retrieve the last node in a generic list */
AjBool ajListPeekLast (const AjPList list, void** Pptr);
```

All the functions return `ajTrue` if the node was retrieved successfully and use an appropriate object pointer to pick up the retrieved node. For example, here the first node in a list of sequences is retrieved. Code to create the list of sequences is not shown:

```
AjPSeq seqtmp = NULL;
AjPList list = NULL;

… /* Code to allocate list of sequences */

if(ajListPeekFirst(list, &seqtmp))
    /* Do something with list */;

else
    /* Error handling */;
```

6.16.8 List iteration

6.16.8.1 List iterator construction

To iterate through a list you must first construct a list iteration object. There are four functions for this:

```
/* Default list iterator constructor, start to end iteration. */
AjIList ajListIterNew (AjPList list);

/* Start to end iteration through read-only list */
AjIList ajListIterNewread (const AjPList list);

/* End to start iteration */
AjIList ajListIterNewBack (AjPList list);

/* End to start iteration through read-only list */
AjIList ajListIterNewreadBack (const AjPList list);
```

You have already come across `ajListIterNew`. The other functions are used in exactly the same way except they set up the iterator for reading from a read-only list (`ajListIterNewread`), to iterate from the end to the start of a list `ajListIterNewBack` or to iterate from end to start through a read-only list (`ajListIterNewreadBack`).

6.16.8.2 List iteration

Flexible control statements can be constructed for list iteration. There are three functions for iterating through a list from start to end:

```
/* Returns the next item in a list                    */
void* ajListIterGet (const AjIList iter);
```

```
/* Returns true if iteration is complete     */
AjBool ajListIterDone (const AjIList iter);
```

You have already come across `ajListIterGet` to return the next item in a list (or `NULL` if the end of the list is reached). `ajListIterDone` will return `ajTrue` if iteration has completed.

Another three functions have the same functionality but are for iteration from the end to the start of a list:

```
/* As above, but end to start iteration */
void* ajListIterGetBack (const AjIList iter);

/* As above, but end to start iteration */
AjBool ajListIterDoneBack (const AjIList iter);
```

6.16.8.3 List iterator destruction

As you have already seen, once you are done with the list iterator, you must destroy the object using:

```
void  ajListIterDel (AjIList* Piter);
```

6.16.9 Querying a list

To search through a list and determine if any nodes are matches according to a given function use:

```
/* Search a generic list */
AjBool ajListMap (const AjPList list, AjBool apply(void** Pptr), void* cl);

/* Search a string list  */
AjBool ajListstrMap (const AjPList list, AjBool apply(AjPStr* Pstr), void* cl);
```

The search function must be provided and is of the same type as the comparison function used for garbage collection (see below). The `cl` argument is usually `NULL`, but can be used to pass external data to the function. Using the same example, if the match function used is:

```
AjBool Compare(const void * ptr, void* cl)
{
   if(ajSeqGetLen((AjPSeq)ptr) > 1000)
        return ajTrue;

   return ajFalse;
}
```

then `ajListMap` will return `ajTrue` if any of the nodes is a sequence whose length is greater than 1000:

```
if(ajListMap(list, Compare, NULL))
/* List contains a sequence > 1000 residues */;
else
/* List does not contain such a sequence */;
```

6.16.10 List editing

Several basic list editing functions are provided. These include:

- Popping a node from the start or end of a list
- Pushing a node onto the start or end of a list
- Push an entire list of nodes onto another list
- Add or remove a list node at any point in the list
- Reverse the order of the nodes in a list.

6.16.10.1 Adding nodes to a list

The following functions add a node to a list (equivalent functions for string lists are available):

```
/* Push a node onto the start of a list */
void   ajListPush (AjPList list, void* ptr);

/* Push a node onto the end of a list */
void   ajListPushAppend (AjPList list, void* ptr);

/* Push a list of nodes onto the start of a list */
void   ajListPushlist (AjPList list, AjPList* Plist);

/* Insert a node into a list using an iterator */
void   ajListIterInsert (AjIList iter, void* ptr);
```

`ajListPushlist` deletes the list from which the nodes were pushed, so you must be careful not to try and delete the node data twice. In the following example, all memory is freed correctly:

```
AjPList   list      = NULL;
AjPList   listnew   = NULL;
AjPSeq    seq1      = NULL;
AjPSeq    seq2      = NULL;
AjPSeq    seqtmp    = NULL;
AjIList   iter      = NULL;

/* Create a list of sequences */
list = ajListNew();
seq1 = ajSeqNewNameC("AAAAAAAAAA", "seqA");
seq2 = ajSeqNewNameC("CCCCCCCCCC", "seqC");

ajListPushAppend(list, seq1);
```

```
ajListPushAppend(list, seq2);

/* Create a new list and add the nodes to it (the original list is deleted) */
listnew = ajListNew();
ajListPushlist(list, &listnew);

/* Free the list and node data */
iter = ajListIterNew(listnew);

while(seqtmp = (AjPSeq) ajListIterGet(iter))
ajSeqDel(&seqtmp);

ajListFree(&listnew);
```

`ajListIterInsert` will insert a node into a list at the current position of the list iterator that is passed. In this example a new sequence is inserted into the middle of the list:

```
AjPList  list  = NULL;
AjPSeq   seq1  = NULL;
AjPSeq   seq2  = NULL;
AjPSeq   seq3  = NULL;
AjIList  iter  = NULL;

/* Create a list from seq1 and seq3 */
list = ajListNew();
seq1 = ajSeqNewNameC("AAAAAAAAAA");
seq3 = ajSeqNewNameC("CCCCCCCCCC");
ajListPushAppend(list, seq1);
ajListPushAppend(list, seq3);

iter = ajListIterNew(list);

while(seqtmp = (AjPSeq) ajListIterGet(iter))
{
    /* Insert a new node once node with sequence "CCCCCCCCCC" is found */
    if(ajStrMatchC(ajSeqGetSeqS(seqtmp), "CCCCCCCCCC"))
    {
          seq2 = ajSeqNewNameC("BBBBBBBBBB");
          ajListIterInsert(iter, (void *) seq2);
          break;
    }
}

/* The order of sequences in the list is now seq1, seq2, seq3 */
```

6.16.10.2 Removing nodes from a list

The following functions remove a node from a list (equivalent functions for string lists are available):

```
/* Pop a node from the start of a list */
AjBool   ajListPop (AjPList list, void** Pptr);

/* Pop a node from the end of a list */
AjBool ajListPopLast (AjPList list, void** Pptr);
```

```
/* Remove current node at a list iterator position */
void   ajListIterRemove (AjIList iter);
```

`ajListIterRemove` simply removes the node at the current position of the list iterator that is passed. You must be careful not to lose your handle on the node data and inadvertently create a memory leak, which is why in this example, which follows from the code above, the sequence `seq3` is deleted explicitly:

```
iter = ajListIterNew(list);
while(seqtmp = (AjPSeq) ajListIterGet(iter))
{
      /* Simply delete the node */
    if(ajStrMatchC(ajSeqGetSeqS(seqtmp), "CCCCCCCCC"))
    {
       ajListIterRemove(iter);    /* Delete the node */
       ajSeqDel(&seqtmp);         /* Delete the node data to avoid a memory leak */
       break;
    }
```

In contrast, when a node is popped using `ajListPop` or `ajListPopLast` the data is caught by the second argument, which should be freed once you are done with it:

```
...

ajListPopLast(list, &seqtmp);

/* Do something with the popped sequence before deleting it */

ajSeqDel(&seqtmp);
```

6.16.10.3 Reversing the order of nodes in a list

To reverse the order of nodes in a list use:

```
void   ajListReverse (AjPList list);       /* Generic list */
void   ajListstrReverse (AjPList list);    /* String list */
```

6.16.10.4 Garbage collection

Garbage collection may be performed on a list to remove nodes whose node data satisfy a certain constraint. `ajListPurge` is used and takes three arguments: the list to be cleaned (`list`), a comparison function (`test`) to check the constraint and a destructor function (`nodedelete`) to delete the node data:

```
void   ajListPurge (AjPList list,
                    AjBool (*test)(const void *),
                    void (*nodedelete)(void **));
```

The comparison and destructor functions must be provided. As you can see from the prototype the comparison function returns a `AjBool` (which is `ajTrue` if the constraint is matched) and takes a generic `void *` argument whereas the destructor function must be of `void` return type and take a generic `void **` argument.

Let us assume you have a list of sequence (`AjPSeq`) objects you want to garbage collect, removing all sequences from the list that are longer than 1000 characters. The comparison function would be as follows:

```
AjBool Compare(const void * ptr)
{
    if(ajSeqGetLen((AjPSeq)ptr) > 1000)
        return ajTrue;

    return ajFalse;
}
```

We cannot use the default sequence destructor function because its parameter is of the wrong type:

```
void  ajSeqDel (AjPSeq* Pseq);
```

A `const` pointer is required, which means you must provide a wrapper function around the default destructor function:

```
void mySeqDelWrap(const void **ptr)
{

    ajSeqDel((AjPSeq *) ptr);

    return;

}
```

To perform garbage collection on a list of sequences (`list`) you would call:

```
void  ajListPurge (list, Compare, mySeqDelWrap);
```

6.16.11 List sorting

A list may be sorted on up to three fields (elements of the C data structure) of the node data. Where more than one field is used the list is first sorted by field 1. Nodes with the same sort order for field 1 are then re-sorted by field 2 such that the relative sort order from field 1 is maintained. Similarly, if a third field is used, nodes with the same sort order for fields 1 and 2 are re-sorted by field 3 such that the relative sort order from sorting by fields 1 and 2 is preserved. The functions are :

```
void    ajListSort (AjPList list,int (*sort1) (const void*, const void*));
void    ajListSortTwo (AjPList list,int (*sort1) (const void*, const void*),
                      int (*sort2) (const void*, const void*));
void    ajListSortTwoThree (AjPList list,int (*sort1) (const void*, const void*),
                           int (*sort2) (const void*, const void*),
                           int (*sort3) (const void*, const void*));
```

Lists may also be sorted (by up to two fields) and duplicate nodes (nodes with the same sort order) removed. The functions are:

```
void    ajListSortUnique (AjPList list,
                         int (*compar) (const void*, const void*),
                         void nodedelete (void**, void*));
void    ajListSortTwoUnique(AjPList list,
                           int (*sort1) (const void*, const void*),
                           int (*sort2) (const void*, const void*),
                           void nodedelete (void**, void*));
```

One or more comparison functions `sort*` must be provided. These take two generic `void *` arguments for the two objects passed, and return the sort order of the objects as an integer: which should be `-1` if the first object sorts higher than the second and `+1` otherwise. For the `ajListSortUnique*` functions, a destructor function is the same as for `ajListMap` (see above): it must be of `void` return type and take a generic `void **` argument and a (usually `NULL`) second argument which can be used to pass external data to the function.

Let us assume you have a list of sequences (`AjPSeq` objects) you want to sort by length (shortest first) and make unique, removing any sequences whose length is the same as another sequence in the list such that the final list of sequences have unique lengths. The comparison function would be as follows:

```
int Compare(const void *ptr1, const void * ptr2)
{
   if(ajSeqGetLen((AjPSeq)ptr1)) < ajSeqGetLen((AjPSeq)ptr2))
       return -1;
   return 1;
}
```

The destructor function is the same as for list deletion collection above.

To sort the list and remove sequences of duplicate length you would call:

```
void  ajListSortUnique (list, Compare, mySeqDelWrap);
```

6.16.11.1 String comparison functions

Various string comparison functions are provided in `ajstr.c` and are useful for sorting lists of strings. The functions perform case-sensitive and case-insensitive comparison, with and without wildcard characters. Variants of the function support string objects (`AjPStr`) and C-type (`char *`) strings (not all shown):

```
/* String to C-type string */
int   ajStrCmpC (const AjPStr thys, const char *text);

/* String to string */
int   ajStrCmpS (const AjPStr str, const AjPStr str2);

/* Case-insensitive string to string */
int   ajStrCmpCaseS (const AjPStr str1, const AjPStr str2);

/* String to C-type string using the first len characters only */
ajint   ajStrCmpLenC (const AjPStr thys, const char *text, ajuint len);

/* String to string using the first len characters only */
int   ajStrCmpLenS (const AjPStr str, const AjPStr str2, ajuint len);

/* String to C-type string with wildcards.*/
int   ajStrCmpWildC (const AjPStr thys, const char* text);

/* String to string with wildcards.*/
int   ajStrCmpWildS (const AjPStr thys, const AjPStr str);

/* Case-insensitive string to C-type string with wildcards.*/
int   ajStrCmpWildCaseC (const AjPStr thys, const char* text);

/* Case-insensitive string to string with wildcards.*/
int   ajStrCmpWildCaseS (const AjPStr thys, const AjPStr str);

/* Same as ajStrCmpS but strings are cast as void */
int   ajStrVcmp (const void* str1, const void* str2);
```

There are also comparison functions for handling C-type strings:

```
/* Case-insensitive C-type string to C-type string */
int   ajCharCmpCase (const char* txt1, const char* txt2);

/* Case-insensitive C-type string to C-
type string using the first len characters only */
int   ajCharCmpCaseLen (const char* txt1, const char* txt2, ajuint len);

/* C-type string to C-type string using wildcards. */
ajint   ajCharCmpWild (const char* txt1, const char* txt2);

/* Case-insensitive C-type string to C-type string using wildcards. */
ajint   ajCharCmpWildCase (const char* txt1, const char* txt2);
```

The functions return an integer representing the sort order which is `-1` if the first string should sort before the second, `+1` if the second string should sort first, or `0` if they are identical.

6.16.12 List conversion

A list may be converted to an array by using:

```
ajint   ajListToarray (const AjPList list, void*** PParray);
ajint   ajListstrToarray (const AjPList list, AjPStr** PParray);
```

A string list may be appended to an existing array by using:

```
ajint  ajListstrToarrayAppend (const AjPList list, AjPStr** PParray);
```

The functions return the size of the array and in all cases memory for the array is (re)allocated as needed and the original list left untouched. The node data are not duplicated, so once these functions return, the list, the array and the node data must be freed. In the example below, code to create the list of sequences is not shown:

```
AjPSeq *seqarray = NULL;
AjPList list = NULL;
AjPSeq seqtmp = NULL;
AjIList iter = NULL;

... /* Code to allocate list of sequences */

ajListToarray(list, &seqarray);

/* Do something with the array */

/* Free the list and node data */
iter = ajListIterNew(list);

while(seqtmp = (AjPSeq) ajListIterGet(iter))
     ajSeqDel(&seqtmp);

ajListFree(&listnew);

/* Delete the iterator */
ajListIterDel(&iter);

/* Free the array */
AJFREE(seqarray);
```

6.16.13 List node functions

To invoke a function on all the nodes in a list use one of:

```
void  ajListMap (AjPList list, void apply(void** Pptr, void* ptr), void* cl);
void  ajListstrMap (AjPList list, void apply(AjPStr* Pstr, void* ptr), void* cl);
void  ajListMapread (const AjPList list, void apply(void* ptr, void* ptr), void* cl);
void  ajListstrMapread (const AjPList list, void apply(AjPStr str, void* ptr), void*
cl);
```

The function that is applied is again of the same type as the comparison function used for garbage collection (see above). The `*MapRead` functions ensure that the function (`apply`) that's applied does not modify the list elements and, as such, takes a list that is `const`. In contrast, any type of operation is permitted with `ajListMap` and `ajListstrMap`.

6.17 Handling arrays

6.17.1 Introduction

Array handling is greatly simplified in EMBOSS. AJAX provides dynamic array objects which grow automatically as required thereby saving you the bother of managing the memory yourself. This makes your code cleaner and less error-prone. Objects are provided for handling arrays of various dimensions (1D, 2D or 3D) of the following C datatypes:

- Character (`char`)
- Short integer (`ajshort`)
- Unsigned integer (`ajuint`)
- Integer (`ajint`)
- Floating point number (`float`)
- Long floating point number (`ajlong`)
- Double-precision floating point number (`double`).

The objects can be allocated to a default or a specified reserved size. Each object keeps track of the reserved space available (in case a function is called which must extend the array) and the actual length of the array. Arrays of higher dimensions can nest arrays of lower dimensions.

Functions are provided for returning the current length and for getting and setting array elements. The array elements should only ever be accessed using these functions. If you assign a value to an element which is beyond the current size of the array, then the array will be dynamically reallocated to a sufficient size to accommodate the new element. In contrast, if you try to retrieve an element beyond the current size of the array an error will be generated.

The C-type array may be retrieved from any of the array objects should this be required. This allows standard array notation to be used to access array elements. Functions are provided to sort C-type arrays of integers, unsigned integers and floating point numbers in various ways.

Array objects are typically created directly within the C source code. A floating point array can however be specified in the ACD file (the `array` ACD datatype) and a pointer to a corresponding AJAX object (`AjPFloat`) retrieved by the program.

6.17.2 AJAX library files

AJAX library files for handling arrays are listed in the table below. Library file documentation, including a complete description of datatypes and functions, is available at: http://emboss.open-bio.org/rel/dev/libs/

Library file documentation	Description
`ajarr`	General array handling
`ajsort`	Array sorting

`ajarr.h/c`. Defines the various array objects and functions for handling dynamic arrays. It also contains static functions for handling them at a low level. You are unlikely to need these unless you plan to extend the core functionality of the library.

`ajsort.h/c`. Functions for sorting arrays (Section 6.17.10, 'Sorting arrays').

6.17.3 AJAX datatypes

For handling an input array defined in the ACD file use:

`AjPFloat`	Array of floating point numbers (for `array` ACD datatype)

AJAX provides dynamic array objects for the fundamental C datatypes (Section 5.1, 'Basic datatypes') below.

6.17.3.1 `char` arrays

`AjPChar`	1D character array

6.17.3.2 `short` arrays

`AjPShort`	1D short array
`AjPShort2d`	2D short array
`AjPShort3d`	3D short array

6.17.3.3 `unsigned int` arrays

`AjPUint`	1D unsigned integer array
`AjPUint2d`	2D unsigned integer array
`AjPUint3d`	3D unsigned integer array

6.17.3.4 `int` arrays

`AjPInt`	1D integer array
`AjPInt2d`	2D integer array
`AjPInt3d`	3D integer array

6.17.3.5 `float` arrays

`AjPFloat`	1D float array
`AjPFloat2d`	2D float array
`AjPFloat3d`	3D float array

6.17.3.6 `long` arrays

`AjPLong`	1D long array
`AjPLong2d`	2D long array
`AjPLong3d`	3D long array

6.17.3.7 `double` arrays

`AjPDouble`	1D double array
`AjPDouble2d`	2D double array
`AjPDouble3d`	3D double array

6.17.4 ACD datatypes

The ACD datatype for handling array input is:

`array`	List of either integer or floating point numbers

6.17.5 ACD data definition

A typical ACD definition for array input:

```
array: thresholds
[
   information: "Values to represent 'identical', 'similar' and 'related'"
   default: "-1.5,0.0,1.5"
   minimum: "0.0"
   size: "3"
   sum: "0"
   sumtest: "Y"
]
```

6.17.5.1 Parameter name

All data definitions for array input should have an intuitive name (Section A.1.3, 'Parameter naming conventions') although no standard name scheme is enforced.

6.17.5.2 Common attributes

Attributes that are typically specified are summarised below. They are datatype-specific (Section A.5, 'Datatype-specific attributes') unless they are indicated as being global attributes (Section A.4, 'Global attributes').

`default:` A global attribute that specifies a default value.
`minimum:` Specifies the minimum permitted value.

size: Specifies the permissible number of elements in an `array` data definition.

sum: Specifies the total of all values in an `array` data definition and is tested for unless the `sumtest:` attribute is false.

sumtest: A boolean attribute which, if turned off, turns off testing for the `sum:` attribute for an array data definition.

6.17.6 ACD file handling

Datatypes and functions for handling arrays via the ACD file are shown below.

Title	To read an array
ACD datatype	`array`
AJAX datatype	`AjPFloat`
To retrieve from ACD	`ajAcdGetArray`

Your application code will call `embInit` to process the ACD file and command line (see Section 6.3, 'Handling ACD files'). All values from the ACD file are read into memory and files are opened as necessary. You have a handle on the files and memory through the `ajAcdGet*` family of functions which return pointers to appropriate objects.

6.17.6.1 Input array retrieval

To retrieve an input array an object pointer is declared and then initialised using `ajAcdGetArray`:

```
AjPFloat thresholds=NULL;
thresholds = ajAcdGetArray("thresholds");
```

6.17.6.2 Processing command line options and ACD attributes

Currently there are no functions for this.

6.17.6.3 Memory management

It is your responsibility to free up memory at the end of the program. You must call the default destructor function (see below) on any array objects returned by calls to `ajAcdGetArray`.

6.17.7 Names of functions

Functions for handling arrays have consistent names with a prefix of the following general form:

```
ajTypeDim
```

`Type` is the type of the array and is one of:

- `Chararr` (`character` array)
- `Short` (`ajshort` array)
- `Uint` (`ajuint` integer array)
- `Int` (`ajint` array)
- `Float` (`float` array)
- `Long` (`ajlong` array)
- `Double` (`double` array)

`Dim` is the dimensionality of the array. It is *not* given for 1D arrays:

- `2d` (two-dimensional array)
- `3d` (three-dimensional array).

For example, all functions for handling 2D arrays of unsigned integers have the prefix `ajUint2d` whereas all functions for handling 1D arrays of floats have the prefix `ajFloat`.

Functions are provided for handling 1D, 2D and 3D arrays; however, only 1D character arrays are supported.

6.17.8 Array object memory management

6.17.8.1 Default object construction

To use an array object that is not defined in the ACD file you must first instantiate the appropriate object pointer. The default construction functions have the suffix `New`:

```
AjPChar       ajChararrNew (void);
AjPShort      ajShortNew (void);
AjPShort2d    ajShort2dNew (void);
AjPShort3d    ajShort3dNew (void);
AjPUint       ajUintNew (void);
AjPUint2d     ajUint2dNew (void);
AjPUint3d     ajUint3dNew (void);
AjPInt        ajIntNew (void);
AjPInt2d      ajInt2dNew (void);
AjPInt3d      ajInt3dNew (void);
AjPFloat      ajFloatNew (void);
AjPFloat2d    ajFloat2dNew (void);
AjPFloat3d    ajFloat3dNew (void);
AjPLong       ajLongNew (void);
AjPLong2d     ajLong2dNew (void);
AjPLong3d     ajLong3dNew (void);
AjPDouble     ajDoubleNew (void);
AjPDouble2d   ajDouble2dNew (void);
AjPDouble3d   ajDouble3dNew (void);
```

All constructors return the address of a new object. The pointers do not need to be initialised to `NULL` beforehand but it is good practice to do so. All the functions above are used in the same way:

```
AjPShort2d      shorts = NULL;
shorts    = ajShort2dNew();
/* The objects are instantiated and ready for use */
```

6.17.8.2 Default object destruction

You must free the memory for an object once your are finished with it. The default destructor functions have the suffix `Del` and take the address of the appropriate object as an argument:

```
void  ajChararrDel    (AjPChar* Parr);
void  ajShortDel      (AjPShort* Parr);
void  ajShort2dDel    (AjPShort2d* Parr);
void  ajShort3dDel    (AjPShort3d* Parr);
void  ajUintDel       (AjPUint* Parr);
void  ajUint2dDel     (AjPUint2d* Parr);
void  ajUint3dDel     (AjPUint3d* Parr);
void  ajIntDel        (AjPInt* Parr);
void  ajInt2dDel      (AjPInt2d* Parr);
void  ajInt3dDel      (AjPInt3d* Parr);
void  ajFloatDel      (AjPFloat* Parr);
void  ajFloat2dDel    (AjPFloat2d* Parr);
void  ajFloat3dDel    (AjPFloat3d* Parr);
void  ajLongDel       (AjPLong* Parr);
void  ajLong2dDel     (AjPLong2d* Parr);
void  ajLong3dDel     (AjPLong3d* Parr);
void  ajDoubleDel     (AjPDouble* Parr);
void  ajDouble2dDel   (AjPDouble2d* Parr);
void  ajDouble3dDel   (AjPDouble3d* Parr);
```

For example to free memory allocated by array input from the ACD file use:

```
void   ajFloatDel (AjPFloat* Parr);
```

It is used as follows:

```
AjPFloat thresholds=NULL;
thresholds = ajAcdGetArray("thresholds");
/* Do something with the array */
ajFloatDel(&thresholds);
```

It is the responsibility of the calling function to destroy any objects once they are finished with:

```
AjPShort2d     shorts = NULL;
shorts     = ajShort2dNew();
/* The object is instantiated and ready for use */
/* Do something with the instantiated objects */
ajShort2dDel(&shorts);
/* The memory is freed and the pointers reset to NULL, ready for re-use. */
shorts = ajShort2dNew();
/* Do something else with the new objects. The pointer variable is reallocated. */
```

```
ajShort2dDel(&shorts);
/* The objects are done with so the memory is freed. */
```

6.17.8.3 Alternative object construction and loading

An array can be constructed with a fixed reserve size by using functions with the suffix `NewRes`. Such functions take an integer argument (`size`) for the reserved size but are otherwise identical to the default constructor. Only a few of the available functions are shown below:

```
AjPChar      ajChararrNewRes (ajint size);
AjPShort     ajShortNewRes (ajint size);
AjPShort2d   ajShort2dNewRes (ajint size);
AjPShort3d   ajShort3dNewRes (ajint size);
AjPUint      ajUintNewRes (ajint size);
AjPUint2d    ajUint2dNewRes (ajint size);
...
```

6.17.9 Getting and setting array elements

If you assign a value to an element array which is beyond the current size of the array, then the array will be dynamically reallocated to a sufficient size to accommodate the new element. If you try to retrieve an array element beyond the current size of the array an error will be generated.

It is therefore handy to know the size of an array in advance. Functions for retrieving the size of an array have the suffix `Len` and either return the array length (for 1D arrays) or write the lengths by argument (2D and 3D arrays). Again, just a few functions are shown:

```
ajint    ajChararrLen (const AjPChar arr);
ajint    ajShortLen (const AjPShort arr);
void     ajShort2dLen (const AjPShort2d arr, ajint* Pn1, ajint* Pn2);
void     ajShort3dLen (const AjPShort3d arr,ajint* Pn1, ajint* Pn2, ajint* Pn3);
ajint    ajUintLen (const AjPUint arr);
void     ajUint2dLen (const AjPUint2d arr, ajint* Pn1, ajint* Pn2);
...
```

For example, to get the sizes of a 3D double array:

```
AjPDouble3d arr = NULL;
ajint dim1;
ajint dim2;
ajint dim3;

/* Allocate array */
ajDouble3dLen(arr, &dim1, &dim2, &dim3);
```

Functions to retrieve an element of an array have the suffix `Get`. They take the index of the element to retrieve and return the appropriate datatype:

```
char    ajChararrGet (const AjPChar arr, ajint n);
short   ajShortGet (const AjPShort arr, ajint n);
short   ajShort2dGet (const AjPShort2d arr, ajint n1, ajint n2);
short   ajShort3dGet (const AjPShort3d arr, ajint n1, ajint n2, ajint n3);
ajuint  ajUintGet (const AjPUint arr, ajint n);
ajuint  ajUint2dGet (const AjPUint2d arr, ajint n1, ajint n2);
...
```

For example, to retrieve the very first element of a 3D float array:

```
AjPFloat3d arr = NULL;
 float val;

 ...
 /* Allocate array */

 val = ajFloat3dGet(arr, 0, 0, 0);
```

Functions to set an element of an array have the suffix `Put`. They take the address of an array, the index of the element to set and of course the value to set. They return `ajTrue` if the element was set successfully:

```
AjBool   ajChararrPut (const AjPChar* Parr, ajint n, char chr);
AjBool   ajShortPut (const AjPShort* Parr, ajint n, short i);
AjBool   ajShort2dPut (const AjPShort2d* Parr, ajint n1, ajint n2, short i);
AjBool   ajShort3dPut (const AjPShort3d* Parr, ajint n1, ajint n2, ajint n3, short i);
AjBool   ajUintPut (const AjPUint* Parr, ajint n, ajuint i);
AjBool   ajUint2dPut (const AjPUint2d* Parr, ajint n1, ajint n2, ajuint i);
...
```

For example, to set the very first element of a 3D float array:

```
AjPFloat3d  arr = NULL;
 float     val = 1.0;

 arr = ajFloat3dNew();

 ajFloat3dPut(&arr, 0, 0, 0, val);
```

6.17.10 Sorting arrays

Functions are provided in `ajsort.h/c` to sort C-type arrays of integers, unsigned integers and floating point numbers. Arrays may be sorted in the following ways:

- Basic sorting of a single array
- Sorting a single array by the order of elements in another array
- Sorting of two arrays simultaneously
- Sorting in ascending or descending order.

Functions for array sorting have the prefix `ajSortInc` (for sorting in incremental ascending order) or prefix `ajSortDec` (for sorting in descending order), followed by the datatype (`Int`, `Uint` or `Float`) of the array(s) that is sorted. The array size must be provided.

The basic sort functions include:

```
extern   void    ajSortIntDec (ajint *a, ajuint n);
extern   void    ajSortUintDec (ajuint *a, ajuint n);
extern   void    ajSortFloatDec (float *a, ajuint n);
extern   void    ajSortIntInc (ajint *a, ajuint n);
extern   void    ajSortUintInc (ajuint *a, ajuint n);
extern   void    ajSortFloatInc (float *a, ajuint n);
```

There are functions for sorting a single array (`p`) by the order of elements in another array (`a`):

```
void    ajSortFloatDecI (const float *a, ajuint *p, ajuint n);
void    ajSortIntDecI (const ajint *a, ajuint *p, ajuint n);
void    ajSortUintDecI (const ajuint *a, ajuint *p, ajuint n);
void    ajSortFloatIncI (const float *a, ajuint *p, ajuint n);
void    ajSortIntIncI (const ajint *a, ajuint *p, ajuint n);
void    ajSortUintIncI (const ajuint *a, ajuint *p, ajuint n);
```

Functions for sorting two arrays simultaneously:

```
void    ajSorttwoIntIncI (ajint *a, ajuint n, ajuint *p);
void    ajSorttwoUintIncI (ajuint *a, ajuint n, ajuint *p);
```

6.17.11 Retrieving a C-type array

The C-type array may be retrieved from any of the array objects using the following functions. The function naming scheme should be obvious. Again, not all variants are shown:

```
char*       ajChararrChararr (const AjPChar arr);

short*      ajShortShort (const AjPShort arr);
short**     ajShort2dShort (const AjPShort2d arr);
short***    ajShort3dShort (const AjPShort3d arr);

ajuint*     ajUintUint (const AjPUint arr);
ajuint**    ajUint2dUint (const AjPUint2d arr);
...
```

Once the C-type array is retrieved then standard array notation can be used to access array elements. It is of course your responsibility to ensure you do not try to access an array element that is out of bounds. To help avoid doing so you can use the `*Len` functions (see above) to return the array size(s).

6.18 Handling tables

6.18.1 Introduction

A hash table is a data structure holding key:value pairs which allows for very efficient retrieval of a value given a key. For example, a key might be the accession number of a sequence and a value might be the sequence itself. Hash tables work by translating the key into a number called a 'hash' using a hashing function. The hash then provides the index of an array element or 'bucket' in the data structure holding the value. Hash tables have the great advantage that the search time is independent of the number of items stored, and that new key:value pairs can be added efficiently.

EMBOSS supports a standard hash table using strings for keys and generic hash tables with keys of arbitrary datatype. All the basic functionality you would expect for creating, manipulating and using hash tables is provided. This includes:

- Hash functions for the construction of string hash functions
- Functions to add and remove key:value pairs or change the value associated with a key
- Functions to return the true key or value associated with a given key
- Ability to apply a function to each key:value pair
- Printing a string hash table.

There is no ACD datatype for hash tables, they are always created directly from within the C source code.

6.18.2 AJAX library files

AJAX library files for handling hash tables are listed in the table below. Library file documentation, including a complete description of datatypes and functions with usage notes is available at: http://emboss.open-bio.org/rel/dev/libs/

Library file documentation	Description
`ajtable`	Hash table functions

`ajtable.h/c`. Defines the hash table object (`AjPTable`) and functions for handling hash tables. They also contains static functions for handling same at a low level. You are unlikely to need the static functions unless you plan to extend the hash table code.

6.18.3 ACD datatypes

There are no ACD datatypes dedicated for tables.

6.18.4 AJAX datatypes

For handling tables use:

`AjPTable`	Hash table – a set of key:value pairs using a simple hash function to provide rapid searching for keys. Tables can hold any data type.

6.18.5 Table object memory management

6.18.5.1 Default object construction

To use a hash table object you must first instantiate the appropriate object pointer. Default constructor functions are provided for generic hash tables and EMBOSS string (`AjPStr`) hash tables:

```
/* Generic hash table */
AjPTable ajTableNewLen(ajuint size)

/* Generic hash table with compare and hash functions */
AjPTable ajTableNewFunctionLen (ajint n,
                     ajint cmp(const void* x, const void* y),
                     unsigned hash(const void* key, unsigned hashsize));

/* String hash table */
AjPTable   ajTablestrNew (ajint n);
AjPTable   ajTablestrNewLen (ajint n);
AjPTable   ajTablestrNewCaseLen (ajint n);
```

Both functions will create, initialise and return a new empty hash table that can hold an arbitrary number of key:value pairs. For a generic hash table, created with `ajTableNewFunctionLen`, an estimate of the number of unique keys (`n`) is passed, and functions for comparison (`cmp`) and for hashing the keys (`hash`) should be provided. If either of the `cmp` or `hash` arguments are `NULL` then it will use a default general-purpose hash function. The `ajTableNewLen` uses the general-purpose functions. Note that these cannot inspect the data, and can only use the addresses or values provided.

All constructors return the address of a new object. In the following code `comp` is the comparison function and `hash` is the hash function. An initial size of 50 is used. The pointer does not need to be initialised to `NULL` but it is good practice to do so:

```
AjPTable table = NULL;

table    = ajTableNewFunctionLen(50, comp, hash);

/* The table is instantiated and ready for use */
```

The comparison function must return the result of comparing two keys as an integer which is `0` if the comparison is equal, `+1` if the `x` is higher than `y` or `-1` otherwise. The hash function must return the hash value for a given key.

Assume that you want a hash table with a case-insensitive C-type character string (`char *`) key. The comparison function may then be defined as `ajCharCmpCase` (available from `ajstr.h`) which returns the sort order of two text strings:

```
ajint comp(const void* x, const void* y)
{
    const char* sx;
    const char* sy;

    sx = (const char*) x;
    sy = (const char*) y;

    return (ajint) ajCharCmpCase(sx, sy);
}
```

The following hash function may be used:

```
ajuint hash(const void* key, ajuint hashsize)
{
   ajuint hash;
   const char* s;

   s = (const char*) key;

   for(hash = 0; *s; s++)
       hash = (hash * 127 + toupper((ajint)*s)) % hashsize;

   return hash;
}
```

6.18.5.2 Default object destruction

You must free the memory for an object once your are finished with it. A default destructor function is provided for generic and string hash tables:

```
void   ajTableFree (AjPTable* Ptable);
void   ajTablestrFree (AjPTable* Ptable);
```

It is the responsibility of the calling function to destroy any objects once they are finished with:

```
AjPTable table = NULL;

table = ajTableNewFunctionLen(50, comp, hash);

/* Do something with the instantiated table */

ajTableFree(&table);

/* The memory is freed and the pointer reset to NULL, ready for re-use. */

table = ajTablestrNew(50);

/* The pointer variable is reallocated. Do something else with the new table. */

ajTableFree(&table);
```

```
/* Done with the object so the memory is freed. */
```

6.18.5.3 Alternative object construction and loading

To create a generic hash table with an explicit (rather than estimated) number of key:value pairs use:

```
AjPTable ajTableNewLen (ajint n,
                        ajint cmp(const void* x, const void* y),
                        unsigned hash(const void* key, unsigned hashsize));
```

6.18.6 Table hash functions

A few table hash functions used during hash table construction (see above) are provided for convenience for string tables only. These provide case-sensitive and case-insensitive hashing for hash tables of EMBOSS strings (`AjPStr`) and C-type (`char *`) strings:

```
/* Case-insensitive hashing of EMBOSS strings */
unsigned   ajTablestrHash (const void* key, unsigned hashsize);

/* Case-sensitive hashing of EMBOSS strings    */
unsigned   ajTablestrHashCase (const void* key, unsigned hashsize);
```

6.18.7 Table comparison functions

A few table comparison functions used during hash table construction (see above) are provided for convenience for string tables only. These provide case-sensitive and case-insensitive comparison for hash tables of EMBOSS strings (`AjPStr`):

```
/* Case-sensitive comparison of EMBOSS strings */
ajint   ajTablestrCmp (const void* ptrx, const void* ptry);

/* Case-insensitive comparison of EMBOSS strings */
ajint   ajTablestrCmpCase (const void* ptrx, const void* ptry);
```

6.18.8 Table edit functions

Key:value pairs may be added to a table, removed from a table or the value associated with a key may be changed.

`ajTablePut` is used to set the value associated with a given key in a hash table. It will return the previous value or, if the table does not hold the specified key, it will add the key and value and return `NULL`.

```
void*  ajTablePut (AjPTable table, const void* key, void* value);
```

ajTableRemove will remove the key:value pair from a table and returns the removed value. If the table does not hold the specified key then the function merely returns NULL:

```
void*  ajTableRemove (AjPTable table, const void* key);
```

6.18.9 Table query functions

The true key or value associated with a given key may be returned by using:

```
/* Return the value */
void*   ajTableFetch (const AjPTable table, const void* key);

/* Return the key   */
void*   ajTableFetchKey (const AjPTable table, const void* key);
```

ajTableFetch is the standard function that returns a value given a query key.

ajTableFetchKey is provided to return the true key (with any capitalisation) in cases where a case-insensitive hash table is used.

To return the number of key:value pairs in a hash table use:

```
ajint   ajTableGetLength (const AjPTable table);
```

6.18.10 Table map functions

It is possible to apply a function to each key:value in a hash table using:

```
void   ajTableMap (AjPTable table, void apply(const void* key, void** Pvalue, void*
ptr), void* ptr);
void   ajTableMapDel (AjPTable table, void apply(const void** Pkey, void** Pvalue,
void* ptr), void* ptr);
```

The function called by ajTableMap should not delete any keys (although values can be modified) whereas the function called by ajTableMapDel may delete keys. In both cases a function must be provided.

Assume that you have a hash table with EMBOSS string keys and integer values and that you want to apply one function to print the key:value pair and another to free a key:value pair.

The print functions will look something like:

```
static void print(const void* key, void** value, void* cl)
{
   AjPStr keystr;
   ajint *valptr;
```

```
    keystr = (AjPStr) key;
    valptr = (ajint *) *value;

    ajUser("type '%S' found %d times", keystr, *valptr);

    return;
}
```

whereas the delete function would be:

```
static void delete(const void** key, void** value, void* cl)
{
    AjPStr keystr;
    ajint *valptr;

    keystr = (AjPStr) *key;
    valptr = (ajint *) *value;

    ajStrDel(&keystr);
    AJFREE(valptr);

    *key = NULL;
    *value = NULL;

    return;
}
```

To call the functions you would use something like:

```
AjPTable table = NULL;

/* Allocate table */

ajTableMap(table, print, NULL);

ajTableMapDel(table, delete, NULL);
```

6.18.11 Print functions

Functions are provided to print a string hash table of EMBOSS strings (`AjPStr`):

```
void ajTablestrPrint (const AjPTable table);
```

6.19 Handling menus

6.19.1 Introduction

Selecting from a list of options is often necessary in molecular biology programs. Two ACD datatypes (`list` and `selection`) provide slightly different styles of list but both present a user with a list of options they can choose from. For the user, the difference between the `list` and `selection` datatype is minimal and lies only in the way the choices are labelled. In a `selection` datatype the choices are numbered automatically from 1 upwards. In a `list` datatype the choices can be labelled by any arbitrary text. The user can choose one of the options by either typing the number (for a `selection` type) or the label text (for a `list` type), or (both list types) a non-ambiguous part of the descriptive text itself given after the number or label. In practice, the `list` data type is preferred.

6.19.2 AJAX library files

There is no specific AJAX library file for handling menus. Selections from menus are handled as strings or arrays of strings. You should consult the documentation for further information about string handling. Library file documentation is available online at: http://emboss.open-bio.org/rel/dev/libs/

Library file documentation	Description
`ajstr`	String handling

`ajstr.h/c`. Defines the string object (`AjPStr`) used for handling strings from the ACD file (see Section 6.4, 'Handling simple ACD datatypes'). They contain most of the functions you will ever need for general string handling (Section 6.5, 'Handling strings').

6.19.3 ACD datatypes

There are two ACD datatypes for handling menus:

`list` A list of options (text descriptions) with text labels. The user is presented with a limited list of options they can choose from. The choices can be labelled by any arbitrary text label. The option descriptions are usually more verbose than for the `selection` datatype.

`selection` A list of options (text descriptions) with automatically generated numerical labels. The user is presented with a limited list of options they can choose from. The choices are numbered automatically from 1 upwards.

6.19.4 ACD data definition

6.19.4.1 `list` menu

A typical ACD definition for a `list` menu:

```
list: frame [
    standard: "Y"
    help: "Allows selection from a set of reading frames"
```

```
    default: "1"
    minimum: "1"
    maximum: "1"
    header: "Translation frames"
    values: "1:1, 2:2, 3:3, F:Forward three frames, -1:-1, -2:-2, -3:-3, R:
Reverse three
    frames, 6:All six frames"
    delimiter: ","
    codedelimiter: ":"
    information: "Frame(s) to translate"
}
```

The user will be presented with the title of the menu from the `header:` attribute. After that will appear the tag/text information from the `values:` attribute. Following that will be the prompt from the `information`: attribute. It will look something like this:

```
Translation frames
    1     1
    2     2
    3     3
    F     Forward three frames
   -1    -1
   -2    -2
   -3    -3
    R     Reverse three frames
    6     All three frames
Frame(s)  to  translate[1]:
```

The user is allowed (generally) to supply a comma-separated list of options from the above; this depends on the minimum and maximum values. In the above example both the minimum number of selections and the maximum number of selections are set to 1, therefore only one selection value is allowed. Selection is done by typing the tag values, therefore if the maximum count had been set to 3 then a user entry of `"-1,F,6"` would be a valid input.

The `codedelimiter` attribute allows you to select any character to separate the tags from the text in the values attribute. The `delimiter` attribute allows you to choose a character to separate the tag/text pairs within the values attribute.

6.19.4.2 `selection` menu

A typical ACD definition for a `selection` menu:

```
select: order
[
     casesensitive: "N"
     default: "score"
     delimiter: ","
     header: "Sort order of results"
     help: "Name of the output file which holds the results of the analysis."
     information: "Sort order of results"
     maximum: "1"
     minimum: "1"
     information: "Select sort order of results"
     standard: "Y"
```

```
    values: "length, position, score"
]
```

Note

The `selection` datatype is very similar to the `list` datatype. The `values:` attribute just contains a delimiter-separated list of text entries (leading whitespace is ignored). Selection takes place on the text values, there are no tags. Also, note that there is a `casesensitive:` attribute which is often set to `"N"` so that shifted and non-shifted characters are equally acceptable. The user can select on the text down to a length of text that has no ambiguity with other values.

6.19.4.3 Parameter name

Parameter names for menus should be intuitive but no standards are enforced.

6.19.4.4 Common attributes

Attributes that are typically specified are summarised below. They are datatype-specific (Section A.5, 'Datatype-specific attributes') unless they are indicated as being global attributes (Section A.4, 'Global attributes').

`values:` For both the `list` and `selection` datatype menus the values that the user can choose from are defined in the `values:` attribute as a string.

`delimiter:` For both the `list` and `selection` datatype menus the values in the `values:` string are delimited by the character given by the `delimiter:` attribute, which defaults to '`;`' (semi-colon).

`codedelimiter:` For the `list` datatype the character defined by the `codedelimiter:` attribute is the delimiter that separates the label from the value, which defaults to '`:`' (colon).

`minimum:` Defines the minimum number of selections a user can make.

`maximum:` Defines the maximum number of selections a user can make.

`header:` Holds the text that is displayed above the list of options.

`information:` This global attribute (defined for all datatypes) defines text to be used as a prompt after the list.

`casesensitive:` Controls whether or not the options are case sensitive. Regardless of the case used by the user, the value of the parameter will be exactly what the corresponding list value is.

`button:` This attribute is provided for interface developers. It is a boolean which is either `"Y"` (Yes) or `"N"` (No), indicating whether a graphical button such as radio-button, checkbox or selection list should be used to select a value, or if the list should be displayed with a text entry box to enter the option.

6.19.5 AJAX datatypes

The AJAX string datatype is used for handling menus defined in the ACD file is:

`AjPStr *` Array of strings (for `list` and `selection` ACD datatypes)

6.19.6 ACD file handling

Datatypes and functions for handling menus via the ACD file are shown below.

ACD datatype	AJAX datatype	To retrieve from ACD
`list` menu		
`list`	`AjPStr *`	`ajAcdGetList`
`selection` menu		
`outfile`	`AjPStr *`	`ajAcdGetSelect`

Your application code will call `embInit` to process the ACD file and command line (see Section 6.3, 'Handling ACD files'). All values from the ACD file are read into memory. You have a handle on the files and memory through the `ajAcdGet*` family of functions which return pointers to appropriate objects.

6.19.6.1 Menu retrieval

To retrieve a menu an object pointer is declared and then initialised using the appropriate `ajAcdGet*` function.

6.19.6.1.1 `list` menu

```
...

AjPStr *flist=NULL;
...

flist = ajAcdGetList("frame");
...
```

The declaration shows that `flist` will be an array of string objects. The values held in the strings are the tags from the values list and the list is terminated by a `NULL` pointer. So, using our example, as only one value is allowed and assuming the user had answered '6' to the prompt the resulting array would be:

```
flist[0]    a string object that contains "6"
flist[1]    a NULL pointer
```

Your code would likely step through the list, if maximum is greater than 1, using something like the following:

```
n = 0;
while(flist[n])
[
        if (ajStrMatchC(flist[n],"6"))
        [
                /* Do something */
        ]

        ...
        ++n;
]
```

6.19.6.1.2 `selection` menu

```
...
AjPStr *oselect=NULL;
...
    oselect = ajAcdGetSelect("order");
...
```

Otherwise the code is the same as for the list datatype with the exception that it would test for the strings ‘length’, ‘position’ and ‘score’.

6.19.6.2 Alternative ACD retrieval functions

```
AjPStr    ajAcdGetListSingle    (const char *token);
AjPStr    ajAcdGetSelectSingle    (const char *token);
```

It is your responsibility to free up memory at the end of the program. You have only retrieved one string; the remainder of the internal data will be cleaned up autiomatically by ACD processing. You should call `ajStrDel` to free the string returned by these functions.

6.19.6.3 Processing command line options and ACD attributes

Currently there are no functions for this.

6.19.6.4 Memory management

It is your responsibility to free up memory at the end of the program. You must call the destructor function on the string array returned by calls to `ajAcdGetList` or `ajAcdGetSelect`.

The destructor for an array of strings is:

```
void   ajStrDelarray   (AjPStr** PPstr);
```

It is used as follows:

```
AjPStr  *flist=NULL;

flist  =  ajAcdGetList("frame");
...
ajStrDelarray(&flist);
```

6.20 Handling graphical output

6.20.1 Introduction

EMBOSS can produce two types of graphics output. First are the 'drawn' graphics. They allow a user to draw in a graphics window using primitives for (e.g.) line and dot plotting. The other, more commonly used, graphic is the 'graph' plot. Several functions are provided to take the sting out of displaying graphs, for example the automatic drawing of axes and the ability to set titles for them. The graphics are currently produced using wrapper functions to **PLplot**. At the time development of EMBOSS began, **PLplot** was the only multi-platform C graphics library available with the LGPL licence.

Graph objects are created through ACD datatypes. The library calls provide an interface to the **PLplot** library which manages output to the screen or to files. **PLplot** defines a graphics viewport to work within. The viewport dimensions depend on the definitions in the **PLplot** drivers for each device and are not identical in size or proportion. Where the application needs to manipulate the scale of objects (for example EMBOSS application **cirdna**) it is recommended to work in units of millimetres (or 4 pixels) rather than arbitrary units which fail to scale when used on a different device.

General features of the graphics output include:

- Graphics functions can define colours, line styles, titles and axis labels and plot scaling factors.
- The objects plotted can be lines, shapes or text.
- Functions are available to calculate scaled character sizes, polar coordinates and distances between plot coordinates.
- The most useful internal data is **PLplot** scaling information including current scaling factors and the plot dimensions in millimetres.
- Where plot output is to a file, there are functions to control the directory and output file names.
- The **PLplot** library has general settings that are independent of any single plot. For this reason you should avoid using more than one active graph object in any EMBOSS application.

6.20.2 AJAX library files

AJAX library files for handling graphics are listed in the table below. Library file documentation, including a complete description of datatypes and functions, is available at: http://emboss.open-bio.org/rel/dev/libs/

Library file documentation	Description
`ajgraph`	General graph handling functions
`ajgraphstruct`	Data structures for use with graphs
`ajgraphxml`	Graph XML file object and functions for manipulating it
`ajhist`	Histograms.
`ajcall`	Graphics package access
`ajdom`	Document Object Model library calls

ajgraph.h/c. Basic functions for plotting and printing graphs. It also contains static data structures and functions for handling graphics at a low level.

ajgraphstruct.h. Objects for handling graphs, including the basic graph object (AjPGraph) and the **PLplot** graph data object (AjPGraphdata).

ajhist.h. Defines the histogram object AjPHist object for handling histograms, which includes the histogram data object (AjPHistdata) as a substructure. It contains basic functions for handling histograms (Section 6.20.17, 'Histograms').

ajcall.h/c. Functions to allow access to different graphics packages, should alternatives to **PLplot** be implemented in the future. They are not currently used or documented here and you should see the online library documentation for further information.

ajdom.h/c. Data structures and functions, and also static functions for handling a document object model. It is under development and is not documented here. See the online library documentation for further information.

You are unlikely to need the static data structures and functions in the above files unless you plan to implement code to extend the core functionality of the library.

6.20.3 ACD datatypes

There are two datatypes for handling graphics output:

graph	Graphical output of any general kind.
graphxy	Graphical output as a simple two dimensional (2D) XY plot with the sequence along the *x*-axis.

6.20.4 ACD data definition

A typical ACD definition for graphics output (graph datatype):

```
graph: graph
[
   gxtitle: "Residue Position"
]
```

A typical ACD definition for graphics output (xygraph datatype):

```
xygraph: graph
[
   gxtitle: "Residue Position"
]
```

6.20.4.1 Parameter name

All data definitions for graphics output should have the standard parameter name `graph`. See Appendix A, *ACD syntax reference*.

6.20.4.2 Common attributes

Attributes that are typically specified are summarised below. These include two datatype-specific attributes (Section A.5, 'Datatype-specific attributes').

`gtitle:` Specifies the graph title for either type of graph. Many other graphical elements can be set using the attributes.

`multiple:` Specifies the number of multiple XY graphs in a single output for a `graph` ACD datatype.

6.20.5 Program initialisation

In earlier releases of EMBOSS, any application that made graphics calls was required to first initialize the graphics library functions. The restructuring of the libraries in EMBOSS 6.2.0 makes development much easier. All applications now simply call `embInit` or `embInitPV`.

6.20.6 AJAX datatypes

For handling graphical output defined in the ACD file use:

`AjPGraph`	Basic graphics object (for `graph` and `graphxy` ACD datatypes)

One further datatype is used for handling histogram graphics:

`AjPHist`	Histogram object

The basic datatypes above include as nested substructures the following AJAX datatypes:

`AjPGraphobj`	**PLplot** graph object (a substructure of `AjPGraph`)
`AjPGraphdata`	**PLplot** graph data object (a substructure of `AjPGraph`)
`AjPHistdata`	Histogram data object (a substructure of `AjPHist`).

You are unlikely to need the static functions unless you plan to extend the graphics library code.

6.20.7 ACD file handling

Datatypes and functions for handling graphics output via the ACD file are shown below.

	To write a graph graph	To write a graphxy graph
ACD datatype	graph	graphxy
AJAX datatype	AjPGraph	AjPGraph
To retrieve from ACD	ajAcdGetGraph	ajAcdGetGraphxy

6.20.7.1 Output graphics retrieval

To retrieve an output graphics object an object pointer is declared and then initialised using the appropriate `ajAcdGet*` function.

6.20.7.1.1 graph

```
AjPGraph graph=NULL;

graph = ajAcdGetGraph("graph");
```

6.20.7.1.2 graphxy

```
AjPGraph graph=NULL;

graph = ajAcdGetGraphxy("graph");
```

6.20.7.2 Processing command line options and ACD attributes

Currently there are no functions for this.

6.20.7.3 Memory and file management

Your application code will call `embInit` or `embInitPV` to process the ACD file and command line. All values from the ACD file are read into memory and files are opened as necessary. You have a handle on the files and memory through the `ajAcdGet*` family of functions which return pointers to appropriate objects. It is your responsibility to call the graphics library to close any files and free up memory at the end of the program.

6.20.7.3.1 Closing graphics files

To close an output graphics file call `ajGraphicsClose`. It closes graphics output through the **PLplot** library, closing any output files and screen displays:

```
AjPGraph graph=NULL;

graph = ajAcdGetGraphxy("graph");

/* Do something with graph */

ajGraphicsClose();
```

6.20.7.3.2 Freeing memory

You must call the default destructor function (see below) on any graphics objects returned by calls to ajAcdGetGraph or ajAcdGetGraphxy.

6.20.8 Graphics object memory management

6.20.8.1 Default object construction

Graphics objects are typically loaded from ACD file processing via a call to ajAcdGetGraph or ajAcdGetGraphxy (see above). You should therefore never need to instantiate the graphics object (AjPGraph) manually. In the unlikely event you do, two constructor functions are provided, the default constructor and ajGraphxyNewI which specifies the number of graphs:

```
AjPGraph ajGraphNew (void);
AjPGraph ajGraphxyNewI (ajint numsets);
```

You may also need to create a graphics data object. Two constructors are provided, the default constructor and ajGraphdataNewI which specifies the number of data points:

```
AjPGraphdata ajGraphdataNew (void);
AjPGraphdata ajGraphdataNewI (ajint numsets);
```

All constructors return the address of a new object. The pointers do not need to be initialised to NULL but it is good practice to do so:

```
AjPGraph          graph    =  NULL;
AjPGraph          graphxy  =  NULL;
AjPGraphdata      gdata1   =  NULL;
AjPGraphdata      gdata2   =  NULL;

graph   =  ajGraphNew();
graphxy =  ajGraphxyNewI(2);
gdata1  =  ajGraphdataNew();
gdata2  =  ajGraphdataNew();

/* The objects are instantiated and ready for use */
```

6.20.8.2 Default object destruction

You must free the memory for an object before the pointer is reused and also once your are finished with it. The default destructor functions are:

```
void   ajGraphxyDel   (AjPGraph* pmult);
void   ajGraphdataDel(AjPGraphdata *thys);
```

They are used as follows:

```
AjPGraph    graph=NULL;

graph   =   ajAcdGetGraphxy("graph");

/* Do something with graph */

ajGraphicsClose();
ajGraphxyDel(&graph);
```

A further example:

```
AjPGraph      graph      = NULL;
AjPGraph      graphxy    = NULL;
AjPGraphdata  gdata1     = NULL;
AjPGraphdata  gdata2     = NULL;

graph   = ajGraphNew();
graphxy = ajGraphxyNewI(2);
gdata1  = ajGraphdataNew();
gdata2  = ajGraphdataNew();

/* Do something with the objects */
ajGraphxyDel(&graph);
ajGraphxyDel(&graphxy);
ajGraphdataDel(&gdata1);
ajGraphdataDel(&gdata2);
```

6.20.9 Graphics file management

Functions for handling graphics output files have the prefix `ajGraph`. The file name and directory can be specified (usually through ACD attributes and command line qualifiers which call these functions). Files are created for **PLplot** output devices that do not write directly to the screen. The library also provides data file output which is used to bypass **PLplot** and directly pass graphics results to the **Staden** package:

```
ajGraphicsSetFilename  /* Set name and extension of output file. */
ajGraphSetOutfileS     /*  Set name of output file. */
ajGraphSetOutdirS      /*  Set directory of output file. */
ajGraphIsData          /*  Checks if the graph is creating a Staden data file. */
ajGraphxyDisplay       /* Displays an xy graph */
```

The following functions set the output file name. You will not normally need to call these as the file will have received a name during ACD file processing.

```
/* Set  name and extension of output file. */
void   ajGraphicsSetFilename (const AjPGraph thys);

/* Set  name of output file.        */
void   ajGraphSetOutfileS (AjPGraph thys, const AjPStr txt);
```

```
/* Set  directory of output file. */
void   ajGraphSetOutdirS(AjPGraph thys, const AjPStr txt);
```

`ajGraphxyDisplay` is used to display the graph(s) in a graph object to the plotting device. It must be called once by any program with graphics output. The second argument (`closeit`) is `ajTrue` if the graphics output file is to be closed once displayed:

```
void ajGraphxyDisplay (AjPGraph thys, AjBool closeit );
```

Finally, `ajGraphIsData` will return `ajTrue` if the graph is creating a data file:

```
AjBool ajGraphIsData (const AjPGraph thys);
```

6.20.10 Initialisation functions

There are various assignment function for initialising and updating graph objects:

```
ajGraphDataAdd            /* Add a graph. */
ajGraphDataAdd            /* Add another graph to a multiple graph. */
ajGraphDataReplace        /* Replace a graph. */
ajGraphDataReplaceI       /* Replace a graph with specified number. */
ajGraphInitSeq            /* Set default values for graph based on a sequence. */
ajGraphxyNewI             /* Create graph object to hold a number of graphs. */
ajGraphxySetRanges        /* Set the min. and max. of data points. */
ajGraphClear              /* Delete drawable objects associated with a graph */
```

`ajGraphSetMulti` is used to add multiple graphs to a graph object. The number of graphs (`numsets`) is specified:

```
void ajGraphSetMulti (AjPGraph thys, ajint numsets);
```

`ajGraphDataAdd` is used to add a graph to a graph object. The graph data (held in a `AjPGraphdata` object) is passed and the function returns `1` on success (`0` otherwise):

```
ajint ajGraphDataAdd (AjPGraph thys, AjPGraphdata graphdata);
```

The function can be called multiple times to add any number of graphs. The graph data object is used by the graph and should not be deleted until you are finished with the graph itself:

```
AjPGraph      graphs  =  NULL;
AjPGraphdata  gdata   =  NULL;

graphs  =  ajAcdGetGraphxy("graphs");
```

```
gdata    =  ajGraphdataNew();

/* Set up graph data */

ajGraphDataAdd(graphs,gdata);

/* Do something with the graph */

ajGraphdataDel(&gdata);
ajGraphxyDel(&graphs);
```

`ajGraphDataReplace` and `ajGraphDataReplaceI` are used to replace one graph in a graph object with another. The new graph data (held in a `AjPGraphdata` object) is passed and graph data object that is replaced is deleted. The function returns `1` on success (`0` otherwise):

```
ajint   ajGraphDataReplace   (AjPGraph thys, AjPGraphdata graphdata);
ajint   ajGraphDataReplaceI   (AjPGraph thys, AjPGraphdata graphdata, ajint num);
```

`ajGraphDataReplace` replaces the first graph whereas `ajGraphDataReplaceI` replaces the numbered graph.

`ajGraphInitSeq` is used to define default values for a graph object based on a sequence that is passed:

```
void ajGraphInitSeq (AjPGraph thys, const AjPSeq seq);
```

The title of the plot is set to the name of the sequence and the *x*-axis is set to the sequence size. It is used as follows:

```
int  main(int argc, char **argv)
{
     AjPSeq seq       =  NULL;
     AjPGraph graph   =  NULL;

     embInit("isochore", argc, argv);

     seq     =  ajAcdGetSeq("sequence");
     graph   =  ajAcdGetGraphxy("graph");

     ajGraphInitSeq(graph, seq);
```

Finally, `ajGraphxySetRanges` will calculate the minimum and maximum of the data points and store the values in the graph object:

```
void   ajGraphxySetRanges    (AjPGraph thys);
```

To delete all the drawable objects connected to the graph object (to reuse it for a new graph page) call:

```
void    ajGraphClear    (AjPGraph thys);
```

6.20.11 Creating windows

The following functions are for managing windows for plotting graphs:

```
ajGraphOpenWin        /* Open a window. */
ajGraphOpenPlotset    /* Open a window with a number of plots. */
ajGraphOpenMm         /* Open a window and return dimensions in mm */
ajGraphicsClose       /* Close a window. */
ajGraphClear          /* Clear screen or new page. */
```

A window is required for each graph (screen or page) to be plotted. Functions with the prefix `ajGraphOpen` will open a window for graph plotting:

```
void    ajGraphOpenWin (AjPGraph thys, float xmin, float xmax, float ymin, float ymax);
void    ajGraphOpenPlotset (AjPGraph thys, ajint numofsets);
void    ajGraphOpenMm (AjPGraph thys, float * xmm, float * ymm);
```

`ajGraphOpenWin` is typically used and should be called once for each graph, typically once per application. The window is defined by the supplied minimum and maximum x and y values. Alternatively, `ajGraphOpenPlotset` will open a window for a number of plots. `ajGraphOpenMm` opens a window and returns the physical size in millimetres which the application can then map to its own plot units. This is the safe way to work with scaling drawable objects as **PLplot** device page sizes are inconsistent.

`ajGraphClear` is used to clear the screen (for screen output) or to create a new page (for plotter/postscript output). It is passed a boolean argument which if `ajTrue` will cause the current pen colour character sizes, plot title and subtitle, etc. to be reset for the next page.

```
void    ajGraphClear (AjPGraph thys, AjBool resetdefaults);
```

Once you are done drawing to the window (see below), you must close it using `ajGraphicsClose`:

```
void    ajGraphicsClose (void)
```

6.20.12 Drawing functions

Functions with the prefix `ajGraphicsDraw` are for drawing graphical elements to the plotter device. The functions are as follows:

```
ajGraphicsDrawposBox                    /* Draw a box.                      */
ajGraphicsDrawposBoxFill                /* Draw a box and fill.             */
ajGraphicsDrawposDia                    /* Draw a diamond.                  */
ajGraphicsDrawposDiaFill                /* Draw a diamond and fill.         */
ajGraphicsDrawsetPoly                   /* Draw a polygon.                  */
ajGraphicsDrawsetPolyFill               /* Draw a polygon and fill.         */
```

```
ajGraphicsDrawposRect                  /* Draw a rectangle.                          */
ajGraphicsDrawposRectFill              /* Draw a rectangle and fill.                 */
ajGraphicsDrawarcRect                  /* Draw a rectangle along a curve.            */
ajGraphicsDrawarcRectFill              /* Draw a rectangle along a curve  and fill.*/
ajGraphicsDrawposTri                   /* Draw a triangle.                           */
ajGraphicsDrawposTriFill               /* Draw a triangle and fill.                  */
ajGraphicsDrawposCircle                /* Draw a circle.                             */
ajGraphicsDrawsetDots                  /* Draw a set of dots.                        */
ajGraphicsDrawbarsHoriz                /* Draw horizontal error bars.                */
ajGraphicsDrawbarsVert                 /* Draw vertical error bars.                  */
ajGraphicsDrawposLine                  /* Draw a line between 2 points.              */
ajGraphicsDrawsetLines                 /* Draw a number of lines.                    */
ajGraphicsDrawsetSymbols               /* Draw a set of dots.                        */
ajGraphicsDrawposTextAtend             /* Draw text at end point.                    */
ajGraphicsDrawposTextAtmid             /* Draw text at mid point.                    */
ajGraphicsDrawposTextAtlineJustify     /* Draw text along line.                      */
ajGraphicsDrawposTextAtstart           /* Draw text at start point.                  */
ajGraphicsDrawarcTextJustify           /* Draw text along a curve.                   */
ajGraphicsDrawarcArc                   /* Draw a portion of a circle (an arc).       */
```

You should see the online documentation for an explanation of how to use these functions.

6.20.13 Setting graphical elements to render

Functions with the prefix `ajGraphicsSet` are for setting graphical elements to be rendered when the graph is plotted. The `ajGraphSetFlag` function is used internally. Developers should prefer to call the individual `ajGraphxyShow` functions for each flag. The functions are as follows:

```
ajGraphicsSetBgcolourBlack /* Background colour to black.                              */
ajGraphicsSetBgcolourWhite /* Background colour to white.                              */
ajGraphicsSetFgcolour      /* Foreground colour to a colour.                           */
ajGraphicsSetLabelsC       /* Labels for the plot.                                     */
ajGraphicsSetRlabelC       /* Right Y-axis label.                                      */
ajGraphSetDescC            /* Graph description.                                       */
ajGraphicsSetLinestyle     /* Line style.                                              */
ajGraphicsSetPortrait      /* Page orientation.                                        */
ajGraphicsSetPenwidth      /* Pen width.                                               */
ajGraphxyShowPointsCircle  /* Draw points as circles.                                  */
ajGraphxyShowPointsJoin    /* Draw lines between the points.                           */
ajGraphSetFlag             /* Set PLplot flags to add or subtract.                     */
ajGraphxySetRanges         /* Max. and min. of the data points for all graphs.         */
ajGraphxySetflagOverlay    /* Sets whether the graphs should lay on top of each other. */
ajGraphSetSubtitleC        /* Subtitle.                                                */
ajGraphShowSubtitle        /* Set whether subtitle is displayed.                       */
ajGraphSetTitleC           /* Title.                                                   */
ajGraphShowTitle           /* Set whether title is displayed.                          */
ajGraphxySetflagGaps       /* Set whether the graphs should enable gaps.               */
ajGraphicsSetDefcharsize   /* Default character size in mm.                            */
ajGraphicsSetCharscale     /* Character scale factor.                                  */
ajGraphicsSetFillpat       /* Fill pattern type.                                       */
```

Functions with the prefix `ajGraphxySetX` and prefix `ajGraphxySetY` are for setting graphical properties of the *x*- or *y*-axis respectively. The functions are as follows:

```
ajGraphxyShowXaxis         /* bottom x-axis. */
ajGraphxyShowUnum          /* labels on the top x-axis. */
```

```
ajGraphxyShowXtick          /* tick marks on the x-axis. */
ajGraphxyShowUaxis          /* left X-axis at the top. */
ajGraphxySetXendF           /* end position for x in the graph. */
ajGraphxyShowXgrid          /* grid the tick marks on the x-axis. */
ajGraphxyShowXinvert        /* tick marks inside the plot on the x-axis. */
ajGraphxyShowXlabel         /* label the x-axis. */
ajGraphxySetXrangeII        /* x-axis range with integers. */
ajGraphxySetXstartF         /* start position for x in the graph. */
ajGraphxyShowYnum           /* labels on the left hand axis. */
ajGraphSetXlabelC           /* title for the x-axis for multiple plots on one graph. */
ajGraphxyShowYaxis          /* left hand y-axis. */
ajGraphxyShowRaxis          /* right hand y-axis. */
ajGraphxyShowYtick          /* tick marks on the y-axis. */
ajGraphxySetYendF           /* end position for y in the graph. */
ajGraphxyShowYgrid          /* grid the tick marks on the y-axis. */
ajGraphxyShowYinvert        /* tick marks inside the plot on the y-axis. */
ajGraphxyShowYlabel         /* label the y-axis. */
ajGraphxySetYrangeII        /* y-axis range with integers. */
ajGraphxySetYstartF         /* start position for y in the graph. */
ajGraphSetYlabelC           /* title for the y-axis for multiple plots on one graph. */
```

The functions above are used in conjunction with functions to add graphical elements to be rendered:

```
ajGraphAddLine          /* Adds a line. */
ajGraphAddRect          /* Adds a rectangle. */
ajGraphAddTextC         /* Adds text. */
ajGraphAddTextScaleC        /* Adds text with scale factor */
```

A more limited set of functions is available for setting elements to be rendered in a graph data object:

```
ajGraphdataSetMinmax        /* Min. and max. of data points for all graphs.  */
ajGraphdataSetTruescale     /* Scale min. and max. of the data points       .*/
ajGraphdataSetTypeC         /* Type of the graph for data output.            */
ajGraphdataSetSubtitleC     /* Subtitle */
ajGraphdataSetTitleC        /* Title */
ajGraphAppendTitleC         /* Title from the description. */
ajGraphdataSetXlabelC       /* x-axis title. */
ajGraphdataSetYlabelC       /* y-axis title. */
ajGraphdataSetColour        /* Colour for the plot on one graph. */
ajGraphdataSetLinetype      /* Line type for the plot on one graph. */
```

These are used in conjunction with functions to add graphical elements to be rendered:

```
ajGraphdataAddposLine       /* Adds a line. */
ajGraphdataAddposRect       /* Adds a rectangle. */
ajGraphdataAddposTextC      /* Adds text. */
ajGraphdataAddXY            /* Adds X,Y data points. */
ajGraphdataCalcXY           /* Adds X,Y data points from an increment. */
```

You should see the online documentation for an explanation of how to use these functions.

6.20.14 Retrieving properties of graphs

Functions for retrieving a property of a graph have the prefix `ajGraphGet`:

```
ajGraphGetSubtitleC           /* Graph subtitle. */
ajGraphGetTitleC              /* Graph title. */
ajGraphGetXlabelC             /* x-axis title */
ajGraphGetYlabelC             /* y-axis title */
ajGraphicsGetFgcolour         /* Current foreground colour. */
ajGraphicsBasecolourNewNuc    /* Base colours array for nucleotide sequence. */
ajGraphicsBasecolourNewProt   /* Base colours array for protein sequence. */
ajGraphicsGetCharsize         /* Character size. */
ajGraphGetParamsPage          /* Output device parameters. */
```

Functions that return the graph titles return the string object itself (not a copy) so you must be careful not to unintentionally change or delete it.

`ajGraphicsGetFgcolour` will return the current foreground colour as an integer:

```
ajint ajGraphicsGetFgcolour (void);
```

`ajGraphicsCheckColourC` takes a string argument for a colour and returns an integer corresponding to that colour, or `-1` if that colour was not found.

```
ajint ajGraphicsCheckColourS (const AjPStr colour);
ajint* ajGraphicsBasecolourNewNuc (const AjPStr codes);
ajint* ajGraphicsBasecolourNewProt (const AjPStr codes);
```

The code below picks up a colour, specified as a string by the user, from ACD file processing. `ajGraphicsCheckColourC` is called to get the integer for the colour, which is then passed to `ajGraphicsSetFgcolour` to set the foreground colour:

```
AjPStr  colour    =   NULL;
ajint   colouri;

colour     = ajAcdGetString("colour");

colouri = ajGraphicsCheckColourC(colour);
if(colouri    == -1)
    colouri        = GREY;

ajGraphicsSetFgcolour(colouri);
```

`ajGraphicsGetCharsize` retrieves the default and current character size. This is useful when setting the size of other graphical elements:

```
void ajGraphicsGetCharsize (float *defheight, float *currentheight);
```

ajGraphicsGetParamsPage gets the output device parameters defined internally in **PLplot**. These are less useful than one might expect as some are hard-coded for each device driver, or simply copies of user values that are ignored in **PLplot**. Where the plot dimensions are needed, use ajGraphOpenMm and work in millimetre units.

6.20.15 Calculated properties of graphs

The following functions return some calculated value (in user coordinates) derived from a graph:

```
ajGraphicsCalcCoord          /* coordinates of a point on a circle knowing the angle. */
ajGraphicsCalcDistance       /* Distances between 2 points. */
ajGraphicsCalcTextlengthC    /* Text length. */
ajGraphicsCalcTextheight     /* Text height. */
ajGraphicsCalcCharsize       /* Text size (mm) to needed for given text length and height. */
```

You should see the online documentation for an explanation of how to use these functions.

6.20.16 Graph debug functions

Functions for debugging graph objects have the prefix ajGraphDebug. See the online documentation for further information.

6.20.17 Histograms

Histograms are a special case of graphics plotting. A dedicated set of functions are provided that handle the underlying graph objects and function calls so that you don't need to.

To plot a histogram you first must create a histogram object. Two constructors are provided and take the number of sets of data (numofsets), all of which must have the same number of data points (numofpoints) per set:

```
AjPHist ajHistNew (ajint numofsets, ajint numofpoints);
AjPHist ajHistNewG (ajint numofsets, ajint numofpoints, AjPGraph graph);
```

ajHistNewG will also assign a graph object (passed as an argument) to the histogram that is returned. A number (numofsets) of graphs are added to the graph object. The graph device and name and extension of output file are also set.

To set the titles for histogram with a single set use one of the functions below:

```
/* Title */
void ajHistSetTitleC (AjPHist hist, const char* strng);

/* x-axis label */
void ajHistSetXlabelC (AjPHist hist, const char* strng);

/* y-axis left label */
void ajHistSetYlabelC (AjPHist hist,const char* strng);

/* y-axis right label */
void ajHistSetRlabelC (AjPHist hist, const char* strng);
```

To set the titles for histogram with multiple sets use one of the functions below. `index` is the number of the histogram in the set to have its title set to `title`:

```
/* Title for histogram with multiple sets */
void ajHistSetmultiTitleC (AjPHist hist, ajint index, const AjPStr title);

/* x-axis title */
void ajHistSetmultiXlabelC (AjPHist hist, ajint index, const AjPStr title);

/* y-axis title */
void ajHistSetmultiYlabelC (AjPHist hist, ajint index, const AjPStr title);
```

`ajHistSetmultiColour` and `ajHistSetmultiPattern` will set the colour or linestyle for the specified set (`index`):

```
/* Set colour for bars in histogram */
void ajHistSetmultiColour(AjPHist hist, ajint index, ajint colour);

/* Set linestyle for set of bars */
void ajHistSetmultiPattern (AjPHist hist, ajint index, ajint style)
```

`ajHistSetMono` is useful for printing in monochrome:

```
/* Set patterns instead of colours */
void ajHistSetMono (AjPHist hist, AjBool set);
```

To display a histogram, call `ajHistDisplay`:

```
void ajHistDisplay (const AjPHist hist);
```

Once you are done with the histogram, you must close the output file and delete the histogram object. The functions are:

```
void ajHistogramClose (void);
void ajHistDel (AjPHist* hist);
```

`ajHistDataAdd` and `ajHistDataCopy` are used to specify a set of data points for a specified histogram (`index`) in a set:

```
void ajHistDataAdd (AjPHist hist, ajint index, PLFLT *data);
void ajHistDataCopy (AjPHist hist, ajint index, const PLFLT *data);
```

`ajHistDataAdd` sets the histogram to point to the data passed whereas `ajHistDataCopy` copies the data.

The example below shows a function for copying a set of data held in an integer array (`results`) into a histogram (`hist`). `ajHistDataCopy` is used to copy the array of data which is why `farray` can be deleted. Had `ajHistDataAdd` been used instead then `AJFREE` would not have been called:

```
void func(AjPHist hist, const ajint * results, ajint npts, ajint hist_num)
{
    ajint i;

    float *farray;

    AJCNEW(farray, npts);
    for(i=0; i<npts; i++)
        farray[i] = (float) results[i];

    ajHistDataCopy(hist, hist_num, farray);
    AJFREE(farray);
```

6.21 Handling maths

6.21.1 Introduction

EMBOSS provides a basic maths library that includes functions for datatype conversion, random number generation, rounding and other basics.

6.21.2 AJAX library files

AJAX library files for handling maths are listed in the table below. Library file documentation, including a complete description of datatypes and functions, is available at: http://emboss.open-bio.org/rel/dev/libs/

Library file documentation	Description
`ajmath`	General maths functions
`ajvector`	Functions for handling vectors

`ajmath.h/c`. General maths functions for rounding numbers, coordinate conversion, random number generation, etc.

`ajvector.h/c`. Data structures and functions for handling vectors in 3D space. The library is not described here and you should see the library documentation online for further information.

6.21.3 ACD datatypes

There are no ACD datatypes dedicated for maths.

6.21.4 AJAX datatypes

There are no AJAX datatypes dedicated for maths.

6.21.5 Conversion functions

Functions are provided for converting one datatype to another. These all have the prefix ajCvt. To interconvert polar and cartesian coordinates, to convert radians to degrees or return a probability given a Gaussian distribution use:

```
float ajCvtDegToRad (float degrees);
void ajCvtPolToRec (float radius, float angle, float* Px, float* Py);
float ajCvtRadToDeg (float radians);
void ajCvtRecToPol (float x, float y, float* Pradius, float* Pangle);
double ajCvtGaussToProb (float mean, float sd, float score);
```

Sequence positions are often specified by the user on the command line. EMBOSS supports negative values for sequence positions which are interpreted as an offset from the *end* of the sequence. Two functions are provided to convert an EMBOSS sequence position to a true position which, for instance, could be used to index into the sequence itself. The functions return an integer between 0 (the start of the sequence) and the length of the sequence minus 1:

```
ajint ajCvtSposToPos (ajint len, ajint ipos);
ajint ajCvtSposToPosStart (ajint len, ajint imin, ajint ipos);
```

In both cases, len is the length of the sequence and ipos is the position. A value of 0 means the start of the sequence whereas negative numbers indicate an offset from the end of the sequence. A minimum relative position (imin) can be specified.

6.21.6 Random number functions

There are three functions for random number generation. These are based on dprand and sdprand and used with the permission of N. M. Maclaren (Copyright, The University of Cambridge). ajRandomSeed initialises random number generation and must be called before ajRandomNumber (returns a random integer between 0 and 32767) and ajRandomDouble (returns a random double-precision value between 0.0 and 1.0):

```
void ajRandomSeed (void);
double ajRandomNumber (void);
double ajRandomDouble (void);
```

6.21.7 Rounding

Two functions are for rounding off numbers, either a float to an int or a float to a float with a specified number of bits left for cumulative addition:

```
ajint ajRound (ajint i, ajint round);
float ajRoundFloat (float f, ajint nbits);
```

6.21.8 Miscellaneous functions

Various miscellaneous functions are provided, for instance, to calculate a geometric mean, the modulo of two numbers or a 32- or 64-bit SwissProt-style checksum for a protein sequence:

```
float ajMathGmean (const float* Parr, ajint n);
ajint ajMathModulo (ajint i1, ajint i2);
ajuint ajMathCrc32 (const AjPStr str);
unsigned long long ajMathCrc64 (const AjPStr str);
```

6.22 Handling system code

6.22.1 Introduction

EMBOSS provides basic functionality for interacting with the system and handling system-specific code. The areas covered include:

- System calls for calling external applications
- Code for parsing the application command line
- Functions for casting one type to another
- Low-level file handling
- Wrapper functions that call (with some modifications) standard C functions, mostly to overcome system-dependent limitations of the standard C functions
- Functions to call the C function `exit` with or without internal clean-up and reporting
- Functions to reverse the byte order in different datatypes.

6.22.2 AJAX library files

AJAX library files for handling system calls are listed in the table below. Library file documentation, including a complete description of datatypes and functions, is available at: http://emboss.open-bio.org/rel/dev/libs/

Library file documentation	Description
`ajsys`	System functions
`ajutil`	System functions

ajsys.h/c. System functions, mostly for low-level handling of files and directories and system-specific functions.

ajutil.h/c. Functions for exiting cleanly, memory clean-up, manipulate bytes and so.

6.22.3 ACD datatypes

There are no ACD datatypes dedicated for system handling.

6.22.4 AJAX datatypes

There are no AJAX datatypes dedicated for system handling.

6.22.5 System calls

There are two functions for calling external applications. `ajSysSystem` will call a user executable file in the current path, whereas `ajSysSystemEnv` will use the environment including the `PATH` that is provided within it:

```
void ajSysSystem (const AjPStr str);
void ajSysSystemEnv (const AjPStr str, char* const env[]);
```

6.22.6 Application arguments

The function `ajSysArglistBuild` extracts the application name (`name`) and list of arguments (`PParglist`) from any application command line:

```
AjBool ajSysArglistBuild (const AjPStr cmdline, char** name, char*** PParglist);
```

The function allocates a string (`name`) and a `NULL`-terminated string array (`PParglist`) which are later freed by calling `AJFREE` and `ajSysArglistFree`:

```
AjPStr cmdline = NULL;
char **arglist = NULL;
char *name;

/* The command line must be assigned to a string */
cmdline = ajStrNewC("ApplicationName arg1 arg1val arg2 arg2val")

ajSysArglistBuild (cmdline, &name, &arglist);

/* Do something with the name and argument list. Individual args are in arglist[0],
arglist[1] etc. */

AJFREE(name);
ajSysArglistFree(&arglist);
```

6.22.7 Cast functions

Functions are available for casting an integer to character types:

```
char ajSysCastItoc (ajint i);
unsigned char ajSysCastItouc (ajint i);
```

6.22.8 File functions

Three functions are available for low-level file handling.

`ajSysFileWhich` will search for a user-executable file in the current path whereas `ajSysFileWhichEnv` will use a `PATH` from the environment that is provided by argument. The functions returns `ajTrue` if the executable was found:

```
AjBool ajSysFileWhich (AjPStr* Pstr);
AjBool ajSysFileWhichEnv (AjPStr* Pstr, char* const env[]);
```

`ajSysFileUnlink` is used to safely delete a file or link:

```
AjBool ajSysFileUnlink (const AjPStr str);
```

6.22.9 C function wrappers

Various functions are provided that wrap (with some modifications) standard C functions. In most cases these are provided to overcome system-dependent limitations of the standard C functions. These functions have the prefix `ajSysFunc` and include:

```
char* ajSysFuncStrtok (const char* txt1, const char* txt2);
char* ajSysFuncStrtokR (const char* txt1, const char* txt2, char** Ptxt, AjPStr* Pbuf);
char* ajSysFuncFgets (char* buf, int n, FILE* fp);
char* ajSysFuncFopen (const char* txt, const char* flags);
FILE* ajSysFuncFdopen (ajint filedes, const char* mode);
```

6.22.10 Functions from `ajutil.c/h`

The following functions call the C function `exit`. The functions `embExit*` (defined in `embexit.h/c`), called from applications code to exit the application, make use of these functions:

```
/* Exits with a success code (zero) after cleanup routines */
__noreturn void ajExit (void);

/* Exits with an unsuccessful code without flushing files or clean-up. */
__noreturn void ajExitAbort (void);

/* Exits with an unsuccessful code without clean-up. */
__noreturn void ajExitBad (void);
```

`ajExit()` calls `exit` with a success code (zero) after calling clean-up routines which report on resource usage, etc.

`ajExitAbort()` calls `exit` with an unsuccessful code (`EXIT_FAILURE` defined in `stdlib.h`) without flushing any files. This is needed to exit from, for example, a failed system call (e.g. ajFileNewInPipe) where the parent process has open output files and the child process needs to exit without affecting them. Failure to exit this way can mean the output buffer is flushed twice. No clean-up or reporting routines are called.

ajExitBad() calls exit() with an unsuccessful code (EXIT_FAILURE). No clean-up or reporting routines are called.

The following functions all reverse the byte order in the datatype indicated:

```
void ajByteRevLen2 (short* i);       /* 2 byte integer      */
void ajByteRevLen4 (ajint* i);       /* 4 byte integer      */
void ajByteRevLen8 (ajlong* i);      /* 8 byte integer      */
void ajByteRevInt (ajint* sval);     /* integer             */
void ajByteRevShort (short* ival);   /* unsigned integer    */
void ajByteRevLong (ajlong* lval);   /* long                */
void ajByteRevUint (ajuint* ival);   /* unsigned integer    */
```

Finally, there are a few miscellaneous useful functions:

```
/* If a log file is in use then this writes run details to end of file. */
void ajUtilLoginfo (void);

/* Dummy function for use in debugging */
void ajUtilCatch (void);

/* Tests whether the host system uses big endian byte order. */
AjBool ajUtilGetBigendian (void);

/* Returns the user's userid. */
AjBool ajUtilGetUid (AjPStr* dest);

/* Returns a command line with all options changed by the user */
const AjPStr ajUtilGetCmdline (void);

/* Returns the user non-default inputs in commandline form */
const AjPStr ajUtilGetInputs (void);

/* Return the program name */
const AjPStr ajUtilGetProgram (void);
```

ajUtilCatch should be called in special cases when debugging in **GDB**. To use, simply put a call to ajUtilCatch() into your code, and use "break ajUtilCatch" in **GDB** to get a traceback.

6.23 Handling exception messages

6.23.1 Introduction

EMBOSS provides a set of useful exception handling functions which format an exception message before calling a function to output the message. The following behaviours are available:

- Call the registered exception handling function (see below)
- Report an informative message
- Report a warning message

- Report an error message
- Dump diagnostic data
- Exit
- Crash (kill) the application
- Debug.

The most widely used are the functions `ajFatal` (crash the application), `ajWarn` (warning message) and `ajDebug` (debug) (see Section 3.3, 'Debugging').

In rare cases it is desirable to write your own customised exception handling functions and functions are provided for this. There are also functions to query or set the internal status of message handling. You will not normally need these but they might come in useful when writing your own exception handling code.

6.23.2 AJAX library files

AJAX library files for handling messages are listed in the table below. Library file documentation, including a complete description of datatypes and functions, is available at: http://emboss.open-bio.org/rel/dev/libs/

Library file documentation	Description
`ajmess`	Message handling
`ajexcept`	Internal exception handling functions
`ajassert`	Functions for handling assertions

ajmess.h/c. Defines the error message object (`AjPError`) and functions for message handling. It also contains static data structures and functions for handling messages at a low level. You are unlikely to need the static datatypes and functions unless you plan to extend the core library functionality.

ajexcept.h/c. Contains the `ajExceptRaise` function for raising an exception.

ajassert.h/c. Contains functions for handling assertions. The simplest use is to add a line:

```
AjPStr mystring = NULL;
 /* ... some code ... */
 assert(mystring);
```

this will raise a fatal exception when `mystring` has a `NULL` value.

6.23.3 ACD datatypes

There are no ACD datatypes dedicated for messages.

6.23.4 AJAX datatypes for handling messages

For handling messages use:

`AjPError`	Error message level information

These are used internally. For handling exceptions define an `Except_T` with a text description, and use the `AJRAISE` macro to report the text string, and the name and line number of the source file. The exception handling datatypes are:

`Except_T`	Types of exceptions
`Except_Frame`	Used for the exception stack

For all intents and purposes you will not need to handle these objects directly.

6.23.5 Message functions

Functions to format messages (provided as a string) all have convenient wrapper functions with short memorable names:

```
void    ajDebug (const char *fmt, ...);
void    ajDie (const char *format, ...);
void    ajErr (const char *format, ...) ;
void    ajUser (const char *format, ...) ;
void    ajWarn (const char *format, ...);
void    ajMessDump (const char *format, ...);
void    ajMessExitmsg(const char *format, ...);
```

All these functions call message output functions that are internally defined.
The functions are invoked with the message text, for example:

```
ajWarn("This is a silly warning message");
```

The call can of course include variables to print:

```
ajint x=0;
...
ajWarn("This is a silly warning message which prints integer value %d", x);
```

`ajDebug` will write a debug message to the debug file if debugging has been turned on. Debugging is typically turned on by adding **`-debug`** to the command line or by defining the environment variable `EMBOSS_DEBUG`. You should avoid using `ajDebug` in any code which might be called before command line processing completes as the debug file has not yet been created.

`ajMessExitmsg` does not give an opportunity to output logged error messages before exiting, but is useful for handling errors that, while being unrecoverable, are not a problem with the EMBOSS code itself.

6.23.6 Querying the status of message handling

A few functions are provided to query the internal status of message handling. These have the prefix `ajMessGet`. For example the number of times the error routines have been called, the current message text, the system error message text from the standard C function `strerror` or the file used for debug output can all be retrieved:

```
ajint  ajMessGetCountError (void);
const char* ajMessGetMessageC (void);
const char* ajMessGetSysmessageC (void);
FILE*  ajMessGetDebugfile (void);
```

6.23.7 Interacting with the user

One function is used for all user prompting in the initialisation on an EMBOSS application. `ajUserGet` writes a prompt to the user and reads a one-line response. `ajMessBeep` generates an audible warning:

```
ajint ajUserGet (AjPStr* pthis, const char* fmt, ...);
void ajMessBeep (void);
```

6.23.8 Miscellaneous functions

To raise an exception manually `ajExceptRaise` (defined in `ajexcept.h/c`) may be used. The function is available as a macro called `(AJRAISE)`:

```
__noreturn void ajExceptRaise(const T* e, const char* file, ajint line);
#define AJRAISE(e) ajExceptRaise(&(e), __FILE__, __LINE__)
```

It will check for an exception and call the registered exception routine if one is caught. Otherwise it will raise an exception message with the filename of the source code and code line number before aborting the program.

The macro version should always be used. For example:

```
static const Except_T Mem_Failed =
{
        "Allocation failed, insufficient memory available"
};

static const Except_T Mem_Badcount =
{
      "Allocation bad byte count"
};
/* Allocates memory using malloc, and fails with an error message if unsuccessful.
   nbytes : number of bytes required
   file   : source file name (generated by macro)
   line   : source line number (generated by a macro)
```

```
void *ajMemAlloc(size_t nbytes, const char* file, ajint line)
{
    void *ptr;

    if(nbytes <= 0)
        ajExceptRaise(&mem_Badcount, file, line);

    ptr = malloc(nbytes);

    if(ptr == NULL)
    {
        if(file == NULL)
            AJRAISE(Mem_Failed);
        else
           ajExceptRaise(&mem_Failed, file, line);

   }
```

Quality assurance 7

7.1 General quality assurance

Various quality assurance (QA) tests are performed on the EMBOSS code and documentation to maintain the quality and integrity of the package.

- Test builds on a variety of platforms to highlight any compiler warnings
- Application test runs on test data (including expected fail conditions)
- Application memory leak tests using **valgrind**
- Validation of C source code documentation and indexing in **SRS**
- Validation and standardisation of application ACD files
- Validation of EMBOSS documentation including tests for completeness
- Automatic updating of application documentation (primarily the command line qualifiers and usage examples including input and output files).

The tests are performed by the EMBOSS developers but there are ways to help. You should provide tests with test data for any applications you develop (see below) and provide adequate documentation for any new library code (see Appendix D, *Code documentation standards*) and applications (see Chapter 8, *Application documentation standards*).

7.2 Application quality assurance

The test data consist of one or more sets of input files, application parameters and the corresponding output files. Each set corresponds to a single run of the application. As many sets of test data as possible should be provided, especially for unusual input conditions, to provide as robust a test as possible.

> **Note**
>
> The QA test definitions are used to generate include files for three sections of the application documentation: usage, input and output. These are created for any QA test that has the suffix `-ex` (example), `-keep` (output reused by another test) or `-check`. All other test names are for QA testing only and do not become part of the application documentation.

The tests are defined in the file:

```
.../emboss/test/qatest.dat
```

An example of an entry in `qatest.dat` is shown below:

```
ID myprogram-ex
AB myemboss
AA myprogram
IN
FI stderr
FC =2
FP 0 /Warning: /
FP 0 /Error: /
FP 0 /Died: /
FI P10932.myprogram
FC =5
FP /^Usa: tembl-id:P10932\n/
FP /^Length: 2167\n/
//
```

The test includes information about the parameters (command line) and what is expected in the output files. Documentation at the start of `qatest.dat` describes the records used to define a test.

7.2.1 Test records

The records used in the file `qatest.dat` are summarised below.

Record	Short description	Requirement
`ID` *`ApplicationName-Tag`*	Test identifier	Mandatory
`AA` *`ApplicationName`*	Application name for EMBASSY applications	Mandatory for EMBASSY applications
`AB` *`PackageName`*	EMBASSY package name	Mandatory where an `AA` record is used
`AP` *`ApplicationName`*	Application name for standard EMBOSS applications	Mandatory for standard EMBOSS applications
`AQ` *`ApplicationName`*	Application name for 'make check' EMBOSS applications	Mandatory for 'make check' EMBOSS applications
`CC` *`comment`*	A comment	Optional
`CL` *`command`*		Optional

	Command line parameters	
`DI` *`directory`*	Output subdirectory	Mandatory where output sub-directories are used
`DF` *`FileName`*	Name of file in output sub-directory	Optional
`DL` *`directive`*	Directive for results directory on test completion	Optional
`ER` *`ErrorCode`*	Error return code	Optional
`FI` *`FileName`*	Name of output file in results directory	Mandatory for all files created in the main results directory
`FC` [<=>] *`NumberLines`*	Test number of lines in the preceding `FI` file	Mandatory for all files other than `stdout` and `stderr` if an `FZ` record is not defined, otherwise optional
`FP` *`count`* / *`regexp`* /	Test for pattern in the preceding `FI` file	Mandatory for all files other than `stdout` and `stderr`
`FZ` [<=>] *`FileSize`*	Test size of the preceding `FI` file	Mandatory for all files other than `stdout` and `stderr` if an `FC` record is not defined, otherwise optional.
`IC` *`text`*	Annotation on the input	Optional
`IN` *`UserInput`*	Value to use in response to an EMBOSS prompt	Mandatory for each application option that will be prompted for by the application
`OC` *`text`*	Annotation on the output	Optional
`PP` *`command`*	Pre-processor command	Optional
`QQ` *`command`*	Post-processor command	Optional
`RQ` *`RequiredApps`*	Required external programs	Mandatory for applications that require other applications to run, e.g. application wrappers
`TI` *`seconds`*	Time limit for test	Optional
`UC` *`text`*	Annotation on the usage examples	Optional
`##`	Comment	Optional

7.2.1.1 `ID` *`ApplicationName-Tag`*

The identifier has the general form *`ApplicationName-Tag`* where *`ApplicationName`* is the name of the application being tested and `Tag` indicate what sort of test is being performed.

When a test is run, a directory is created under `/emboss/test/` to hold the output of the test run. This includes all the application output files and files with any output written by the application to the UNIX file streams `stdout` and `stderr`. The identifier defines the name of the directory. For example, if the identifier was `myprogram-ex` then a directory called `test/qa/myprogram-ex` would be created.

Tag may be any of the following:

`-ex`	The test is used to create the first usage example in the application HTML documentation. A number following `-ex` may be given to indicate the second and subsequent tests (see below).
`-ex2`, `-ex3`, etc.	Further tests that are used for usage examples in the documentation. In general, all such examples should correspond to a test defined in `qatest.dat` to ensure the documentation is correct.
`-keep`	Used where the test is used in the documentation and the test output is retained once the test completes. The test should include a `DL keep` record. This is used where the test output is read by a subsequent test, as is the case for database indexing. It is important that this test appears in the test file before any tests that use the results.
`-keep2`, `-keep3`, etc.	A number following `-keep` may be given to indicate the second and subsequent tests.
`-fail`	The test for a failure condition. Such tests typically have an `ER` record to indicate the error return code.

7.2.1.2 **AA** *ApplicationName*

This is used to generate the command line and for reporting statistics. If an `AA` record is given for an EMBASSY package it must be identified with an `AB` record in the test definition. The application (*ApplicationName*) must be built by using **make** on the package. If the application is not found, **qatest** will assume this EMBASSY package was not installed and will not run the test.

7.2.1.3 **AB** *PackageName*

This informs the **qatest** script that a test depends on installation of an EMBASSY package (*PackageName*) and may fail to find the binary if the package is not installed.

7.2.1.4 **AP** *ApplicationName*

This is used to generate the command line and for reporting statistics. The application (*ApplicationName*) must be built by using **make**. The test is always performed by the **qatest** script.

7.2.1.5 **AQ** ApplicationName

This is used to generate the command line and for reporting statistics. The application (*ApplicationName*) must be built by using **make check**. The **qatest** script will look in the build directory for the executable as this application will not be installed. If the application is not found, **qatest** will assume **make check** was not run and omit the test.

7.2.1.6 **CC** *`comment`*

Used in commenting on failed tests. The comment (*`comment`*) will appear in the test output on failure as a guide to an acceptable failure condition. In such cases the text of the error message would be validated using e.g. `FP` records (see below).

The record `##` (see below) is used for more general comments within the test definition.

7.2.1.7 **CL** *`command`*

This is the command line (*`command`*) without the application name. The command line may span multiple `CL` records, in which case text following the `CL` tokens is appended and a space inserted between the text from each line.

7.2.1.8 **DI** *`directory`*

This is the name of a directory (*`directory`*) created in the results directory (defined by the `ID` record) that may be used to hold output where an application opens a new output directory. Currently, tests check each file is named in a `DF` line but tests cannot be defined directly for file contents in those subdirectories. Howver, a `QQ` post-processing command (see below) could be used to copy the contents to another file (in the main results directory), which could then be tested, e.g. for size and patterns.

7.2.1.9 **DF** *`FileName`*

This is the name of a file (*`FileName`*) created in a sub-directory of the results directory. The sub-directory the file belongs to is identified by the preceding `DI` record; `DF` records must be listed under the appropriate `DI` record. The files must be written for the test to pass.

7.2.1.10 **DL** *`directive`*

This controls what happens to the results directory on completion of the test. *`directive`* can be one of:

`success`	Delete the test directory on success. This is the default behaviour.
`all`	Delete the test directory always.
`keep`	Keep the test results directory always. This is used where another test reuses the results, as is the case in database indexing.

7.2.1.11 **ER** *`ErrorCode`*

The expected error code (*`ErrorCode`*). This is used for tests that are testing failure conditions. The test will report an error if the specified error code is not returned by the application. All applications which succeed return 0, those that fail return 1.

7.2.1.12 **FI** *`FileName`*

Currently, an `FI` record can only reference a file in the main results directory and not in a subdirectory of the results directory. The filename (*`FileName`*) needn't be given `stdout` and `stderr` which are assumed to exist and may be empty, unless defined otherwise.

7.2.1.13 FC [<=>] *NumberLines*

For the test to pass, the number of lines (*NumberLines*) may be checked as follows:

=	There must be *NumberLines* lines
<	There must be more than *NumberLines* lines
>	There must be less than *NumberLines* lines

7.2.1.14 FP *count* / *regexp* /

For the test to pass the Perl regular expression *regexp* given between forward-slashes must be found in the preceding `FI` file. The expression may be preceded by an optional integer (*count*) to check the exact number of the pattern.

This record can be used to check `stderr` (which often contains the user prompts) for the absence of exception messages (warning, error or fatal).

7.2.1.15 FZ [<=>] *FileSize*

For the test to pass, the size (*FileSize*) of the file may be checked as follows:

=	File must be exactly *FileSize*
<	File must be bigger than *FileSize*
>	File must be smaller than *FileSize*

There is an implicit test that `stdout` and `stderr` must be of size zero unless otherwise stated.

7.2.1.16 IC *text*

This is annotation (*text*) for the input and is used by **makeexample.pl** when creating the HTML documentation files. The information is not used in QA testing.

7.2.1.17 IN *UserInput*

The value *UserInput* will be used. If there is nothing on the line then an empty line is input to the application which will use the default value for the option (if defined). Care is needed where options are conditionally prompted for.

7.2.1.18 OC *text*

This is annotation (*text*) for the output and is used by **makeexample.pl** when creating the HTML documentation files. The information is not used in QA testing.

7.2.1.19 PP *command*

This is a command executed (by `/bin/sh`) before the test is run. Each `PP` line defines a single command: long commands may *not* be concatenated over multiple lines as is allowed for the `CL` records.

A typical use is to set an environment variable required by the application. The variable should always be exported (this is `/bin/sh`), for example:

```
PP EMBOSS_ACDROOT=../../acd
PP export EMBOSS_ACDROOT
```

7.2.1.20 **QQ** *command*

This is a command executed (by `/bin/sh`) *after* the test is run. Each `QQ` line defines a single command: long commands may *not* be concatenated over multiple lines as was allowed for the `CL` records.

This is not used at present but the most likely application is to list the contents of a directory to another file which can then be tested for size and patterns.

7.2.1.21 **RQ** *RequiredApps*

This is an application the test is dependent upon. A single required 'helper' application should be given per `RQ` record. For example, **srs** is required for tests that use **getz**, or **clustalw** for **emma**.

7.2.1.22 **TI** *seconds*

This the time limit at which point the test times out. The default is 60 seconds. Some examples can take longer on a heavily loaded system.

7.2.1.23 **UC** *text*

This is annotation (*text*) for the QA test itself and is used by **makeexample.pl** when creating the HTML documentation files. The information is not used in QA testing.

7.2.1.24

This is a general comment in the test definition and is not reported. You should use `CC` records to comment on tests for failure.

7.2.2 Writing an application quality assurance test

Quality assurance tests for new applications must be added to the appropriate place in the file `qatest.dat`. For example, if you were writing a test for an application in the EMBASSY package **myemboss** you would search for the line `"AB myemboss"` and add the test under there. The test, as a minimum, should include the following records:

- An `ID` record with an identifier code for the test
- An `AA` (EMBASSY) or `AP` or `AQ` (EMBOSS) record for the application name and an `AB` record with the name of the EMBASSY package (if appropriate)
- A `CL` record giving any options on the command line for the test and/or one or more `IN` lines to give responses to any requests for input from the program. One `IN` record, which may be blank if the default response is acceptable, is required for each prompt
- An `FI` for the name of each output file

- An FC record under each FI file to test for correct line count or an FZ record to test for file size
- One or more FP tests under each FI record to test for file contents
- One or more DI records for output subdirectories where they are used
- An RQ record for the name of required applications if there are any
- An FI, FZ (or FC) and an FP record should be given for each of stderr and stdout.

7.2.2.1 Location of test data

Any test data, i.e. the input files required by the application, should be added to the directory:

```
emboss/test/data
```

Where many input files are required then, to keep things tidy, these may be put under a subdirectory of the above directory. In either case, you should not create files unnecessarily: it is possible, likely even, that a file already exists under emboss/test/data that is suitable for your needs. More information on the contents of this directory is given below (Section 7.2.4, 'Useful files and directories').

7.2.3 Running an application quality assurance test

To perform tests, you must edit your .embossrc file (in your home directory) or the emboss.default file to set the EMBOSS environment variable emboss_qadata to the test directory, e.g.

```
/home/auser/emboss/emboss/test
```

To run a test you must run the script:

```
/emboss/scripts/qatest.pl
```

from the directory /emboss/test/qa. The command has the following form:

```
qatest.pl TestIdentifier
```

where *TestIdentifier* is the test identifier given on the ID record of the appropriate entry in qatest.dat.

7.2.3.1 Example test

Let's assume you want to run the example entry in qatest.dat shown previously:

```
ID myprogram-ex
AB myemboss
AA myprogram
IN
FI stderr
FC =2
FP 0 /Warning: /
FP 0 /Error: /
FP 0 /Died: /
FI P10932.myprogram
FC =5
FP /^Usa: tembl-id:P10932\n/
FP /^Length: 2167\n/
//
```

Let's also assume you are in the directory `emboss/test`. To run the test `myprogram-ex` you would type:

```
cd qa
../../scripts/qatest.pl myprogram-ex -keep
```

If the output files are to be retained once the test completes, for tests that include no `DL keep` record, then **qatest.pl** must be invoked with the **`-keep`** qualifier. If it is not, the output files will be deleted. So, had the test included these lines:

```
ID myprogram-keep
DL keep
```

It would be invoked thus:

```
../../scripts/qatest.pl myprogram-ex
```

In either case, output files are created in the main results directory. If the test identifier line is

```
ID myprogram-ex
```

then the results directory will be:

```
emboss/qa/myprogram-ex
```

For other tests, files might also be written to subdirectories in the main results directory (see `DI` and `DF` records).

If **qatest.pl** is run on something not defined in `qatest.dat` it will report:

```
Tests total: 0.
```

If it succeeds, all files are deleted unless the test entry included a `DL keep` line, or **`-keep`** was specified on the command line.

If it fails, it will report why and all results will be saved in the results directory for inspection. You would check, for example, `myprogram-ex`, then identify the problem, update the test definition and try again until it works.

A typical session looks something like this:

```
../../scripts/qatest.pl -without=srs
Tests total: 1586 pass: 1586 fail: 0
Skipped: 12 check: 1 embassy: 0 requirements: 11
Missing documentation html: 0 text: 0 sourceforge: 0
Time: 677 seconds
```

7.2.4 Useful files and directories

There are some useful files held under `/emboss/test/`. The directories are as follows:

`acd`	ACD files for test applications. These are used for testing ACD file parsing. You will not need this directory unless you extend ACD.
`data`	The directory for test data that has already been mentioned. Your test data input files should go in here, or in a sub-directory beneath it.
`gb`	Some *GenBank* data files in NBRF/GCG database format.
`qa`	The directory from which all tests must be run. Application output files are written to their own results directory underneath this, and are deleted on successful test completion unless otherwise stated.
`rc`	This directory is used for database and resource definition tests. You will not need it.
`swnew`	Files from the *SwissProt* database used to make the `tsw` test database.
`embl`	Files from the *EMBL* database used to make the `tembl` test database.
`genbank`	Some *GenBank* data files used to make the `tgenbank` test database.
`memtest`	Files for tracing memory leaks: you will not need this.
`pir`	Data files from the *PIR* database used to make the `tpir` test database.
`swiss`	Data files from the *SwissProt* database.
`wormpep`	Data files from the *WormPep* (worm peptide) database.

Application documentation standards 8

8.1 Introduction

The guidelines here describe how to write documentation for new applications conforming to the EMBOSS style that is used online:

```
http://emboss.open-bio.org/rel/dev/apps/
```

They also cover documentation for EMBASSY packages:

```
http://emboss.open-bio.org/rel/dev/apps/EMBASSYPackagesTable.html
```

To get a feel for the documentation, look for example at the documentation for the **seqret** application:

```
http://emboss.open-bio.org/rel/dev/apps/seqret.html
```

and for the HMMER package:

```
http://emboss.open-bio.org/rel/dev/apps/hmmer.html
```

You will notice that the application and package documentation is organised into sections for easy navigation. The sections, and processes for generating the documentation, are explained below.

8.2 Application documentation

8.2.1 ACD file documentation

An application ACD file (see Chapter 5, *C Programming*) does not require explicit documentation but in some cases, however, it is helpful to add comments. Comment lines begin with # and continue to the end of the line (see Section 4.1, 'Introduction to ACD file development').

An ACD file includes an application definition describing the application itself, and one or more data definitions describing the parameters. Definitions have attributes which describe the application or parameter in detail. Data definitions are organised into sections in the file such as "Input","Output" and so on. The ACD file is parsed when generating the application documentation. The components used include:

- The application name (from the application definition) and short description (from the `documentation:` attribute of the application definition)
- Example data files (from the "Input" section)
- Command line synopsis
- Table of application options (from the data definitions)
- List of links to related applications (from the `groups:` attribute of the application definition).

Therefore you should ensure that the application ACD file is valid and properly formatted by using utilities provided for this purpose.

8.2.1.1 Application quality assurance test

The command line sessions from the application QA tests are included in the application documentation. You should therefore write these tests and run them (see Section 7.2, 'Application quality assurance') before documenting the application to make sure that the example files will be correctly generated.

If you did not write the QA test before documenting the application, rerun **autodoc.pl** and it will create the remaining three include files (usage examples and input and output files), with complete HTML documentation created in `embassy/myemboss/emboss_doc/html/myprogram.html`

8.2.2 Sections of application documentation

Sections in the application documentation are described in the table below.

8.2.3 Application documentation process

The process for generating the application documentation combines manually written components with parts that are generated automatically (mostly from parsing the ACD file and from the QA tests). The process is largely automated and involves running the **autodoc.pl** script which generates a part-completed 'application template' which you complete by hand.

The application name and index terms are written directly to the application template. However, most of the automatically generated content is kept in 'include files' that are generated from template XML include files.

autodoc.pl is included in the EMBOSS distribution under the `scripts` directory, e.g.

```
/home/auser/emboss/emboss/scripts/autodoc.pl
```

Section	Description	Generation
Name	Application name, short description and version number.	Generated automatically from the application C source code and ACD file.
Function	Terse, formal description of application inputs, outputs and functionality.	Written by the application author.
Description	Description of the application suitable for the biologist end-user.	Written by the application author.
Command line arguments	Synopsis of how to use the application at the command line including command line synopsis and arguments table.	Generated automatically.
Usage	Usage example(s) which are command line sessions, or the equivalent in e.g. **wEMBOSS**, **Jemboss** or **SOAPLAB**.	Generated automatically from the quality assurance test cases.
Input Files	Example input files. Optional description of input files (e.g. file formats) or URL linking to same.	Example input files are generated automatically from the test cases. Description is written by the application author.
Data Files	Examples of data files. Optional description of data files (e.g. file formats) or URL linking to same.	Example data files are generated automatically from the ACD file. Description is written by the application author.
Output Files	Example output files. Optional description of output files (e.g. file formats) or URL linking to same.	Example output files are generated automatically from the test cases. Description is written by the application author.
Algorithm	Technical description of the algorithm suitable for computer scientists.	Written by the application author.
Notes	Peripheral documentation such as usage notes.	Written by the application author.
Warnings	Cautionary usage advice, known bugs, etc.	Written by the application author. The application-specific messages can be found by searching the code.
Diagnostics	Diagnostic warning and error messages the application might generate.	Written by the application author. The application-specific messages can be found by searching the code.
Authors	Author contact details.	Generated automatically from the source code.
Revision History	Application revision history, including date of last revision, name of revising author and brief remarks about the revision.	Used for major revisions. Detailed history is available in CVS for EMBOSS code.

References	Literature references and instructions on how to cite the application.	Instructions on how to cite EMBOSS are included automatically. Otherwise, written by the application author.
See Also	List of links to applications in the same group.	Generated automatically from the ACD file.
Target Users	The group of users (such as biologists, bioinformaticians, system administrators, etc.) the application is intended for.	Written by the application author.
Exit Status	Values the application might return to the operating system on exit.	Generated automatically from the source code.

To document a new program, first ensure you have an up-to-date set of programs compiled, and that any programs you've written but which are not part of the main EMBOSS package have had their executable deleted. Otherwise references to them might occur in the automatically generated "See Also" sections (see above). The ACD file should be complete and validated and QA tests for the application should have been run.

To generate the documentation, **autodoc.pl** is run on each application you wish to document in turn.

For EMBOSS applications:

```
autodoc.pl ApplicationName
```

For EMBASSY applications:

```
autodoc.pl -embassy= PackageName ApplicationName
```

You should replace `PackageName` and `ApplicationName` with something sensible.

The following example assumes you are working in the EMBASSY package **myemboss** and are writing a program called `myprogram`. To document the application:

1. Move to the documentation directory:

```
cd embassy/myemboss/emboss_doc/master
```

2. The `emboss_doc/master` directory has a template file for documenting a **myemboss** application. Copy the application template to a file called **myprogram**`.html` in the same directory:

```
cp template.html.save myprogram .html
```

3. Edit this file, replacing 'ProgramNameToBeReplaced' with the application name. This defines named include files for the auto-generated elements of the documentation.
4. Run **autodoc.pl**:

```
autodoc.pl -embassy=myemboss myprogram
```

5. Complete `myprogram.html` writing documentation text in the places indicated (see below).

The script will run **wossname** to check that `myprogram` really exists, then generate the application template documentation file (for you to fill in) with include directives, plus include files for the following sections:

- Name (entire section)
- Synopsis (entire section)
- Data Files (data file examples)
- Authors (entire section)
- Revision History (entire section)
- References (instructions on how to cite EMBOSS only).

The include files for the following parts are generated from running the QA test which is why (typically) the QA test should be written before documenting your application:

- Usage (entire section)
- Input Files (input file examples only)
- Output Files (output file examples only).

The application name and index terms in the application template are also set correctly. When all is done the HTML application template is created in:

```
embassy/myemboss/emboss_doc/html/myprogram.html
```

To complete the application template `myprogram.html` you add documentation text as follows:

1. Write "Function" section.
2. Write description of input files in "Input Files" section (optional).
3. Write description of data files in "Data Files" section (optional).
4. Write description of output files in "Output Files" section (optional).
5. Write description of application in "Description" section.
6. Write description of algorithm in "Algorithm" section.
7. Write peripheral documentation in "Notes" section.
8. Write usage advice, known bugs, etc. in "Warnings" section.

9. Write warning and error messages in "Diagnostics" section.
10. Write references in "References" section (if appropriate).

The template is commented enough for you to see how to fill it out.

autodoc.pl also generates documentation in plain text format (with all the text from the include files) that is used for manual pages and is displayed when running:

```
application_name -help
```

Once you complete the template and save it, the documentation (in HTML and plain text formats, plus include files) is ready for submission to the EMBOSS authors for publication online. Other online files that reference the new documentation, such as indices of applications (http://emboss.open-bio.org/rel/dev/apps/), will be updated automatically.

8.2.3.1 Documenting applications when not using myemboss

When documenting EMBOSS applications work in the directory:

```
doc/programs/master/emboss/apps/
```

Leave out the `-embassy=myemboss` qualifier from the **autodoc.pl** command line:

```
autodoc.pl ApplicationName
```

The final documentation is written to:

```
doc/programs/html/myprogram.html
```

8.2.3.2 Worked example

The example below is for **seqret**. All paths are relative to the documentation directory, e.g. `/home/auser/emboss/emboss/doc/`.

The application template file with include directives is:

```
./programs/master/emboss/apps/seqret.html
```

`Files generated by autodoc.pl`
The include files are below. All files are written to `/programs/master/inc/`:

`/seqret.itable`	Generated by running **acdtable**
`.../seqret.ione`	Generated from the ACD file by **autodoc.pl**
`.../seqret.ihelp`	Generated from the ACD file by **autodoc.pl**
`.../seqret.itable`	Generated from the ACD file by **autodoc.pl**
`.../seqret.isee`	Generated from **seealso** using the groups in the ACD file by `autodoc.pl`
`.../seqret.usage`	Generated from QA tests (using UsageExampleTemplate.xml)
`.../seqret.input`	Generated from QA tests (using `InputFilesTemplate.xml`)
`.../seqret.output`	Generated from QA tests (using `OutputFilesTemplate.xml`)
`.../seqret.comment`	Generated by `autodoc.pl` for editing by hand
`.../seqret.history`	Generated by `autodoc.pl` for editing by hand

autodoc.pl generates the final HTML file, with all data included, that is used online:

```
./programs/html/seqret.html
```

autodoc.pl also writes a plain text file of documentation, containing exactly the same documentation as in the completed application template, but with all included text:

```
./programs/text/seqret.txt
```

8.3 EMBASSY package documentation

8.3.1 Sections of EMBASSY package documentation

The sections in the EMBASSY package documentation are described in the table below.

8.3.2 EMBASSY package documentation process

The process for generating the EMBASSY package documentation is similar to that for applications. Manually written components are combined with automatically generated text. The process again involves running the **autodoc.pl** script which generates a part-completed package template which you complete by hand.

To document a new package, first ensure you have an up-to-date set of programs compiled. To generate the documentation run **autodoc.pl** on the package:

```
autodoc.pl PackageName
```

Section	Description	Generation
Name	Package name, short description and version number.	Generated automatically from the application C source code and ACD file.
Contents	Links to sections in the documentation.	Generated automatically.
List of Applications	Names, short description and links to full documentation for the applications in the package.	Generated automatically from the application ACD files.
Function	Terse, formal description of the package and its main inputs, outputs and functionality using terms from a controlled vocabulary.	Written by the application author.
Description	Description of the package suitable for the biologist end-user.	Written by the application author.
Installation Notes	Installation instructions for the package.	General instructions are generated automatically. Information, such as required files or versions, that are specific to the package should be written by the application author.
Notes	Peripheral documentation such as usage notes.	Written by the application author.
Warnings	Cautionary usage advice, known bugs, etc.	Written by the application author.
Authors	Author contact details.	Generated automatically from the source code.
Revision History	Package revision history, including date of last revision, name of revising author and brief remarks about the revision.	Written by the application author.
References	Literature references and instructions on how to cite the application.	Instructions on how to cite EMBOSS are generated automatically. Otherwise, written by the application author.

The following example assumes you have created a new EMBASSY package called `mypackage`. To document the package:

1. Move to the documentation directory:

```
cd embassy/mypackage/emboss_doc/master
```

2. Copy the application template to a file called *mypackage*`.html` in the same directory. An example template is provided in the EMBASSY package **myemboss**. Change MYEMBOSS to the name of your package when you edit the template file:

```
cp template.html.save mypackage .html
```

3. Run **autodoc.pl**:

```
autodoc.pl -embassy=mypackage
```

4. Complete *mypackage*.html writing documentation text in the places indicated (see below).

The script will generate a template documentation file (for you to fill in) with include directives, plus include files for the following sections:

- Name (entire section)
- List of Applications (entire section)
- Authors (entire section)
- Revision History (entire section)
- Installation Notes (general instructions only)
- References (instructions on how to cite EMBOSS only).

The EMBASSY package name and index terms in the template are also set correctly. The HTML application template is created in:

```
embassy/mypackage/emboss_doc/html/mypackage.html
```

To complete the template mypackage.html add text in the places indicated:

1. Write "Function" section.
2. Write description of package in "Description" section.
3. Write package-specific installation information in "Installation Notes" section (optional).
4. Write peripheral documentation in "Notes" section.
5. Write usage advice, known bugs, etc. in "Warnings" section.
6. Write peripheral documentation in "Notes" section.
7. Write references in "References" section (if appropriate).

Again, the template is commented so it is clear where to add the text.

autodoc.pl also generates documentation in plain text format (with all included text) that is included in the README file for the package.

Once you complete the template and save it, the documentation (in HTML and plain text formats, plus include files) is ready for submission to the EMBOSS authors for publication online. Other files that reference the new package, such as the master table of EMBASSY packages (http://emboss.open-bio.org/rel/dev/apps/EMBASSYPackagesTable.html) will be updated automatically.

A complete application: seqret

9

seqret was the first EMBOSS application and got its name partly from the its function (it can be used to retrieve sequences from databases) and partly from how the application source belies the complexity of the underlying library code. In simple terms, the program reads in a sequence and writes it out again, but there are many options which make it rather powerful. Its command line interface will prompt for an input stream of one or more sequences of any type and prompt for an output stream. **seqret** reads and writes any sequence format that EMBOSS supports and uses the Uniform Sequence Address (USA) (see the *EMBOSS User's Guide*) for specifying the location and format of the sequence data.

The C source code, ACD file, quality assurance tests and documentation for **seqret** are exposed below, to give you an appreciation of all the basic components of a fully functioning EMBOSS application.

9.1 The ACD file (`seqret.acd`)

The ACD file specifies an input sequence stream (`sequence`) and an output sequence stream (`outseq`).

```
application: seqret [
  documentation: "Reads and writes (returns) sequences"
 groups: "Edit"
]

section: input [
 information: "Input section"
 type: "page"
]

 boolean: feature [
   information: "Use feature information"
 ]

 seqall: sequence [
   parameter: "Y"
   type: "gapany"
   features: "$(feature)"
 ]
endsection: input

section: advanced [
```

```
   information: "Advanced section"
   type: "page"
]

   boolean: firstonly [
     information: "Read one sequence and stop"
 ]

endsection: advanced

section: output [
   information: "Output section"
   type: "page"
]

   seqoutall: outseq [
     parameter: "Y"
     features: "$(feature)"
 ]

endsection: output
```

Data definitions for `sequence` and `outseq` are specified as parameters, which means that a value for them (i.e. a USA) can be given on the command line without the requirement of specifying a flag (the parameter name, i.e. **`-sequence`** or **`-outseq`**) on the command line. The flags, however, may still be given.

So, **seqret** can be invoked like this:

```
seqret InputSequenceUSA OutputSequenceUSA
```

in which case the parameters must be in the order shown (*`InputSequenceUSA`* and *`OutputSequenceUSA`* are the USAs of the input and output sequences).

It can also be invoked like this:

```
seqret -sequence InputSequenceUSA -outseq OutputSequenceUSA
```

in which case the parameters can be given in any order, so this is also acceptable:

```
seqret -outseq OutputSequenceUSA -sequence InputSequenceUSA
```

9.1.1 Application definition

The application definition is given at the top of the file:

```
application: seqret
[
   documentation: "Reads and writes (returns) sequences"
```

```
   groups: "Edit"
]
```

The application name (seqret) is given after the application: token. A succinct description of the application function ("Reads and writes (returns) sequences") is given after the documentation: attribute and will be printed to screen when the program is run. It also appears when the program is run with the **-help** qualifier. This text is also searched by the **wossname** utility, which displays applications after searching for keywords.

The groups: attribute has the value of "Edit", indicating that **seqret** belongs to the group "Edit", i.e. applications for sequence editing. All EMBOSS and EMBASSY applications are put into groups of related functionality (Section 4.2.4.1, 'Application group names file (groups.standard)').

Other attributes (see Section 4.2, 'Application definition') may be given in the application definition.

9.1.2 ACD file sections

The ACD file is organised into sections for Input, Advanced and Output sections, which are enclosed within text tokens (section: input, endsection: input etc):

```
section: input
[
 information: "Input section"
 type: "page"
]
.
.
.
endsection: input
```

Clearly, the Input and Output sections are used for application inputs and outputs respectively. Advanced is used for application options which are never prompted for at the command line (none in this case). In practice there are other types of application parameters and corresponding sections that might appear in the ACD file (see Section 4.1, 'Introduction to ACD file development').

9.1.3 Data definitions

These sections contain the ACD data definitions which define the application input and output data and other parameters. Conceptually the data definitions can be thought of as two basic types, either basic input and output files or other types that control the application. For example, sequence and outseq are data definitions for input and output files:

```
seqall: sequence
[
  parameter: "Y"
  type: "gapany"
  features: "$(feature)"
]
.
.
```

```
.
seqoutall: outseq
[
  parameter: "Y"
  features: "$(feature)"
]
```

In contrast, `boolean: feature` is a control definition which, if set by the user, will cause **seqret** to read sequence feature information, if available, from the input file.

`boolean: firstonly` is another control definition which, if set, will cause only a single sequence to be read:

```
boolean: feature
[
    information: "Use feature information"
]
.
.
.
boolean: firstonly
[
    information: "Read one sequence and stop"
]
```

In practice, many different ACD datatypes (Section A.2, 'Datatypes') are available. Programming these data definitions is covered in greater detail in Section 5.5, 'Programming with objects'.

9.1.4 Attributes

Each ACD definition has attributes which are `name: value` pairs. Attributes are of two basic types. *Global attributes* are available for all datatypes whereas *datatype-specific attributes* are available for individual or groups of related datatypes.

Consider the definitions for `feature` and `sequence`:

```
boolean: feature
[
   information: "Use feature information"
]
seqall: sequence
[
   parameter: "Y"
   type: "gapany"
   features: "$(feature)"
]
```

`parameter:` and `information:` are global attributes whereas `type:` and `features:` are datatype-specific.

`parameter: "Y"` defines a data definition to be a *parameter* as opposed to some kind of *qualifier*. A value for a parameter can be given on the command line with or without a flag (parameter name) as described above. All parameters are required by the application and a value will be prompted for if one is not given on the command line.

parameter: "Y" is not given for the boolean definitions which means, in the absence of standard: "Y" or additional: "Y" attributes, that they will default to being an advanced qualifier. An advanced qualifier is never prompted for. Had standard: "Y" or additional: "Y" been defined, then the data definitions would be *standard* or *additional* qualifiers. More information on the behaviour of qualifiers and parameters is available (Section 4.1, 'Introduction to ACD file development').

information: gives a very succinct description of the data definition. The text is used in web forms and other GUIs. It is also the text used to prompt the user for a value at the command line. You'll notice information: is not defined for the sequence inputs and outputs. A default prompt is provided by EMBOSS for these datatypes. All boolean: datatypes have the default value of False which means that a default needn't be defined explicitly in the ACD file.

The type: attribute is specific to the sequence datatypes. type: "gapany" indicates that the input sequence can be of any type and may contain gap characters. In practice many different sequence types are supported (Section A.7, 'Sequence types').

The features: attribute is also sequence-specific. The attribute definition looks a little strange:

```
features: "$(feature)"
```

The $ syntax means 'retrieve the value of ...', in this case 'retrieve the value of the feature ACD definition', such that the features: attribute of outseq is set to that value. In other words, the input sequence will include features if the user turned the feature advanced qualifier on by giving it on the command line, e.g. **seqret -feature**. Whilst it is not strictly necessary to have features: "$(feature)" (the information is after all specified in boolean: feature) it makes the application source code a good deal cleaner.

In addition to global and datatype-specific attributes there are two other types of attribute that may appear in an ACD file. *Calculated attributes* receive a value once the ACD file has been processed and application input files have been read. Finally, there are *datatype-specific command line qualifiers* which are defined for single or groups of ACD datatypes and may be hard-coded as attributes within an ACD file. See Section A.4 ('Global attributes'), Section A.5 ('Datatype-specific attributes'), and Section A.6 ('Calculated attributes').

9.2 The C source code (seqret.c)

The application C source code (see below) is very simple. Its basic functions are:

- Declare variables for holding values from the ACD file (AjP* type declarations)
- Process the ACD file and command line (embInit)
- Read the values (input and output sequence streams) from the ACD file into memory (prefix ajAcdGet family of functions)
- Iterate through the input sequences and load a sequence into memory (ajSeqallNext)
- Write the sequence out (ajSeqoutWriteSeq)
- Close the output file

- Exit cleanly (embExit).

The source code is shown below:

```
/* @source seqret application
**
** Return a sequence
**
** @author Copyright (C) Peter Rice
** @@
**
** This program is free software; you can redistribute it and/or
** modify it under the terms of the GNU General Public License
** as published by the Free Software Foundation; either version 2
** of the License, or (at your option) any later version.
**
** This program is distributed in the hope that it will be useful,
** but WITHOUT ANY WARRANTY; without even the implied warranty of
** MERCHANTABILITY or FITNESS FOR A PARTICULAR PURPOSE. See the
** GNU General Public License for more details.
**
** You should have received a copy of the GNU General Public License
** along with this program; if not, write to the Free Software
** Foundation, Inc., 59 Temple Place - Suite 330, Boston, MA 02111-1307, USA.
******************************************************************************/

#include "emboss.h"



/* @prog seqret **************************************************************
**
** Reads and writes (returns) sequences
**
******************************************************************************/

int main(int argc, char **argv)
{
   AjPSeqall seqall;
   AjPSeqout seqout;
   AjPSeq seq = NULL;
   AjBool firstonly;

   embInit("seqret", argc, argv);

   seqout = ajAcdGetSeqoutall("outseq");
   seqall = ajAcdGetSeqall("sequence");
   firstonly = ajAcdGetBoolean("firstonly");

   while(ajSeqallNext(seqall, &seq))
   {
       ajSeqoutWriteSeq(seqout, seq);
       if(firstonly)
          break;
   }
   ajSeqoutClose(seqout);

   embExit();

   return 0;
}
```

9.2.1 Variable declarations

The first block of code in `main()` declares variables for holding values from the ACD file:

```
AjPSeqall seqall=NULL;
AjPSeqout seqout=NULL;
AjPSeq seq = NULL;
AjBool firstonly;
```

The variables beginning with `AjP` are all C pointers to EMBOSS objects (C data structures for the corresponding types). These include `AjPSeqall`, `AjPSeqout` and `AjPSeq`. Many EMBOSS objects, for both complex biological and other types, are defined in the AJAX and NUCLEUS programming libraries. `AjBool`, in contrast, is the EMBOSS datatype for a simple boolean variable. You'll notice that the pointer variables are initialised to `NULL`. It is good practice to always initialise pointers to `NULL` where they're first defined. For an explanation of why, and detailed information on programming with objects, see Section 5.3, 'Objects (C data structures)'.

`AjPSeq` is for single sequence input, `AjPSeqall` for multiple sequence input and `AjPSeqout` for single sequence output. For the **seqret** application `AjPSeqall seqall` is used for the input sequence stream, `AjPSeq seq` is used to hold the data for a single sequence from that stream and `AjPSeqout seqout` is used for the output sequence stream. `AjBool firstonly` is used to hold the value of the `firstonly` control attribute from the ACD file.

You'll notice that no variable is required for the `feature` data definition. The value for this option is set on the input sequence stream, within the ACD file itself, by the ACD code `features: "$(feature)"`. In other words whether feature information will or will not be included with the sequences is set within the ACD file and stored within the `seqall` object, therefore no additional variable is needed for it in the C code.

9.2.2 ACD file and command line processing

The code:

```
embInit("seqret", argc, argv);
```

is used to process the ACD file and command line. It handles all of the user input processing which is why it's called first. `embInit` reads in local database definitions, finds the right ACD file to use (the first argument is `"seqret"` so it looks for `seqret.acd` in the ACD directory), reads the ACD file and processes the command line using `argc` and `argv` from `main`.

By the time `embInit` returns, the input sequence stream (`sequence`) will be opened for reading and the first sequence read into memory, the boolean variables `feature` and `firstonly` will have received values (possibly the default) and an output file will be opened for `outseq`. Memory is allocated for these objects and is available for use by the program.

`embInit` handles prompting of the user for values that are not entered on the command line, including functionality such as reprompting the user for values that are out of range.

9.2.3 Retrieving values from the ACD file

To retrieve C pointers to these data items the following code is used:

```
seqout = ajAcdGetSeqoutall("outseq");
seqall = ajAcdGetSeqall("sequence");
```

Similarly, to retrieve the value of the simple data type variable `firstonly`:

```
firstonly = ajAcdGetBoolean("firstonly");
```

You can see that the argument to the `ajAcdGet*` functions is the name of the ACD definition which is to be retrieved.

9.2.4 Sequence handling

To iterate through the input sequences and load a sequence into memory the following code is used:

```
while(ajSeqallNext(seqall, &seq))
{
    ajSeqoutWriteSeq(seqout, seq);
    if(firstonly)
       break;
}
```

`ajSeqallNext` is called in a loop to retrieve consecutive sequences in turn from the input stream. The second argument (`&seq`) sets the pointer to the current sequence in the stream. `ajSeqoutWriteSeq` is used to write this sequence to the output stream `seqout`. The loop will terminate after the first sequence if `firstonly` has been set.

9.2.5 Exiting cleanly

The output stream is closed by calling `ajSeqoutClose()`. The application terminates cleanly with the call `embExit()` before returning `0` to the operating system:

```
  ajSeqoutClose(seqout);
   embExit();
   return 0;
}
```

9.3 Inbuilt functionality

seqret has rich inbuilt functionality:

- It can read and write any sequence format that EMBOSS supports (see the *EMBOSS User's Guide*), which includes all the common formats.

- It uses the powerful USA syntax (see the *EMBOSS User's Guide*) for specifying the location and format of the sequence data.
- **seqret** can read one or more sequences from a database, file, listfile (file of USAs), the command line or the output of other programs and can then write them to a file, database or to `stdout` which can be redirected to another program. It's therefore useful for extracting sequences from databases and displaying them.
- **seqret** supports many command line options which are not listed in the ACD file. These include global and datatype-specific qualifiers. Many others are available and may be used in combination, providing greater flexibility without the requirement for any additional programming.
- It supports global qualifiers (see the *EMBOSS User's Guide*) which are available to all EMBOSS applications.
- It supports various datatype-specific qualifiers (see the *EMBOSS User's Guide*); in this case, the sequence datatypes which include the `seqall: sequence` input and `seqoutall: outseq` output definitions from the ACD file.

For example, you can specify the input and output formats by using the ***-sformat FormatName*** (input) and **-osformat *FormatName*** (output) options. If you don't specify the input format it will try all known formats until one succeeds, and if you don't specify the output format then FASTA will be used by default. **seqret** is useful for reformatting sequences, perhaps in preparation for input to another program.

The **-sbegin** and **-send** options specify the start and end position of a subsequence in the input sequence. This fragment is written to the output file, therefore **seqret** is useful for simple extraction of sequence regions. The **-sreverse** switch will use the reverse complement of a nucleic acid sequence.

As you've seen from `seqret.c`, no application code is needed to benefit from this inbuilt functionality. What's more, as new sequence formats are added to EMBOSS, **seqret** will automatically be able to use them; no application code needs to change. It should be obvious then that this inbuilt functionality saves you, as a software developer, a great deal of time.

9.3.1 Example usage

Examples illustrating a few of the many use cases of **seqret** are shown below.

Here **seqret** is being used to retrieve the entry 5HT1D_FUGRU from the *SwissProt* database and write the entry out in FASTA format:

```
% seqret
Reads and writes (returns) sequences
Input (gapped) sequence(s): swissprot:5HT1D_FUGRU
Output sequence [5ht1d_fugru.fasta]:
% more 5ht1d_fugru.fasta
>5HT1D_FUGRU P79748 5-hydroxytryptamine receptor 1D (5-HT-1D)
MELDNNSLDYFSSNFTDIPSNTTVAHWTEATLLGLQISVSVVLAIVTLATMLSNAFVIAT
IFLTRKLHTPANFLIGSLAVTDMLVSILVMPISIVYTVSKTWSLGQIVCDIWLSSDITFC
TASILHLCVIALDRYWAITDALEYSKRRTMRRAAVMVAVVWVISISISMPPLFWRQAKAH
EELKECMVNTDQISYTLYSTFGAFYVPTVLLIILYGRIYVAARSRIFKTPSYSGKRFTTA
QLIQTSAGSSLCSLNSASNQEAHLHSGAGGEGGGSPLFVNSVKVKLADNVLERKRLCAAR
```

```
ERKATKTLGIILGAFIICWLPFFVVTLVWAICKECSFDPLLFDVFTWLGYLNSLINPVIY
TVFNDEFKQAFQKLIKFRR
```

The same thing could be achieved by using the USA (see the *EMBOSS User's Guide*) mechanism to specify the same input and output sequences on the command line:

```
seqret swissprot:5Ht1D_FUGRU fasta:5ht1d_fugru.fasta
```

This example retrieves *all* of the sequences from the input stream. Not a very sensible thing to do, but it illustrates that the USA mechanism supports wildcard specification of sequences:

```
% seqret swissprot:"*"
Reads and writes (returns) sequences
Output sequence [ubr5_rat.fasta]: all.fasta
% more all.fasta
>UBR5_RAT Q62671 E3 ubiquitin-protein ligase UBR5 (6.3.2.-)
MMSARGDFLNYALSLMRSHNDEHSDVLPVLDVCSLKHVAYVFQALIYWIKAMNQQTTLDT
PQLERKRTRELLELGIDNEDSEHENDDDTSQSATLNDKDDESLPAETGQNHPFFRRSDSM
TFLGCIPPNPFEVPLAEAIPLADQPHLLQPNARKEDLFGRPSQGLYSSSAGSGKCLVEVT
MDRNCLEVLPTKMSYAANLKNVMNMQNRQKKAGEDQSMLAEEADSSKPGPSAHDVAAQLK
SSLLAEIGLTESEGPPLTSFRPQCSFMGMVISHDMLLGRWRLSLELFGRVFMEDVGAEPG
SILTELGGFEVKESKFRREMEKLRNQQSRDLSLEVDRDRDLLIQQTMRQLNNHFGRRCAT
TPMAVHRVKVTFKDEPGEGSGVARSFYTAIAQAFLSNEKLPNLDCIQNANKGTHTSLMQR
LRNRGERDREREREREMRRSSGLRAGSRRDRDRDFRRQLSIDTRPFRPASEGNPSDDPDP
LPAHRQALGERLYPRVQAMQPAFASKITGMLLELSPAQLLLLLASEDSLRARVEEAMELI
VAHGRENGADSILDLGLLDSSEKVQENRKRHGSSRSVVDMDLDDTDDGDDNAPLFYQPGK
RGFYTPRPGKNTEARLNCFRNIGRILGLCLLQNELCPITLNRHVIKVLLGRKVNWHDFAF
FDPVMYESLRQLILASQSSDADAVFSAMDLAFAVDLCKEEGGGQVELIPNGVNIPVTPQN
VYEYVRKYAEHRMLVVAEQPLHAMRKGLLDVLPKNSLEDLTAEDFRLLVNGCGEVNVQML
ISFTSFNDESGENAEKLLQFKRWFWSIVERMSMTERQDLVYFWTSSPSLPASEEGFQPMP
SITIRPPDDQHLPTANTCISRLYVPLYSSKQILKQKLLLAIKTKNFGFV
>5HT1D_FUGRU P79748 5-hydroxytryptamine receptor 1D (5-HT-1D) (5HT1D)
MELDNNSLDYFSSNFTDIPSNTTVAHWTEATLLGLQISVSVVLAIVTLATMLSNAFVIAT
IFLTRKLHTPANFLIGSLAVTDMLVSILVMPISIVYTVSKTWSLGQIVCDIWLSSDITFC
TASILHLCVIALDRYWAITDALEYSKRRTMRRAAVMVAVVWVISISISMPPLFWRQAKAH
EELKECMVNTDQISYTLYSTFGAFYVPTVLLIILYGRIYVAARSRIFKTPSYSGKRFTTA
QLIQTSAGSSLCSLNSASNQEAHLHSGAGGEGGGSPLFVNSVKVKLADNVLERKRLCAAR
ERKATKTLGIILGAFIICWLPFFVVTLVWAICKECSFDPLLFDVFTWLGYLNSLINPVIY
TVFNDEFKQAFQKLIKFRR
>ACTB1_FUGRU P68142 Actin, cytoplasmic 1 (Beta-actin A)
MEDEIAALVVDNGSGMCKAGFAGDDAPRAVFPSIVGRPRHQGVMVGMGQKDSYVGDEAQS
KRGILTLKYPIEHGIVTNWDDMEKIWHHTFYNELRVAPEEHPVLLTEAPLNPKANREKMT
QIMFETFNTPAMYVAIQAVLSLYASGRTTGIVMDSGDGVTHTVPIYEGYALPHAILRLDL
... data omitted
```

The following example illustrates the use of the **`-firstonly`** option. This is an advanced option therefore will never be prompted for; whenever used it has to be specified on the command line:

```
% seqret swissprot:"*" -outseq all.fasta -firstonly
Reads and writes (returns) sequences
%more all.fasta
>UBR5_RAT Q62671 E3 ubiquitin-protein ligase UBR5 (6.3.2.-)
```

```
MMSARGDFLNYALSLMRSHNDEHSDVLPVLDVCSLKHVAYVFQALIYWIKAMNQQTTLDT
PQLERKRTRELLELGIDNEDSEHENDDDTSQSATLNDKDDESLPAETGQNHPFFRRSDSM
TFLGCIPPNPFEVPLAEAIPLADQPHLLQPNARKEDLFGRPSQGLYSSSAGSGKCLVEVT
MDRNCLEVLPTKMSYAANLKNVMNMQNRQKKAGEDQSMLAEEADSSKPGPSAHDVAAQLK
SSLLAEIGLTESEGPPLTSFRPQCSFMGMVISHDMLLGRWRLSLELFGRVFMEDVGAEPG
SILTELGGFEVKESKFRREMEKLRNQQSRDLSLEVDRDRDLLIQQTMRQLNNHFGRRCAT
TPMAVHRVKVTFKDEPGEGSGVARSFYTAIAQAFLSNEKLPNLDCIQNANKGTHTSLMQR
LRNRGERDREREREREMRRSSGLRAGSRRDRDRDFRRQLSIDTRPFRPASEGNPSDDPDP
LPAHRQALGERLYPRVQAMQPAFASKITGMLLELSPAQLLLLLASEDSLRARVEEAMELI
VAHGRENGADSILDLGLLDSSEKVQENRKRHGSSRSVVDMDLDDTDDGDDNAPLFYQPGK
RGFYTPRPGKNTEARLNCFRNIGRILGLCLLQNELCPITLNRHVIKVLLGRKVNWHDFAF
FDPVMYESLRQLILASQSSDADAVFSAMDLAFAVDLCKEEGGGQVELIPNGVNIPVTPQN
VYEYVRKYAEHRMLVVAEQPLHAMRKGLLDVLPKNSLEDLTAEDFRLLVNGCGEVNVQML
ISFTSFNDESGENAEKLLQFKRWFWSIVERMSMTERQDLVYFWTSSPSLPASEEGFQPMP
```

Here, usage information is being generated by using the **-help** global qualifier:

```
% seqret -help
  Standard (Mandatory) qualifiers:
 [-sequence]          seqall     (Gapped) sequence(s) filename and optional
                                 format, or reference (input USA)
 [-outseq]            seqoutall  Sequence set(s) filename and optional format
                                 (output USA)

  Additional    (Optional)   qualifiers:    (none)
  Advanced    (Unprompted)   qualifiers:
  -feature                boolean  Use feature information
  -firstonly              boolean  Read one sequence and stop

  General   qualifiers:
  -help                   boolean  Report command line options. More
                                   information on associated and general
                                   qualifiers can be found with -help -verbose
```

In the example below, the sequence-specific command line options **-sbegin** and **-send** are used to specify a sequence region:

```
% seqret swissprot:5HT1D_FUGRU -sbegin 10 -send 20 fasta:5ht1d_fugru.fasta
Reads and writes (returns) sequences
%more 5ht1d_fugru.fasta
>5HT1D_FUGRU P79748 5-HYDROXYTRYPTAMINE 1D RECEPTOR (5-HT-1D) (SEROTONIN RECEPTOR).
YFSSNFTDIPS
```

9.4 Documentation (`seqret.html`)

The end-user documentation is organised into sections for easy navigation. You can see this for yourself in the online application documentation:

```
http://emboss.open-bio.org/rel/dev/apps/seqret.html
```

For more information see Chapter 8, *application documentation standards*.

9.5 Quality assurance tests

Quality assurance tests for **seqret** are defined in the file:

```
.../emboss/emboss/test/qatest.dat
```

The test data for the application is given under the `/test` directory and consists of input files, application parameters and the corresponding output files.

You'll see from inspecting the file that many tests are defined for the application, testing it under all sorts of input conditions in order to be as robust as possible. These are performed frequently to ensure the applications work as advertised and are not broken by (e.g.) recent changes to the library code.

Quality assurance testing, including how to write entries in `qatest.dat`, is covered in detail elsewhere (Chapter 7, *Quality assurance*).

Test 1

This test extracts an entry from a database and writes it to a file:

```
ID seqret-ex1
UC Extract an entry from a database and write it to a file:
AP seqret
IN tembl:x65923
IN
FI stderr
FC = 2
FP 0 /Warning: /
FP 0 /Error: /
FP 0 /Died: /
FI x65923.fasta
FP /X65923/
//
```

The command line session is below:

```
% qatest.pl -simple -testfile=seqret_1.dat seqret-ex1
Tests total: 1 pass: 1 fail: 0
Time: 0 seconds
```

Test 2

Display the contents of the sequence on the screen:

```
ID seqret-ex2
UC Display the contents of the sequence on the screen:
AP seqret
IN tembl:x65923
IN stdout
FI stdout
```

```
FP /X65923/
FI stderr
FC = 2
FP 0 /Warning: /
FP 0 /Error: /
FP 0 /Died: /
//
```

The command line session is below:

```
% qatest.pl -simple -testfile=seqret_2.dat seqret-ex2
Tests total: 1 pass: 1 fail: 0
Time: 0 seconds
```

Test 3
Write the result in GCG format by using the qualifier **-osformat**:

```
ID seqret-ex3
UC Write the result in GCG format by using the qualifier '-osformat'.
AP seqret
CL -osf gcg
IN tembl:x65923
IN
FI stderr
FC = 2
FP 0 /Warning: /
FP 0 /Error: /
FP 0 /Died: /
FI x65923.gcg
FP /X65923/
//
```

The command line session is below:

```
% qatest.pl -simple -testfile=seqret_3.dat seqret-ex3
Tests total: 1 pass: 1 fail: 0
Time: 0 seconds
```

Test 4
Write the result in GCG format by specifying the format in the output USA on the command line:

```
ID seqret-ex4
UC Write the result in GCG format by specifying the format
UC in the output USA on the command line.
AP seqret
CL -outseq gcg::x65923.gcg
IN tembl:x65923
FI stderr
FC = 2
```

```
FP 0 /Warning: /
FP 0 /Error: /
FP 0 /Died: /
FI x65923.gcg
FP /X65923/
//
```

The command line session is below:

```
% qatest.pl -simple -testfile=seqret_4.dat seqret-ex4
Tests total: 1 pass: 1 fail: 0
Time: 0 seconds
```

Test 5

Write the result in GCG format by specifying the format in the output USA at the prompt:

```
ID seqret-ex5
UC Write the result in GCG format by specifying the format
UC in the output USA at the prompt.
AP seqret
IN tembl:x65923
IN gcg::x65923.gcg
FI stderr
FC = 2
FP 0 /Warning: /
FP 0 /Error: /
FP 0 /Died: /
FI x65923.gcg
FP /X65923/
//
```

The command line session is below:

```
% qatest.pl -simple -testfile=seqret_5.dat seqret-ex5
Tests total: 1 pass: 1 fail: 0
Time: 0 seconds
```

Test 6

Write the reverse complement of a sequence:

```
ID seqret-ex6
UC Write the reverse-complement of a sequence:
AP seqret
CL -srev
IN tembl:x65923
IN
FI stderr
FC = 2
FP 0 /Warning: /
FP 0 /Error: /
FP 0 /Died: /
```

```
FI x65923.fasta
FP /X65923/
```

The command line session is below:

```
% qatest.pl -simple -testfile=seqret_6.dat seqret-ex6
Tests total: 1 pass: 1 fail: 0
Time: 0 seconds
```

Test 7

Extract the bases between the positions starting at 5 and ending at 25:

```
ID seqret-ex7
UC Extract the bases between the positions starting at 5 and ending at 25:
AP seqret
CL -sbegin 5 -send 25
IN tembl:x65923
IN
FI stderr
FC = 2
FP 0 /Warning: /
FP 0 /Error: /
FP 0 /Died: /
FI x65923.fasta
FP /X65923/
//
```

The command line session is below:

```
% qatest.pl -simple -testfile=seqret_7.dat seqret-ex7
Tests total: 1 pass: 1 fail: 0
Time: 0 seconds
```

Test 8

Extract the bases between the positions starting at 5 and ending at 5 bases before the end of the sequence:

```
ID seqret-ex8
UC Extract the bases between the positions starting at 5
UC and ending at 5 bases before the end of the sequence:
AP seqret
CL -sbegin 5 -send -5
IN tembl:x65923
IN
FI stderr
FC = 2
FP 0 /Warning: /
FP 0 /Error: /
FP 0 /Died: /
FI x65923.fasta
```

```
FP /X65923/
//
```

The command line session is below:

```
% qatest.pl -simple -testfile=seqret_8.dat seqret-ex8
Tests total: 1 pass: 1 fail: 0
Time: 0 seconds
```

Test 9

Read all entries in the database `tembl` that start with `h` and write them to a file:

```
ID seqret-ex9
UC Read all entries in the database 'tembl' that start with 'h'
UC and write them to a file:
AP seqret
IN tembl:h*
IN hall.seq
FI stderr
FC = 2
FP 0 /Warning: /
FP 0 /Error: /
FP 0 /Died: /
FI hall.seq
FP /H45989/
//
```

The command line session is below:

```
% qatest.pl -simple -testfile=seqret_9.dat seqret-ex9
Tests total: 1 pass: 1 fail: 0
Time: 0 seconds
```

Test 10

This test reads all entries in the database `tembl` that start with `ab` and writes them to a file. In this example the specification is all done on the command line. To stop UNIX getting confused by the `*` character, it has a backslash (\) before it:

```
ID seqret-ex10
UC Read all entries in the database 'tembl' that start with 'ab'
UC and write them to a file.
UC In this example the specification is all done in the command line
UC and to stop Unix getting confused by the '*' character,
UC it has to have a backslash ('\') before it:
AP seqret
CL 'tembl:ab*' aball.seq
FI stderr
FC = 1
FP 0 /Warning: /
FP 0 /Error: /
FP 0 /Died: /
```

```
FI aball.seq
FP /^>AB000095/
FP 4 /^>/
//
```

The command line session is below:

```
% qatest.pl -simple -testfile=seqret_10.dat seqret-ex10
Tests total: 1 pass: 1 fail: 0
Time: 0 seconds
```

Test 11

seqret does not read in features by default as this results in slightly faster performance. If, however, you wish to read in features with your sequence and also write them out then using **-feature** will change the default behaviour. Note EMBL format is used for the output file. The default FASTA format would report the features in GFF format (file *seqname*.gff where *seqname* is the name of the input sequence):

```
ID seqret-ex11
UC seqret does not read in features by default because this results in
UC slightly faster performance. If however you wish to read in features
UC with your sequence and write them out on output, using '-feature' will
UC change the default behaviour to use any features present in the sequence.
UC N.B. use embl format for the output file as the default format 'fasta'
UC reports the features in gff (file "<seqname>.gff")
AP seqret
CL -feature
IN tembl:x65923
IN embl::x65923.embl
FI stderr
FC = 2
FP 0 /Warning: /
FP 0 /Error: /
FP 0 /Died: /
FI x65923.embl
FP /X65923/
//
```

The command line session is below:

```
% qatest.pl -simple -testfile=seqret_11.dat seqret-ex11
Tests total: 1 pass: 1 fail: 0
Time: 0 seconds
```

Incorporating third-party applications 10

10.1 Benefits to incorporating applications into EMBOSS

There are lots of good reasons for making the effort to incorporate third-party applications into EMBOSS. Just a few are listed below.

Distribution It's clearly desirable to reach as many users as possible. EMBOSS has thousands of users and many sites regularly download the latest release, so any software that's incorporated will be widely distributed.

Training cost Incorporated software is used in exactly the same way as all other EMBOSS applications. Someone familiar with EMBOSS will be able to use any application that's been ported or wrapped. There's no additional training requirement beyond, of course, learning about the specific methods the application implements. That is a requirement regardless of whether or not an application is in EMBOSS.

Support cost Incorporated applications will benefit from a basic level of support provided by EMBOSS through the mailing lists. The level of support depends on the circumstances. An application that's been fully ported with the understanding of it being adopted by EMBOSS would be fully supported. At the very least a user would get a reply to any email query and forwarded to where they could get more help. This is especially valuable in cases where the original author of an application has since moved away from the institute that now maintains it.

Maintenance cost There might be savings on the maintenance costs. This again depends on the level of integration and the specific arrangements. At the very least EMBOSS would investigate any bugs and forward appropriate reports to the authors of the original software. There is an additional cost to maintain the port or wrapper as well as the original code, but this is more than offset by the benefits.

Web services, workflows and interfaces EMBOSS has been incorporated into many graphical interfaces and workflow software such as **SRS**, **Pipeline Pilot**, **Taverna**, **wEMBOSS** and **Jemboss**. Further, thanks to ongoing developments in **SOAPLAB**, it is relatively easy to deploy an EMBOSS application as a web service. If your software is incorporated then there's little or no additional cost in setting it up in these contexts.

Database support EMBOSS supports all the common formats for input and output of sequences, sequence features and sequence alignments. Many other common data formats are supported. Any software you incorporate will be able to use these formats too. What's more, when EMBOSS is updated to support changes to the formats, or entirely new formats, then your software will automatically benefit from these changes.

10.2 Wrappers and ports

There are two basic approaches for incorporating third-party applications. You can write a 'wrapper' to an application or you can port the application source code itself. A wrapper is merely a program whose primary purpose is to call another program, the latter is usually (but not always) not native to EMBOSS. In contrast, a 'port' is where the source code of the application being incorporated is modified such that it's suitable for the native system (EMBOSS).

Basic methods are outlined below for wrapping and porting applications. It is pointless being too prescriptive as what's required very much depends on the application in question. The pros and cons of wrapping and porting applications are summarised. The methods used, and C source code and ACD files, are illustrated based on experience of incorporating the **HMMER** package into EMBOSS.

Note

The **HMMER** package is a collection of tools for hidden Markov model analysis of protein sequences written by Sean Eddy. Version 2.2.1 was ported into EMBOSS as the **HMMER** EMBASSY package. **HMMER** is under the GPL licence so there were no licensing issues to consider when modifying the source code. This version proved hard to maintain, however, so for the latest release, v2.3.2, an EMBASSY package (**HMMERNEW**) of application wrappers was developed.

When porting an application it's necessary to identify the functions that read input from the user, write an ACD file that defines the same values, then add code to pick up the values from the ACD file instead of processing them as in the original.

When wrapping an application an ACD file is still needed and this is usually written to follow closely the original interface. There are drawbacks in the naming of parameters as you'll see later. The wrapper must construct an appropriate command line based on the user specifications. The main difficulty is ensuring that sequences and other data are in the format required by the third-party application. The wrapper may need to read the input and convert it as necessary before the underlying program is launched.

In both cases any output files may be left in whatever format is produced by the original application or the data might be converted into a format specified by the user. For a port this could involve modifying the original source code and replacing the output function with appropriate calls to the EMBOSS libraries. Alternatively, and this is the approach taken when developing some wrappers, temporary files may be used to achieve the conversion.

10.3 Pros and cons

A subjective view of the various costs involved in wrapping and porting applications under EMBOSS follows:

Category	Wrappers	Port
Development cost	Low to medium	Medium to high
Flexibility	Low to medium	High
Limitations	Medium	Low
Maintenance cost	Low to medium	Medium to high
Support cost	Low	High
Perceived risk	Low to medium	Medium to high

10.3.1 Development cost

More or less the same ACD file must be written for a wrapper or a port. The main difficulty is that third-party software is typically not as flexible as EMBOSS and requires input files in a specific format, for example sequences in FASTA format. In such cases there's a decision to make – either to support all input datatypes that EMBOSS supports or to stick within the constraints of the original program.

Fully supported input requires additional code for manipulating temporary files to convert the input data into a format acceptable to third-party software. If this isn't done it's necessary to enforce any input constraints at the level of the ACD file (if possible), or otherwise document them and raise an exception if the application receives data in the wrong format. Constraining the permissible formats of sequence input within ACD would go against the whole ethos of EMBOSS, so there's no alternative but to add code for reformatting or to raise an exception.

The big advantage to writing a wrapper is that you don't need to worry about the third-party source code itself. All the wrapper code must do (after reformatting the input files and processing the ACD file) is to construct an appropriate command line, invoke the command, then (possibly) reformat the output (using temporary files). Building the correct command line usually involves some comparison of parameter values and therefore some housekeeping code, but that's more or less trivial.

When porting software it's essential to consider the third-party source code carefully. No new files of code are usually needed, but the `main()` function will need rewriting to handle the ACD file processing. Also, any other functions that read input data, and these might live in different files, will need rewriting so that they work with the data read via ACD.

Therefore it might be necessary to edit multiple files. The edits themselves most probably will not be too difficult but could present a barrier. Furthermore, knowing where to edit certainly does require knowledge of the third-party code, possibly quite deep knowledge. For these reasons the initial development cost for a port tends to be significantly higher than it is for a wrapper.

10.3.2 Flexibility

For the reasons just explained, ported software provides the greatest flexibility in terms of support for input and output formats. In fact support is as complete as it is for any EMBOSS application. In contrast there is potentially less flexibility for wrappers, though this might be mitigated with extra coding involving the use of temporary files.

10.3.3 Limitations

In addition to greater flexibility, ports may have fewer intrinsic limitations owing to technical reasons. For example, difficulties in getting one program to execute another have been reported under MS Windows. Issues can also arise with inter-process communication.

Note

In principle, a call to **`system()`** or **`exec()`** could be used to invoke the third-party application. Both **`system()`** and **`exec()`** create a new process. The difference is that **`system()`** spawns a shell to invoke the program whereas **`exec()`** just invokes the program. When using **`system()`** the shell can get in the way of setting up inter-process communication.

10.3.4 Maintenance cost

To maintain a wrapper you only need to worry about changes to the interface. It's relatively easy to add one or two new options to the ACD file and corresponding wrapper code. Upon major releases, however, many of the options might have changed and it might be simpler to start from scratch, rather than incrementally modify what's already there. In either case the package documentation must be updated.

With a port, in addition to changes to the interface, any changes to the application code must be integrated. As there is no convenient automatic mechanism for doing this, for example by using a common CVS repository, then one has to either use **`diff`** on the code to detect changes, or otherwise start afresh by adding the EMBOSS-specific code in the port to the code to the new release. The latter may well be the simpler and safer option. This is easier if all the insertions of EMBOSS-specific code are well documented. Nonetheless it may not be obvious where the changes should be made, requiring code inspection even if the code is well documented. For these reasons porting software is likely to be more error-prone.

10.3.5 Support cost

The cost of supporting a wrapper is substantially lower than a port. This is because you didn't write the third-party application, only a wrapper to it. Therefore you can reasonably forward any queries to the original authors, so long as you're certain errors have not arisen as a result of the wrapper code. In the case of a port you've modified the original source code and so may be reasonably expected to support it.

10.3.6 Perceived risk

'Perceived risk' is subjective, but it boils down to whether the end-user will trust your software enough to use it. Inevitably, ported software is treated with some suspicion because someone other than the original author has modified the code, regardless of whether they have in fact fixed bugs or improved it in some other way. Therefore a port could be overlooked if a user is being cautious and is trying to avoid any possibility of discrepancy in results. They might also stick with the original, with warts and all, simply because it's what they know and have used in the past. In contrast, the perceived risk is lower with a wrapper where they are understood to merely call the wrapped software.

10.3.7 Summary

Whether you should port or wrap software depends on the case in question. Generally, wrapper applications are preferred as they maintain the separation between the original and EMBOSS code and are easier to develop, maintain and support. If you are the author of the original code, however, or you see EMBOSS as the main access point to the software, it may well be preferable to port the application.

10.4 How to wrap third-party applications

Steps that were taken to wrap the **HMMER** package are described below. It is impossible to give entirely generic guidelines because the requirements depend on the software being wrapped. Nonetheless all the basic steps you are likely to take are illustrated here.

1. Planning and design
2. ACD file development
3. C source code development
4. Quality assurance tests
5. Documentation
6. Integration.

10.4.1 Planning and design

The steps taken were:

1. Download the source code and documentation
2. Read the documentation
3. Decide which options to keep in the EMBOSS version
4. Decide if new parameters are needed e.g. for application output (normally to `stdout`).

10.4.1.1 Documentation

HMMER includes an excellent *User's Guide*. It was necessary to read the Introduction, work through the Tutorial and then work through the manual pages for each application in turn. Not all applications and packages are documented to the same high standard! It's essential that you familiarise yourself with the package as a whole, and in particular identify all of the possible parameters for all the applications and their interactions. You should not start coding until you have this information.

10.4.1.2 Application options

The first design step is to decide which application options to keep in the EMBOSS version. An option should be discarded if it is:

- Redundant to inbuilt EMBOSS functionality
- Sensibly subsumed by a new EMBOSS qualifier
- Always set so need not be defined in the ACD file.

You should familiarise yourself with the functionality that is built into EMBOSS (see Section 3.1, 'EMBOSS programming') to help decide what options are redundant. For example the **HMMER** help option `-h` is not needed because **`-help`** is an inbuilt qualifier for all EMBOSS applications.

One or more options might sensibly be covered by a single EMBOSS qualifier; for example there are five options in **hmmbuild** for setting sequence weighting which are handled by a single weighting option in the EMBASSY wrapper equivalent.

Certain options should always be set in the EMBOSS version and so needn't be defined in the ACD file; for example the **`-F`** option to force overwrite of files is always set.

10.4.1.3 New options

The second step is to decide whether any new parameters are required. Typically a parameter for an output file is needed to catch output written to `stdout` by default.

10.4.2 ACD file development

The key things to consider are:

- Application name
- Application short description
- Documentation for program options
- Qualifier names
- Validating and reformatting the ACD.

10.4.2.1 Application name

For **HMMER** the original application names were used except that the EMBOSS versions are prefixed with an '`e`'. You should use the original names or some simple derivative except in unavoidable cases, for example because an EMBOSS or system application with that name already exists.

10.4.2.2 Application short description

The application short description was taken directly from the *User Guide* and pasted into the `documentation:` attribute in the application definition. A description and documentation for each option was again taken from the *User Guide* and pasted into the `help:` and `information:` attributes as appropriate. This is vital documentation and cannot be omitted.

10.4.2.3 Application options

The qualifier names chosen were identical to the option names in the original wherever possible. There is however a compromise between consistency with the original and consistency with EMBOSS. If in doubt, ask the EMBOSS developers.

10.4.2.4 ACD file

The ACD file was tested and reformatted using the EMBOSS utilities **acdc** and **acdpretty**. You should routinely use these tools when developing ACD files.

More information is available on ACD file development (Chapter 5, *C programming*) and on the ACD utilities (Section 4.6, 'ACD utilities').

10.4.3 C source code development

The application C source code was implemented in the following order:

1. Application header documentation
2. `main()` function
3. Variables to handle ACD data items
4. Call to `embInitP`
5. Calls to `ajAcdGet*` functions to retrieve objects for ACD data definitions
6. Code to reformat input files (if necessary)
7. Code to construct and call the **HMMER** command line
8. Code to reformat output files (if necessary)
9. Code to clean up the ACD variables.

10.4.3.1 Application header

The application header documentation (Section D.3, 'Application code documentation') was pasted in from another EMBOSS application. Then an empty `main()` function and variables to handle ACD data items were added. A call to `embInitP` was added to process the ACD file and the `ajAcdGet*` functions used to retrieve ACD values.

10.4.3.2 File reformatting and housekeeping

Code was added, where necessary and possible, to reformat the input files by using temporary files. Code to reformat the output files, again by using temporary files, was again added where necessary. Finally code to clean up memory for the ACD variables was added.

10.4.3.3 Command line generation

The hardest part of the code was to construct the call to the **HMMER** command line, but this is quite trivial once all the options are properly understood. A few tricky issues arose in generating the command line and you'll see these in the code later. These were documented in the code to save others time in the future. You should always document such tricky steps in your own code.

10.4.4 Quality assurance testing

Quality assurance tests (Section 7.2, 'Application quality assurance') were written for the applications. These were based on the examples in the tutorial which use files from the **HMMER** distribution. In cases where test data are not already available then these would have to be collected.

10.4.5 Documentation

The documentation consisted of a `README` file for the package distribution and the standard EMBASSY package and application documentation.

A `README` file for the package distribution was written to cover all the basics and included the following topics:

- How to download the original and EMBASSY versions of **HMMER**
- Where to get installation instructions and documentation
- Requirements, caveats, etc.
- Differences in the application between the two versions (see below).

A note was made for each application to describe:

- Which **HMMER** options are supported as ACD qualifiers
- Any new qualifiers and parameters in the EMBOSS version
- If the order of parameters was changed.

Formal documentation for the package was then generated following the guidelines (Section 8.2, 'Application documentation'). For many of the sections in the application documentation, text could be pasted in directly from the original documentation. No new documentation, other than the `README` file already described, was written. Once the text was inserted the EMBOSS-provided scripts were used to generate full documentation files automatically.

10.4.6 Integration

The last steps were to commit the new package code to the EMBOSS CVS server (see Section 1.5, 'Contributing software to EMBOSS') and update the EMBOSS ftp and web sites. Such integration issues are handled by the EMBOSS developers.

10.5 HMMER wrapper: overview

The EMBASSY **HMMER** package of wrappers to the original **HMMER** v2.3.2 applications is summarised below.

- **HMMER** v2.3.2 must be installed on the same system as EMBOSS and the location of the **HMMER** executables must be defined in your path for EMBASSY **HMMER** to work.
- Where possible the same command line qualifier names and parameter order are used as in the original **HMMER**. There are however several unavoidable differences and these are clearly documented.
- More or less all options documented as 'expert' in the original **HMMER** user guide are given in ACD as 'advanced' options. **`-options`** must be specified on the command line to be prompted for a value for them.
- The original **HMMER** uses the BLAST environment variables, if defined, to locate files whereas the EMBASSY **HMMER** does not.
- Input and output of alignments and sequences is limited to the formats that the original **HMMER** supports. It would be fairly straightforward to adapt the code to support all EMBOSS-supported formats.
- Automatic processing of **gzip** files is not supported.
- **HMMER** v2.3.2 and therefore EMBASSY **HMMER** is only recommended for use with protein sequences. If you provide a non-protein sequence you will be reprompted for a protein sequence – this is achieved by setting the `type:` attribute of sequence inputs to `protein` in the ACD file. To accept nucleic acid sequences you'll need to replace `protein` with `nucleic`.

10.6 HMMER wrapper: hmmbuild

The basic function of **hmmbuild** is to read a multiple sequence alignment file, build a new profile HMM and save the HMM to file. It is called as follows:

```
ehmmbuild [options] alignfile hmmfile
```

By default the model is configured to find one or more non-overlapping alignments that are global with respect to the model and local with respect to the sequence. Various other alignment strategies can be set by using the appropriate option.

One limitation is that the user must provide the full filename of an alignment for the `alignfile` option and not an indirect reference to a set of sequences, so a USA (see the *EMBOSS User's Guide*) is *not* acceptable. This is because **hmmbuild** (which **ehmmbuild** wraps) requires an alignment and does not support USAs.

Differences between the wrapper and the original software are as follows:

- Note that the `alignfile` parameter (used for input) and the `hmmfile` parameter (used for output) are specified in the reverse order in the original **HMMER**. This is because input parameters are always specified before output parameters in EMBOSS.
- Several of the original **HMMER** options are not supported. **`-help`** is used instead of the **`-h`** **HMMER** option, and the three **HMMER** options for setting the alignment strategy are replaced with the **`-strategy`** qualifier which is a list in ACD.
- An option is not needed to specify whether or not to append to the output file as this can be set in the ACD file by using the `append:` attribute. Arguably this should be a supported option.
- The **`-F`** option, to force an existing HMM file to be overwritten, is always set because an application should always generate output if possible and EMBOSS trusts its users to specify files correctly.
- The **`-amino`** and **`-nucleic`** options are not needed because they are set in the ACD file by using the `type:` attribute.
- **`-informat`** is not supported because it would confuse the user, who should not have to specify the format of input files.
- The six options for specifying the sequence weighting algorithm are all handled by a single **`-weighting`** qualifier defined as a list in the ACD file. This is purely for convenience.
- Finally, **`-verbosity`** is used instead of **`-verbose`**.

10.6.1 HMMER wrapper: `ehmmbuild.acd`

10.6.1.1 Application definition and inputs

The start of the ACD file is shown below. Text for the `help:` attribute is not shown but is given in the ACD files.

You can see that the `alignfile` option is handled by a `seqset` sequence input type. As mentioned before, this has to be an alignment file and not a USA referring indirectly to a set of sequences. This limitation could be overcome by first translating the USA into a local file, but this wasn't done for this version of the wrapper:

```
application: ehmmbuild
[
# EMBOSS wrapper to hmmbuild from Sean Eddy's HMMER package
# v.2.3.2
    documentation: "Build a profile HMM from an alignment."
    groups: "HMM"
    gui: "yes"
    batch: "yes"
    cpu: "medium"
    embassy: "hmmernew"
```

```
]

section: input
[
    information: "Input section"
    type: "page"
]

seqset: alignfile
[
# User must provide the full filename of an alignment, not an
# indirect reference to a set of sequences, e.g. a USA is NOT
# acceptable.
    parameter: "Y"
    type: "gapstopprotein"
    aligned: "Y"
]

...
```

There are then three `infile` ACD definitions to handle various **HMMER** input files. All of these are advanced ACD options not normally set by the user. Note that a `NULL` default value is set for them which requires the `nullok:` attribute to be specified and set to `True`.

```
infile: prior
[
#   Advanced input file
    information: "Dirichlet prior file."
    knowntype: "dirichlet prior"
    default: ""
    nullok: "Y"
]

infile: null
[
#   Advanced input file
    information: "NULL model file"
    knowntype: "hmmer null model"
    default: ""
    nullok: "Y"
]

infile: pam
[
#   Advanced input file
    information: "PAM file"
    knowntype: "hmmer matrix file"
    default: ""
    nullok: "Y"
]

float: pamwgt
[
    default: "20.0"
    information: "Weighting for PAM."
]

endsection: input
```

10.6.1.2 Required section

The required section is shown below. As you can see it is quite sparse. It contains a string to specify the name of the HMM and a list which is used to set the alignment strategy. The list replaces three individual **HMMER** options. There is also a default setting so this list has four entries in total:

```
section: required
[
    information: "Required section"
    type: "page"
 ]

string: n
[
    standard: "Y"
    default: ""
    information: "Name for this HMM."
    word: "Y"
    knowntype: "name"
 ]

list: strategy
[
    standard: "Y"
    default: "D"
    minimum: "1"
    maximum: "1"
    values: "D:global-multidomain,F:local-multidomain,G:global-singledomain,S:
local-singledomain"
    delimiter: ","
    codedelimiter: ":"
    header: "Alignment preference"
    information: "Select preference"
    button: "Y"
]

endsection: required
```

10.6.1.3 Advanced section

The bulk of the **HMMER** options are defined as 'expert' options in the original **HMMER** documentation and so are given in the advanced section of the ACD file. These options are not normally set by the user and a default value, taken from the **HMMER** documentation, is given:

```
section: advanced
[
    information: "Advanced section"
    type: "page"
 ]

integer: pbswitch
[
    default: "1000"
```

```
    information: "Threshold to switch to position-based weights."
  ]

float: archpri
[
    default: "0.85"
    information: "Architecture prior"
  ]

boolean: binary
[
    default: "N"
    information: "Write HMM as binary."
  ]

boolean: fast
[
    default: "N"
    information: "Work in fast mode"
  ]

float: gapmax
[
    default: "0.5"
    information: "Fast mode control"
  ]

boolean: hand
[
    default: "N"
    information: "Specify model by hand."
  ]

float: sidlevel
[
    default: "0.62"
    information: "Cutoff ID threhold"
  ]
```

The sequence weighting algorithm is also specified as an advanced ACD qualifier. This one list replaces the six command line options given in the original **HMMER**:

```
boolean: noeff
[
    default: "N"
    information: "Turn off the effective sequence number calculation."
  ]

float: swentry
[
    default: "0.5"
    information: "Probability control for local entries"
  ]

float: swexit
[
    default: "0.5"
    information: "Probability control for exits"
  ]
```

```
boolean: verbosity
[
    default: "N"
    information: "Verbosity."
 ]

list: weighting
[
    default: "G"
    minimum: "1"
    maximum: "1"
    values: "B:Blosum, G:Gerstein/Sonnhammer/Chothia, K:Krogh/Mitchison, W:
Henikoff, V:Sibbald/Argos Voronoi, N:None"
    delimiter: ","
    codedelimiter: ":"
    header: "Weighting method"
    information: "Select weighting"
    button: "Y"
]

endsection: advanced
```

10.6.1.4 Output section

The output section is shown here. This contains the new parameter defined for the HMM output file, which was written directly to `stdout`, and two other output files used by **HMMER**.

```
section: output:
[
    information: "Output section"
    type: "page"
 ]

outfile: hmmfile
[
    parameter: "Y"
    knowntype: "hmm file"
    append: "Y"
 ]

outfile: o
[
    nullok: "Yes"
    nulldefault: "Yes"
    information: "Resave starting alignment."
    knowntype: "selex file"
 ]

outfile: cfile
[
    nullok: "Yes"
    nulldefault: "Yes"
    information: "Emission and transition count file"
    knowntype: "hmmer count file"
 ]

endsection: output
```

10.6.2 HMMER wrapper: `ehmmbuild.c`

10.6.2.1 Header documentation

The start of the file of C source code is shown below. This just shows the standard documentation that should be given for any EMBOSS application. There is also a line (`#include emboss.h`) to import the AJAX and NUCLEUS library interfaces:

```
/* @source ehmmbuild application
**
** EMBOSS wrapper to hmmbuild from Sean Eddy's HMMER package v.2.3.2
** Build a profle HMM from an alignment.
**
** @author Copyright (C) Jon Ison
** @@
**
** This program is free software; you can redistribute it and/or
** modify it under the terms of the GNU General Public License
** as published by the Free Software Foundation; either version 2
** of the License, or (at your option) any later version.
**
** This program is distributed in the hope that it will be useful,
** but WITHOUT ANY WARRANTY; without even the implied warranty of
** MERCHANTABILITY or FITNESS FOR A PARTICULAR PURPOSE. See the
** GNU General Public License for more details.
**
** You should have received a copy of the GNU General Public License
** along with this program; if not, write to the Free Software
** Foundation, Inc., 59 Temple Place - Suite 330, Boston, MA 02111-1307, USA.
******************************************************************************/

#include "emboss.h"
```

10.6.2.2 `main()` function

This shows the `main()` function and the variable declarations. All the variables for handling the ACD data items have the same name as the corresponding qualifier. This style is not enforced but it is recommended because it makes reading the source code much easier.

Housekeeping variables are given sensible names. All variables are initialised to `NULL` or zero. It is good practice to do that, in fact dangerous not to, because some parts of the EMBOSS libraries assume that non-`NULL` pointers have had memory assigned to them; if there is a junk value assigned to them at run-time then you may be heading for a segmentation fault.

```
/* @prog ehmmbuild ***********************************************************
**
** EMBOSS wrapper to hmmbuild from Sean Eddy's HMMER package v.2.3.2
** Build a profle HMM from an alignment.
**
******************************************************************************/
int main(int argc, char **argv)
{
    /* ACD data item variables */
```

```
    AjPSeqset alignfile  = NULL;
    AjPFile       prior = NULL;
    AjPFile        null = NULL;
    AjPFile         pam = NULL;
    float       pamwgt  = 0.0;
    AjPStr            n = NULL;
    AjPStr    *strategy = NULL;
    ajint      pbswitch = 0;
    float      archpri  = 0.0;
    AjBool       binary = ajFalse;
    AjBool         fast = ajFalse;
    float       gapmax  = 0.0;
    AjBool         hand = ajFalse;
    float      idlevel  = 0.0;
    AjBool        noeff = ajFalse;
    float      swentry  = 0.0;
    float       swexit  = 0.0;
    AjBool    verbosity = ajFalse;
    AjPStr   *weighting = NULL;
    AjPFile     hmmfile  = NULL;
    AjPFile           o = NULL;
    AjPFile       cfile  = NULL;

    /* Housekeeping variables */
    AjPStr          cmd = NULL;
    AjPStr         rnd1 = NULL;
    AjPStr         rnd2 = NULL;
    AjPStr          tmp = NULL;
    AjPStr          fmt = NULL;
    char          option;
    AjBool        fmtok = ajFalse;
    AjPStr  hmmfilename  = NULL;
```

10.6.2.3 Processing the ACD file

The code below shows the function calls for processing the ACD file. `embInitP` processes the ACD file and prompts the user for any required values that are not specified on the command line. The prefix `ajAcdGet` family of functions are used to retrieve values from the ACD data definitions and store them in the variables defined earlier:

```
  /* ACD file processing */
    embInitP("ehmmbuild",argc,argv,"HMMERNEW");

    alignfile  = ajAcdGetSeqset("alignfile");
    prior      = ajAcdGetInfile("prior");
    null       = ajAcdGetInfile("null");
    pam        = ajAcdGetInfile("pam");
    pamwgt     = ajAcdGetFloat("pamwgt");
    n          = ajAcdGetString("n");
    strategy   = ajAcdGetList("strategy");
    pbswitch   = ajAcdGetInt("pbswitch");
    archpri    = ajAcdGetFloat("archpri");
    binary     = ajAcdGetBoolean("binary");
    fast       = ajAcdGetBoolean("fast");
```

```
gapmax     = ajAcdGetFloat("gapmax");
hand       = ajAcdGetBoolean("hand");
idlevel    = ajAcdGetFloat("sidlevel");
noeff      = ajAcdGetBoolean("noeff");
swentry    = ajAcdGetFloat("swentry");
swexit     = ajAcdGetFloat("swexit");
verbosity = ajAcdGetBoolean("verbosity");
weighting = ajAcdGetList("weighting");
hmmfile    =  ajAcdGetOutfile("hmmfile");
o          = ajAcdGetOutfile("o");
cfile      =  ajAcdGetOutfile("cfile");
```

10.6.2.4 Housekeeping and file format handling

The start of the application code proper is shown below. First of all there is some housekeeping code. Then there is a block of code to check that the sequence alignment input file is in a format that **HMMER** can understand. An exception is raised if an unsupported format is specified. This could be replaced in the future with code to reformat the alignment file into an appropriate format. At the time of writing, it was not fully tested whether all alignment formats, including SELEX and Stockholm, could be interconverted without any loss of data or annotation, so the safe option was chosen:

```
/* MAIN APPLICATION CODE */
  /* 1. Housekeeping */
  cmd  = ajStrNew();
  rnd1 = ajStrNew();
  rnd2 = ajStrNew();
  tmp  = ajStrNew();
  fmt  = ajStrNew();
  hmmfilename = ajStrNew();


  ajStrAssignC(&hmmfilename, ajFileGetNameC(hmmfile));
  /* 2. Ensure alignfile is in format HMMER can understand. These include
     FASTA, GENBANK,EMBL, GCG, PIR, STOCKHOLM, SELEX, MSF,CLUSTAL and PHYLIP.
     EMBOSS name definitions are taken from seqInFormatDef in ajseqread.c and
     seqOutFormat in ajseqwrite.c */
  fmtok=ajFalse;
  ajStrAssignS(&fmt, ajSeqsetGetFormat(alignfile));
  if(ajStrMatchC(fmt, "fasta")     ||
     ajStrMatchC(fmt, "genbank")   ||
     ajStrMatchC(fmt, "embl")      ||
     ajStrMatchC(fmt, "gcg")       ||
     ajStrMatchC(fmt, "pir")       ||
     ajStrMatchC(fmt, "stockholm") ||
     ajStrMatchC(fmt, "selex")     ||
     ajStrMatchC(fmt, "msf")       ||
     ajStrMatchC(fmt, "clustal")   ||
     ajStrMatchC(fmt, "phylip"))
     fmtok = ajTrue;
  /* This could be replaced with code to reformat the file. */
  if(!fmtok)
      ajFatal("Input alignment ('alignfile' ACD option) is not in a format "
      "HMMER understands. Please use a file in FASTA, GENBANK,"
      "EMBL, GCG, PIR, STOCKHOLM, SELEX, MSF,CLUSTAL or PHYLIP format.");
```

10.6.2.5 Building the command line

The first part of the code for building the command line is shown below. The command line is constructed in a specific order to make updating the wrapper for new releases easier. First the application name is pasted into a string, then the original **HMMER** options are given in the order they appear in the ACD file. Next the **HMMER** options that do not have any parallel in the ACD file are given. Finally, new parameters and options that are specific to the EMBASSY wrapper are given:

```
/* 3. Build hmmbuild command line */
   /* Command line is built in this order:
      i. Application name.
      ii. HMMER 'options' (in order they appear in ACD file)
      iii.HMMER 'options' (that don't appear in ACD file)
      iv. HMMER and new parameters.
      */
   ajFmtPrintS(&cmd, "hmmbuild ");

   if(prior)
       ajFmtPrintAppS(&cmd, " --prior %s ", ajFileGetNameC(prior));

   if(null)
       ajFmtPrintS(&cmd, " --null %s ", ajFileGetNameC(null));

   if(pam)
       ajFmtPrintAppS(&cmd, " --pam %s --pamwgt %f ",
                       ajFileGetNameC(pam), pamwgt);

   ajFmtPrintAppS(&cmd, " -n %S ", n);

   /* ACD option only allows one selection */
   option = ajStrGetCharFirst(strategy[0]);

   if(option == 'F')
       ajStrAppendC(&cmd, " -f ");
   else if(option == 'G')
       ajStrAppendC(&cmd, " -g ");
   else if(option == 'S')
       ajStrAppendC(&cmd, " -s ");

   /* else go with default ('D' option in ACD file) */
   ajFmtPrintAppS(&cmd, " --pbswitch %d ", pbswitch);
   ajFmtPrintAppS(&cmd, " --archpri %f ", archpri);

   if(binary)
       ajStrAppendC(&cmd, " --binary ");

   if(fast)
       ajFmtPrintAppS(&cmd, " --fast --gapmax %f ", gapmax);

   if(hand)
       ajStrAppendC(&cmd, " --hand ");

   ajFmtPrintAppS(&cmd, " --idlevel %f ", idlevel);

   if(noeff)
       ajStrAppendC(&cmd, " --noeff ");

   ajFmtPrintAppS(&cmd, " --swentry %f ", swentry);
```

```
    ajFmtPrintAppS(&cmd, " --swexit %f ", swexit);

    if(verbosity)
        ajStrAppendC(&cmd, " --verbose ");
```

The rest of the code for building the command line is below. The only thing to point out is that the **append** option is always set. This means that output should always be appended to whatever is given in the specified output file. EMBOSS clears its output files by default though, so for this to work the `append:` attribute of the `hmmfile` ACD data item must be set to `True`:

```
/* ACD option only allows one selection */
    option = ajStrGetCharFirst(weighting[0]);

    if(option == 'B')
         ajStrAppendC(&cmd, " --wblosum ");
    else if(option == 'G')
         ajStrAppendC(&cmd, " --wgsc ");
    else if(option == 'K')
         ajStrAppendC(&cmd, " --wme ");
    else if(option == 'W')
         ajStrAppendC(&cmd, " --wpb ");
    else if(option == 'V')
         ajStrAppendC(&cmd, " --wvoronoi ");
    else if(option == 'N')
         ajStrAppendC(&cmd, " --wnone ");

    if(o)
         ajFmtPrintAppS(&cmd, " -o %s ", ajFileGetNameC(o));

    if(cfile)
         ajFmtPrintAppS(&cmd, " --cfile %s ", ajFileGetNameC(cfile));

    /* -A (append) always set but file will be wiped by EMBOSS first unless
    ** append: "Y" is set for "hmmfile" in the ACD file.
    */
    ajStrAppendC(&cmd, " -A -F ");
    ajFmtPrintAppS(&cmd, " %S %S", hmmfilename, ajSeqsetGetFilename(alignfile));
```

10.6.2.6 Invoking the application

The code below shows the system call to invoke the **hmmbuild** application using the command line just constructed. Note that `system()` is used here but that should probably be replaced with a call to `exec()` for reasons explained earlier. There is also some housekeeping code for memory management to ensure that the application can close cleanly:

```
/* 4. Close ACD files */
    ajSeqsetDel(&alignfile);
    ajFileClose(&prior);
    ajFileClose(&null);
    ajFileClose(&pam);
    ajFileClose(&hmmfile);
    ajFileClose(&o);
    ajFileClose(&cfile);

    /* 5. Call hmmbuild */
    ajFmtPrint("\n%S\n", cmd);
```

```
    system(ajStrGetPtr(cmd));
    /* 6. Exit cleanly */
    ajStrDel(&n);
    ajStrDel(&cmd);
    ajStrDel(&rnd1);
    ajStrDel(&rnd2);
    ajStrDel(&tmp);
    ajStrDel(&fmt);
    ajStrDel(&hmmfilename);

    embExit();

    return 0;
}
```

10.7 HMMER wrapper: hmmalign

We'll now look at another **HMMER** application, **hmmalign**. Its basic function is to read an HMM profile and a set of sequences, align the sequences to the profile and output a multiple sequence alignment. It is called as follows:

```
ehmmalign [options] hmmfile seqfile outfile
```

The set of sequences may be unaligned or aligned. If aligned the existing alignment is ignored and **hmmalign** will align them in the way it wants.

hmmalign is covered in basic detail because nearly everything that's been said about **hmmbuild** applies to all the other **HMMER** applications.

The **`-outfile`** parameter is new to EMBASSY **HMMER**. The multiple sequence alignment is always written to `outfile` rather than to `stdout`.

In contrast to **hmmbuild** the user may specify a USA for sequence input. This is because any alignment is ignored by **HMMER**, therefore the wrapper can treat the file as unaligned sequences which can be converted if necessary into a format that will be understood by **HMMER**. The application will make a temporary local copy of its input sequence data. It's down to the user to ensure that there's enough disk space in the directory it's run in.

A few of the original **HMMER** options are not supported. Again **`-h`** is redundant. **`-informat`**, **`-oneline`** and **`-outformat`** were provided for the user to specify the format of the input sequence file and the output alignment. None are needed in the wrapper.

More or less any sequence format will be understood, whereas the alignment format can be specified in the ACD file or by using the inbuilt **`-aformat`** command line qualifier.

10.7.1 HMMER wrapper: `hmmalign.acd`

The ACD file is very simple. It only contains an input and output section.

10.7.1.1 Input section

An excerpt from the input section is shown here. Note that an `infile` is used for the HMM file, whereas a `seqset` is used for sequence input. As mentioned before, all sequence formats that EMBOSS normally supports are fully supported.

```
application: ehmmalign
[
# EMBOSS wrapper to hmmalign from Sean Eddy's HMMER package
# v.2.3.2
    documentation: "Align sequences to an HMM profile"
    groups: "HMM"
    gui: "yes"
    batch: "yes"
    cpu: "medium"
    embassy: "hmmernew"
]

section: input
[
    information: "Input section"
    type: "page"
]

infile: hmmfile
[
    parameter: "Y"
    information: "HMM file"
    knowntype: "hmm file"
    help: "File containing a HMM profile"
]

seqset: seqfile
[
    parameter: "Y"
    type: "gapstopprotein"
    help: "File containing a (set of) sequence(s)"
    aligned: "N"
]
...
endsection: input
```

10.7.1.2 Output section

The output section is shown here. The only things to point out are that the output file is handled by an `align` data item and that the format of the alignment is set by the `aformat:` attribute:

```
section: output
[
    information: "Output section"
    type: "page"
]

align: o
[
    parameter: "Y"
    help: "Multiple sequence alignment output file."
    aformat: "fasta"
]

boolean: m
[
    additional: "Y"
    default: "N"
    information: "Only show match state alignment symbols."
]
```

```
boolean: q
[
    additional: "Y"
    default: "N"
    information: "Suppress all output except the alignment."
 ]

endsection: output
```

10.7.2 HMMER wrapper: `ehmmalign.c`

10.7.2.1 Documentation header

The start of the C source code is shown here. The documentation is just the same as it was for **hmmbuild**.

```
/* @source ehmmalign application
**
** EMBOSS wrapper to hmmalign from Sean Eddy's HMMER package v.2.3.2
** Align sequences to an HMM profile.
**
** @author Copyright (C) Jon Ison
** @@
**
** This program is free software; you can redistribute it and/or
** modify it under the terms of the GNU General Public License
** as published by the Free Software Foundation; either version 2
** of the License, or (at your option) any later version.
**
** This program is distributed in the hope that it will be useful,
** but WITHOUT ANY WARRANTY; without even the implied warranty of
** MERCHANTABILITY or FITNESS FOR A PARTICULAR PURPOSE. See the
** GNU General Public License for more details.
**
** You should have received a copy of the GNU General Public License
** along with this program; if not, write to the Free Software
** Foundation, Inc., 59 Temple Place - Suite 330, Boston, MA 02111-1307, USA.
******************************************************************************/

#include "emboss.h"

 /* @prog ehmmalign ************************************************************
**
** EMBOSS wrapper to hmmalign from Sean Eddy's HMMER package v.2.3.2
** Align sequences to an HMM profile.
**
******************************************************************************/
```

10.7.2.2 `main()` function, housekeeping and ACD file processing

This shows the `main()` function, the variable declarations, the code to process the ACD file and some housekeeping code.

```
int main(int argc, char **argv)
{
    /* ACD data item variables */
    AjPFile   hmmfile   = NULL;
    AjPSeqset seqfile   = NULL;
    AjPFile    mapali   = NULL;
    AjPFile   withali   = NULL;
    AjPAlign        o   = NULL;
    AjBool          m   = ajFalse;
    AjBool          q   = ajFalse;

    /* Housekeeping variables */
    AjPStr         cmd = NULL;
    AjPStr         tmp = NULL;
    AjPStr         fmt = NULL;
    AjBool       fmtok = ajFalse;
    AjPStr         rnd = NULL;
    AjPSeqout     rndo = NULL;


    /* ACD file processing */
    embInitP("ehmmalign",argc,argv,"HMMERNEW");

    hmmfile  = ajAcdGetInfile("hmmfile");
    seqfile  = ajAcdGetSeqset("seqfile");
    mapali   = ajAcdGetInfile("mapali");
    withali  = ajAcdGetInfile("withali");
    o        = ajAcdGetAlign("o");
    m        = ajAcdGetBoolean("m");
    q        = ajAcdGetBoolean("q");

    /* MAIN APPLICATION CODE */
    /* 1. Housekeeping */
    cmd  = ajStrNew();
    tmp  = ajStrNew();
    fmt  = ajStrNew();
    rnd  = ajStrNew();
```

10.7.2.3 File format handling

This shows the code required to reformat the input file into a format suitable for **HMMER**, i.e. FASTA. You have to do this because **hmmalign** only understands FASTA format on input, and besides the sequence may have been specified by a USA which will need transforming into a file of sequences.

`ajFilenameSetTempname` is used to set an available random filename, and then `ajSeqoutOpenFilename` to initialise a `seqout` object with that filename.

The output format is set by using `ajSeqoutSetFormatS`. Sequences are written by using `ajSeqoutWriteSeq`. Finally the file is closed by using `ajSeqoutClose` and the `seqout` object is deleted:

```
/* 2. Re-write seqfile to a temporary file in a format (FASTA) HMMER can understand.
      We cannot just pass the name of seqfile to HMMER as the name provided might be
a     USA which HMMER would not understand. */
   rnd = ajStrNew();
   ajFilenameSetTempname(&rnd);
   rndo = ajSeqoutNew();
```

```
    if(!ajSeqoutOpenFilename(rndo, rnd))
        ajFatal("Failed to open file '%S'", rnd);
    ajSeqoutSetFormatC(rndo, "fasta");
    ajSeqoutWriteSet(rndo, seqfile);
    ajSeqoutClose(rndo);
    ajSeqoutDel(&rndo);
```

10.7.2.4 Building the command line

Here's the code for building the command line. Once again the command line is built in a particular order to make maintenance easier in the future.

The thing to point out here is that EMBOSS supports certain alignment formats that the original **HMMER** does not, and **HMMER** supports certain formats that EMBOSS doesn't (or didn't at the time of writing).

If the user-specified format is not supported then an exception is raised and the format is set to Stockholm. In the future this could be replaced by code to reformat the output file as appropriate.

```
/* 3. Build hmmalign command line */
    /* Command line is built in this order:
       i. Application name.
       ii. HMMER 'options' (in order they appear in ACD file)
       iii.HMMER 'options' (that don't appear in ACD file)
       iv. HMMER and new parameters.
       */
    ajFmtPrintS(&cmd, "hmmalign ");

    if(mapali)
        ajFmtPrintAppS(&cmd, " --mapali %s ", ajFileGetNameC(mapali));

    if(withali)
        ajFmtPrintAppS(&cmd, " --withali %s ", ajFileGetNameC(withali));

    if(m)
        ajStrAppendC(&cmd, " -m ");

    if(q)
        ajStrAppendC(&cmd, " -q ");

    /* Ensure output alignment is in user-specified format. */
    fmtok=ajTrue;
    ajStrAssignS(&fmt, ajAlignGetFormat(o));
    /* fasta and a2m are identical formats. */

    if(ajStrMatchC(fmt, "fasta"))
        ajStrAssignC(&fmt, "A2M");
    else if(ajStrMatchC(fmt, "a2m"))
        ajStrAssignC(&fmt, "A2M");
    else if(ajStrMatchC(fmt, "msf"))
        ajStrAssignC(&fmt, "MSF");
    else if(ajStrMatchC(fmt, "phylip"))
        ajStrAssignC(&fmt, "PHYLIP");
    /* hmmer also supports stockholm, SELEX and Clustal output, EMBOSS does not.
    ** EMBOSS supports unknown/multiple/simple and srs output, hmmer does not.
    */
    else
```

```
        fmtok = ajFalse;
    if(!fmtok)
    {
        /* This could be replaced with code to reformat the file. */
        ajWarn("Specified output alignment format ('o' ACD option) is "
                "not understood by HMMER. Using stockholm format instead.");
        ajStrAssignC(&fmt, "Stockholm");
    }
```

10.7.2.5 Invoking the application

This shows the code for calling the **hmmalign** application. Again the call to `system()` should probably be replaced by one to `exec()`.

You can see that a temporary variable called `rnd` is used for the name of the rewritten sequence input file. The FASTA format has to be specified explicitly by using the **`-informat`** option.

```
/* rnd is the name of the rewritten seqfile. MUST specify FASTA format explicitly. */
    ajFmtPrintAppS(&cmd, " --informat FASTA --outformat %S -o %s %s %S",
                   fmt,
                   ajAlignGetFilename(o),
                   ajFileGetNameC(hmmfile),
                   rnd);

    /* 4. Close ACD files */
    ajFileClose(&hmmfile);
    ajSeqsetDel(&seqfile);
    ajFileClose(&mapali);
    ajFileClose(&withali);
    ajAlignDel(&o);

    /* 5. Call hmmalign */
    ajFmtPrint("\n%S\n\n", cmd);
    system(ajStrGetPtr(cmd));
    /* 6. Exit cleanly */
    ajFmtPrintS(&tmp, "rm %S", rnd);
    system(ajStrGetPtr(tmp));

    ajStrDel(&cmd);
    ajStrDel(&tmp);
    ajStrDel(&fmt);
    ajStrDel(&rnd);

    embExit();

    return 0;
}
```

10.8 Reducing the length of the command line

A problem can arise when wrapping applications that have many possible command line qualifiers. An example of this is the **MIRA** package which has of the order of a hundred. For a package like this one cannot sensibly take the approach of constructing a command line with all the possible qualifiers being specified; the command line might be too long for the shell and it'd certainly be hard to read and debug.

Two approaches are taken by the **MIRA** wrapper to avoid command line clutter. The first is to make use of the `ajAcdIsUserdefinedC` library function. This is demonstrated by the code snippet below:

```
...
AjPStr cl = NULL;
AjPStr squal = NULL;

...

if(ajAcdIsUserdefinedC("genome"))
{
     squal = ajAcdGetListSingle("genome");
     ajFmtPrintAppS(&cl," -genome%S",squal);
     ajStrDel(&squal);
}
```

The `ajAcdIsUserdefinedC` function call checks whether a user has typed anything in response to the named ACD qualifier name or has specified a value for that qualifier on the command line. If either case is true then the code above retrieves the value from ACD and adds the qualifier and value to the command line. If the user has not provided a value then the (correct) assumption is that the **MIRA** program will use a default value and so the command line is left unchanged.

Now that approach works quite nicely but there is a drawback. The `ajAcdIsUserdefinedC` call will return a true value if the user has typed anything; that includes typing the default value held in ACD for a given qualifier. It would be preferable if the command line was kept clear for unnecessarily specified default values. That is what the rest of the code, the second approach, in the **MIRA** wrapper does. It is a useful exercise to study the code.

The code is very slightly complicated by the fact that **MIRA** allows a user to optionally supply a prefix which can be attached to a command line qualifier. For example, the qualifier **`-project`** can also be specified as **`-GE:project`** as it belongs to a 'GEneral' class of input options. The wrapper therefore first lists all the qualifiers and their prefixes in a static array. It then loads them all into an `AjPTable` lookup table as one of the first jobs within `main()` (the code is trivial and therefore not shown here)

```
typedef struct MiraSPrefix
{
    const char* qname;
    const char* prefix;
} MiraOPrefix;
#define MiraPPrefix MiraOPrefix*

static MiraOPrefix miraprefix[] = {
    {"project", "GE:"},
    {"lj", "GE:"},
    {"fo", "GE:"},
    {"mxti", "GE:"},
    {"rns", "GE:"},
    {"eq", "GE:"},
    {"eqo", "GE:"},
    {"droeqe", "GE:"},
```

```
    {"uti", "GE:"},
    {"ess", "GE:"},
    {"ps", "GE:"},
    ...
    {NULL, NULL}
};
```

Having explained that complication the strategy used by the wrapper can now be described. The wrapper programmer looks through the ACD file and, for the C code, clusters all the different datatypes into groups. In other words the wrapper will deal with all the string datatypes as one code section, then all the input file datatypes, etc. Taking the string datatype as an example the wrapper defines these in another static block.

```
typedef struct MiraSQuals
{
    const char* qname;
    const char* mname;
    const char* def;
} MiraOQuals;
#define MiraPQuals MiraOQuals*

static MiraOQuals mirastrings[] =
{
    {"project", "project", "mira"},
    {"bsn", "bsn", ""},
    {"np", "np", "mira"},
    {"gapfda", "gap4da", "gap4da"},
    {"log", "log", "miralog"},
    {"co", "co", "mira_out.caf"},
    ...
    {NULL, NULL, NULL}
};
```

The `MiraSQuals` structure is common to defining all the other datatype name blocks as well as the strings. For each definition block the variable `qname` is the qualifier name specified in ACD, the variable `mname` is the name to be printed out on the command line (as you can see they're different for `gapfdna`) and `def` is the default value for the qualifier as specified in the ACD file. Therein lies the slight drawback to this method, i.e. you have the specify the default value both within the ACD file and in the C code. You also obviously should make sure that they match, although no great harm will arise if you don't – you'd just get an unexpected qualifier appearing on the command line. The advantages outweigh the disadvantage in this case. All that then needs to be done is to write a simple function to handle each of the datatypes and to call that function from within `main()`. The function for handling strings is shown here:

```
static void emira_dostrings(AjPStr *cl, AjPTable table)
{
    ajuint i;
    AjPStr squal = NULL;
    AjPStr prefix = NULL;
    AjPStr key = NULL;
    AjPStr value = NULL;

    prefix = ajStrNew();
    key = ajStrNew();
```

```
    i = 0;
    while(mirastrings[i].qname)
    {
        squal = ajAcdGetString(mirastrings[i].qname);
        ajStrAssignC(&key,mirastrings[i].qname);
        ajStrAssignC(&prefix,"");

        value = ajTableFetch(table, key);

        if(value)
            ajStrAssignS(&prefix,value);

        if(!ajStrMatchC(squal,mirastrings[i].def))
            ajFmtPrintAppS(cl," -%S%s=%S",prefix,mirastrings[i].mname,squal);
        ajStrDel(&squal);
        ++i;
    }
    ajStrDel(&key);
    ajStrDel(&prefix);

    return;
}
```

The code looks sequentially through the static table of defined strings. First the code retrieves the associated value from ACD. It then performs a lookup in the qualifier prefix table for any associated prefix value and sets a variable accordingly (this need not be done for most packages so the code can be simplified). It then compares the value retrieved from ACD to the default value given in the static definition block and only adds the qualifier to the command line if the two values don't match.

The result of this code is a nice clean command line. The code also has the advantage that it avoids having to define a long unsightly block of `ajAcdGet*` function calls at the start of the wrapper. Though you may consider this approach to be over-engineered for wrapping packages with simpler interfaces it is nevertheless worth considering as the resulting wrapper is much easier to maintain.

10.9 How to port third-party applications

Most of the steps done in developing a wrapper also need to be done when porting software. An ACD file is still required. The main differences are in writing the source code. The basic steps are the same:

1. Implement the C source code
2. Header documentation
3. Empty `main()` function
4. Call to `embInitPV`
5. Variables to handle ACD data items
6. Calls to `ajAcdGet*` to retrieve ACD values
7. Code to clean up the ACD variables.

For a port there are significantly more things to consider. The `main()` function will need rewriting to handle the ACD file processing as follows:

1. Add code to pick up application options from ACD file processing rather than processing them as in the original
2. Identify functions that read input from user and update the function calls
3. Replace the output functions with appropriate calls to the EMBOSS libraries (possibly).

You must identify any functions that read input from the user, then add code to pick up values from the ACD file instead of processing them as in the original. The output can be left in whatever format is produced by the original application. You might add code to use the output functions provided by the EMBOSS libraries.

Clearly it's necessary to carefully familiarise yourself with the third-party source code. It might be necessary to make edits across multiple files. This is the main disadvantage to porting software.

The remaining sections illustrate the approach taken for the port **ohmmalign**.

10.10 HMMER port

10.10.1 HMMER port: `ohmmalign.acd`

The ACD file is essentially the same as in the wrapper apart from a few changes resulting from differences in options between the two releases. One thing to point out is in the output section. In the port the alignment output is handled by an `outfile` ACD data item. The `align` datatype that the wrapper used wasn't available at the time.

```
section: output
[
    information: "Output section"
    type: "page"
]

outfile: outfile
[
    parameter: "Y"
    knowntype: "hmmalign output"
]

endsection: output
```

10.10.2 HMMER port: `ohmmalign.c`

10.10.2.1 Heading code

This shows the top of the C source code. You can see straight away that there's an amalgamation of the original **HMMER** code and additions for EMBOSS. For example the EMBOSS interface is imported before the first include directive.

```
/************************************************************
 * HMMER - Biological sequence analysis with profile HMMs
 * Copyright (C) 1992-1998 Washington University School of Medicine
 * All Rights Reserved
```

```
 *
 *     This source code is distributed under the terms of the
 *     GNU General Public License. See the files COPYING and LICENSE
 *     for details.
 ***********************************************************/

/* hmmalign.c
 * SRE, Thu Dec 18 16:05:29 1997 [St. Louis]
 *
 * main() for aligning a set of sequences to an HMM.
 * RCS $Id: DevWrappingPorting.xml,v 1.30 2010/07/08 14:21:05 jonsourceforge Exp $
 * Modified for EMBOSS by Alan Bleasby (ISMB 2001)
 */

#include "emboss.h"
#include <stdio.h>
#include <stdlib.h>

#include "structs.h"            /* data structures, macros, #define's   */
#include "config.h"             /* compile-time configuration constants */
#include "funcs.h"              /* function declarations                 */
#include "globals.h"            /* alphabet global variables             */
#include "squid.h"              /* general sequence analysis library    */

#ifdef MEMDEBUG
#include "dbmalloc.h"
#endif


extern void emboss_rseqs(AjPSeqset seqset, char ***ret_rseqs,SQINFO **ret_sqinfo,int
*ret_num);

static void include_alignment(char *seqfile, struct plan7_s *hmm, int do_mapped,
char ***rseq, char ***dsq, SQINFO
**sqinfo, struct p7trace_s ***tr, int *nseq);
```

10.10.2.2 `main()` function variable declarations

The `main()` function and the variable declarations are shown here. So far as possible the original code is left unchanged, but it's necessary to add variables for AJAX-specific types to hold values to be retrieved from the ACD file. These are set to `NULL` as before.

```
int main(int argc, char **argv)
{
  const char      *hmmfile;       /* file to read HMMs from                  */
  HMMFILE         *hmmfp;         /* opened hmmfile for reading              */
  const char      *seqfile;       /* file to read target sequence from       */
  char           **rseq;          /* raw, unaligned sequences                */
  SQINFO         *sqinfo;         /* info associated with sequences          */
  char         **dsq;             /* digitized raw sequences                 */
  int            nseq;            /* number of sequences                     */
  char **aseq;                    /* aligned sequences                       */
  AINFO ainfo;                    /* alignment information                   */
  float *wgt;                     /* per-sequence weights                    */
  int i;
  struct plan7_s *hmm;            /* HMM to align to                         */
  struct p7trace_s **tr;          /* traces for aligned sequences            */
```

```
int be_quiet;                   /* TRUE to suppress verbose banner          */
int matchonly;                  /* TRUE to show only match state syms       */
const char *outfile;            /* optional alignment output file           */
FILE *ofp;                      /* handle on alignment output file          */
AjPFile ajwithali;               /* name of additional alignment file to align */
AjPFile ajmapali;                /* name of additional alignment file to map   */
AjBool ajmatch     = ajFalse;
AjPFile outf       = NULL;
AjPStr outfname    = NULL;
AjPFile inf        = NULL;
AjPStr infname     = NULL;
AjPSeqset seqset   = NULL;
AjPStr ajseqfile   = NULL;
char* mapali       = NULL;
char* withali      = NULL;
```

10.10.2.3 ACD file processing

This shows the code for retrieving the values from ACD. A few housekeeping variables are needed for that. `mapali` and `withali` are C type strings (`char *`) used to hold the names of the EMBOSS input files. The names are retrieved by a call to `ajFileGetNameS`. The files are then closed, leaving them ready for **HMMER**. A similar thing is done for the other **HMMER** input and output files:

```
 /**********************************************
  * Parse command line
  **********************************************/
matchonly = FALSE;
outfile   = NULL;
be_quiet  = FALSE;
withali   = NULL;
mapali    = NULL;

embInitP("ohmmalign",argc,argv,"HMMER");

ajmatch = ajAcdGetBoolean("matchonly");

if(ajmatch)
    matchonly=TRUE;
else
    matchonly=FALSE;

ajmapali = ajAcdGetInfile("mapalifile");

if (ajmapali)
    mapali = ajCharNewS(ajFileGetNameS(ajmapali));

ajFileClose(&ajmapali);
ajwithali = ajAcdGetInfile("withalifile");

if (ajwithali)
    withali = ajCharNewS(ajFileGetNameS(ajwithali));

ajFileClose(&ajwithali);

be_quiet=TRUE;
```

```
outf = ajAcdGetOutfile("outfile");
outfname = ajStrNewC((char *)ajFileGetNameC(outf));

if(*ajStrGetPtr(outfname)>31)
    ajFileClose(&outf);
outfile = ajStrGetPtr(outfname);

inf = ajAcdGetInfile("hmmfile");
infname = ajStrNewC((char *)ajFileGetNameC(inf));
ajFileClose(&inf);
hmmfile = ajStrGetPtr(infname);

seqset = ajAcdGetSeqset("sequences");
ajseqfile = ajStrNewC(ajStrGetPtr(seqset->Filename));
seqfile = ajStrGetPtr(ajseqfile);
```

10.10.2.4 Handling input data

The code for managing the data input is here. You can see that the housekeeping variables used for holding the ACD values are passed into the **HMMER** functions. The only thing to point out here is that exception handling is dealt with by calls to the EMBOSS function `ajFatal`. Clearly it's necessary to work through the original source code when writing a port:

```
/***********************************************
 * Open HMM file (might be in HMMERDB or current directory).
 * Read a single HMM from it.
 *
 * Currently hmmalign disallows the J state and
 * only allows one domain per sequence. To preserve
 * the S/W entry information, the J state is explicitly
 * disallowed, rather than calling a Plan7*Config() function.
 * this is a workaround in 2.1 for the 2.0.x "yo!" bug.
 ***********************************************/
 if ((hmmfp = HMMFileOpen(hmmfile, "HMMERDB")) == NULL)
   ajFatal("Failed to open HMM file %s\n", hmmfile);

 if (!HMMFileRead(hmmfp, &hmm))
   ajFatal("Failed to read any HMMs from %s\n", hmmfile);

 HMMFileClose(hmmfp);

 if (hmm == NULL)
   ajFatal("HMM file %s corrupt or in incorrect format? Parse failed", hmmfile);

 hmm->xt[XTE][MOVE] = 1.;      /* only 1 domain/sequence ("global" alignment)
 */hmm->xt[XTE][LOOP] = 0.;
 P7Logoddsify(hmm, TRUE);

 /* do we have the map we might need? */
 if (mapali != NULL && ! (hmm->flags & PLAN7_MAP))
   ajFatal("HMMER: HMM file %s has no map; you can't use --mapali.", hmmfile);

 /***********************************************
  * Open sequence file in current directory.
  * Read all seqs from it.
  ***********************************************/
```

```
/*
  if (! SeqfileFormat(seqfile, &format, NULL))
     switch (squid_errno) {
     case SQERR_NOFILE:
       ajFatal("Sequence file %s could not be opened for reading", seqfile);
     case SQERR_FORMAT:
     default:
       ajFatal("Failed to determine format of sequence file %s", seqfile);
  }

  if (! ReadMultipleRseqs(seqfile, format, &rseq, &sqinfo, &nseq))
     ajFatal("Failed to read any sequences from file %s", seqfile);
*/

 emboss_rseqs(seqset,&rseq,&sqinfo,&nseq);
```

10.10.2.5 Exiting cleanly

There's then lots of native code which isn't shown. Finally the program must exit with a call to `embExit()`:

```
if (outfile != NULL && (ofp = fopen(outfile, "w")) != NULL)
     {
       WriteSELEX(ofp, aseq, &ainfo, 50);
       printf("Alignment saved in file %s\n", outfile);
       fclose(ofp);
     }
 else
   WriteSELEX(stdout, aseq, &ainfo, 50);

 /**********************************************
   * Cleanup and exit
   **********************************************/
 for (i = 0; i < nseq; i++)
    {
      P7FreeTrace(tr[i]);
      FreeSequence(rseq[i], &(sqinfo[i]));
      free(dsq[i]);
    }
 FreeAlignment(aseq, &ainfo);
 FreePlan7(hmm);
 free(sqinfo);
 free(rseq);
 free(dsq);
 free(wgt);
 free(tr);

 SqdClean();

 ajStrDel(&outfname);
 ajStrDel(&infname);
 ajStrDel(&ajseqfile);

#ifdef MEMDEBUG
  current_size = malloc_inuse(&histid2);
  if (current_size != orig_size) malloc_list(2, histid1, histid2);
  else fprintf(stderr, "[No memory leaks.]\n");
#endif
```

```
    embExit();

    return 0;
}
```

10.11 Summary

To incorporate third-party applications into EMBOSS takes significant work, regardless of whether an application is wrapped or ported. This work is very worthwhile though because of benefits in terms of distribution (the software will reach more users) and, in some cases, because of lowered costs of training, support and maintenance. Moreover, any incorporated software can make use of inbuilt EMBOSS functionality such as support for common sequence formats and will be integrated into the many available EMBOSS interfaces. These include graphical user interfaces, workflow software and web services.

Fully ported software provides the most flexibility but this comes at a high relative maintenance cost. Nonetheless, this is the preferred route where the code is to be essentially adopted by EMBOSS. More or less the same flexibility can be achieved in a wrapper by the use of temporary files, although there are drawbacks and potential complications to this. For example, code for interconverting file formats is needed.

Appendix A ACD syntax reference

A.1 Introduction to ACD syntax

A.1.1 General syntax

The Ajax Command Definition (ACD) language was designed for writing ACD files for EMBOSS applications. The ACD syntax allows for very flexible descriptions of an application's parameters and its command line interface. It can specify everything that can appear on the command line or in another interface such as a web page.

ACD files are plain ASCII text files and *must* have the extension `.acd`. Typically they have the same name as the application, but this is not mandatory.

A.1.1.1 Whitespace

During ACD file parsing, the entire file contents are effectively treated as a single string which is parsed into tokens delimited by *space* characters. A single space between individual tokens is required: extraneous whitespaces are ignored.

A.1.1.2 Comments

Comment lines can be added and begin with "#" and continue to the end of the line.

A.1.2 ACD definitions

An ACD file contains a single *application definition* and a *data definition* for each parameter. The application definition is given first, followed by the data definitions. Data definitions are organised into sections (Section A.1.6, 'ACD file sections').

Application and data definitions have the following general form: a single text token followed by a colon `':'` (or `'='`) and a whitespace, followed by a second token. The definition body follows, which is one or more attributes delimited by a mandatory pair of square brackets `[ ]`, which can span multiple lines. Each attribute is a `name: value` pair with the attribute value given between quotes (`" "`):

Either:

```
token: token
{
   Attribute1Name: "Attribute1Value"
   Attribute2Name: "Attribute2Value"
}
```

Or:

```
token=token
[
   Attribute1Name: "Attribute1Value"
   Attribute2Name: "Attribute2Value"
]
```

The first token is either `application:` (for the application definition) or an AJAX datatype (e.g. `sequence`) for data definitions. The second token is either the name of the application (e.g. `seqret`) or the name of parameter (e.g. `asequence`).

Application definition:

```
application: ApplicationName
{
  ApplicationAttribute1Name: "ApplicationAttribute1Value"
  ApplicationAttribute2Name: "ApplicationAttribute2Value"
}
```

Data definition:

```
Datatype: ParameterName
{
   DataAttribute1Name: "DataAttribute1Value"
   DataAttribute2Name: "DataAttribute2Value"
}
```

The `application` token and tokens for the datatype and attribute names can be can be abbreviated up to the point where they are not ambiguous. Such abbreviations are *not* recommended however because they tend to make the ACD file more difficult to read.

Attribute values are normally enclosed in double quotes, although this is only mandatory for values (typically strings) which include whitespace.

A.1.2.1 Application definition

The application definition must be the first definition in the file:

```
application: ApplicationName
{
  ApplicationAttribute1Name: "ApplicationAttribute1Value"
  ApplicationAttribute2Name: "ApplicationAttribute2Value"
}
```

The application name is arbitrary but is typically the same as that used for the ACD file name. It is the ACD file name (not *`ApplicationName`*, if different) that's used from within the application C source code to associate it with an ACD file. This allows multiple ACD files (and therefore command line interfaces) to be developed for a single file of application C source code.

For complete description of the available application attributes see Section A.3, 'Application attributes'.

A.1.2.2 Data definition

All application parameters must have a data definition. Data definitions follow the application definition and must be placed in an appropriate file section (Section A.1.6, 'ACD file sections'):

```
Datatype: ParameterName
{
  DataAttribute1Name: "DataAttribute1Value"
  DataAttribute2Name: "DataAttribute2Value"
}
```

`Datatype` must be a valid ACD datatype (see Section A.2, 'Datatypes').

`ParameterName` is the name of the parameter. It is a string that must conform to certain conventions (Section A.1, 'Introduction to ACD syntax'). This name is used to refer to the data definition from the command line and from within the C source code (see Section 6.3, 'Handling ACD files').

For a complete description of the available attributes see: Section A.4 ('Global attributes'), Section A.5 ('Datatype-specific attributes') and Section A.6 ('Calculated attributes').

A.1.3 Parameter naming conventions

A.1.3.1 General conventions

Parameter and qualifier names:

- Must not contain whitespace characters
- Should not normally be single characters
- Should be meaningful words and indicate the function of the option so far as possible
- Are not case-sensitive.

A.1.3.2 Datatype-specific conventions

Some conventions for parameter names apply to individual datatypes. Where more than one instance of a datatype is specified in an ACD file, then the character `a`, `b`, etc. can be appended to the flag: `asequence`, `bsequence`, etc. This is indicated in the table by an asterisk in the parameter name, for example `*sequence`.

Datatype	Name	Usage
sequence	sequence, *sequence	Primary input sequence, generally required
seqall	sequence, *sequence, seqall	Primary input sequence database, generally required

seqset	sequence, *sequence, sequences	Primary input sequences, generally required
seqsetall	sequence, *sequence, sequences	Primary input sequences, generally required
seqout, seqoutset, seqoutall	outseq, *outseq, *outfile	Primary output sequence, generally required, generally should default to the primary input sequence name, extension defaults to the name of the output sequence format
outfile	outfile, *file	Primary output non-sequence results file, generally required. The file extension should be allowed to default to the application name. outfile should be used for the first output file. outfile or *file is acceptable for the second and subsequent output files
report	outfile, *file	Report output file. outfile should be used for the first report file. outfile or *file is acceptable for the second and subsequent report files
align	outfile, *file	Alignment output file. outfile should be used for the first output alignment. outfile or *file is acceptable for the second and subsequent output alignments
infile	infile, *file	Primary input non-sequence file
infile	data	Primary auxiliary input data file, generally optional
infile	patterns	File of patterns to search for in sequence
integer	minlen	Minimal length of sequence feature to be found
integer	maxlen	Maximum length of sequence feature to be found
integer	wordsize	Word size for hash tables, etc. Generally minimum value = 2 for protein, 4 for DNA
integer	window	Window length for calculating dotplots, features, etc.
integer	shift	Amount by which window is shifted in each iteration
boolean	consensus	Flag for whether consensus sequence should be output
float	gap	Gap penalty

float	gapext	Gap extension penalty
integer	from	Position of start of input sequence to specify for an operation (e.g. deletion), defaults to start of sequence, minimum value = 1, maximum value = <sequence length >
integer	to	Position of end of input sequence to specify for an operation (e.g.: deletion), defaults to the from value, minimum value = from value, maximum value = <sequence length >
float or integer	threshold	Threshold for various operations
boolean	left	Operation should be done at the start of the sequence
boolean	right	Operation should be done at the end of the sequence
string	pattern	Pattern to search for in sequence
graph	graph	Graphical output
xygraph	graph	Graphical output
directory	directory, *dir, *path	Directory of files
outdir	outdir, *outdir	Output directory of files
dirlist	directory	Directory of files
filelist	*files	List of files
matrix	matrix	Matrices
datafile	datafile	Datafiles
feature	feature, *feature	Feature input
featout	outfeat, *outfeat	Feature output
regexp	pattern	Regular expressions

A.1.3.3 Validated parameter names

For some datatypes, conventions are more strongly enforced: a warning will be generated during ACD processing if a standard name is not used for the following datatypes:

- Sequence inputs (any data definition of the type sequence, seqall, seqsetall or seqset) and sequence outputs (seqout, seqoutall and seqoutset datatypes)
- Feature inputs (any data definition of the type feature) and feature outputs (featout datatype)
- Alignments (align datatype)
- File inputs and outputs (any data definition of the type infile, filelist, directory, dirlist or outfile)

• Report output (`report` datatype).

A.1.4 Types of attributes

Application attributes may be defined for an application definition (Section A.3, 'Application attributes').

There are three basic types of attributes that may be defined for a data definition:

- Global attributes (Section A.4, 'Global attributes')
- Datatype-specific attributes (Section A.5, 'Datatype-specific attributes')
- Calculated attributes (Section A.6, 'Calculated attributes').

Additionally, there are various 'datatype associated' command line qualifiers (or simply 'associated qualifiers') that are inbuilt for certain ACD datatypes may also be defined as attributes in the appropriate data definition. These are listed in the datatype descriptions (Section A.2, 'Datatypes').

A.1.5 Parameters and qualifiers

Every data definition in the ACD file can be defined via an appropriate attribute to be one of the following:

- Parameter
- Standard qualifier
- Additional qualifier

with the default being:

- Advanced qualifier.

They differ in terms of how they are prompted for, how they may be specified on the command line and whether help information for them appears.

This behaviour is summarised in the table below. 'Flag' indicates whether the flag (parameter or qualifier name) must be given on the command line. 'Prompt' indicates whether a value will be prompted for if one is not specified on the command line. Additional qualifiers will only be prompted for if **`-options`** is specified. 'Help' indicates where the

Type	Attribute	Flag	Prompt	Help
parameter	`parameter: "Y"`	No	Yes	Required section
standard	`standard: "Y"`	Yes	Yes	Required section
additional	`additional: "Y"`	Yes	Yes (with **`-options`**) or No (default needed)	Advanced section
advanced (default)	No attribute needed	Yes	No	Advanced section

information from the built-in **-help** qualifier is shown. For more information, see Section 4.5, 'Controlling the prompt'.

A.1.6 ACD file sections

Any data definitions in an ACD file *must* be contained within an appropriate Section and given in the correct order. The sections must appear in this order:

1. Input
2. Required
3. Additional
4. Advanced
5. Output.

Subsections with arbitrary names can also be defined. They can appear in any order but must be nested in a major section.

Sections and subsections have the following general syntax:

```
section: SectionName
[
    information: "SectionName section"
    type: "page"
]
.
.   (data definitions go here)
.
   section: NestedSectionName
   [
       information: "NestedSectionName section"
       type: "page"
   ]
   .
   . (data definitions go here)
   .
   endsection: NestedSectionName
.
endsection: SectionName
```

For example:

```
section: input
[
  information: "Input section"
  type: "page"
]
.
. (input data definitions go here)
.
   section: inputsubSection
   [
   information: "Input sub-section"
   type: "page"
```

```
    ]
.
. (input sub-Section data definitions go here)
.
endsection: inputsubsection

endsection: input
```

The section contents is summarised in the table below.

A.1.6.1 Validation of sections

Restrictions on the order of sections and what data definitions can appear in what sections are defined in the EMBOSS system file `sections.standard` (see Section 4.1, 'Introduction to ACD file development'). The restrictions are enforced during ACD processing and an error will be generated in the following circumstances:

- If major sections appear in the wrong order
- If subsections appear in the wrong major sections
- If a parameter (data definition with a `parameter: "Y"` attribute) or a standard qualifier (`standard: "Y"` attribute) occurs in the 'Advanced' or 'Additional' sections
- If an additional qualifier (`additional: "Y"` attribute) occurs in the 'Advanced' or 'Required' sections
- If an advanced qualifier (no `parameter: "Y"`, `standard: "Y"` or `additional: "Y"` attribute) occurs in the 'Additional' or 'Required' sections

Section name	Description
Input	Simple input values and any ACD type that will read input, including `infile`, `sequence`, `seqset`, `seqall`, `matrix`, `fmatrix` and `codon`. Any other parameters and qualifiers related to input can also be placed in this section. At present `datafile` is also included.
Required	Parameters and standard qualifiers, including any whose `standard:` attribute can be true but depends on a conditional operation. Any `toggle:` definitions that are used by the parameters and standard qualifiers. Note that input and output parameters and qualifiers must be in their respective sections.
Additional	Additional qualifiers, including any whose `additional` attribute can be true but depends on a conditional operation. Any `toggle:` definitions that are used by additional qualifiers. Input and output parameters and qualifiers must be in their respective sections.
Advanced	Any qualifiers (except input and output qualifiers) which have no `standard:` or `additional:` attribute defined.
Output	Any datatype that will write output, including any `outfile`, `outdata`, `seqout`, `seqoutall`, `seqoutset` and `outtree`. Other qualifiers related to output can also be placed in this section.

A.2 Datatypes

For convenience, the available ACD datatypes are organised into five groupings reflecting similar properties or modes of usage as follows:

- Simple
- Input
- Selection
- Output
- Graphics.

The available datatypes are described in detail below.

A.2.1 Description of simple ACD datatypes

A.2.1.1 `array`

A list of either integer or floating point numbers.

Data value The data value is a list of numbers separated by spaces or commas. For example:

```
"1 2 3 4 5"
"1.5, 2.0, 2.5, 3.0"
```

Default value A default value is set using the `default:` global attribute.

Key attributes The ACD attributes control validation, for example the permissible number of values (`size:` attribute), or a total value the list of numbers must add up to `sumtest:`, within a certain tolerance `tolerance:`, which is only tested if the boolean `sum:` attribute is set.

A.2.1.2 `boolean`

Simple boolean value.

Data value The data value has a 'true' or 'false' value which may be specified as follows:

```
"Y"
"yes"
"true"
"N"
"no"
"false"
```

The value will be "Y" if the parameter name is entered on the command line as a flag, for example `-BooleanOption`. If the qualifier is absent from the command line the default value is used. The flag can also be prefixed by **no**, for example `-noBooleanOption`, to force the value to be "N". This is needed if the default value is "Y".

Default value A default value is set using the `default:` global attribute.

Key attributes None.

A.2.1.3 integer

Simple integer number.

Data value The data value is any integer value.
For example:

```
"100"
```

Default value A default value is set using the `default:` global attribute.

Key attributes Many applications will stipulate a minimum and / or maximum value, e.g. a minimum value of 0 or 1. The permissible value range is controlled by the `minimum:` and `maximum:` attributes.

A.2.1.4 float

Simple floating point number.

Data value The data value is any valid floating point number.
For example:

```
"100.24"
```

Default value A default value is set using the `default:` global attribute.

Key attributes The value range is controlled by `minimum:` and `maximum:` attributes and the maximum precision by `precision:`.

A.2.1.5 range

Range(s) of sequence positions.

Data value One or more ranges may be defined on the command line or in a range file.
On the command line, a range is defined by a pair of integer numbers and multiple ranges may be given. The numbers may be delimited by any non-digit, non-alphabetic character. For example:

```
"24-45, 56-78"
"1:45, 67=99;765..888"
"1,5,8,10,23,45,57,99"
```

A range file contains a list of pairs of numbers with optional text comments. One pair of numbers must be given per line and the file can contain comment lines which are preceded with a # character. For example:

```
# A set of ranges in a range file.
 12       23
  4       5        This is an optional comment.
 67       10348    Another comment.
```

Range files are specified on the command line by preceding the filename with **@filename**. For example:

```
@filename RangeFileName
```

In cases where the numbers are sequence positions, the upper and lower bounds will in practice depend on the length of the sequence to which they are applied. You should bear in mind that sequence positions can be negative, in which case they count back from the *end* of the sequence.

Default value A default value is set using the `default:` global attribute.

Key attributes None.

A.2.1.6 `regexp`

A regular expression pattern.

EMBOSS uses the 'Perl-compatible regular expression library' (PCRE) to process regular expressions.

Data value Any regular expression that is valid in Perl 5.0 (http://search.cpan.org/~nwclark/perl-5.8.7/pod/perlre.pod) should be valid here.

Default value A default value is set using the `default:` global attribute.

Key attributes Attributes provide validation, for example, to control the length (`minlength:` and `maxlength` attributes) and case of the regular expression, which can be set to upper (`upper: "Y"`) or lower case (`lower: "Y"`) only.

A.2.1.7 `pattern`

A sequence pattern.

Data value The standard IUPAC one-letter codes for the amino acids are used. The symbol x is used for a position where any amino acid is accepted. Ambiguities are indicated by listing the acceptable amino acids for a given position, between square parentheses `[ ]`. For example:

```
[ALT]
```

stands for Ala or Leu or Thr. Ambiguities are also indicated by listing between a pair of curly brackets { } the amino acids that are not accepted at a given position. For example:

```
{AM}
```

stands for any amino acid except Ala and Met. Each element in a pattern is separated from its neighbour by a dash (–). Repetition of an element of the pattern can be indicated by following that element with a numerical value or a numerical range between parenthesis. For example:

```
x(3) corresponds to x-x-x
x(2,4) corresponds to x-x or x-x-x or x-x-x-x
```

When a pattern is restricted to either the N- or C-terminal of a sequence, that pattern either starts with a < symbol or respectively ends with a > symbol. A period ends the pattern. For example:

```
[DE](2)HS{P}X(2)PX(2,4)C.
```

Default value A default value is set using the `default:` global attribute.

Key attributes Attributes provide validation, for example, to control the length (`minlength:` and `maxlength` attributes) and case of the pattern, which can be set to upper (`upper: "Y"`) or lower case (`lower: "Y"`) only. The type of pattern (nucleotide or protein) can be set using `type:`.

A.2.1.8 `string`

Simple string.

Most string values are free text, although strings can be used by a program for any input that is not covered by the other ACD datatypes so long as the string is subsequently parsed.

Data value The data value is any valid ASCII text string which should be enclosed in double quotes. For example:

```
"This is a valid text string"
```

Default value A default value is set using the `default:` global attribute.

Key attributes Whenever a `string` datatype is defined, a type for it should be specified with the `knowntype:` attribute: a warning message will be generated during ACD processing otherwise.

ACD attributes are available to control the length or to provide a regular expression pattern to validate the string if necessary.

A.2.1.9 `toggle`

Simple boolean switch for controlling other parameters.

`toggle` parameters work in the same way as `boolean` parameters but are intended for use in turning the prompting for other parameters on or off (see Section 4.5, 'Controlling the prompt'). Typically this is done by using a calculation to determine the value of a `standard:` or `additional:` attribute of the data definition that is being controlled. In such cases the `toggle` parameters are used in the calculated values, and can be placed in the "Required" section of an ACD file even if not themselves defined as standard qualifiers (using the `standard:` attribute).

Data value The data value has a 'true' or 'false' value which may be specified as follows:

```
"Y"
"yes"
"true"
"N"
"no"
"false"
```

Exactly like `boolean` parameters, the value will be `"Y"` if the flag for the parameter is entered on the command line (for example **-*ToggleOption***). If the qualifier is absent from the command line the value will be the default value. The flag can also be prefixed by **no** (for example **-no*ToggleOption***) to force the value to be `"N"` (false). This is needed if the default value is `"Y"`.

Default value A default value is set using the `default:` global attribute.

Key attributes None.

A.2.2 Description of input ACD datatypes

A.2.2.1 `codon`

Codon usage table file.

Codon usage table files are ASCII text files and can be read in several formats including GCG. Codon usage files are distributed in the EMBOSS data directory.

Data value The data value is the name of a codon usage table file in the EMBOSS data search path (see the *EMBOSS User's Guide*).

Default value EMBOSS uses the human codon usage table `Ehum.cut` provided in the EMBOSS distribution by default but this would typically be overridden by the user. Codon usage tables are species-specific and in some cases specific to a class of genes within a species, so it is useful to be able to set the codon usage table on an application-specific basis.

A default value is set using the `default:` global attribute.

Key attributes None.

A.2.2.2 `cpdb`

Protein coordinate data in CCF (clean coordinate file) format.

CCF format is a simple 'clean' file format for protein and domain coordinate data. See the documentation for **pdbparse**, part of the EMBASSY **domainatrix** package, which generates CCF files from PDB file input.

Data value The data value is the name of a CCF file.

Default value A default value is set using the `default:` global attribute.
There is an internally defined default value (`"1azu"`) although it is not normally appropriate to use it.

Key attributes None.

A.2.2.3 `datafile`

A formatted data file read from the standard EMBOSS data search path (see the *EMBOSS User's Guide*).

Many data files already have their own ACD datatype, for example, `matrix`, `matrixf` and `codon`. Other data files do not have or need their own ACD definition and `datafile` is used for these.

Data value The data value is the name of a data file in the EMBOSS data search path (see the *EMBOSS User's Guide*).

Default value A default value is set using the `default:` global attribute.
The default data file name may also be defined by two ACD attributes, for the file base name (`name:`) and file extension (`extension:`).

Key attributes Data files often have a hard-coded filename. You are free to use `name:` to override that name.

A.2.2.4 `directory`

A directory that can be used for input or output.

Data value The data value is the name of any valid directory. For example:

```
"."
"/data"
"/data/sequences"
```

Default value A default value is set using the `default:` global attribute.

Key attributes The attributes provide additional validation of user input.

A.2.2.5 `dirlist`

A list of file names that are read from a directory.

Data value The data value is the name of any valid directory. For example:

```
"."
"/data"
"/data/sequences"
```

Default value A default value is set using the `default:` global attribute.

Key attributes The attributes provide additional validation of the user input.
The type of data in the files can be identified by specifying a value for the `knowntype:` attribute. This allows inputs to be matched to outputs where the `knowntype:` attribute is set, for example, for an `outfile` definition.

A.2.2.6 `discretestates`

Discrete states file.

`discretestates` was implemented for the **phylipnew** EMBASSY package. `discretestates` input is used by the **phylip** 'discrete character' applications.

`discretestates` could be replaced by a simple input file in GUIs, with the user required to provide the correct data format.

Data value The data value is the name of a **phylip** 'discrete states' file.

Default value A default value is set using the `default:` global attribute.

Key attributes The attributes provide detailed type checking, and can automatically detect and validate the various alternative formats that **phylip** supports without the need for complex extra command line options.

A.2.2.7 `distances`

Distance matrix.

`distances` is specific to the **phylipnew** EMBASSY package. `distances` input is used by the **phylip** 'distance matrix' applications.

The distances datatype can be replaced by a simple input file in GUIs, with the user required to provide the correct data format.

Data value The data value is the name of a distance matrix file. The accepted file formats includes all the formats read by **phylip**, with automatic interconversion.

Default value A default value is set using the `default:` global attribute.

Key attributes The attributes provide detailed type checking, and can automatically detect and validate the various alternative formats that **phylip** supports without the need for complex extra command line options.

A.2.2.8 `features`

Sequence feature annotation in any known feature format.

Data value The data value is the name of a features file. A features file contains sequence feature information. Several feature formats are supported (see the *EMBOSS User's Guide*).

Default value A default value is set using the `default:` global attribute.

Key attributes The type of features can be restricted by setting the `type:` attribute, for example, so that the program accepts only DNA features. The feature type must be one of `protein` or `nucleotide`. There is a default based on the type of an input sequence (where used), but a value should be specified so that the application can validate that the input is of the specified type. If no type is specified for input features and there is no sequence input from which to take a default type, then an error will be generated during ACD processing.

Features can also be read from an input sequence (`sequence`, `seqall`, `seqset` and `seqsetall` datatypes) and written alongside an output sequence (`seqout`, `seqoutall` and `seqoutset` datatypes) if their `features:` attribute is set.

A.2.2.9 `filelist`

A list of input files.

Data value The data value is a list of filenames separated by commas.

For example:

```
"../data/file1.dat, file2.dat"
```

`Filelist` is equivalent to the `infile` datatype, but allows the user to specify one or more input files.

Default value A default value is set using the `default:` global attribute.

Key attributes The attributes provides additional validation of the user input.

The type of data can be identified by specifying a value for the `knowntype:` attribute. This allows inputs to be matched to outputs where the `knowntype:` attribute is set, for example, for an `outfile` definition.

A.2.2.10 `frequencies`

Frequency value(s).

`frequencies` is specific to the **phylipnew** EMBASSY package and is used by the **phylip** 'gene frequency and continuous character' applications.

The frequencies datatype can be replaced by a simple input file in GUIs, with the user required to provide the correct data format.

Data value The data value is the name of a frequencies file. The accepted file formats include all the formats read by **phylip**, with automatic interconversion.

Default value A default value is set using the `default:` global attribute.

Key attributes The attributes provide detailed type checking, and can automatically detect and validate the various alternative formats that **phylip** supports without the need for complex extra command line options.

A.2.2.11 `infile`

General input file.

`infile` is used for files of data not catered for by some other ACD datatype. For example, an `infile` would not normally contain sequence data.

Data value The data value is the name of an input file.
For example:

```
"data.in"
"/data/infile.1"
```

Default value A default value is set using the `default:` global attribute.

Key attributes The type of data can be identified by specifying a value for the `knowntype:` attribute. This allows inputs to be matched to outputs where the `knowntype:` attribute is also set for the `outfile` definition. A directory containing the file can be specified, via an environment variable, by using `directory:`.

A.2.2.12 `matrix`

Comparison matrix file (integer values).

These are typically amino acid or nucleotide substitution matrices. The matrix files distributed with **BLAST** are distributed with EMBOSS in the EMBOSS data directory.

The `matrix` datatype defines integer matrices which are usually faster than floating point matrices. Floating point matrices (`matrixf` datatype) are available if needed, and an integer matrix file can of course also be read as floating point.

Typically where a comparison matrix is specified, gap penalties will also be required. These must be specified separately in one or more other data definitions.

Data value The data value is the name of an integer comparison matrix file in the EMBOSS data search path (see the *EMBOSS User's Guide*).

Default value A default value is set using the `default:` global attribute.

Key attributes Attributes of the `matrix` datatype define characteristics and allow validation of matrices of integer numbers for biological data.

The matrix datatype has a `protein:` attribute to force selection of a nucleic acid or protein comparison matrix. In ACD files, the type of the input sequence is often used to set the type of matrix.

A.2.2.13 `matrixf`

Comparison matrix file (floating point values).

The `matrixf` datatype defines floating point matrices which are usually slower than floating point matrices. An integer matrix file can of course also be read as floating point.

These are typically amino acid or nucleotide substitution matrices. The matrix files distributed with **BLAST** are distributed with EMBOSS in the EMBOSS data directory.

Typically where a comparison matrix is specified, gap penalties will also be required. These must be specified separately in one or more other data definitions.

Data value The data value is the name of a floating point comparison matrix file in the EMBOSS data search path (see the *EMBOSS User's Guide*).

Default value A default value is set using the `default:` global attribute.

Key attributes Attributes of the `matrixf` datatype define characteristics and allow validation of matrices of floating point numbers for biological data. The `matrixf` datatype has a `protein:` attribute to force selection of a nucleic acid or protein comparison matrix. In ACD files, the type of the input sequence is often used to set the type of matrix.

A.2.2.14 `properties`

Property value(s).

`properties` is specific to the **phylipnew** EMBASSY package. Properties input is used by the **phylip** applications to define weights, ancestral states and factors (multi-state characters).

The properties datatype can be replaced by a simple input file in GUIs, with the user required to provide the correct data format.

Data value The data value is the name of a properties file. The accepted formats include all the formats read by **phylip**, with automatic interconversion.

Default value A default value is set using the `default:` global attribute.

Key attributes The attributes provide detailed type checking and can automatically detect and validate the various alternative formats that **phylip** supports without the need for complex extra command line options.

A.2.2.15 `scop`

SCOP and CATH domain classification data in DCF (domain classification file) format.

DCF (domain classification file) format is a simple 'clean' file format for domain classification data. See the documentation for **domainer**, part of the EMBASSY **domainatrix** package, which generates DCF files from SCOP and CATH file input.

Data value The data value is the name of a DCF file.

Default value A default value is set using the `default:` global attribute.

Domain classification file input has an internally defined default value (`"d3sdha"`) although it is not normally appropriate to use this default.

Key attributes None.

A.2.2.16. `sequence`

A single sequence for reading.

Data value The data value is the USA (see the *EMBOSS User's Guide*) of a single sequence. For example, the USA might be a database reference or file.

Default value A default value is set using the `default:` global attribute.

Key attributes The type of sequence can be restricted by setting the `type:` attribute, for example, so that the program accepts only DNA sequences. The sequence type must be a standard sequence type (Section A.7, 'Sequence types').

Sequence features (Section 6.9, 'Handling features') can be read if the `features:` ACD attribute is set.

A.2.2.17 `seqall`

A set of single sequences that are addressed one after another.

Data value The data value is the USA of a set of single sequences. For example, the USA (see the *EMBOSS User's Guide*) might specify a sequence database for sequential reading of entries.

Default value A default value is set using the `default:` global attribute.

Key attributes The type of sequence can be restricted by setting the `type:` attribute, for example, so that the program accepts only DNA sequences. The sequence type must be a standard sequence type (Section A.7, 'Sequence types').

Sequence features (Section 6.9, 'Handling features') can be read if the `features:` ACD attribute is set.

A.2.2.18 `seqset`

A set of single sequences that can be used all at the same time.

Data value. The data value is the USA (see the *EMBOSS User's Guide*) of a set of single sequences. For example, set of sequences from a multiple alignment file, or sequences from a database.

Default value A default value is set using the `default:` global attribute.

Key attributes The type of sequence can be restricted by setting the `type:` attribute, for example, so that the program accepts only DNA sequences. The sequence type must be a standard sequence type (Section A.7, 'Sequence types').

Sequence features (Section 6.9, 'Handling features') can be read if the `features:` ACD attribute is set.

The `aligned:` attribute *must* be set: an error will be generated during ACD processing otherwise.

A.2.2.19 `seqsetall`

One or more sets of single sequences that can be used all at the same time.

Data value The data value is the USA (see the *EMBOSS User's Guide*) of one or more sets of single sequences. For example, sets of sequences from two databases or two alignment files. The data value would typically be a 'list file' (a file containing a list of USAs).

Default value A default value is set using the `default:` global attribute.

Key attributes The type of sequence can be restricted by setting the `type:` attribute, for example, so that the program accepts only DNA sequences. The sequence type must be a standard sequence type (Section A.7, 'Sequence types').

Sequence features (Section 6.9, 'Handling features') can be read if the `features:` ACD attribute is set.

The `aligned:` attribute *must* be set: an error will be generated during ACD processing otherwise.

A.2.2.20 `tree`

Phylogenetic tree.

The `tree` datatype is specific to the **phylipnew** package. Tree input is used by the **phylip** applications to define one or more phylogenetic trees.

The `tree` datatype can be replaced by a simple input file in GUIs, with the user required to provide the correct data format.

The trees are currently parsed by **phylip** itself, but in the future native parsing methods might be implemented.

Data value The data value is the name of a tree file. The formats accepted include all the formats read by **phylip**, with automatic interconversion.

Default value A default value is set using the `default:` global attribute.

Key attributes The attributes provide detailed type checking, and can automatically detect and validate the various alternative formats that **phylip** supports without the need for complex extra command line options.

A.2.3 Description of output ACD datatypes

A.2.3.1 `align`

Output file for sequence alignments.

The data is stored as sequences and all of the common alignment formats are supported (see the *EMBOSS User's Guide*).

Data value The data value is any valid filename.

Default value An alignment filename with the format *`name.extension`* is constructed if the datatype-specific qualifiers **`-aname`** and **`-aextension`** are specified. Values may be hard-coded with the corresponding `aname:` and `aextension:` attributes.

A default value is also set by defining the `default:` global attribute.

Key attributes An alignment output file is defined in the same way as a plain output file (`outfile` datatype) but has extra qualifiers to allow a choice of alignment formats and attributes to specify whether the alignment will have two or more sequences (which limits the possible formats).

The `multiple:` boolean attribute should be set to `"Y"` if the output can contain more than one alignment from the same input.

The output format is normally set at the command line but a default may be hard-coded with `aformat:`.

A.2.3.2 `featout`

Output file for sequence feature annotation.

The data is stored as a feature table and most common sequence feature formats are supported (see the *EMBOSS User's Guide*).

Data value The data value is any valid filename.

Default value The output filename has the format `name.extension` and is constructed from the `name:` and `extension:` attributes (which the **`-ofname`** and

offormat built-in command line qualifiers override (see the *EMBOSS User's Guide*). If the `name:` attribute is not defined in the ACD file, it will default to the calculated attribute `name:` of the *first* sequence that is read in. The ACD operation to get this value is `$(asequence.name)` if the sequence parameter was named `asequence`. The `extension:` attribute will default to the output feature format, e.g. `.gff`.

A default value is also set by defining the `default:` global attribute.

Key attributes Features can be read from an input sequence (`sequence`, `seqall`, `seqset` and `seqsetall` datatypes) and written alongside an output sequence (`seqout`, `seqoutall` and `seqoutset` datatypes) if their `features:` attribute is set.

GFF format is used by default for the output feature(s). The format is normally set at the command line but a default may be hard-coded with `offormat:`.

The type of features can be restricted by setting the `type:` attribute, for example, so that the program accepts only DNA features. The feature type must be one of `protein` or `nucleotide`. There is a default based on the type of an input sequence (where used), but a value should be specified so that the application can validate that the input is of the specified type. If no type is specified for input features and there is no sequence input from which to take a default type, then an error will be generated during ACD processing.

A.2.3.3 outcodon

Output file for codon usage data.

Data value The data value is any valid filename.

The data is stored as a codon usage table. Codon usage table files are ASCII text files and can be read in several formats (see the *EMBOSS User's Guide*).

Default value The output filename has the format *name.extension* and is constructed from the `name:` and `extension:` attributes. The `name:` attribute will default to `outfile`. The `extension:` attribute will default to the format, with `cut` defined as the default format to match the usual codon usage file naming convention. This format is also called *EMBOSS codon format*.

A default value is also set by defining the `default:` global attribute.

Key attributes The default data format can be specified by an `oformat:` attribute which the **-oformat** associated qualifier can override.

A.2.3.4 outcpdb

Output file for protein coordinate data in CCF (clean coordinate file) format.

CCF (clean coordinate file) format is a simple 'clean' file format for protein and domain coordinate data. See the documentation for **pdbparse**, part of the EMBASSY **domainatrix** package, which generates CCF files from PDB file input.

Data value The data value is any valid filename.

Default value A default value is set using the `default:` global attribute.

Key attributes The default data format can be specified by an `oformat:` attribute which the **-oformat** associated qualifier can override.

A.2.3.5 outdata

Output file for data formatted cleanly as a table or list.

The output corresponding to multiple outdata definitions in an ACD file is appended to a single file. The individual ACD definitions allow the format of each file Section to be defined.

Data value The data value is any valid filename.

Default value A default value is set using the default: global attribute.

Key attributes The default data format can be specified by an oformat: attribute which the **-oformat** associated qualifier can override.

A.2.3.6 outdir

Output directory for writing of multiple output files.

Data value The data value is the name of any valid directory.
For example:

```
"."
"/data"
"/data/sequences"
```

Default value A default value is set using the default: global attribute.

Key attributes The default file extension can be set with the extension: attribute.

A.2.3.7 outdiscrete

Output file for phylogenetics discrete characteristics data.

Data value The data value is any valid filename.

Default value A default value is set using the default: global attribute.

Key attributes The default data format can be specified by an oformat: attribute which the **-oformat** associated qualifier can override.

A.2.3.8 outdistance

Output file for phylogenetics distance matrix data.

Data value The data value is any valid filename.

Default value A default value is set using the default: global attribute.

Key attributes The default data format can be specified by an oformat: attribute which the **-oformat** associated qualifier can override.

A.2.3.9 outfile

General output file.

`outfile` is used for data not catered for by some other output ACD datatype. It is suitable for general application output in plain text. For example, the output file would not normally contain sequence data.

Data value The data value is any valid filename.

Default value The output filename has the format *`name.extension`* and is constructed from the `name:` and `extension:` attributes. The `extension:` attribute will default to the program name, and is usually left as the default value.
A default value is also set by defining the `default:` global attribute.

Key attributes The type of data can be identified by a `knowntype:` attribute and matched to an standard type of an `infile` data definition for use as input to another program. The standard EMBOSS known types are described elsewhere (Section 4.3, 'Data definition').

A.2.3.10 outfileall

Multiple general output files.
`outfileall` is used for data not catered for by some other output ACD datatype. It is suitable for general application output in plain text. For example, the output files would not normally contain sequence data.

Data value The data value is the base filename of the output files.

Default value The output filename has the format `name.extension` and is constructed from the `name:` and `extension:` attributes. The `extension:` attribute will default to the program name, and is usually left as the default value.
A default value is also set by defining the `default:` global attribute.

Key attributes The type of data can be identified as a standard types (Section A.4, 'Global attributes') using the `knowntype:` attribute. The type can be matched to that for an `infile` data definition for use as input to another program.

A.2.3.11 outfreq

Output file for phylogenetics character frequency data.

Data value The data value is any valid filename.

Default value A default value is set using the `default:` global attribute.

Key attributes The default data format can be specified by an `oformat:` attribute which the **`-oformat`** associated qualifier can override.

A.2.3.12 outmatrix

Output file for integer comparison matrix data.

Data value The data value is any valid filename.

Default value A default value is set using the `default:` global attribute.

Key attributes The default data format can be specified by an `oformat:` attribute which the **`-oformat`** associated qualifier can override.

A.2.3.13 `outmatrixf`

Output file for floating point comparison matrix data.

Data value The data value is any valid filename.

Default value A default value is set using the `default:` global attribute.

Key attributes The default data format can be specified by an `oformat:` attribute which the **`-oformat`** associated qualifier can override.

A.2.3.14 `outproperties`

Output file for phylogenetics property data.

Data value The data value is any valid filename.

Default value A default value is set using the `default:` global attribute.

Key attributes The default data format can be specified by an `oformat:` attribute which the **`-oformat`** associated qualifier can override.

A.2.3.15 `outscop`

Output file for SCOP and CATH domain classification information in DCF (domain classification file) format.

DCF (domain classification file) format is a simple 'clean' file format for domain classification data. See the documentation for **domainer**, part of the EMBASSY **domainatrix** package, which generates DCF files from SCOP and CATH file input.

Data value The data value is any valid filename.

Default value A default value is set using the `default:` global attribute.

Key attributes The default data format can be specified by an `oformat:` attribute which the **`-oformat`** associated qualifier can override.

A.2.3.16 `outtree`

Output file for phylogenetic tree data.

Data value The data value is any valid filename.

Default value The output filename has the format *`name.extension`* and is constructed from the `name:` and `extension:` attributes. The `extension:` attribute will default to the output file format, and is usually left as the default value.

A default value is also set by defining the `default:` global attribute.

Key attributes The default data format can be specified by an `oformat:` attribute which the **`-oformat`** associated qualifier can override.

A.2.3.17 `report`

Output file for sequence annotation.

Report data is stored internally as a feature table, so the supported formats (see the *EMBOSS User's Guide*) include the most common feature formats.

Data value The data value is any valid filename.

Default value A report filename with the format *name.extension* is constructed if the datatype-specific qualifiers **-rname** and **-rextension** are specified. Values may be hard-coded with the corresponding rname: and rextension: attributes.
A default value is also set by defining the default: global attribute.

Key attributes A report file is defined in the same way as a plain output file (outfile) but has extra qualifiers to allow a choice of report formats.

- rformat: specifies the report format to use, which must be one of the supported report formats (see the *EMBOSS User's Guide*).
- multiple: is a boolean attribute which should be set to "Y" if the output can contain more than one report from the same input.
- type: is defined as one of "protein" or "nucleotide" where the report format is one of the standard feature table formats (see the *EMBOSS User's Guide*).
- taglist: defines the tag:value pairs from the internal feature table to be reported in the output.

A.2.3.18 seqout

Output file for a single sequence.

Data value The data value is a USA (see the *EMBOSS User's Guide*) of sequence output stream. For example, the USA might be a database reference or file.

Default value The output filename has the format *name.extension* and is constructed from the name: and extension: attributes. If the name: attribute is not defined in the ACD file it will default to the calculated attribute name: of the *first* sequence that is read in. The ACD operation to get this value is $(asequence.name) if the sequence parameter was named asequence.
A default value is also set by defining the default: global attribute.

Key attributes The type of sequence can be restricted by setting the type: attribute. The sequence type must be a standard type (Section A.7, 'Sequence types').
Sequence features (Section 6.9, 'Handling features') can be written if the features: ACD attribute is set.
FASTA format is used by default for the output sequence(s). The format is normally set at the command line but a default may be hard-coded with osformat:.

A.2.3.19 seqoutall

Output file for multiple sequences.

Data value The data value is a USA (see the *EMBOSS User's Guide*) of a sequence output stream. For example, the USA might be a database reference or file.

Default value The output filename has the format *name.extension* and is constructed from the name: and extension: attributes. If the name: attribute is not defined in the ACD file it will default to the calculated attribute name: of the *first* sequence that is read in. The ACD operation to get this value is $(asequence.name) if the sequence parameter was named asequence.

A default value is also set by defining the `default:` global attribute.

Key attributes The type of sequence can be restricted by setting the `type:` attribute. The sequence type must be a standard type (Section A.7, 'Sequence types').

Sequence features (Section 6.9, 'Handling features') can be written if the `features:` ACD attribute is set.

FASTA format is used by default for the output sequence(s). The format is normally set at the command line but a default may be hard-coded with `osformat:`.

A.2.3.20 `seqoutset`

Output file for a set of sequences.

Data value The data value is a USA (see the *EMBOSS User's Guide*) of a sequence output stream on a set of single sequences stored in memory together, to be written to file. For example, the USA might be a database reference or file.

Default value The output filename has the format `name.extension` and is constructed from the `name:` and `extension:` attributes. If the `name:` attribute is not defined in the ACD file. It will default to the calculated attribute `name:` of the *first* sequence that is read in. The ACD operation to get this value is `$(asequence.name)` if the sequence parameter was named `asequence`.

A default value is also set by defining the `default:` global attribute.

Key attributes The type of sequence can be restricted by setting the `type:` attribute. The sequence type must be a standard type (Section A.7, 'Sequence types').

Sequence features (Section 6.9, 'Handling features') can be written if the `features:` ACD attribute is set.

FASTA format is used by default for the output sequence(s). The format is normally set at the command line but a default may be hard-coded with `osformat:`.

A.2.4 Description of selection ACD datatypes

A.2.4.1 `list`

A list of options (text descriptions) with text labels.

The user is presented with a limited list of options they can choose from. The choices can be labelled by any arbitrary text label. The option descriptions are usually more verbose than for the `selection` datatype.

Data value The data value is one (or more) of the valid options.

An option is specified by the label text or a non-ambiguous part of the descriptive text itself given after the label. If multiple selections are allowed, the user must supply a comma-separated list of labels (options).

For example, the following ACD definition:

```
list: frame [
  standard:"Y"
  help: "Allows selection from a set of reading frames"
  default: "1"
  minimum: "1"
```

```
  maximum: "1"
  header: "Translation frames"
  values: "1:1, 2:2, 3:3, F:Forward three frames, -1:-1, -2:-2, -3:-3, R:
Reverse three
frames, 6:All six frames"
  delimiter: ","
  codedelimiter: ":"
  information: "Frame(s) to translate"
]
```

would present to the user something like:

```
Translation frames
   1      1
   2      2
   3      3
   F      Forward three frames
  -1     -1
  -2     -2
  -3     -3
   R      Reverse three frames
   6      All three frames
Frame(s) to translate[1]:
```

To select from the list, the user specifies one (or sometimes more) labels, or partial text descriptions. The program is given a list of text labels as input. In this example, the minimum and maximum number of selections are set to one in the ACD definition, therefore only one selection value is allowed. For example, these are all valid selections:

```
"1"
"F"
"Forward"
"For"
"R"
"Reverse"
"Rev"
```

If the maximum count had been set to 3, say, then the following would be valid:

```
"-1,F,6"
```

Default value A default value is set using the `default:` global attribute.

Key attributes None.

A.2.4.2 `selection`

A list of options (text descriptions) with automatically generated numerical labels.

The user is presented with a limited list of options they can choose from. The choices are numbered automatically from 1 up. The option descriptions are typically more verbose than for `list` definitions.

Data value The data value is one (or more) of the valid options.

An option is specified by number or a non-ambiguous part of the descriptive text itself given after the label. If multiple selections are allowed, the user must supply a comma-separated list of numbers (options).

For example, the following ACD definition:

```
selection: reject [
    default: "3, 5, 6"
    minimum: "1"
    maximum: "6"
    values: "None, AAINDEX, CVS, CODONS, PRINTS, PROSITE, REBASE"
    delimiter: ","
    header: "Directories to ignore"
    information: "Select directories"
    help: "This specifies the names of the sub-directories of the
           EMBOSS data directory that should be ignored when displaying data
           directories."
    button: "Y"
 ]
```

would present to the user something like:

```
Directories to ignore
1         None
2         AAINDEX
3         CVS
4         CODONS
5         PRINTS
6         PROSITE
7         REBASE
Select directories[3, 5, 6]:
```

To select from the list, the user specifies one (or sometimes more) numbers, or partial text descriptions. The program is given a list of text labels as input. In this example, a minimum of one and maximum of six selections are set in the ACD definition. Here are some valid selections:

```
"3,5,6"
"3"
"CVS"
"5"
"PRINTS"
"PRI"
```

Default value A default value is set using the `default:` global attribute.

Key attributes None.

The `list` datatype is preferred to the `selection:`.

A.2.5 Description of graphics ACD datatypes

A.2.5.1 Graph

Graphical output of any general kind, e.g. dotplots may be generated.

Data value The data value is the graphics device, as specified by the **PLplot** graphics library used in EMBOSS at present.
The currently supported devices include:

- `ps` (Postscript)
- `png` (PNG files)
- `X11` (X-Windows).

A value of `?` in answer to the prompt will list the available graphics devices on your installation. Some permissible values therefore are:

```
"ps"
"png"
"X11"
"?"
```

Default value A default value is set using the `default:` global attribute.

Key attributes `gtitle:` specifies the graph title (many other graphical elements can be set).

A.2.5.2 `graphxy`

Graphical output as a simple two-dimensional (2D) XY plot with the sequence along the *x*-axis.

Data value The data value is the graphics as specified by the **PLplot** graphics library used in EMBOSS at present.
The currently supported devices include:

- `ps` (Postscript)
- `png` (PNG files)
- `X11` (X-Windows).

A value of `?` in answer to the prompt will list the available graphics devices on your installation. Some permissible values therefore are:

```
"ps"
"png"
"X11"
"?"
```

Default value A default value is set using the `default:` global attribute.
Key attributes `multiple:` specifies the number of multiple XY graphs in a single output.
`gtitle:` specifies the graph title (many other graphical elements can be set).

A.3 Application attributes

Application attributes describe the application itself and should only be given in an `application:` definition. The available application attributes are described in detail below.

Their general format is:

```
ApplicationAttributeName: "ApplicationAttributeValue"
```

A.3.1 `documentation:`

This is a succinct description of application function and will be printed to screen when the program is run, and when the program is run with the **`-help`** qualifier. The `documentation:` string is also searched by the **wossname** utility, which finds applications by searching for keywords in the `documentation:` text and application group. If the `documentation:` attribute is missing, a warning will be issued when you run the program.

The `documentation:` string must conform to these conventions:

- The text should start with an upper-case letter.
- The length should be kept to 63 characters or shorter in order to allow the **wossname** utility to display each program name and its documentation on one 80-character line.

Attribute definition	Description	Default value
`documentation:` VT_STR	Short description of the application function	""
`groups:` VT_STR	Standard application group(s) for **wossname** and GUIs	""
`keywords:` VT_STR	Standard application group(s) for **wossname** and GUIs	""
`gui:` VT_STR	Suitability for launching in a GUI	""
`batch:` VT_STR	Suitability for running in batch	""
`embassy:` VT_STR	EMBASSY package name	""
`external:` VT_STR	Third-party tool(s) required by this program	""
`cpu:` VT_STR	Estimated maximum CPU usage	""
`supplier:` VT_STR	Supplier name	""
`version:` VT_STR	Version number	""
`nonemboss:` VT_STR	Non-EMBOSS application name for **SoapLab**	""
`executable:` VT_STR	Non-EMBOSS executable for **SoapLab**	""
`template:` VT_STR	Command line template for **SoapLab**'s ACD files	""
`comment:` VT_STR	Comment for **SoapLab**'s ACD files	""

- The text should *not* end with a '.' character
- Any acronyms or abbreviations in the text should be written in upper-case (e.g. "SNPs", "EST", "DNA", "ABI", "SRS", "ASCII", "CDS", "mRNA", "B-DNA", "RNA", "CpG", "ORFs", "MAR/SAR", "PCR", "STS", "REBASE", "SCOP", "PROSITE", "PRINTS", "EMBL", "TRANSFAC", "AAINDEX", "BLAST", "GCG", "EMBOSS").

A.3.2 groups:

This allows the EMBOSS programs to be placed into groups or sub-groups based on their functionality. Programs in the same sub-group generally do similar things whereas programs in the same group but different sub-group do different things but in the same general area. The `groups:` string is searched by the **wossname** program, which, given a keyword, will return all the applications containing that keyword in the `groups:` text or the `documentation:` text. The `groups:` string is also used by the **seealso** application which takes the name of an EMBOSS program and returns a list of the programs which share some functionality with it.

The `groups:` attribute is followed by a string value, containing the name of one or more groups and one or more sub-groups to which the application belongs. The sub-group(s) are optional, but if specified the group(s) and sub-group(s) must be separated by a colon (`:`). When an application belongs to more than one group (or sub-group), the group names must be separated by either a comma (`,`) or semicolon (`;`).

The general format for the `groups:` attribute is as follows:

```
application: ApplicationName
[
    groups: "GroupName1, GroupName2, … : SubgroupName1, SubgroupName2, … "
]
```

The `groups:` string must conform to these conventions:

- The group name should, ideally, be taken from a list of standard group names defined in the EMBOSS system file `groups.standard` in the application ACD file directory (see Section 4.2.4.1, 'Application group names file (`groups.standard`)'). New groups can be defined, but the definitions should be sent to the EMBOSS developers so they can be added to the standard set.
- Group names can have spaces in them.
- Sub-group(s) are optional, but if specified the group(s) and sub-group(s) must be separated by a colon (`:`).
- EMBOSS supports arbitrary numbers of levels of sub-group; however, several third-party interfaces rely upon there being a maximum of two levels in total, so do not use more than one.
- When an application belongs to more than one group (or sub-group), the group names must be separated by either a comma (`,`) or semicolon (`;`).

A.3.3 `keywords:`

A set of keywords describing the application functionality. These are used by the application **wossname** and by some GUIs to EMBOSS.

A.3.4 `gui:`

This attribute has a value of `"Yes"` or `"No"` and reflects the suitability for launching the application in a GUI.

A.3.5 `batch:`

This attribute has a value of `"Yes"` or `"No"` and reflects the suitability for launching the application in a batch queue.

A.3.6 `embassy:`

The EMBASSY package name (if appropriate).

A.3.7 `external:`

Third-party tool(s) required by this program. This is used by applications which wrap (call) third-party applications and therefore rely on this software being present on the system that is running EMBOSS.

A.3.8 `cpu:`

Estimated maximum CPU usage.

A.3.9 `supplier:`

Supplier name, in cases where the application was provided by a body other than the EMBOSS core developers.

A.3.10 `version:`

Version number of the application in case a different version number than the EMBOSS or EMBASSY package version number is required.

A.3.11 `nonemboss:`

This boolean attribute specifies whether an application is part of EMBOSS or EMBASSY. It is only used by **SoapLab**.

A.3.12 `executable:`

Name of an application that is not part of EMBOSS or EMBASSY. This is used in environments (for example, **SoapLab**) where applications are defined using the ACD syntax but are not part of EMBOSS or EMBASSY itself.

A.3.13 `template:`

This is a command line template for **SoapLab's** ACD files.

A.3.14 `comment:`

This is used for a comment about the application and is used by **SoapLab's** ACD files.

A.4 Global attributes

A.4.1 Introduction

Global attributes are available for all datatypes and can be defined in any ACD data definition as follows:

```
GlobalAttributeName: "GlobalAttributeValue"
```

Most global attributes have string or boolean values which are set explicitly in the ACD file. The booleans have a hard-coded default value which can be overridden by stating a value explicitly. Any of `"Y"`, `"Yes"`, `"N"` or `"No"` (the strings are case-insensitive) are acceptable.

The global attributes are described below and, for convenience, they are grouped by function as follows:

- Parameters and qualifiers
- User prompting
- Datatype definition
- Help information and documentation
- Hints for GUIs
- For use by **SoapLab**.

In the descriptions of the attributes, the value type and default value are indicated as follows:

```
GlobalAttributeName: Type ("Default")
```

For example, the following indicates the `parameter:` global attribute has a boolean value with a default value of `"N"`:

```
parameter: Boolean ("N")
```

In contrast, this indicates the `help:` global attribute has a string value with a default value of `""` (an empty string):

```
help: "String" ("")
```

A.4.2 Global attributes table

The global attributes are tabulated below:

A.4.3 Parameters and qualifiers

Each ACD data definition can be defined to be one of the following:

- Parameter
- Standard qualifier
- Additional qualifier

with the default being:

- Advanced qualifier.

Global attribute	Description	Default value
`default:` *string*	Default value	""
`information:` *string*	Information for menus, etc., and default prompt	""
`prompt:` *string*	Prompt (if `information:` string is unclear)	""
`code:` *string*	Code name for information/prompt which is searched for in `codes, english`	""
`help:` *string*	Text for help documentation	""
`parameter:` *boolean*	Command line parameter. Can be specified with no qualifier name. Implies standard qualifier	"N"
`standard:` *boolean*	Standard qualifier, value required. Interactive prompt if missing	"N"
`additional:` *boolean*	Additional qualifier. Value required if **`-options`** is on the command line	"N"
`missing:` *boolean*	Allow a `NULL` value on the command line	"N"
`valid:` *string*	`help:` *String* description of allowed values for **`-help`** output, used if the default help is nuclear	""
`expected:` *string*	`help:` *String* description of the expected value for **`-help`** output, used if the default help is nuclear	""
`needed:` *boolean*	Include in GUI form, used to hide options if they are unclear in GUIs	"Y"
`knowntype:` *string*	Known standard type, used to define input and output types for workflows	""
`relations:` *string*	Relationships between this ACD item and others, defined as specially formatted text	""

outputmodifier: *boolean*	Modifies the output in ways that can break parsers	"N"
style: *string*	Style for **SoapLab**'s ACD files	""
qualifier: *string*	Qualifier name for **SoapLab**'s ACD files	""
template: *string*	Commandline template for **SoapLab**'s ACD files	""
comment: *string*	Comment for **SoapLab**'s ACD files	""
pformat: "string"	File format	""
pname: "string"	Pattern base name	""

The attributes to use are as follows:

```
Parameter: "Y"
Standard:   "Y"
Additional: "Y"
```

Only one of `parameter: "Y"`, `standard: "Y"` or `additional: "Y"` should ever be given in a data definition. If more than one is used an error message will be generated during ACD processing.

You should never explicitly specify `parameter: "N"` `standard: "N"` or `additional: "N"`: an error will be generated during ACD processing if you do. The `"Y"` in the previous definitions is given for consistency because every ACD attribute, being a label:value pair, has to have a value. In practice, calculated values of `"N"` are in fact supported and can be used, in exceptional circumstances, to override the default behaviour of these attributes (see Section 4.5, 'Controlling the prompt').

For more information, see Section A.1.5, 'Parameters and qualifiers'.

A.4.3.1 `parameter:`

Defines the data definition as a *parameter*.

A value for a parameter can be given on the command line with or without a flag (parameter name). Any parameters that are specified without their flag must appear in the same order as they do in the ACD file. All parameters are required by the application and a value will be prompted for if one is not given on the command line.

For example:

```
application: acddemo
[
    documentation: "Test application"
    groups: "Test"
]
sequence: asequence
```

```
[
    parameter: "Y"
]
```

The above application could be run with a command line flag:

```
acddemo -asequence filename.seq
```

or without it:

```
acddemo filename.seq
```

In both examples *filename.seq* is the value of the parameter named asequence for the EMBOSS program **acddemo**.

In the example below, two parameters are defined:

```
application: acddemo
[
    documentation: "Test appplication"
    groups: "Test"
]
sequence: asequence
[
    parameter: Y
]
outseq: outseq
[
    parameter: Y
]
```

Unless the command line flags are used, the order in which they appear on the command line must be the same as in the ACD file:

```
acddemo infilename.seq outfilename.seq
```

This will assign the name *infilename.seq* to parameter asequence, and *outfilename.seq* to parameter outseq.

If the flags are used however the values may be given in any order:

```
acddemo -outseq outfilename.seq -asequence infilename.seq
```

Certain ACD datatypes (see below) are typically a primary input or output and as such, should be defined as parameters. The first such datatype in an ACD file *must* be defined as a parameter (an error will be generated during ACD processing otherwise). It's recommended that subsequent definitions of that datatype are also parameters (a warning will be generated if they are not). The exception is if the `default:` or `nullok:` attributes are set, in which case no warning or error messages are generated as the application can run with a default or without any value for the definition.

The datatypes are as follows:

- File input (`infile`, `filelist`, `directory` or `dirlist`)
- Sequence inputs (`sequence`, `seqall`, `seqsetall` or `seqset`)
- Sequence feature input (`feature`)
- Alignment output (`align`)
- Report output (`report:`)
- Sequence feature output (`featout`)
- File output (`outdir`).

A.4.3.2 `standard:`

Defines the data definition as a *standard qualifier*.

The behaviour of standard qualifiers is identical to parameters except that the command line flag must always be used when values are given on the command line. All standard qualifiers are required by the application and a value will be prompted for if one is not given on the command line.

Any program is expected to have one or more required inputs. An ACD data definition that is defined as a `parameter:` is automatically counted as required. All other required inputs should have the `standard:` attribute set. When the program runs, the user will be prompted for any 'required' values that are not already on the command line.

In the example below, two standard qualifiers are defined:

```
application: acddemo
[
    documentation: "Test appplication"
    groups: "Test"
]
sequence: asequence
[
    standard: Y
]
outseq: outseq
[
    standard: Y
]
```

The flags must be used on the command line and the values may be given in any order:

```
acddemo -asequence infilename.seq -outseq outfilename.seq
acddemo -outseq outfilename.seq -asequence infilename.seq
```

If either value had been omitted, the user would be prompted for it.

A.4.3.3 `additional:`"*Boolean*" (`"N"`)

Defines the data definition as an *additional qualifier*.

The behaviour of additional qualifiers is identical to standard qualifiers except that the value will *not* be prompted for if not given on the command line, *unless* the application is run with the **`-options`** global qualifier, in which case all additional qualifiers are prompted for.

Additional qualifiers are taken to be not absolutely required by the application. The command line flag must be always be used when values are given on the command line.

In the example below, two additional qualifiers are defined:

```
application: acddemo
[
    documentation: "Test appplication"
    groups: "Test"
]
sequence: asequence
[
    additional: Y
]
outseq: outseq
[
    additional: Y
]
```

The flags must be used on the command line and the values may be given in any order:

```
acddemo -asequence infilename.seq -outseq outfilename.seq
acddemo -outseq outfilename.seq -asequence infilename.seq
```

If either value had been omitted, the user would *not* be prompted for it, unless the application was run with **`-options`**:

```
acddemo -asequence infilename.seq -options
```

In this example, a value is specified for `asequence` but omitted for `outseq`. The **`-options`** flag is set which means the user will be prompted for a value for `outseq`.

A.4.4 User prompting

A.4.4.1 `information:`"*String*" (`""`)

Very succinct description of the parameter.

The text is used in web forms and other GUIs. It is also the text used to prompt the user for a value at the command line. Only one of `code:`, `prompt:` or `information` should ever be defined. The use of `information:` (with a standard name, see below) is preferred instead of `code:` or `prompt:`.

To provide standard prompts, a default value for the `information:` string is defined for most common datatypes. The defaults are in the EMBOSS system file `codes.english` in the application ACD file directory (see Section 4.3.5.2.1, 'Standard prompts file (`codes.english`)').

The standard practice is to use the default prompt where possible. A warning is generated during ACD processing if either the `information:` attribute is missing or, if there is standard value available, the value is not the standard value.

If a non-standard prompt is used, the text given after the `information:` attribute should conform to the following conventions (a warning will be generated during ACD processing otherwise):

- The string should start with an upper-case, alphabetic character
- The string should not end with a full stop.

A.4.4.2 `code:`"*String*" ("")

A code word which is searched for in the EMBOSS system file `codes.english` (Section 4.3.5.2.1, 'Standard prompts file (codes english)') to give a standard prompt. `codes english` gives the standard prompts for all common ACD datatypes.

For example, gap insertion and extension penalties may be defined as either integer or floating point values. The standard prompt for these is identified by the code `GAP` or `GAPEXT` in `codes.english`. These prompts can be used with the `code:` attribute, for example `code: "GAP"`. The code word is not case-sensitive, so `code: "gap"` is also acceptable.

In practice, GUI developers found the `code:` attributes hard to use, so the `information:` attribute tends to be used instead.

A.4.4.3 `prompt:` "*String*" ("")

The string used to prompt the user for a value at the command line.

The use of `prompt:` is now deprecated; `information`: is preferred. In rare cases where the `information:` string is misleading, however, a `prompt:` string can be defined.

The text given after the `prompt:` attribute should conform to the following conventions (a warning will be generated during ACD processing otherwise):

- The string should start with an upper-case, alphabetic character
- The string should not end with a full stop.

A.4.5 Datatype definition

A.4.5.1 `knowntype:` "*String*" ("")

The known type of a data definition taken from a controlled vocabulary of known value types.

`knowntype:` should be defined where the type is not already clear from the datatype itself. It is typically defined for `string`, `infile`, `outfile` and `outfileall` datatypes but not, for example, for a `sequence:`.

A warning is generated for the `string`, `outfile` and `outdir` ACD datatypes if it is not specified.

The standard values (known types) are listed in the EMBOSS system file `knowntypes.standard` in the application ACD file directory.

A few other values not listed in `knowntypes.standard` are accepted, for example "*ApplicationName* output" for an `outfile:` datatype. These are documented with the datatypes (Section A.5, 'Datatype-specific attributes'). If a value is given that is not a standard known type or other accepted value, a warning message will be generated during ACD processing. The **acdvalid** utility will check all `knowntype` values in an ACD file, and report any missing values for data definitions that require a known type.

A.4.5.2 `default:` "*Value*" ("")

Default value for the parameter.

For many but not all data definitions it is appropriate to define a default value. For some datatypes, a default is hard-coded or generated by a naming scheme from datatype-specific attributes. See the datatypes descriptions (Section A.5, 'Datatype-specific attributes') for further information.

The default value can be dependent on the values of parameters defined earlier; this is done using an ACD operation to set the value (see Section A.8, 'Operations').

A.4.5.3 `relations:` "*String*" ("")

Text to describe the relationship between this ACD data definition and others.

A.4.5.4 `outputmodifier:` "*boolean*" ("N")

This is set to `"True"` for data definitions which control the fields that are written in the application output.

In practice `outputmodifier:` is used exclusively for `boolean` and `toggle` datatypes which modify the output in a way that might be problematic in parsing software.

A.4.5.5 `missing:` "*boolean*" ("N")

Indicates whether a null value can be specific on the command line for the parameter. This is only ever required for output files for which there is a default file naming scheme available but a default value is also specified in the ACD file. In such cases, the default value in the ACD file can be overridden and the naming scheme used if a null value (`""`) for the parameter is given on the command line.

A.4.6 Help information and documentation

A.4.6.1 `help:` "*String*" ("")

Helpful information for the parameter which is displayed when the application is run with the **`-help`** qualifier on the command line, and used in the application documentation.

`help:` is usually only defined if a deeper explanation of an application parameter is needed. If `help:` is not defined, the value of the `information:` attribute (if available) or the default help string will be used instead. Default help strings are given for each datatype in the EMBOSS system file `codes.english` file (Section 4.3.5.4.1, 'Standard help strings file (`codes.english` file)').

The text given after the `help:` attribute should conform to the following conventions (a warning will be generated during ACD processing otherwise):

- The string should start with an upper-case, alphabetic character
- The string should not end with a full stop.

A.4.6.2 `valid:` "*String*" ("")

A description of the allowed values of the parameter.

Used in the 'Allowed values' column of the command line syntax table of the application documentation. This table is automatically generated from the ACD file. In most cases a reasonable value is generated automatically but where there is no suitable value, this attribute should be used to provide one.

A.4.6.3 `expected:` "*String*" ("")

A description of the expected values of the parameter.

Used in the 'Expected values' column of the command line syntax table of the application documentation. This table is automatically generated from the ACD file. In most cases a reasonable value is generated automatically but where there is no suitable value, this attribute should be used to provide one.

A.4.7 Hints for GUIs

A.4.7.1 `needed:` *boolean* ("Y")

Indicates whether a parameter is expected to be included in a GUI form. This is a hint for use by GUI developers. Some parameters are available on the command line, but are not generally useful to users, or can cause confusion when presented in a GUI form with all other options.

A.4.8 For use by SoapLab

A.4.8.1 `qualifier:` "*String*" ("")

Qualifier name for SoapLab's ACD files. It should not be used in standard EMBOSS application ACD files.

A.4.8.2 `template:` "*String*" ("")

Command line template for **SoapLab**'s ACD files. It should not be used in standard EMBOSS application ACD files.

A.4.8.3 `comment:` "*String*" ("")

A comment for **SoapLab**'s ACD files. It should not be used in standard EMBOSS application ACD files.

A.5 Datatype-specific attributes

Datatype-specific attributes are available (can be defined) for certain ACD datatypes only. Each datatype has its own set of these attributes. Additionally, there are various datatype-associated command line qualifiers (or simply 'associated qualifiers') that are inbuilt for certain ACD datatypes and may also be hard-coded as attributes in the appropriate data definition. Both types of attribute are described in detail below. Their value is set explicitly in an ACD data definition as follows:

```
DataAttributeName: "DataAttributeValue"
```

In the descriptions of the attributes, the value type and default value are indicated as follows:

```
DataAttributeName: Type ("Default")
```

A.5.1 Simple attributes description

A.5.1.1 array

size: "*integer*" ("1") The number of elements in the array.

minimum: "*float*" ("-FLT_MIN") Lower value limit. Defaults to the boundaries as specified by the systems setup.

maximum: "*float*" ("+FLT_MAX") Upper value limit. Defaults to the boundaries as specified by the systems setup.

sum: "*float*" ("1.0") Total for all values in the array (tested unless the sumtest: attribute is false).

sumtest: "*boolean*" ("Y") If false, turns off testing for the sum: attribute.

tolerance: "*float*" ("0.01") Specifies how closely the sum: should match the total. Most floating point fractions cannot be represented accurately in binary form.

warnrange: "*boolean*" ("Y") Controls whether a warning is issued if values are given that are out of range.

increment: "*integer*" ("0") Defines the steps the array values may take when iterated. The increment: can be any valid float value. Implemented for GUIs to EMBOSS only and is not used by EMBOSS itself (it is disregarded during ACD file processing).

precision: "*integer*" ("0") Defines the maximum number of significant decimal places that will be taken into account for this value. Implemented for GUIs to EMBOSS only and is not used by EMBOSS itself (it is disregarded during ACD file processing).

A.5.1.1.1 Associated qualifiers

The array datatype has no associated qualifiers.

A.5.1.2 `boolean`

`boolean` currently has no specific attributes.

A.5.1.2.1 Associated qualifiers

The `boolean` datatype has no associated qualifiers.

> **Caution**
>
> Care should be taken over the definition of the **`information:`** and **`help:`** global attributes. These are used to prompt the user (interactively or via a GUI), and to provide help text. The text provided in each case should reflect the expected default value of the boolean option, which may be the opposite of what the name implies. For example, if set to **`"Y"`** by default, then the command line option would typically be **`-no`*`Flag`*** where ***`Flag`*** is the qualifier. If set to **`"N"`** by default, then the default action may be the opposite of what the information or help text implies. If the value is calculated, the user may need some extra guidance.

A.5.1.3 `float`

`minimum:` **`"`*`float`*`"`** **`("-FLT_MIN")`** Lower value limit. Defaults to the boundaries as specified by the systems setup.

`maximum:` **`"`*`float`*`"`** **`("+FLT_MAX")`** Upper value limit. Defaults to the boundaries as specified by the systems setup.

`increment:` **`"`*`float`*`"`** **`("1")`** The steps that this parameter is allowed to take, in case there is a need to iterate this parameter. The `increment:` attribute can be any valid float value.

`precision:` **`"`*`integer`*`"`** **`("3")`** The maximum number of significant decimal places that will be taken into account for this value.

`warnrange:` **`"`*`boolean`*`"`** **`("Y")`** Controls whether a warning is issued if values are given that are out of range.

A.5.1.3.1 Associated qualifiers

The `float` datatype has no associated qualifiers.

A.5.1.4 `integer`

`minimum:` **`"`*`integer`*`"`** **`("INT_MIN")`** Lower value limit. Defaults to the boundaries as specified by the systems setup.

`maximum:` **`"`*`integer`*`"`** **`("INT_MAX")`** Upper value limit. Defaults to the boundaries as specified by the systems setup.

`increment:` **`"`*`integer`*`"`** **`("0")`** Defines the steps that this parameter is allowed to take, in case there is a need to iterate this parameter.

`warnrange:` **`"`*`boolean`*`"`** **`("Y")`** Controls whether a warning is issued if values are given that are out of range.

A.5.1.4.1 Associated qualifiers

The `integer` datatype has no associated qualifiers.

A.5.1.5 `range`

`minimum:` **`"integer"`** **`("1")`** Lower value limit.

`maximum:` **`"integer"`** **`("INT_MAX")`** Upper value limit. Defaults to the boundaries as specified by the systems setup.

`minsize:` **`"integer"`** **`("0")`** Where multiple ranges are defined `minsize:` defines the minimum number of ranges that must be defined.

`size:` **`"integer"`** **`("0")`** Exact number of values required.
Only one of `size:` or `minsize:` should be defined for any range.

A.5.1.5.1 Associated qualifiers
The `range` datatype has no associated qualifiers.

A.5.1.6 `regexp`

`minlength:` **`"integer"`** **`("1")`** The minimum length the regular expression string can be.

`maxlength:` **`"integer"`** **`("INT_MAX")`** The maximum length the regular expression string can be.

`upper:` **`"boolean"`** **`("N")`** Convert the regular expression to upper case before compiling.

`lower:` **`"boolean"`** **`("N")`** Convert the regular expression to lower case before compiling.

`maxsize:` **`"integer"`** **`("INT_MAX")`** The maximum number of patterns.

`type:` The type of pattern, currently one of `string`, `nucleotide` or `protein`.

A.5.1.6.1 Associated qualifiers
The `regexp` datatype has the following associated qualifiers which may be hard-coded as attributes in an ACD file. These qualifiers are described in the *EMBOSS User's Guide*:

A.5.1.7 `pattern`

`minlength:` **`"integer"`** **`("1")`** The minimum length the pattern can be.

`maxlength:` **`"integer"`** **`("INT_MAX")`** The maximum length the pattern can be.

`upper:` **`"boolean"`** **`("N")`** Convert the pattern to upper-case before compiling.

`lower:` **`"boolean"`** **`("N")`** Convert the pattern to lower-case before compiling.

`maxsize:` **`"integer"`** **`("INT_MAX")`** The maximum number of patterns.

`type:` The type of pattern, currently one of `string`, `nucleotide` or `protein`.

A.5.1.7.1 Associated qualifiers
The `pattern` datatype has the following associated qualifiers which may be hard-coded as attributes in an ACD file. These qualifiers are described in the *EMBOSS User's Guide*:

Value	Type(s)	Gaps	Ambiguity codes	Conversions	Description
any	Nucleotide or protein	Removed	Yes	'?'=>'X'	Any valid sequence
gapany	Nucleotide or protein	Kept	Yes	'?'=>'X'	Any valid sequence with gaps
dna	Nucleotide only	Removed	Yes	'?'=>'N' 'X'=>'N' 'U'=>'T'	DNA sequence
puredna	Nucleotide only	Removed	No	'U'=>'T'	DNA sequence, bases ACGT only
gapdna	Nucleotide only	Kept	Yes	'?'=>'N' 'X'=>'N' 'U'=>'T'	DNA sequence with gaps
gapdnaphylo	Nucleotide only	Kept	Yes	'U'=>'T'	DNA sequence with gaps and queries
rna	Nucleotide only	Removed	Yes	'?'=>'N' 'X'=>'N' 'T'=>'U'	RNA sequence
purerna	Nucleotide only	Removed	No	'T'=>'U'	RNA sequence, bases ACGU only
gaprna	Nucleotide only	Kept	Yes	'?'=>'N' 'X'=>'N' 'T'=>'U'	RNA sequence with gaps
gaprnaphylo	Nucleotide only	Kept	Yes	'T'=>'U'	RNA sequence with gaps and queries
nucleotide	Nucleotide only	Removed	Yes	'?'=>'N' 'X'=>'N'	nucleotide sequence
purenucleotide	Nucleotide only	Removed	No		Nucleotide sequence, bases ACGTU only
gapnucleotide	Nucleotide only	Kept	Yes	'?'=>'N' 'X'=>'N'	Nucleotide sequence with gaps
gapnucleotidephylo	Nucleotide only	Kept	Yes		Nucleotide sequence with gaps and queries
protein	Protein only	Removed	Yes	'?'=>'X' '*'=>'X'	Protein sequence

pureprotein	Protein only	Removed	No		Protein sequence without BZ U X or *
stopprotein	Protein only	Removed	Yes	'?pp '=>'X'	Protein sequence with possible stops
gapprotein	Protein only	Kept	Yes	'?'=>'X' '*'=>'X'	Protein sequence with gaps
gapstopprotein	Protein only	Kept	Yes	'?'=>'X'	Protein sequence with gaps and possible stops
gapproteinphylo	Protein only	Kept	Yes		Protein sequence with gaps, stops and queries
proteinstandard	Protein only	Removed	Yes	'?'=>'X' '*'=>'X' 'U'=>'X'	Protein sequence with no selenocysteine
stopproteinstandard	Protein only	Removed	Yes	'?'=>'X' 'U'=>'X'	Protein sequence with a possible stop but no selenocysteine
gapproteinstandard	Protein only	Kept	Yes	'?'=>'X' '*'=>'X' 'U'=>'X'	Protein sequence with gaps but no selenocysteine

pformat: "string" ("")	File format
pmismatch: "integer" ("")	Pattern mismatch
pname: "string" ("")	Pattern base name

A.5.1.8 string

minlength: "*integer*" ("0") Minimum length of the string. The default minimum length is zero.

maxlength: "*integer*" ("INT_MAX") Maximum length of the string. There is no default maximum.

pattern: "*string*" ("") Defines a regular expression used to validate the string value: an invalid string will be reprompted for. EMBOSS uses the Perl-compatible regular expression library (PCRE) so any Perl-compatible regular expression should be usable.

word: "*boolean*" ("N") If set requires the result to be a valid word with no whitespace. Note that the default minimum length of zero allows an empty string; however, this is *not* accepted as a word.

upper: "*boolean*" ("N") Converts the case of the string internally to upper-case.

lower: "*boolean*" ("N") Converts the case of the string internally to lower-case. Only one of `upper:` or `lower:` should be specified.

A.5.1.8.1 Associated qualifiers

The `string` datatype has no associated qualifiers.

A.5.1.9 `toggle`

Although there are (currently) no specific attributes for a `toggle` ACD type, care should be taken over the definition of the `information:` and `help:` attributes. These are used to prompt the user (interactively or via a GUI), and to provide help text. The text provided in each case should reflect the expected default value of the toggle option, which may be the opposite of what the name implies. For example, if set to `"Y"` by default, then the command line option would typically be **-no*Flag*** where *`Flag`* is the qualifier. If set to `"N"` by default, then the default action may be the opposite of what the information or help text implies. If the value is calculated, the user may need some extra guidance.

A.5.1.9.1 Associated qualifiers

The `toggle` datatype has no associated qualifiers.

A.5.2 Input attributes description

A.5.2.1 `codon`

name: "*string*" ("Ehum.cut") Individual codon data definitions can set their own default codon usage tables with the `name:` attribute. Alternatively, this can be set via the global attribute `default:`.

nullok: "*boolean*" ("N") If the application can accept a null value for this definition and can run without a codon usage table, the `nullok:` attribute must be set to `"Y"`. In such cases the codon usage file might not be specified (in the ACD file or on the command line) for the data definition. This might occur when the codon usage table is only required by the application under certain circumstances.

A.5.2.1.1 Associated qualifiers

The `codon` datatype has the following associated qualifiers which may be hard-coded as attributes in an ACD file. These qualifiers are described in the *EMBOSS User's Guide*:

format: "*string*" ("") Data format.

A.5.2.2 cpdb

name: "*string*" ("") Individual cpdb inputs can set their own default names with the name: attribute which in the current version has the same effect as setting the default: global attribute.

nullok: "*boolean*" ("N") If the application can accept a null value for this definition and can run without clean PDB file input, the nullok: attribute must be set to "Y". In such cases a clean coordinate file might not be specified (in the ACD file or on the command line) for the data definition. This might occur when the file is only required by the application under certain circumstances.

A.5.2.2.1 Associated qualifiers

The cpdb datatype has the following associated qualifiers which may be hard-coded as attributes in an ACD file. These qualifiers are described in the *EMBOSS User's Guide*:

format: "*string*" ("") Data format.

A.5.2.3 datafile

name: "*string*" ("") Base name of the default data file.

extension: "*string*" ("") File extension of the default data file.

directory: "*string*" ("") The EMBOSS data sub-directory to be searched.

nullok: "*boolean*" ("N") If the application can accept a null value for this definition and can run without a data file, the nullok: attribute must be set to "Y". This allows a default value to be omitted, or **-no*Flag*** (where *Flag* is the name of the definition), to be specified on the command line.

A.5.2.3.1 Associated qualifiers

The datafile datatype has no associated qualifiers.

A.5.2.4 directory

extension: "*string*" ("") The file extension for all files read from the directory. Files with other extensions will not be read.

fullpath: "*boolean*" ("N") If set to "Y" will interpret the specified directory as a full path rather than a relative path.

nullok: "*boolean*" ("N") If the application can accept a null value for this definition and can run without a directory the nullok: attribute must be set to "Y". In such cases a default directory might not be set, or **-no*Flag***, where *Flag* is the name of the data definition flag, might be given on the command line.

nulldefault: "*boolean*" ("N") By default, names for files in the directory are constructed from the path and extension as described above. The nulldefault: attribute overrides the default name generation, and uses an empty string (no directory) by default. This is useful for programs which only occasionally require a directory. If an empty string is specified for the directory on the command line, the default name generation will be used. In combination with the `nullok:`, this allows qualifiers to be null by default, and turned on from the command line.

A.5.2.4.1 Associated qualifiers

The `directory` datatype has no associated qualifiers.

A.5.2.5 `dirlist`

`extension: "`*`string`*`" ("")` The extension for all files read from the directories. Files with other extensions will not be read.

`fullpath: "`*`boolean`*`" ("N")` If set to `"Y"` will interpret the given path as a full rather than a relative path specification.

`nullok: "`*`boolean`*`" ("N")` If the application can accept a null value for this definition and can run without a directory the `nullok:` attribute must be set to `"Y"`. In such cases a default directory might not be set, or **-no*Flag***, where *`Flag`* is the name of the data definition flag, might be given on the command line.

A.5.2.5.1 Associated qualifiers

The `dirlist` datatype has no associated qualifiers.

A.5.2.6 `discretestates`

`length: "`*`integer`*`" ("0")` The number of state values (the length of the discrete characters string) in each set.

`size: "`*`integer`*`" ("1")` The number of sets of values (usually 1) but some programs will accept multiple sets.

`characters: "`*`string`*`" ("01")` Defines which discrete state characters can be specified. This is defined as a string containing all possible characters.

`nullok: "`*`boolean`*`" ("N")` If the application can accept a null value for this definition and can run without a `discretestates` file, the `nullok:` attribute must be set to `"Y"`. This allows the application to run without a default value or to be run with - **no*Flag*** (where *`Flag`* is the name of the data definition) on the command line.

A.5.2.6.1 Associated qualifiers

The `discretestates` datatype has no associated qualifiers.

A.5.2.7 `distances`

`size: "`*`integer`*`" ("1")` The number of rows in the distance matrix.

`missval: "`*`boolean`*`" ("N")` If set to `"Y"` means the matrix can have missing values.

`nullok: "`*`boolean`*`" ("N")` If the application can accept a null value for this definition and can run without a distance matrix file, the `nullok:` attribute must be set to `"Y"`. This allows the application to run without a default value or to be run with **-no*Flag*** (where *`Flag`* is the data definition flag) on the command line.

A.5.2.7.1 Associated qualifiers

The `distances:` datatype has no associated qualifiers.

A.5.2.8 features

type: "*string*" ("") Whether the feature input is 'protein' or 'nucleotide'. There is a default based on the type of any input sequence, but a value should always be specified.

nullok: "*boolean*" ("N") If the application can accept a null value for this definition and can run without features input, the nullok: attribute must be set to "Y". This allows the application to run without a default value or to be run with **-no*Flag*** (where *Flag* is the data definition flag) on the command line.

A.5.2.8.1 Associated qualifiers

The features: datatype has the following associated qualifiers which may be hard-coded as attributes in an ACD file. These qualifiers are described in the *EMBOSS User's Guide*:

fformat: "string" ("")	Features format
fopenfile: "string" ("")	Features file name
fask: "boolean" ("N")	Prompt for begin/end/reverse
fbegin: "integer" ("0")	Start of the features to be used
fend: "integer" ("0")	End of the features to be used
freverse: "boolean" ("N")	Reverse (if DNA)

A.5.2.9 filelist

nullok: "*boolean*" ("N") If the application can accept a null value for this definition and can run without a filelist input, the nullok: attribute must be set to "Y". This specifies that a missing input file is acceptable to the application, and that **-no*Flag*** (where *Flag* is the data definition flag) can be used on the command line.

A.5.2.9.1 Associated qualifiers

The filelist: datatype has no associated qualifiers.

A.5.2.10 frequencies

length: "*integer*" ("0") Defines the number of loci (or values) in the frequencies file.

size: "*integer*" ("1") The number of sets of values, usually 1 but some programs will accept multiple sets.

continuous: "*boolean*" ("N") Specifies a frequencies file with continuous character data values.

genedata: "*boolean*" ("N") Specifies a frequencies file with genetic locus data values.

within: "*boolean*" ("N") Specifies a frequencies file with continuous data for multiple individuals (additional values on each line).

nullok: "*boolean*" ("N") If the application can accept a null value for this definition and can run without a frequencies file, the nullok: attribute must be set to "Y". This allows the application to be run without a default value or with **-no*Flag*** (where *Flag* is the of the data definition flag) on the command line.

A.5.2.10.1 Associated qualifiers

The frequencies: datatype has no associated qualifiers.

A.5.2.11 `infile`

`nullok: "`*`boolean`*`" ("N")` Specifies that a missing input file is acceptable to the application. The application can then run without a default value or with **`-no`*`Flag`*** (where *`Flag`* is the name of the data definition) on the command line.

`directory: "@(value:`*`ENVARNAME`*`"` Specifies a directory in which to search for the file. The directory is given by the environment variable *ENVARNAME*, which is resolved first as `EMBOSS_`*`ENVARNAME`* (in `emboss.default` or set by a **`setenv`**) call. It then checks for *ENVARNAME* (set using **`setenv`**). If an explicit path is given in the filename by the user, the directory is ignored.

A.5.2.11.1 Associated qualifiers

The `infile` datatype has no associated qualifiers.

A.5.2.12 `matrix`

`protein: "`*`boolean`*`" ("Y")` If set to `"Y"` will specify that the scoring matrix is a protein matrix. If not set the matrix is presumed to be for nucleic acids.

`pname: "`*`string`*`" ("EBLOSUM62")` Defines the default protein matrix to use. The system default is `EBLOSUM62` for protein matrices.

`nname: "`*`string`*`" ("EDNAFULL")` Defines the default nucleotide matrix to use. The system default is `EDNAFULL` for nucleotide matrices.

A.5.2.12.1 Associated qualifiers

The `matrix:` datatype has no associated qualifiers.

A.5.2.13 `matrixf`

`protein: "`*`boolean`*`" ("Y")` If set to `"Y"` will specify that the scoring matrix is a protein matrix. If not set the matrix is presumed to be for nucleic acids.

`pname: "`*`string`*`" ("EBLOSUM62")` Defines the default protein matrix to use. The system default is `EBLOSUM62` for protein matrices.

`nname: "`*`string`*`" ("EDNAFULL")` Defines the default nucleotide matrix to use. The system default is `EDNAFULL` for nucleotide matrices.

A.5.2.13.1 Associated qualifiers

The `matrixf:` datatype has no associated qualifiers.

A.5.2.14 `properties`

`length: "`*`integer`*`" ("0")` The number of values in the properties file.

`size: "`*`integer`*`" ("1")` The number of sets of values, usually 1 but some programs will accept multiple sets.

`characters: "`*`string`*`" ("")` Defines which property characters can be specified. This is defined as a string containing all possible characters.

nullok: "*boolean*" ("N") If the application can accept a null value for this definition and can run without a properties file, the `nullok:` attribute must be set to `"Y"`. This allows a default value to be omitted or the application to be run with **-no*Flag*** on the command line (where *Flag* is the data definition flag).

A.5.2.14.1 Associated qualifiers

The `properties:` datatype has no associated qualifiers.

A.5.2.15 scop

nullok: "*boolean*" ("N") Used to set defaults for individual `scop` inputs and has the same effect as setting the `default:` attribute.

A.5.2.15.1 Associated qualifiers

The `scop` datatype has the following associated qualifiers which may be hard-coded as attributes in an ACD file. These qualifiers are described in the *EMBOSS User's Guide*:

format: "*string*" ("") Data format.

A.5.2.16 sequence

type: "*string*" ("") Forces the sequence to be of the given type. By default any sequence type is accepted. Ideally the `type:` attribute will always be set so that the accepted input sequence type is clear. The sequence type must be a standard sequence type (Section A.7, 'Sequence types').

features: "*boolean*" ("N") If the `features:` attribute is set, the sequence input will include feature information either in the same file (if the sequence format supports it) or in a separate file, which by default is in GFF format.

entry: "*boolean*" ("N") When set, the sequence input will include the full original text of the input sequence or database entry.

nullok: "*boolean*" ("N") If the application can accept a null value for this definition and can run without sequence input, the `nullok:` attribute must be set to `"Y"`. This allows a default value to be omitted or the application to be run with **-no*Flag*** on the command line (where *Flag* is the data definition flag).

A.5.2.16.1 Associated qualifiers

The `sequence` datatype has the following associated qualifiers which may be hard-coded as attributes in an ACD file. These qualifiers are described in the *EMBOSS User's Guide*:

`sbegin: "integer" ("0")`	Start of the sequence to be used
`send: "integer" ("0")`	End of the sequence to be used
`sreverse: "boolean" ("N")`	Reverse (if DNA)
`sask: "boolean" ("N")`	Ask for begin/end/reverse

`snucleotide: "boolean" ("N")`	**Sequence is nucleotide**
`sprotein: "boolean" ("N")`	**Sequence is protein**
`slower: "boolean" ("N")`	**Make lower-case**
`supper: "boolean" ("N")`	**Make upper-case**
`sformat: "string" ("")`	**Input sequence format**
`sdbname: "string" ("")`	**Database name**
`sid: "string" ("")`	**Entryname**
`ufo: "string" ("")`	**UFO features**
`fformat: "string" ("")`	**Features format**
`fopenfile: "string" ("")`	**Features filename**

A.5.2.17 `seqall`

`type: "`*`string`*`" ("")` Forces the sequence(s) to be of the given type. By default any sequence type is accepted. The `type:` attribute is normally set so that the accepted input sequence type is clear. The sequence type must be a standard sequence type (Section A.7, 'Sequence types').

`features: "`*`boolean`*`" ("N")` If the `features:` attribute is set, the sequence input will include feature information either in the same file (if the sequence format supports it) or in a separate file (by default in GFF format).

`entry: "`*`boolean`*`" ("N")` If set, the sequence input will include the full original text of the input sequence or database entry.

`nullok: "`*`boolean`*`" ("N")` If the application can accept a null value for this definition and can run without sequence input, the `nullok:` must be set to `"Y"`. This allows a default value to be omitted or the application to be run with **`-no`*`Flag`*** on the command line (where *`Flag`* is the data definition flag).

`minseqs: "`*`integer`*`" ("1")` Minimum number of sequences to be read. By default, a minimum of one sequence is acceptable.

`maxseqs: "`*`integer`*`" ("INT_MAX")` Maximum number of sequences to be read. By default, there is no maximum.

A.5.2.17.1 Associated qualifiers

The `seqall` datatype has the following associated qualifiers which may be hard-coded as attributes in an ACD file. These qualifiers are described in the *EMBOSS User's Guide*:

`sbegin: "integer" ("0")`	Start of the sequence to be used
`send: "integer" ("0")`	End of the sequence to be used
`sreverse: "boolean" ("N")`	Reverse (if DNA)

`sask: "boolean" ("N")`	**Ask for begin/end/reverse**
`snucleotide: "boolean" ("N")`	**Sequence is nucleotide**
`sprotein: "boolean" ("N")`	**Sequence is protein**
`slower: "boolean" ("N")`	**Make lower-case**
`supper: "boolean" ("N")`	**Make upper-case**
`sformat: "string" ("")`	**Input sequence format**
`sdbname: "string" ("")`	**Database name**
`sid: "string" ("")`	**Entryname**
`ufo: "string" ("")`	**UFO features**
`fformat: "string" ("")`	**Features format**
`fopenfile: "string" ("")`	**Features filename**

A.5.2.18 `seqset`

`type:` *"string"* `("")` Forces the sequence set to be of the given type. By default any sequence type is accepted. Ideally the `type:` attribute will always be set so that the accepted input sequence type is always clear. The sequence type must be a standard sequence type (Section A.7, 'Sequence types').

`aligned:` *"boolean"* `("N")` If set, specifies that all sequences in the input are expected to be aligned. If false then the sequences are assumed to be unaligned, and are simply read into memory together for processing. We recommend always defining the `aligned:` attribute so that the nature of the sequence set is clearly defined.

`features:` *"boolean"* `("N")` If the `features:` attribute is set, the sequence input will include feature information either in the same file (if the sequence format supports it) or in a separate file (by default in GFF format).

`entry:` *"boolean"* `("N")` If set, the sequence input will include the full original text of the input sequence or database entry.

`nullok:` *"boolean"* `("N")` If the application can accept a null value for this definition and can run without sequence input, the `nullok:` attribute must be set to `"Y"`. This allows a default value to be omitted or the application to be run with **-no*Flag*** on the command line (where *`Flag`* is the data definition flag).

`nulldefault:` *"boolean"* `("N")` The `nulldefault:` attribute overrides the default name generation, and uses an empty string (no sequence input) as the default for programs where sequence input is only occasionally required. If an empty string is specified on the command line, the standard default value will be generated instead. In combination with the `nullok:` and `missing:` attributes, this allows qualifiers to be null by default, and turned on from the command line.

minseqs: "*integer*" ("1") Minimum number of sequences to be read. By default, a minimum of one sequence is acceptable.

maxseqs: "*integer*" ("INT_MAX") Maximum number of sequences to be read. By default, there is no maximum.

A.5.2.18.1 Associated qualifiers

The seqset datatype has the following associated qualifiers which may be hard-coded as attributes in an ACD file. These qualifiers are described in the *EMBOSS User's Guide*:

sbegin: "integer" ("0")	Start of the sequence to be used
send: "integer" ("0")	End of the sequence to be used
sreverse: "boolean" ("N")	Reverse (if DNA)
sask: "boolean" ("N")	Ask for begin/end/reverse
snucleotide: "boolean" ("N")	Sequence is nucleotide
sprotein: "boolean" ("N")	Sequence is protein
slower: "boolean" ("N")	Make lower-case
supper: "boolean" ("N")	Make upper-case
sformat: "string" ("")	Input sequence format
sdbname: "string" ("")	Database name
sid: "string" ("")	Entryname
ufo: "string" ("")	UFO features
fformat: "string" ("")	Features format
fopenfile: "string" ("")	Features filename

A.5.2.19 seqsetall

type: "*string*" ("") Forces the sequence set(s) to be of the given type. By default any sequence type is accepted. Ideally the type: attribute will be set so that the accepted input sequence type is clear. The sequence type must be a standard sequence type (Section A.7, 'Sequence types').

aligned: "*boolean*" ("N") If true, it specifies that all sequences in the input are expected to be aligned. If false then the sequences are assumed to be unaligned, and are simply read into memory together for processing. Ideally the aligned: attribute should always be set so that the nature of the sequence set is clearly defined.

features: "*boolean*" ("N") If set, will make the sequence input include feature information either in the same file (if the sequence format supports it) or in a separate file (by default in GFF format).

entry: "*boolean*" ("N") If set, the sequence input will include the full original text of the input sequence or database entry.

nullok: "*boolean*" ("N") If the application can accept a null value for this definition and can run without sequence input, the `nullok:` attribute must be set to `"Y"`. This allows a default value to be omitted or the application to be run with **-no*Flag*** on the command line (where *Flag* is the data definition flag).

nulldefault: "*boolean*" ("N") The `nulldefault:` attribute overrides the default name generation, and uses an empty string (no sequence input) as the default for programs where sequence input is only occasionally required. If an empty string is specified on the command line, the standard default value will be generated instead. In combination with the `nullok:` and `missing:` attributes, this allows qualifiers to be null by default, and turned on from the command line.

minsets: "*integer*" ("1") Minimum number of sequence sets to be read. By default, a minimum of one sequence set is acceptable.

maxsets: "*integer*" ("INT_MAX") Maximum number of sequence sets to be read. By default, there is no maximum (other than system-imposed limits on the size of an integer).

minseqs: "*integer*" ("1") Minimum number of sequences to be read. By default, a minimum of one sequence is acceptable.

maxseqs: "*integer*" ("INT_MAX") Maximum number of sequences to be read. By default, there is no maximum.

A.5.2.19.1 Associated qualifiers

The `seqsetall` datatype has the following associated qualifiers which may be hard-coded as attributes in an ACD file. These qualifiers are described in the *EMBOSS User's Guide*:

`sbegin: "integer" ("0")`	Start of the sequence to be used
`send: "integer" ("0")`	End of the sequence to be used
`sreverse: "boolean" ("N")`	Reverse (if DNA)
`sask: "boolean" ("N")`	Ask for begin/end/reverse
`snucleotide: "boolean" ("N")`	Sequence is nucleotide
`sprotein: "boolean" ("N")`	Sequence is protein
`slower: "boolean" ("N")`	Make lower-case
`supper: "boolean" ("N")`	Make upper-case
`sformat: "string" ("")`	Input sequence format
`sdbname: "string" ("")`	Database name
`sid: "string" ("")`	Entryname
`ufo: "string" ("")`	UFO features
`fformat: "string" ("")`	Features format
`fopenfile: "string" ("")`	Features filename

A.5.2.20 tree

size: "*integer*" ("0") The number of trees in the input file. Some can only accept a single tree so the value should be set to "1" for these. A value of "0" means that any number of trees is acceptable.

nullok: "*boolean*" ("N") If the application can accept a null value for this definition and can run without tree input, the nullok: attribute must be set to "Y". This allows a default value to be omitted or the application to be run with **-no*Flag*** on the command line (where *Flag* is the data definition flag).

A.5.2.20.1 Associated qualifiers

The tree datatype has no associated qualifiers.

A.5.3 Output attributes description

A.5.3.1 align

minseqs: "*integer*" ("1") The minimum number of sequences that are expected in the output alignment.

The minseqs: and maxseqs: attributes can be used to validate application output, for example, when selecting formats on the command line with the **-aformat** qualifier.

maxseqs: "*integer*" ("INT_MAX") The maximum number of sequences that are expected in the output alignment.

The minseqs: and maxseqs: attributes can be used to validate application output, for example, when selecting formats on the command line with the **-aformat** qualifier.

multiple: "*boolean*" ("N") Should be set to "Y" if the output can contain more than one alignment from the same input.

type: The sequence type. Although this will default to the type of the first output sequence, it is recommended that a value is always defined to make the output sequence type clear.

taglist: "*string*" ("") Defines additional tags to be reported from the internal feature table for the aligned sequences. The tag names and types must match the source code of the application. Each tag is in the format type:tagname[=columnname], for example int:length or string:gc=GC%.

nullok: "*boolean*" ("N") If the application can accept a null value for this definition and can run without generating alignment output, the nullok: attribute must be set to "Y". This allows a default value to be omitted or the application to be run with **-no*Flag*** on the command line (where *Flag* is the data definition flag).

nulldefault: "*boolean*" ("N") The nulldefault: attribute overrides the default name generation, and uses an empty string (no alignment file) as the default for programs where an alignment file is only occasionally required. If an empty string is specified on the command line, the standard default value will be generated instead.

A.5.3.1.1 Associated qualifiers

The `align:` datatype has the following associated qualifiers which may be hard-coded as attributes in an ACD file. These qualifiers are described in the *EMBOSS User's Guide*:

`aformat: "string" ("")`	Alignment format
`aextension: "string" ("")`	File name extension
`adirectory: "string" ("")`	Output directory
`aname: "string" ("")`	Base file name
`awidth: "integer" ("0")`	Alignment width
`aaccshow: "boolean" ("N")`	Show accession number in the header
`adesshow: "boolean" ("N")`	Show description in the header
`ausashow: "boolean" ("N")`	Show the full USA in the alignment
`aglobal: "boolean" ("N")`	Show the full sequence in alignment

The `align` datatype has two associated qualifiers (**`-aformat`** and **`-aglobal`**) which which are often hard-coded as attributes in an ACD file (see the *EMBOSS User's Guide*). `aformat:` defines the default value for the **`-aformat`** qualifier, which is the default alignment format. `aglobal:` defines the default value for the **`-aglobal`** qualifier, and should be set to `"Y"` for programs that produce a global alignment.

A.5.3.2 featout

`name: "string" ("")` Base name of output file name. Use of **`-ofname`** (see below) is preferred.

`extension: "string" ("")` Extension of output file name. Use of **`-offormat`** (see below) is preferred.

`type: "string" ("")` Defines whether the feature output is `"protein"` or `"nucleotide"`. There is a default based on the type of any input sequence, but a value should always be specified.

`multiple: "boolean" ("N")` Should be set to `"Y"` if the feature annotation is for multiple sequences.

`nullok: "boolean" ("N")` If the application can accept a null value for this definition and can run without generating feature output, the `nullok:` attribute must be set to `"Y"`. This allows a default value to be omitted or the application to be run with **-no*Flag*** on the command line (where *`Flag`* is the data definition flag).

`nulldefault: "boolean" ("N")` The `nulldefault:` attribute overrides the default name generation, and uses an empty string (no feature output) as the default for programs where feature output is only occasionally required. If an empty string is specified on the command line, the standard default value will be generated instead. In combination with the `nullok:` and `missing:` attributes, this allows qualifiers to be null by default, and turned on from the command line.

A.5.3.2.1 Associated qualifiers

The `featout` datatype has the following associated qualifiers which may be hard-coded as attributes in an ACD file. These qualifiers are described in the *EMBOSS User's Guide*:

`offormat: "string" ("")`	Output feature format
`ofopenfile: "string" ("")`	Features file name
`ofextension: "string" ("")`	File name extension
`ofdirectory: "string" ("")`	Output directory
`ofname: "string" ("")`	Base filename
`ofsingle: "boolean" ("N")`	Separate file for each entry

The `featout` datatype has two associated qualifiers (**`-offormat`** and **`-ofname`**) which are often hard-coded as attributes in an ACD file (see the *EMBOSS User's Guide*). `offormat:` defines the default value for the **`-offormat`** qualifier which is the feature format and default file extension. `ofname:` defines the default value for the **`-ofname`** qualifier which is the default base file name

A.5.3.3 `outcodon`

`name: "`*`string`*`" ("")` Base name of output filename.

`extension: "`*`string`*`" ("")` Extension of output filename.

`nullok: "`*`boolean`*`" ("N")` If the application can accept a null value for this definition and can run without generating codon output, the `nullok:` attribute must be set to `"Y"`. This allows a default value to be omitted or the application to be run with **`-noFlag`** on the command line (where *`Flag`* is the data definition flag).

`nulldefault: "`*`boolean`*`" ("N")` The `nulldefault:` attribute overrides the default name generation, and uses an empty string (no data output) as the default for programs where data output is only occasionally required. If an empty string is specified on the command line, the standard default value will be generated instead. In combination with the `nullok:` and `missing:` attributes, this allows qualifiers to be null by default, and turned on from the command line.

A.5.3.3.1 Associated qualifiers

The `outcodon` datatype has the following associated qualifiers which may be hard-coded as attributes in an ACD file. These qualifiers are described in the *EMBOSS User's Guide*:

`odirectory: "string" ("")`	Output directory
`oformat: "string" ("")`	Output format specific to this datatype

The `outcodon` datatype has a datatype-associated command line qualifier (**`-oformat`**) which is often hard-coded as an attribute in an ACD file (see the *EMBOSS User's Guide*). `oformat:` defines the default value for the **`-oformat`** qualifier which is the file format.

A.5.3.4 outcpdb

extension: "*string*" ("") Extension of output file name. The extension: attribute will default to the output file format.

nullok: "*boolean*" ("N") If the application can accept a null value for this definition and can run without generating coordinate file output, the nullok: attribute must be set to "Y". This allows a default value to be omitted or the application to be run with **-no*Flag*** on the command line (where *Flag* is the data definition flag).

nulldefault: "*boolean*" ("N") The nulldefault: attribute overrides the default name generation, and uses an empty string (no data output) as the default for programs where data output is only occasionally required. If an empty string is specified on the command line, the standard default value will be generated instead. In combination with the nullok: and missing: attributes, this allows qualifiers to be null by default, and turned on from the command line.

A.5.3.4.1 Associated qualifiers

The outcpdb datatype has the following associated qualifiers which may be hard-coded as attributes in an ACD file. These qualifiers are described in the *EMBOSS User's Guide*:

odirectory: "string" ("")	Output directory
oformat: "string" ("")	Output format specific to this datatype

The outcpdb datatype has a datatype-associated command line qualifier (**-oformat**) which is often hard-coded as an attribute in an ACD file (see the *EMBOSS User's Guide*). oformat: defines the default value for the **-oformat** qualifier which is the file format and default file extension.

A.5.3.5 outdata

type: "*string*" ("") Defines the output datatype. It is recommended that a value is always defined to make the output datatype clear.

nullok: "*boolean*" ("N") If the application can accept a null value for this definition and can run without generating formatted data output, the nullok: attribute must be set to "Y". This allows a default value to be omitted or the application to be run with **-no*Flag*** on the command line (where *Flag* is the data definition flag).

nulldefault: "*boolean*" ("N") The nulldefault: attribute overrides the default name generation, and uses an empty string (no data output) as the default for programs where data output is only occasionally required. If an empty string is specified on the command line, the standard default value will be generated instead. In combination with the nullok: and missing: attributes, this allows qualifiers to be null by default, and turned on from the command line.

A.5.3.5.1 Associated qualifiers

The outdata datatype has the following associated qualifiers which may be hard-coded as attributes in an ACD file. These qualifiers are described in the *EMBOSS User's Guide*:

`odirectory: "string" ("")`	Output directory
`oformat: "string" ("")`	Output format specific to this datatype

The `outdata` datatype has a datatype-associated command line qualifier (**`-oformat`**) which is often hard-coded as an attribute in an ACD file (see the *EMBOSS User's Guide*). `oformat:` defines the default value for the **`-oformat`** qualifier which is the file format and default file extension.

A.5.3.6 outdir

`fullpath: "`*`boolean`*`" ("N")` If set to `"Y"` requires the path to be specified in full when passed to the program, although the user may provide a path from the current working directory.

`extension: "`*`string`*`" ("")` The default extension for all files written to the directory.

`nullok: "`*`boolean`*`" ("N")` If the application can accept a null value for this definition and can run without an output directory, the `nullok:` attribute must be set to `"Y"`. This allows a default value to be omitted or the application to be run with **`-noFlag`** on the command line (where *`Flag`* is the data definition flag).

`nulldefault: "`*`boolean`*`" ("N")` The `nulldefault:` attribute overrides the default name generation, and uses an empty string (no data output) as the default for programs where data output is only occasionally required. If an empty string is specified on the command line, the standard default value will be generated instead. In combination with the `nullok:` and `missing:` attributes, this allows qualifiers to be null by default, and turned on from the command line.

A.5.3.6.1 Associated qualifiers

The `outdir` datatype has no associated qualifiers.

A.5.3.7 outdiscrete

`nullok: "`*`boolean`*`" ("N")` If the application can accept a null value for this definition and can run without generating an output file, the `nullok:` attribute must be set to `"Y"`. This allows a default value to be omitted or the application to be run with **`-no`*`Flag`*** on the command line (where *`Flag`* is the data definition flag).

`nulldefault: "`*`boolean`*`" ("N")` The `nulldefault:` attribute overrides the default name generation, and uses an empty string (no data output) as the default for programs where data output is only occasionally required. If an empty string is specified on the command line, the standard default value will be generated instead. In combination with the `nullok:` and `missing:` attributes, this allows qualifiers to be null by default, and turned on from the command line.

A.5.3.7.1 Associated qualifiers

The `outdiscrete` datatype has the following associated qualifiers which may be hard-coded as attributes in an ACD file. These qualifiers are described in the *EMBOSS User's Guide*:

`odirectory: "string" ("")`	Output directory
`oformat: "string" ("")`	Output format specific to this datatype

The `outdiscrete` datatype has a datatype-associated command line qualifier (**`-oformat`**) which is often hard-coded as an attribute in an ACD file (see the *EMBOSS User's Guide*). `oformat:` defines the default value for the **`-oformat`** qualifier which is the file format and default file extension.

A.5.3.8 `outdistance`

`nullok: "boolean" ("N")` If the application can accept a null value for this definition and can run without generating a distance matrix file, the `nullok:` attribute must be set to `"Y"`. This allows a default value to be omitted or the application to be run with **-no*Flag*** on the command line (where *`Flag`* is the data definition flag).

`nulldefault: "boolean" ("N")` The `nulldefault:` attribute overrides the default name generation, and uses an empty string (no data output) as the default for programs where data output is only occasionally required. If an empty string is specified on the command line, the standard default value will be generated instead. In combination with the `nullok:` and `missing:` attributes, this allows qualifiers to be null by default, and turned on from the command line.

A.5.3.8.1 Associated qualifiers

The `outdistance` datatype has the following associated qualifiers which may be hard-coded as attributes in an ACD file. These qualifiers are described in the *EMBOSS User's Guide*:

`odirectory: "string" ("")`	Output directory
`oformat: "string" ("")`	Output format specific to this datatype

The `outdistance` datatype has a datatype-associated command line qualifier (**`-oformat`**) which is often hard-coded as an attribute in an ACD file (see the *EMBOSS User's Guide*). `oformat:` defines the default value for the **`-oformat`** qualifier which is the file format and default file extension.

A.5.3.9 `outfreq`

`nullok: "boolean" ("N")` If the application can accept a null value for this definition and can run without generating an output frequencies file, the `nullok:` attribute must be set to `"Y"`. This allows a default value to be omitted or the application to be run with **`-noFlag`** on the command line (where *`Flag`* is the data definition flag).

`nulldefault: "boolean" ("N")` The `nulldefault:` attribute overrides the default name generation, and uses an empty string (no data output) as the default for programs where data output is only occasionally required. If an empty string is specified on the command line, the standard default value will be generated instead. In combination with the `nullok:` and `missing:` attributes, this allows qualifiers to be null by default, and turned on from the command line.

A.5.3.9.1 Associated qualifiers

The `outfreq` datatype has the following associated qualifiers which may be hard-coded as attributes in an ACD file. These qualifiers are described elsewhere (see the *EMBOSS User's Guide*):

`odirectory: "string" ("")`	Output directory
`oformat: "string" ("")`	Output format specific to this datatype

The `outfreq` datatype has a datatype-associated command line qualifier (**`-oformat`**) which is often hard-coded as an attribute in an ACD file (see the *EMBOSS User's Guide*). `oformat:` defines the default value for the **`-oformat`** qualifier which is the file format and default file extension.

A.5.3.10 outfile

`name: "`*`string`*`" ("")` Base name of output filename. By default, any input file or sequence name will be used. If there is no input file processed so far, the default name is `"outfile"`.

`extension: "`*`string`*`" ("")` Extension of output filename.

`append: "`*`boolean`*`" ("N")` If set, specifies that output is appended to the end of an existing output file. By default the output file will be overwritten.

`nullok: "`*`boolean`*`" ("N")` If the application can accept a null value for this definition and can run without generating an output file, the `nullok:` attribute must be set to `"Y"`. This allows a default value to be omitted or the application to be run with **-no*Flag*** on the command line (where *`Flag`* is the data definition flag).

`nulldefault: "`*`boolean`*`" ("N")` The `nulldefault:` attribute overrides the default name generation, and uses an empty string (no output file) as the default for programs where an output file is only occasionally required. Examples include programs where the original output format is available, usually for users that still require it for parsing in automated scripts. In combination with the `nullok:` and `missing:` attributes, this allows qualifiers to be null by default, and turned on from the command line.

The `knowntype:` global attribute (Section A.4, 'Global attributes') should always be defined. If the output is not of any of the standard EMBOSS known types, *`ApplicationName`* `output` is the recommended value.

A.5.3.10.1 Associated qualifiers

The `outfile` datatype has the following associated qualifiers which may be hard-coded as attributes in an ACD file. These qualifiers are described in the *EMBOSS User's Guide*:

`odirectory: "string" ("")`	Output directory

A.5.3.11 outfileall

`name: "`*`string`*`" ("")` Base name of output filename. By default, any input file or sequence name will be used. If there is no input file processed so far, the default name is `"outfile"`.

`extension: "`*`string`*`" ("")` Extension of output filename.

`nullok: "`*`boolean`*`" ("N")` If the application can accept a null value for this definition and can run without generating an output file, the `nullok:` attribute must be set to

"Y". This allows a default value to be omitted or the application to be run with **-noFlag** on the command line (where *Flag* is the data definition flag).

nulldefault: "*boolean*" ("N") The nulldefault: attribute overrides the default name generation, and uses an empty string (no output file) as the default for programs where an output file is only occasionally required. Examples include programs where the original output format is available, usually for users that still require it for parsing in automated scripts. In combination with the nullok: and missing: attributes, this allows qualifiers to be null by default, and turned on from the command line. The knowntype: global attribute (Section A.4, 'Global attributes') should always be defined. If the output is not of any of the standard EMBOSS known types (Section A.4, 'Global attributes'), "ApplicationName output" is the recommended value.

A.5.3.11.1 Associated qualifiers

The outfileall datatype has the following associated qualifiers which may be hard-coded as attributes in an ACD file. These qualifiers are described in the *EMBOSS User's Guide*:

odirectory: "string" ("")	Output directory

A.5.3.12 outmatrix

nullok: "*boolean*" ("N") If the application can accept a null value for this definition and can run without generating an matrix file, the nullok: attribute must be set to "Y". This allows a default value to be omitted or the application to be run with **-noFlag** on the command line (where *Flag* is the data definition flag).

nulldefault: "*boolean*" ("N") The nulldefault: attribute overrides the default name generation, and uses an empty string (no data output) as the default for programs where data output is only occasionally required. If an empty string is specified on the command line, the standard default value will be generated instead. In combination with the nullok: and missing: attributes, this allows qualifiers to be null by default, and turned on from the command line.

A.5.3.12.1 Associated qualifiers

The outmatrix: datatype has the following associated qualifiers which may be hard-coded as attributes in an ACD file. These qualifiers are described in the *EMBOSS User's Guide*:

odirectory: "string" ("")	Output directory
oformat: "string" ("")	Output format specific to this datatype

The outmatrix datatype has a datatype-associated command line qualifier (**-oformat**) which is often hard-coded as an attribute in an ACD file (see the *EMBOSS User's Guide*). oformat: defines the default value for the **-oformat** qualifier which is the file format and default file extension.

A.5.3.13 outmatrixf

nullok: "*boolean*" ("N") If the application can accept a null value for this definition and can run without generating an matrix file, the nullok: attribute must be set to

"Y". This allows a default value to be omitted or the application to be run with **-no*Flag*** on the command line (where *Flag* is the data definition flag).

nulldefault: "*boolean*" ("N") The nulldefault: attribute overrides the default name generation, and uses an empty string (no data output) as the default for programs where data output is only occasionally required. If an empty string is specified on the command line, the standard default value will be generated instead. In combination with the nullok: and missing: attributes, this allows qualifiers to be null by default, and turned on from the command line.

A.5.3.13.1 Associated qualifiers

The outmatrixf: datatype has the following associated qualifiers which may be hard-coded as attributes in an ACD file. These qualifiers are described in the *EMBOSS User's Guide*:

odirectory: "*string*" ("")	Output directory
oformat: "*string*" ("")	Output format specific to this datatype

The outmatrixf datatype has a datatype-associated command line qualifier (**-oformat**) which is often hard-coded as an attribute in an ACD file (see the *EMBOSS User's Guide*). oformat: defines the default value for the **-oformat** qualifier which is the file format and default file extension.

A.5.3.14 outproperties

nullok: "*boolean*" ("N") If the application can accept a null value for this definition and can run without generating a properties file, the nullok: attribute must be set to "Y". This allows a default value to be omitted or the application to be run with **-no*Flag*** on the command line (where *Flag* is the data definition flag).

nulldefault: "*boolean*" ("N") The nulldefault: attribute overrides the default name generation, and uses an empty string (no data output) as the default for programs where data output is only occasionally required. If an empty string is specified on the command line, the standard default value will be generated instead. In combination with the nullok: and missing: attributes, this allows qualifiers to be null by default, and turned on from the command line.

A.5.3.14.1 Associated qualifiers

The outproperties: datatype has the following associated qualifiers which may be hard-coded as attributes in an ACD file. These qualifiers are described in the *EMBOSS User's Guide*:

odirectory: "string" ("")	Output directory
oformat: "string" ("")	Output format specific to this datatype

The outproperties datatype has a datatype-associated command line qualifier (**-oformat**) which is often hard-coded as an attribute in an ACD file (see the *EMBOSS User's Guide*). oformat: defines the default value for the **-oformat** qualifier which is the file format and default file extension.

A.5.3.15 outscop

nullok: "*boolean*" ("N") If the application can accept a null value for this definition and can run without generating an output file, the `nullok:` attribute must be set to `"Y"`. This allows a default value to be omitted or the application to be run with **-no*Flag*** on the command line (where *Flag* is the data definition flag).

nulldefault: "*boolean*" ("N") The `nulldefault:` attribute overrides the default name generation, and uses an empty string (no data output) as the default for programs where data output is only occasionally required. If an empty string is specified on the command line, the standard default value will be generated instead. In combination with the `nullok:` and `missing:` attributes, this allows qualifiers to be null by default, and turned on from the command line.

A.5.3.15.1 Associated qualifiers

The `outscop:` datatype has the following associated qualifiers which may be hard-coded as attributes in an ACD file. These qualifiers are described in the *EMBOSS User's Guide*:

odirectory: "string" ("")	Output directory
oformat: "string" ("")	Output format specific to this datatype

The `outscop` datatype has a datatype-associated command line qualifier (**-oformat**) which is often hard-coded as an attribute in an ACD file (see the *EMBOSS User's Guide*). `oformat:` defines the default value for the **-oformat** qualifier which is the file format and default file extension.

A.5.3.16 outtree

name: "*string*" ("") Base name of output filename.

extension: "*string*" ("") Extension of output filename.

nullok: "*boolean*" ("N") If the application can accept a null value for this definition and can run without generating a phylogenetic tree, the `nullok:` attribute must be set to `"Y"`. This allows a default value to be omitted or the application to be run with **-no*Flag*** on the command line (where *Flag* is the data definition flag).

nulldefault: "*boolean*" ("N") The `nulldefault:` attribute overrides the default name generation, and uses an empty string (no data output) as the default for programs where data output is only occasionally required. If an empty string is specified on the command line, the standard default value will be generated instead. In combination with the `nullok:` and `missing:` attributes, this allows qualifiers to be null by default, and turned on from the command line.

A.5.3.16.1 Associated qualifiers

The `outtree:` datatype has the following associated qualifiers which may be hard-coded as attributes in an ACD file. These qualifiers are described in the *EMBOSS User's Guide*:

odirectory: "string" ("")	Output directory
oformat: "string" ("")	Output format specific to this datatype

The `outtree` datatype has a datatype-associated command line qualifier (**-oformat**) which is often hard-coded as an attribute in an ACD file (see the *EMBOSS User's Guide*).

oformat: defines the default value for the **-oformat** qualifier which is the file format and default file extension.

A.5.3.17 report

multiple: "*boolean*" ("N") Should be set to "Y" if the output can contain more than one report from the same input.

precision: "*integer*" ("3") The floating point precision of any reported score values. For integer score values this can be set to "0".

type: "*string*" ("") Where the report format is one of the standard feature table formats (see the *EMBOSS User's Guide*), the type: attribute defines whether the report output is "protein" or "nucleotide". There is a default based on the type of any input sequence, but a value should always be specified.

taglist: "*string*" ("") Defines the additional tags to be reported from the internal feature table. The tag names and types must match the source code of the application. Each tag is in the format type:tagname[=columnname] for example int:length or string:gc=GC%.

nullok: "*boolean*" ("N") If the application can accept a null value for this definition and can run without generating report output, the nullok: attribute must be set to "Y". This allows a default value to be omitted or the application to be run with **-no*Flag*** on the command line (where *Flag* is the data definition flag).

nulldefault: "*boolean*" ("N") The nulldefault: attribute overrides the default name generation, and uses an empty string (no report output) as the default for programs where report output is only occasionally required. If an empty string is specified on the command line, the standard default value will be generated instead. In combination with the nullok: and missing: attributes, this allows qualifiers to be null by default, and turned on from the command line.

A.5.3.17.1 Associated qualifiers

The report: datatype has the following associated qualifiers which may be hard-coded as attributes in an ACD file. These qualifiers are described in the *EMBOSS User's Guide*:

rformat: "string" ("")	Report format
rname: "string" ("")	Base filename
rextension: "string" ("")	Filename extension
rdirectory: "string" ("")	Output directory
raccshow: "boolean" ("N")	Show accession number in the report
rdesshow: "boolean" ("N")	Show description in the report
rscoreshow: "boolean" ("N")	Show the score in the report
rusashow: "boolean" ("N")	Show the full USA in the report
rmaxall: "integer" ("0")	Maximum total hits to report
rmaxseq: "integer" ("0")	Maximum hits to report for one sequence

The report datatype has three associated qualifiers (**-rextension**, **-rscoreshow** and **-rformat**) which are often hard-coded as attributes in an ACD file (see the *EMBOSS User's Guide*). The rextension: attribute sets the default value for the **-rextension** qualifier, which is normally the program name. The `rscoreshow:` boolean attribute specifies whether the feature score value should appear in the report. Where a score is not set by the program, the default 0.0 score can be excluded by setting the value to `"N"`. The `rformat:` attribute is required. It defines the default value for the **`-rformat`** qualifier which is the report format.

A.5.3.18 `seqout`

`name: "`*`string`*`" ("")` Base name of output filename.

`extension: "`*`string`*`" ("")` Extension of output filename.

`features: "`*`boolean`*`" ("N")` If set, the sequence output will include feature information either in the same file (if the sequence format supports it) or in a separate file (by default in GFF format).

`type: "`*`string`*`" ("")` The output sequence type. Although this will default to the type of the first output sequence, it is recommended that a value is always defined to make the output sequence type clear. The sequence type must be a standard sequence type (Section A.7, 'Sequence types').

`nullok: "`*`boolean`*`" ("N")` If the application can accept a null value for this definition and can run without generating sequence output, the `nullok:` attribute must be set to `"Y"`. This allows a default value to be omitted or the application to be run with **`-no`*`Flag`*** on the command line (where *`Flag`* is the data definition flag).

`nulldefault: "`*`boolean`*`" ("N")` The `nulldefault:` attribute overrides the default name generation, and uses an empty string (no sequence output) as the default for programs where sequence output is only occasionally required. If an empty string is specified on the command line, the standard default value will be generated instead. In combination with the `nullok:` and `missing:` attributes, this allows qualifiers to be null by default, and turned on from the command line.

A.5.3.18.1 Associated qualifiers

The `seqout` datatype has the following associated qualifiers which may be hard-coded as attributes in an ACD file. These qualifiers are described in the *EMBOSS User's Guide*:

`osformat: "string" ("")`	Output seq format
`osextension: "string" ("")`	Filename extension
`osname: "string" ("")`	Base filename
`osdirectory: "string" ("")`	Output directory
`osdbname: "string" ("")`	Database name to add
`ossingle: "boolean" ("N")`	Separate file for each entry
`oufo: "string" ("")`	UFO features
`offormat: "string" ("")`	Features format
`ofname: "string" ("")`	Features filename
`ofdirectory: "string" ("")`	Output directory

A.5.3.19 `seqoutall`

`name: "`*`string`*`" ("")` Base name of output filename.

extension: "*string*" ("") Extension of output filename.

features: "*boolean*" ("N") If set, the sequence output will include feature information either in the same file (if the sequence format supports it) or in a separate file (by default in GFF format).

type: "*string*" ("") The output sequence type. Although this will default to the type of the first output sequence, it is recommended that a value is always defined to make the output sequence type clear. The sequence type must be a standard sequence type (Section A.7, 'Sequence types').

minseqs: "*integer*" ("1") The minimum number of sequences that are expected in the output file.

maxseqs: "*integer*" ("INT_MAX") The maximum number of sequences that are expected in the output file.

nullok: "*boolean*" ("N") If the application can accept a null value for this definition and can run without generating sequence output, the `nullok:` attribute must be set to `"Y"`. This allows a default value to be omitted or the application to be run with **-no*Flag*** on the command line (where *`Flag`* is the data definition flag).

nulldefault: "*boolean*" ("N") The `nulldefault:` attribute overrides the default name generation, and uses an empty string (no sequence output) as the default for programs where sequence output is only occasionally required. If an empty string is specified on the command line, the standard default value will be generated instead. In combination with the `nullok:` and `missing:` attributes, this allows qualifiers to be null by default, and turned on from the command line.

A.5.3.19.1 Associated qualifiers

The `seqoutall` datatype has the following associated qualifiers which may be hard-coded as attributes in an ACD file. These qualifiers are described in the *EMBOSS User's Guide*:

`osformat: "string" ("")`	Output seq format
`osextension: "string" ("")`	Filename extension
`osname: "string" ("")`	Base filename
`osdirectory: "string" ("")`	Output directory
`osdbname: "string" ("")`	Database name to add
`ossingle: "boolean" ("N")`	Separate file for each entry
`oufo: "string" ("")`	UFO features
`offormat: "string" ("")`	Features format
`ofname: "string" ("")`	Features filename
`ofdirectory: "string" ("")`	Output directory

The `seqoutall` datatype has a datatype-associated command line qualifier (**`-ossingle`**) which is often hard-coded as an attribute in an ACD file (see the *EMBOSS User's Guide*). `ossingle:` defines the default value for the **`-ossingle`** qualifier and can be set to `"Y"` to direct output to multiple sequence files. For example, the EMBOSS program **seqretsplit** splits an input sequence input multiple files using this attribute.

A.5.3.20 seqoutset

name: "*string*" ("") Base name of output filename.

extension: "*string*" ("") Extension of output filename.

features: "*boolean*" ("N") If set, the sequence output will include feature information either in the same file (if the sequence format supports it) or in a separate file (by default in GFF format).

type: "*string*" ("") Output sequence type. Although this will default to the type of the first output sequence, it is recommended that a value is always defined to make the output sequence type clear. The sequence type must be a standard sequence type (Section A.7, 'Sequence types').

minseqs: "*integer*" ("1") The minimum number of sequences that are expected in the output file.

maxseqs: "*integer*" ("INT_MAX") The maximum number of sequences that are expected in the output file.

aligned: "*boolean*" ("N") If true, specifies that all sequences in the output are expected to be aligned. If false then the sequences are assumed to be unaligned. We recommend always defining the aligned: attribute so that the nature of the sequence set is clearly defined.

nullok: "*boolean*" ("N") If the application can accept a null value for this definition and can run without generating sequence output, the nullok: attribute must be set to "Y". This allows a default value to be omitted or the application to be run with **-no*Flag*** on the command line (where *Flag* is the data definition flag).

nulldefault: "*boolean*" ("N") The nulldefault: attribute overrides the default name generation, and uses an empty string (no sequence output) as the default for programs where sequence output is only occasionally required. If an empty string is specified on the command line, the standard default value will be generated instead. In combination with the nullok: and missing: attributes, this allows qualifiers to be null by default, and turned on from the command line.

A.5.3.20.1 Associated qualifiers

The seqoutset datatype has the following associated qualifiers which may be hard-coded as attributes in an ACD file. These qualifiers are described in the *EMBOSS User's Guide*:

osformat: "string" ("")	Output seq format
osextension: "string" ("")	File name extension
osname: "string" ("")	Base filename
osdirectory: "string" ("")	Output directory
osdbname: "string" ("")	Database name to add
ossingle: "boolean" ("N")	Separate file for each entry
oufo: "string" ("")	UFO features
offormat: "string" ("")	Features format
ofname: "string" ("")	Features filename
ofdirectory: "string" ("")	Output directory

A.5.4 Selection attributes description

A.5.4.1 `list`

`values: "string"` (`""`) A string containing a list of option labels and descriptions which define the options the user can choose from. The `delimiter:` and `codedelimiter:` attributes specify how to parse this string into individual list items.

`delimiter: "string"` (`";"`) The character used in the `values:` string to separate list items. Defaults to a semi-colon (`;`).

`codedelimiter: "string"` (`":"`) The character used in the `values:` string to separate option labels from their descriptions. Defaults to a colon `:`.

`minimum: "integer"` (`"1"`) The minimum number of selections required. By default one selection is required.

`maximum: "integer"` (`"1"`) The maximum number of selections allowed. By default exactly one selection is required. A higher value allows multiple selections.

`header: "string"` (`""`) Text that is displayed to the user above the list of options.

`button: "boolean"` (`"N"`) Provided for interface developers. It has a boolean value used to suggest whether a list is best represented as check-boxes or radio-buttons in an interface (value of `"Y"`) or if a text box should be used to enter a value (`"N"`).

`casesensitive: "boolean"` (`"N"`) Defines whether user input must match the exact case of the list item. Regardless of the case used by the user, the value of the parameter will be exactly what the corresponding list value is.

A.5.4.1.1 Associated qualifiers
The `list` datatype has no associated qualifiers.

A.5.4.2 `select`

`values: "string"` (`""`) A string containing a list of option descriptions which define the options the user can choose from. The `delimiter:` attribute specifies how to parse this string into individual selection list items.

`delimiter: "string"` (`":"`) The character used in the `values:` string to separate list items. Defaults to a semicolon (`;`).

`minimum: "integer"` (`"1"`) The minimum number of selections required. By default one selection is required.

`maximum: "integer"` (`"1"`) The maximum number of selections allowed. By default exactly one selection is required. A higher value allows multiple selections.

`header: "string"` (`""`) Text that is displayed to the user above the list of options.

`button: "boolean"` (`"N"`) Provided for interface developers. It has a boolean value used to suggest whether a list is best represented as check-boxes or radio-buttons in an interface (value of `"Y"`) or if a text box should be used to enter a value (`"N"`).

`casesensitive: "boolean"` (`"N"`) Defines whether the input must match the exact case of the list item. Regardless of the case used by the user, the value of the parameter will be exactly what the corresponding list value is.

Note

The **`information:`** attribute, defined for all datatypes, defines text to be used as a prompt after a **`list`** or **`select`** type list.

A.5.4.2.1 Associated qualifiers

The `selection` datatype has no associated qualifiers.

A.5.5 Graphics attributes description

A.5.5.1 `graph`

`nullok: "`*`boolean`*`" ("N")` If the application can accept a null value for this definition and can run without generating graphical output, the `nullok:` attribute must be set to `"Y"`. This allows a default value to be omitted or the application to be run with **`-noFlag`** on the command line (where *`Flag`* is the data definition flag).

`nulldefault: "`*`boolean`*`" ("N")` Overrides the default name generation, and uses an empty string (no graph) as the default for programs where a graph is only occasionally required. If an empty string is specified on the command line, the standard default value will be generated instead.

A.5.5.1.1 Associated qualifiers

The `graph` datatype has the following associated qualifiers which may be hard-coded as attributes in an ACD file. These qualifiers are described in the *EMBOSS User's Guide*:

`gprompt: "boolean" ("N")`	Graph prompting
`gdesc: "string" ("")`	Graph description
`gtitle: "string" ("")`	Graph title
`gsubtitle: "string" ("")`	Graph subtitle
`gxtitle: "string" ("")`	Graph *x*-axis title
`gytitle: "string" ("")`	Graph *y*-axis title
`goutfile: "string" ("")`	Output file for non-interactive displays
`gdirectory: "string" ("")`	Output directory

The `graph` datatype has four associated qualifiers which are often hard-coded as attributes in an ACD file (see the *EMBOSS User's Guide*). **`-gtitle`**, **`-gxtitle`** and **`-gytitle`** are used to control the titles of the graph and its axes and are set with the attributes `gtitle:`, `gxtitle:` and `gytitle:`. The `goutfile:` attribute specifies the base file name for output. It is used, for example by the EMBOSS program **tmap**, to direct output to a named file rather than the default name, which is the name of first input sequence read (if any).

A.5.5.2 xygraph

multiple: "*integer*" ("1") The multiple: attribute specifies the number of multiple XY graphs in a single output. The default value is 1, but any integer value may be defined in the ACD file.

nullok: "*boolean*" ("N") If the application can accept a null value for this definition and can run without generating graphical output, the nullok: attribute must be set to "Y". This allows a default value to be omitted or the application to be run with **-no*Flag*** on the command line (where *Flag* is the data definition flag).

nulldefault: "*boolean*" ("N") The nulldefault: attribute overrides the default name generation, and uses an empty string (no graph) as the default for programs where a graph is only occasionally required. If an empty string is specified on the command line, the standard default value will be generated instead.

A.5.5.2.1 Associated qualifiers

The xygraph datatype has the following associated qualifiers which may be hard-coded as attributes in an ACD file. These qualifiers are described in the *EMBOSS User's Guide*:

gprompt: "boolean" ("N")	Graph prompting
gdesc: "string" ("")	Graph description
gtitle: "string" ("")	Graph title
gsubtitle: "string" ("")	Graph subtitle
gxtitle: "string" ("")	Graph *x*-axis title
gytitle: "string" ("")	Graph *y*-axis title
goutfile: "string" ("")	Output file for non-interactive displays
gdirectory: "string" ("")	Output directory

The graph datatype has four associated qualifiers which are often hard-coded as attributes in an ACD file (see the *EMBOSS User's Guide*). **-gtitle**, **-gxtitle** and **-gytitle** are used to control the titles of the graph and its axes and are set with the attributes gtitle:, gxtitle: and gytitle:. The goutfile: attribute specifies the base file name for output. It is used to direct output to a named file, otherwise a default of the name of first input sequence read (if any) is used.

A.6 Calculated attributes

Calculated attributes are datatype-specific attributes that are assigned values *after* the data definition has been processed, for example, once a sequence has been read in from file. Their value is calculated automatically during ACD file processing. This allows data definitions to refer to attributes whose value depends on the input data. The available calculated attributes are summarised in the table and described in detail below.

Values of calculated attributes for ACD data definitions are retrieved by an operation from within the ACD file. This is done using the $ ('get value of') syntax before a pair of parentheses enclosing the parameter and attribute name as follows:

```
$(ParameterName.CalculatedAttributeName)
```

A.6.1 Simple ACD datatypes

A.6.1.1 `string`

`length:` *integer*
The length of the string.

A.6.2 Input ACD datatypes

A.6.2.1 Sequences

A.6.2.1.1 `seqall`

`begin:` *integer*
Start of the first sequence used.

`end:` *integer*
End of the first sequence used.

`length:` *integer*
Total length of the first sequence.

`protein:` *Y/N*
Boolean, indicates if sequence is protein.

`nucleic:` *Y/N*
Boolean, indicates if sequence is DNA.

`name:` *string*
The name/ID/accession of the sequence.

A.6.2.1.2 `seqset`

`begin:` *integer*
The beginning of the selection of the sequence.

`end:` *integer*
The end of the selection of the sequence.

`length:` *integer*
The maximum length of the sequence set.

`protein:` *Y/N*
Boolean, indicates if sequence set is protein.

`nucleic:` *Y/N*
Boolean, indicates if sequence set is DNA.

`name:` *string*
The name of the sequence set.

`totweight:` *float*
Float, total sequence weight for a set.

`count:` *integer*
Integer, number of sequences in the set.

A.6.2.1.3 `seqsetall`

`begin:` *integer*
The beginning of the selection of the sequence.

`end:` *integer*
The end of the selection of the sequence.

`length:` *integer*
The maximum length of the sequence set.

`protein:` *Y/N*
Boolean, indicates if sequence set is protein.

`nucleic:` *Y/N*
Boolean, indicates if sequence set is DNA.

`name:` *string*
The name of the sequence set.

`totweight:` *float*
Float, total sequence weight for each set.

`count:` *integer*
Integer, number of sequences in each set.

`multicount:` *integer*
Integer, number of sets of sequences.

A.6.2.1.4 `sequence`

`begin:` *`integer`*
Start of the sequence used.

`end:` *`integer`*
End of the sequence used.

`length:` *`integer`*
Total length of the sequence.

`protein:` *`Y/N`*
Boolean, indicates if sequence is protein.

`nucleic:` *`Y/N`*
Boolean, indicates if sequence is DNA.

`name:` *`string`*
The name/ID/accession of the sequence.

A.6.2.2 Sequence features

A.6.2.2.1 `features`

`fbegin:` *`integer`*
Start of the features to be used.

`fend:` *`integer`*
End of the features to be used.

`flength:` *`integer`*
Total length of sequence (`fsize` is feature count).

`fprotein:` *`Y/N`*
Feature table is protein.

`fnucleic:` *`Y/N`*
Feature table is nucleotide.

`fname:` *`string`*
The name of the feature table.

`fsize:` *`string`*
Integer, number of features.

A.6.2.3 Other inputs (not sequences or features)

A.6.2.3.1 `distances`

`distancesize:` ***`integer`***
Number of distance rows.

`replicates:` ***`Y/N`***
Replicates data found in input.

`hasmissing:` ***`Y/N`***
Missing values found (replicates = N).

A.6.2.3.2 `frequencies`

`freqlength:` ***`integer`***
Number of frequency values per set.

`freqsize:` ***`integer`***
Number of frequency sets.

`freqloci:` ***`integer`***
Number of frequency loci.

`freqgenedata:` ***`Y/N`***
Gene frequency data.

`freqcontinuous:` ***`Y/N`***
Continuous frequency data.

`freqwithin:` ***`Y/N`***
Individual within species frequency data.

A.6.2.3.3 `properties`

`propertylength:` ***`integer`***
Number of property values per set.

`propertysize:` ***`integer`***
Number of property sets.

A.6.2.3.4 `regexp`

`length:` ***`integer`***
The length of the regular expression.

A.6.2.3.5 `tree`

`treecount:` ***`integer`***
Number of trees.

speciescount: ***integer***
Number of species.

haslengths: ***Y/N***
Branch lengths defined.

A.7 Sequence types

All of the sequence ACD datatypes have an attribute called `type:`. Not surprisingly, this is used to limit the kind of sequence that EMBOSS will accept. The `type:` attribute describes the type of the sequence as a single token. The EMBOSS initialisation routines will try to establish the type, by reading the (first) sequence and examining the contents. Possible values for the `type:` attribute are listed in the table below.

Table A.6. ACD Sequence Types

Value	Type(s)	Gaps	Ambiguity codes	Conversions	Description
any	Nucleotide or protein	Removed	Yes	'?'=>'X'	Any valid sequence
gapany	Nucleotide or protein	Kept	Yes	'?'=>'X'	Any valid sequence with gaps
dna	Nucleotide only	Removed	Yes	'?'=>'N' 'X'=>'N' 'U'=>'T'	DNA sequence
puredna	Nucleotide only	Removed	No	'U'=>'T'	DNA sequence, bases ACGT only
gapdna	Nucleotide only	Kept	Yes	'?'=>'N' 'X'=>'N' 'U'=>'T'	DNA sequence with gaps
gapdnaphylo	Nucleotide only	Kept	Yes	'U'=>'T'	DNA sequence with gaps and queries
rna	Nucleotide only	Removed	Yes	'?'=>'N' 'X'=>'N' 'T'=>'U'	RNA sequence
purerna	Nucleotide only	Removed	No	'T'=>'U'	RNA sequence, bases ACGU only
gaprna	Nucleotide only	Kept	Yes	'?'=>'N' 'X'=>'N' 'T'=>'U'	RNA sequence with gaps
gaprnaphylo	Nucleotide only	Kept	Yes	'T'=>'U'	RNA sequence with gaps and queries

`nucleotide`	Nucleotide only	Removed	Yes	'?'=>'N' 'X'=>'N'	nucleotide sequence
`purenucleotide`	Nucleotide only	Removed	No		Nucleotide sequence, bases ACGTU only
`gapnucleotide`	Nucleotide only	Kept	Yes	'?'=>'N' 'X'=>'N'	Nucleotide sequence with gaps
`gapnucleotidephylo`	Nucleotide only	Kept	Yes		Nucleotide sequence with gaps and queries
`protein`	Protein only	Removed	Yes	'?'=>'X' '*'=>'X'	Protein sequence
`pureprotein`	Protein only	Removed	No		Protein sequence without BZ U X or *
`stopprotein`	Protein only	Removed	Yes	'?'=>'X'	Protein sequence with possible stops
`gapprotein`	Protein only	Kept	Yes	'?'=>'X' '*'=>'X'	Protein sequence with gaps
`gapstopprotein`	Protein only	Kept	Yes	'?'=>'X'	Protein sequence with gaps and possible stops
`gapproteinphylo`	Protein only	Kept	Yes		Protein sequence with gaps, stops and queries
`proteinstandard`	Protein only	Removed	Yes	'?'=>'X' '*'=>'X' 'U'=>'X'	Protein sequence with no selenocysteine
`stopproteinstandard`	Protein only	Removed	Yes	'?'=>'X' 'U'=>'X'	Protein sequence with a possible stop but no selenocysteine
`gapproteinstandard`	Protein only	Kept	Yes	'?'=>'X' '*'=>'X' 'U'=>'X'	Protein sequence with gaps but no selenocysteine

A.8 Operations

A.8.1 Types of operation

The arithmetic operators are:

- `@(a + b)` (Addition)
- `@(a - b)` (Subtraction)
- `@(a * b)` (Multiplication)
- `@(a / b)` (Division)

The logical operators are:

- `@(!a)` (Not boolean)
- `@(a|b)` (Or)
- `@(a&b)` (And)

The equality operators are:

- `@(token1==token2)` (Equality)
- `@(token1!=token2)` (Non-equality)
- `@(token1<token2)` (Less-than)
- `@(token1>token2)` (Greater-than)

The conditional operators are:

- `@(boolval ? iftrue : iffalse)` (If)
- `@(testval = A : 1 B : 2 else : 0)` (Case)

A.8.2 Retrieving data values

The attribute values for a given data definition in an ACD file can depend on the values from other data definitions. It is possible to retrieve the value of:

- A data definition (application parameter)
- An attribute of a data definition
- A calculated attribute of a data definition
- An ACD variable definition.

Such values are retrieved using the ACD 'get the value of' syntax which consists of a term *ParameterName.AttributeName* term surrounded by parentheses with a dollar sign ($) at the front:

```
$(ParameterName.AttributeName)
```

If just the value of the data definition is to be retrieved, the *AttributeName* component is omitted:

```
$(ParameterName)
```

A variable name may also be given. Variables do not have attributes therefore the syntax is:

```
$(VariableName)
```

A.8.3 Types of calculations and tests

Calculations can be performed in ACD using the `@` syntax. Calculations can be used to add, subtract, multiply or divide, or test for equality, inequality, 'greater than' or 'less than' tests. The test values can be integers, floats and strings.

Only a single operation is allowed per set of parentheses. This means that that in some cases two or more sets of parentheses are required. In this example of a nested operations, two sets of parentheses are required, the first around the addition of `a` and `b` which gives `@(a + b)`. The second around the addition of `@(a + b)` to `c`, which gives `@(@(a + b)+c)`:

```
integer: n
{
default: "@(@(a + b)+c)"
}
```

Where more than one operation is required, however, one would typically use an internal ACD variable to hold the intermediate results.

A.8.3.1 Arithmetic operations

The supported arithmetic operations are addition, subtraction, multiplication and division. The standard characters for the arithmetic operations are used: `+`, `-`, `*` and `/`:

```
@(a+b) (Addition)
@(a-b) (Subtraction)
@(a*b) (Multiplication)
@(a/b) (Division)
```

The operands `a` and `b` *must* resolve to a numerical (integer or a floating point): the result is undefined otherwise but will most probably result in an error during ACD processing.

A.8.3.2 Tests for equality

The supported equality tests (symbols in parenthesis) are 'equality' (==), 'non-equality' (`!=`), 'less than ' (<) and 'greater than ' (>):

```
@(token1 == token2)  (Equality)
@(token1  != token2)    (Non-equality)
@(token1  < token2) (Less-than)
@(token1  > token2) (Greater-than)
```

The above equality tests can be used on strings in which case the lexicographical sorting order of the string is used.

A.8.3.3 Boolean tests

The supported boolean operations are logical AND, logical OR and logical NOT. Again, the standard characters are used: &, | and !:

```
@(!a) (NOT)
@(a|b) (OR)
@(a&b) (AND)
```

In the following example, the integer rtfm will only be prompted for if the value of fubar is either 3 or 7. Each of the equality tests is a calculation and the boolean test is another calculation. There are therefore three instances of @():

```
integer: rtfm
{
   standard: "@(@($(fubar)==3) | @($(fubar)==7))"
   etc
}
```

A.8.4 Conditional statements

There are three kinds of conditional statements in ACD: unary, ternary and case-type.

A.8.4.1 Unary conditional statements

A typical use for unary conditionals is to switch prompts on or off.

In the following example, if the sequence is a protein then the prompt for window is effectively turned off (see Section 4.5, 'Controlling the prompt' for more information about negation):

```
sequence: sequence
{
  parameter: "Y"
  type: gapany
}
integer: window
{
  standard: "$(sequence.protein)"
  etc
}
```

A.8.4.2 Ternary conditional statements

Ternary conditional statements have the general form:

```
@(conditional ? value-if-true : value-if-false)
```

They are useful when setting up the application for two distinct modes of usage, for example when setting gap penalty values differently for proteins and nucleic acids in alignment programs.

In the following example, `penalty` is set to 14 for proteins and 16 for nucleic acids:

```
integer: penalty
{
  standard: "N"
  default: "@($(sequence.protein) ? 14 : 16)"
  etc
}
```

A.8.4.3 Case conditional statements

These have the general form:

```
@(testval = poss_valA : ass_valA
            poss_valB : ass_valB
            else : default_val)
```

The test value (*`testval`*) is compared with a list of possible values (*`poss_valA`* etc). If a match is found, the operation resolves to the result (*`ass_valB`* etc) associated with that possible value. The test value is parsed as a string and if none of the possible values match, the operation will resolve to the default result (*`default_val`*). The `else :` default value pair is optional and the default of `NULL` is used as required where it is not defined.

If the test value cannot unambiguously be assigned to a single associated value, the operation will resolve to the *last* associated value that matches its possible value.

A.8.5 Use of variables

Variables are useful for holding partial calculations or values and can keep your ACD files looking neat and tidy. The general syntax for them is:

```
VariableName : Variable value
```

As an example, here is a calculation to determine the maximum size of a sequence window:

```
integer: start
{
  standard: "Y"
  maximum:  "@(@($(sequence.length) - $(window)) + 1)"
}
```

This can be tidied by storing one of the calculations in a partial result as follows:

```
variable: lminusw "@($(sequence.
length) - $(window))"
integer: start
```

```
{
  standard: "Y"
  maximum: "@($(lminusw) + 1)"
}
```

A.8.5.1 Automatic variables

Currently there is just one of these (`acdprotein`) which is set to true or false depending upon the type of the first sequence read.

Appendix B C programming libraries

EMBOSS includes extensive C programming libraries (AJAX and NUCLEUS). These contain C data structures and functions for application development and static data structures and functions that may be used to extend the core functionality of the libraries themselves. AJAX provides low-level functionality such as array and string handling. NUCLEUS contains code for higher-level tasks, for example sequence alignment algorithms.

B.1 Library documentation

Documentation for AJAX and NUCLEUS is available on the EMBOSS website, for the CVS (Developer's) Release and major versions of the Stable Release.

CVS (Developer's) Release Documentation	http://emboss.open-bio.org/rel/dev/libs/
Stable Release 6 Documentation	http://emboss.open-bio.org/rel/rel6/libs

B.2 Source code in SRS

The library source code is indexed in SRS at the EBI SRS Server: http://srs.ebi.ac.uk/
There are separate SRS databases for objects (C data structures) and functions:

EFUNC: Functions in CVS (Developer's) Release	http://srs.ebi.ac.uk/srs7bin/cgi-bin/wgetz?-newId+-page+LibInfo+-lib+EFUNC
EDATA: Objects in CVS (Developer's) Release	http://srs.ebi.ac.uk/srs7bin/cgi-bin/wgetz?-newId+-page+LibInfo+-lib+EDATA
EFUNCREL: Functions in Stable Release 6.0.0	http://srs.ebi.ac.uk/srs7bin/cgi-bin/wgetz?-newId+-page+LibInfo+-lib+EFUNCREL
EDATAREL: Objects in Stable Release 6.0.0	http://srs.ebi.ac.uk/srs7bin/cgi-bin/wgetz?-newId+-page+LibInfo+-lib+EDATAREL

B.3 AJAX library files

Topic	Library file	Short description
AJAX database reading library		
Sequence database	ajseqdb	Defines a sequence database reading object (SeqSAccess) and static data structures and functions for handling all aspects of sequence database access.
AJAX graphics library		
Graphical output	ajgraph	All the basic functions you are likely to need for plotting and printing graphs.
Histogram graphical functions	ajhist	Defines the histogram object (AjPHist) for handling histograms, which includes the histogram data object (AjPHistdata). It contains basic functions for handling histograms.
Graphical output	ajgraphstruct	Objects for handling graphs, including the basic graph object (AjPGraph) and the **PLplot** graph data object (AjPGraphdata).
AJAX ACD library		
AJAX command definitions	ajacd	The ajAcdGet* family of functions that return AJAX objects for data definitions in the application ACD file.
AJAX core library files		
Alignments	ajalign	Defines the main alignment object (AjPAlign). It can be used for retrieving an input sequence alignment via ACD file processing. Most of the functions you will ever need for general handling of sequence alignments.
Array handling	ajarr	Various objects and functions for handling dynamic arrays.
Assert function	ajassert	Functions for handling assertions.
Base code functions	ajbase	Functions for character conversion of nucleotide bases and other character conversion functions for use with molecular sequences.
Call registration functions	ajcall	Functions to allow access to different graphics packages, should alternatives to **PLplot** be implemented in the future.

Codon analysis functions	ajcod	Defines the AjPCod object and functions for handling codon usage tables.
DNA melting functions	ajdan	Functions for calculating properties of nucleotide sequences, mostly related to DNA melting/annealing.
Domainatrix functions	ajdmx	Objects and functions for handling structure as used by some of the EMBASSY applications.
DOM parser	ajdom	DOM XML parser functions
Domain functions	ajdomain	Defines the scop object (AjPScop), cath object (AjPCath), and generic domain classification object (AjPDomain) object for storing protein structural domain data. Functions for handling protein domain data, including reading the SCOP and CATH domain classification database parsable files and for reading and writing the DCF (domain classification file) format files.
Exception handling	ajexcept	Contains the ajExceptRaise function for raising an exception.
Feature table functions	ajfeat	Most of the functions you will ever need for general feature handling.
File handling	ajfile	Objects and functions for handling of data files, general files and directories and static functions for manipulating them at a low level.
Data file handling	ajfiledata	Data file handling.
File I/O handling	ajfileio	File input/output handling.
Formatted output	ajfmt	Functions for string formatting. The functions are similar to the C functions printf, fprintf, etc., but the set of conversion specifiers and other functionality is extended.
Indexing functions	ajindex	Database indexing functions.
Java functions	ajjava	Java functions.
Lists	ajlist	All the functions you are likely to ever need for the creation and control of linked lists. They define the basic list object (AjPList) which includes the list node object (AjPListNode) as a nested substructure, and a list iteration object (AjIList).
Maths utilities	ajmath	General maths functions for rounding numbers, coordinate conversion, random number generation, etc.

Comparison matrix functions	ajmatrices	Defines the AjPMatrix and AjPMatrixf objects and functions for handling comparison matrices.
Memory allocation	ajmem	Memory allocation.
Message handling	ajmess	Defines the error message object (AjPError) and functions for message handling.
Names: Variables and database definitions	ajnam	Objects and functions for EMBOSS system variables and database definitions.
Nexus data functions	ajnexus	Functions and objects (including static data structures and functions) for parsing the NEXUS file format.
Pattern functions	ajpat	Defines the sequence pattern list object (AjPPatlistSeq) and general pattern list object (AjPPatlistRegex) and functions for handling lists of regular expression patterns.
Protein databank data functions	ajpdb	Protein databank data functions.
Protein databank data I/O functions	ajpdbio	Protein databank data I/O functions.
Phylogenetic data functions	ajphylo	Defines the objects and functions for handling phylogenetic data.
Sequence range functions	ajrange	Defines the range specification object (AjPRange) and functions for handling of range specifications.
Regular expressions	ajreg	Defines the regular expression object (AjPRegexp) and functions for handling of regular expressions.
Feature report functions	ajreport	Defines the AjPReport object and functions for handling of reports.
Sequence	ajseq	Most of the functions you will ever need for general sequence handling. Defines the AjPSeqCvt object used for sequence conversion.
Sequence ABI traces	ajseqabi	Provides functions for parsing sequences (and other data) from ABI trace files.
Sequence reading	ajseqread	Functions for handling the sequence query and sequence input objects and a few general sequence handling functions.
Sequence types	ajseqtype	Functions for handling of sequence types. Static data structures and functions for

		validation and processing of sequence types at a low level.
Sequence writing	ajseqwrite	Defines the sequence output object (AjPSeqout) and functions for writing sequences to file. Defines static data structures and functions for handling sequence output, in the supported formats, at a low level.
Sort utility functions	ajsort	Functions for sorting arrays (Section 6.17.10, 'Sorting arrays').
SQL	ajsql	SQL database access.
String manipulation	ajstr	Functions for general string handling. Defines the basic string object AjPStr, string iteration object (AjIStr and string token parser object (AjPStrTok) for use with the functions.
System utility functions	ajsys	System functions, mostly for low level handling of files and directories and system-specific functions.
Tables	ajtable	Defines the hash table object (AjPTable) and functions for handling hash tables.
Time functions	ajtime	Time functions.
Sequence translation functions	ajtranslate	Sequence translation object (AjPTrn) and include functions for handling sequence translation.
Phylogenetic tree functions	ajtree	Create and manipulate phylogenetic trees.
Utilities	ajutil	Exit cleanly, memory clean-up, manipulate bytes and so on.
Mathematical vector functions	ajvector	Data structures and functions for handling vectors in 3D space.
System specific definitions	ajarch	System-specific type definitions.
General data definitions	ajax	Master AJAX include file.
General data definitions	ajdefine	General type definitions.
Feature table data	ajfeatdata	Basic feature objects (AjPFeattable, AjPFeature and AjPFeattabOut) for general use, e.g. retrieving features via ACD file processing. It also defines a feature input object (AjPFeattabIn) used for low-level feature input handling.
Sequence data	ajseqdata	Define the basic sequence objects (AjPSeq, AjPSeqset and AjPSeqall) for general use e.g. retrieving sequences

		via ACD file processing. Sequence input object (AjPSeqin) and a sequence query object (AjPSeqQuery) used for low level sequence retrieval.

B.4 NUCLEUS library files

Description	Library file	Subject
Alignments	embaln	Sequence alignment
Comments	embcom	Sequence complexity calculation
Consensus sequences	embcons	Consensus
Data file utilities	embdata	Data file handling
Database indexing using EMBL-CD indices	embdbi	Database index
Domainatrix package algorithms	embdmx	**Domainatrix** methods
Domain algorithms	embdomain	Domain methods
EST algorithms	embest	EST methods
Exit calls for applications	embexit	Exit methods
Groups of applications	embgroup	Groups
Isoelectric point utilities	embiep	Protein properties
Indexing databases using B+ trees	embindex	Indexing
Initialising internals	embinit	Initialization
Comparison matrices	embmat	Comparison matrices
Miscellaneous methods	embmisc	Miscellaneous methods
Molecular fragment algorithms	embmol	Molecular fragments
N-mer (word) algorithms	embnmer	Word (n-mer) methods
Pattern matching algorithms	embpat	Pattern matching methods
Pattern list file reading	embpatlist	Pattern list input methods
Protein databank (PDB) structure file utilities	embpdb	Protein databank
Properties of protein sequences	embprop	Sequence properties
Reading data files	embread	Reading data files
Showing sequences in pretty formats	embshow	Show (display) methods
Signatures of protein domains	embsig	Signature methods
Word matching algorithms	embword	Word methods
Include file for all applications	emboss	Master include file

Appendix C C coding standards

C.1 General guidelines

The EMBOSS C coding standards are summarised below. Mostly they concern layout of the code, some well established principles of C programming, and tips for programming EMBOSS objects (C data structures) and functions. There are a few guidelines which you should try to follow when writing code for EMBOSS.

C.1.1 Use of libraries

It is very wasteful to write code unnecessarily; often the functionality is available in AJAX or NUCLEUS. Check the libraries before coding and contribute any new code so that it can be incorporated into the libraries.

C.1.2 Over-engineered code

A common mistake is to waste time implementing functionality that you think you'll need one day, but never actually do. Over-engineered code is confusing and difficult to maintain. Only program what you need today, but design your code so that it can, if necessary, be extended in the future.

C.1.3 Duplicated code

Duplicated code is error-prone and difficult to maintain. Do not duplicate blocks of code, write a function instead. Where two functions do essentially the same thing but have different arguments, make one function simply call the other.

C.1.4 Managing change to code

Your code should be easy to modify for new functionality. Where you find yourself modifying multiple objects or functions to implement a single change it's likely your data model or program structure is not ideal. Consider defining a new object containing the elements you need or new functions as appropriate.

C.1.5 Keep objects clean

The purpose of each element in an object should be obvious. Objects containing variables that are only rarely used or which are used for housekeeping, for instance to hold temporary values, can be difficult to understand. Review your code and establish whether the variable really needs to be in the object or whether it's best moved somewhere else.

C.1.6 Avoid redundancy

Where two or more different objects share common elements there is likely scope for removing redundancy throughout your code. Consider whether a new object encapsulating the common elements would make your code easier to understand and maintain.

C.1.7 Manage your variables

Functions with very long lists of variables are difficult to understand and maintain. Where a group of variables are always used together, consider encapsulating them in a new object, especially where the group occurs elsewhere in your code.

C.1.8 Long functions

Big functions are difficult to understand. Smaller functions are easier to document and therefore easier for the programmer to identify. Functionality split into smaller functions is more likely to be reused. Consider breaking big functions down into smaller ones. If necessary, retain the function with the original name which can call the new, smaller functions. Do not overdo it though and try to avoid too many levels of function calls (see Section C.1.10, 'Nesting of functions').

C.1.9 Long parameter lists

Functions with many parameters are difficult to understand, use and maintain. Where possible, you should pass an object pointer rather than the individual elements of a single object. If the parameters do not belong to an object, consider defining a new object to encapsulate them and pass a pointer to that instead.

C.1.10 Nesting of functions

Code which uses deeply nested chains of functions can be extremely difficult to understand. Design your code to avoid this.

C.2 Code layout

C.2.1 Application code layout

Use the following order for laying out the application C source code:

1. Application standard header
2. Inclusion of system header files
3. Inclusion of local header files

4. Definition of constants
5. Definition of data types
6. Global variables (try to avoid these altogether)
7. Function prototypes
8. `main()` function
9. Functions.

A template application C source code file is provided in the **myemboss** applications directory, e.g.:

```
…/embassy/myemboss/src/template.c
```

The file is shown below:

```
** @source __ApplicationName__
**
** __ShortDescription__
**
** @author Copyright (C) __Year__ __AuthorName__
** @version __VersionNumber__ __description of version__
** @modified __EditDate__ __EditorName__ __description of edit__
** @@
**
** This program is free software; you can redistribute it and/or
** modify it under the terms of the GNU General Public License
** as published by the Free Software Foundation; either version 2
** of the License, or (at your option) any later version.
**
** This program is distributed in the hope that it will be useful,
** but WITHOUT ANY WARRANTY; without even the implied warranty of
** MERCHANTABILITY or FITNESS FOR A PARTICULAR PURPOSE. See the
** GNU General Public License for more details.
**
** You should have received a copy of the GNU General Public License
** along with this program; if not, write to the Free Software
** Foundation, Inc., 59 Temple Place - Suite 330, Boston, MA 02111-1307, USA.
******************************************************************/


/* =============================================================== */
/* ========================== include files ====================== */
/* =============================================================== */

#include "emboss.h"

/* Inclusion of system and local header files goes here */


/* =============================================================== */
/* =========================== constants ========================= */
/* =============================================================== */

/* #define and enum statements go here */
```

```
/* ====================================================================== */
/* ========================= global variables =========================== */
/* ====================================================================== */

/* Global variables definitions go here */


/* ====================================================================== */
/* =============================== data ================================= */
/* ====================================================================== */

/* Definition of datatypes go here */


/* ====================================================================== */
/* ===================== function prototypes ============================ */
/* ====================================================================== */

/* Function prototypes for public (external) functions go here */


/* @prog __ApplicationName ******************************************************
**
** __ShortDescription__
**
********************************************************************************/

int main(int argc, char **argv)
{
}


/* ====================================================================== */
/* ============================= functions =============================== */
/* ====================================================================== */
```

C.2.1.1 Standard file sections

Standard sections in the file .../embassy/myemboss/src/template.c are summarised below.

C.2.1.1.1 Application standard header

All application source code files must begin with the standard documentation header (see Appendix D, *Code documentation standards*).

C.2.1.1.2 Inclusion of header files

The EMBOSS header file emboss.h is included first followed by includes for any required system files. For example:

```
#include "emboss.h"
#include <limits.h>
#include <float.h>
```

emboss.h includes the EMBOSS header file ajax.h which has the following includes:

```
#include <stdio.h>
#include <stdarg.h>
#include <stdio.h>
#include <stdlib.h>
#include <string.h>
```

There is therefore no need to include these files explicitly in the application source code.

C.2.1.1.3 Definition of constants

These should be given in a block and documented where necessary. For example:

```
#define TIMEOUT 30        /* Arbitrary pipe timeout (secs)                    */
#define TIMEBUFFER 256    /* Arbitrary length buffer for time printing        */
#define PUTTIMEOUT 120    /* Max no. of secs to write a file                  */
#define R_BUFFER 2048     /* Arbitrary length buffer for reentrant syscalls */
```

Or:

```
enum blastdbtype [BLAST1P, BLAST1N, BLAST2P, BLAST2N];
```

C.2.1.1.4 Definition of global variables

You should not use global variables unless absolutely necessary. Where they are used they should be documented.

C.2.1.1.5 Definition of data types

Any unions and C data structure definitions that are specific to the application should be given next. All datatypes should be defined in the EMBOSS style. It is recommended, but not strictly required, that application data structures are documented in a standard way (see Appendix D, *Code Documentation Standards*).

C.2.1.1.6 Function prototypes

All functions must be prototyped using the full ANSI C style.

C.2.1.1.7 `main()` function

The `main` function is preceded with a standard documentation header (see Appendix D, *Code documentation standards*). The `main()` function should be given before other functions. This saves people from having to wade through many functions before they find it, and also helps avoid accidental implicit function declarations.

C.2.1.1.8 Functions

Functions should be listed in the order they are prototyped and separated by whitespace (four newlines). All application-specific functions should be declared `static` and have a name that begins with the application name. All application functions should, ideally, be documented in a standard way (see Appendix D, *Code documentation standards*).

C.2.2 Library code layout

C.2.2.1 Library header file

In header files, use the following order for the code:

1. Standard pre-processor directives
2. Inclusion of system header files
3. Inclusion of local header files
4. Definition of constants
5. Definition of public (external) data types
6. Function prototypes for public (external) functions.

All exported prototype definitions must be bounded by the following comment definitions:

```
/*
** Prototype definitions
*/

… exported function prototypes here …
/*
** End of prototype definitions
*/
```

This is to allow the information to be extracted for Windows compilation. The function names of exported prototypes must be on the same line as their return datatype.

A template header file is provided in the AJAX and NUCLEUS directories, e.g.:

```
…/emboss/ajax/core/template.h
```

The file (which is the same for AJAX and NUCLEUS) is shown below:

```
#ifdef __cplusplus
extern "C"
{
#endif

#ifndef LibraryFileName_h
#define LibraryFileName_h


/* ================================================================== */
/* ========================== include files ========================= */
/* ================================================================== */

/* Inclusion of system and local header files goes here */

/* ================================================================== */
/* =========================== constants ============================ */
/* ================================================================== */
```

```
/* #define and enum statements go here */

/* ================================================================ */
/* ========================= public data =========================== */
/* ================================================================ */

/* Definition of public (external) data types go here */


/* ================================================================ */
/* ====================== public functions ======================== */
/* ================================================================ */

/* Function prototypes for public (external) functions go here */


#endif
#ifdef __cplusplus
}
#endif
```

C.2.2.2 Library source file

Use the following order for the source file:

1. Standard library documentation header
2. Inclusion of system header files
3. Inclusion of local header files
4. Definition of constants
5. Definition of global variables (avoid them!)
6. Definition of private data types
7. Function prototypes for private (static) functions
8. Definition of functions.

A template source file is provided in the AJAX and NUCLEUS directories, e.g.:

```
.../emboss/ajax/core/template.c
```

The file is shown below:

```
** @source __AJAX or NUCLEUS__ __LibraryArea__ functions
**
** __ShortDescription__
**
** @author Copyright (C) __Year__ __AuthorName__
** @version __VersionNumber__ __description of version__
```

```
** @modified __EditDate__ __EditorName__ __description of edit__
** @@
**
** This program is free software; you can redistribute it and/or
** modify it under the terms of the GNU General Public License
** as published by the Free Software Foundation; either version 2
** of the License, or (at your option) any later version.
**
** This program is distributed in the hope that it will be useful,
** but WITHOUT ANY WARRANTY; without even the implied warranty of
** MERCHANTABILITY or FITNESS FOR A PARTICULAR PURPOSE. See the
** GNU General Public License for more details.
**
** You should have received a copy of the GNU General Public License
** along with this program; if not, write to the Free Software
** Foundation, Inc., 59 Temple Place - Suite 330, Boston, MA 02111-1307, USA.
*********************************************************************/

/* ====================================================================== */
/* =========================== include files ============================ */
/* ====================================================================== */

/* Inclusion of system and local header files goes here */


/* ====================================================================== */
/* ============================= constants ============================== */
/* ====================================================================== */

/* #define and enum statements go here */


/* ====================================================================== */
/* ========================= global variables =========================== */
/* ====================================================================== */

/* Global variables definitions go here */


/* ====================================================================== */
/* =========================== private data ============================= */
/* ====================================================================== */

/* Definition of private datatypes go here */



/* ====================================================================== */
/* ======================= private functions ============================ */
/* ====================================================================== */

/* Function prototypes for private (static) functions go here */


/* ====================================================================== */
/* ====================== All functions by section ====================== */
/* ====================================================================== */

/* Functions definitions go here, organised into sections */
```

C.2.2.3 Standard file sections

Standard sections in the library header and C source files are summarised below.

C.2.2.3.1 Standard pre-processor directives

The contents of every header (`*.h`) file must be enclosed in a standard pre-processor directive (below). `LibraryFileName` is the name of the library file, e.g. `str` for `ajstr.h/c`:

```
#ifdef __cplusplus
extern "C"
{
#endif

#ifndef ajLibraryFileName_h
#define ajLibraryFileName_h

/* Everything else goes here */

#endif
#ifdef __cplusplus
}
#endif
```

C.2.2.3.2 Standard library documentation header

A standard documentation header (see Appendix D, *Code documentation standards*) should go at the top of every library source (`*.c`) file.

C.2.2.3.3 Inclusion of header files

Any files that are required for compilation are included before any code proper. Inclusions of system files are given before local files. You should never use nested includes.

C.2.2.3.4 Definition of constants

These include `#define` and `enum` statements and should be given in a block and documented where necessary.

C.2.2.3.5 Definition of global variables

Global variables follow constants in the library source (`*.c`) file. Global variables should normally be commented if their function is not obvious from their name. They should be avoided wherever possible.

C.2.2.3.6 Definition of datatypes

Any unions and C data structures (objects) that are private (and should be declared as `static`) are given in the library source (`*.c`) file. Any public (external) unions and structures are given in the library header (`*.h`) file. All datatypes should be defined in the EMBOSS style (Section C.8, 'Structures and unions'). All public datatypes must be documented in a standard way (see Appendix D, *Code documentation standards*). Such documentation is not required for private structures and unions, although it is recommended.

C.2.2.3.7 Function prototypes

Avoid exporting names outside individual C source files, i.e. declare as `static` (in the library source file `*.c`) every function that you possibly can (i.e. any which are not known to be externally required). All functions must be prototyped using the full ANSI C style.

C.2.2.3.8 Definition of functions

The functions are given last of all in the library source (`*.c`) file. All public functions must be documented in a standard way (see Appendix D, *Code documentation standards* below). Such documentation is not required for private functions, although it is recommended.

C.2.3 Ease of reading

Your code should be easy to read. This is perhaps more important than the code actually working. If it's easy to read then at least someone else stands a chance of fixing it, whereas if it's difficult to read, no one is ever easily going to be able to modify or improve it.

C.2.3.1 Line length

Lines should typically be no longer than 79/80 characters to avoid linewrap on screen (or premature line termination on printouts), which can be confusing. Exceptions are allowed where it renders the code easier to read.

C.2.3.2 Variable names

Variable names should not be too short or too long. Overly long variable names can make the code unreadable, names that are too short can make the code incomprehensible. That said, single characters (`i`, `j`, `k`, etc.) are fine for standard loop counters. Use common sense to strike a balance. If for example you're coding a mathematical formula or algorithm from a paper, then use variables with the same name as the symbols. If a variable is a temporary one then make that clear in its name. Consider:

```
for (elementindex = 0; elementindex < DIMENSION; ++elementindex)
    printf("%d\n", element[elementindex]);

for (i =0; i < DIMENSION; ++i)
    printf("%d\n", element[i]);
```

In the first example, you have to read more text before you can recognise the `for` loop idiom, and then you have more hard work to parse the loop body. Since clarity is the goal, a name should contain only necessary and sufficient information to identify the variable. Informative names are unnecessary, however, if the declaration and use of the variable is limited to a small, local section of code. Longer, more informative names are often suitable for variables with larger scope.

C.2.3.3 Braces

Matching braces should appear in the same column and indentation should be consistent. This is good:

```
int fred(void)
{
    ...

    for(i=0; i<LIMIT; ++i)
    {
        ...
    }

    return 0;
}
```

whereas this is messy:

```
int fred(void) {
    ...
    for(i=0; i<LIMIT; ++i) {
        ...
    }

    return 0;
}
```

and so is this:

```
int fred(void)
{
     ...
     for(i=0;  i<LIMIT; ++i)
          {
               ...
          }

     return 0;
}
```

Don't use braces unnecessarily. The following is good:

```
for(i=0; i<LIMIT; ++i)
    sum += i;
```

but this code is cluttered:

```
for(i=0; i<LIMIT; ++i)
{
    sum += i;
}
```

The body of a `for` statement should appear in its own block distinct from the `for` statement itself. The following could be confusing if nested in a larger body of code:

```
for(i=0; i<LIMIT; ++i) sum += i;
```

C.2.3.4 Indentation

Indentation of four characters is recommended but, more importantly, be consistent. If you find that indentation within nested loops results in many of the lines wrapping then you should check whether the code structure can be improved.

There are some lines you can add to your `.emacs` file which will (usually) automatically place braces in the correct places and indent properly. They are available in the file http://emboss.open-bio.org/downloads/EmacsConfig and are shown below:

```
(defun set-if-void (symbol value)
  (or (boundp symbol)
      (set symbol value)))

(require 'cc-mode)
(setq auto-mode-alist
  (append
    '(("\\.C$" .     c++-mode)
      ("\\.H$" .     c++-mode)
      ("\\.cc$" .    c++-mode)
      ("\\.hh$" .    c++-mode)
      ("\\.c$" .     c-mode)
      ("\\.h$" .     c-mode)
      ("\\.m$" .     objc-mode)
      ("\\.java$" . java-mode)
     ) auto-mode-alist))


(defconst emboss-c-style
  '((c-tab-always-indent              . t)
    (c-comment-only-line-offset       . 0)
    (c-hanging-braces-alist           . ((brace-list-open)))
    (c-hanging-colons-alist           . ((member-init-intro before)
                                          (inher-intro)
                                          (case-label after)
                                          (label after)
                                          (access-label after)))
    (c-cleanup-list                   . (scope-operator
                                          empty-defun-braces
                                          defun-close-semi))
;; Uncomment the following line if you don't want newlines after semicolons
;;      (c-hanging-semi&;comma-criteria)
    (c-offsets-alist                  . ((arglist-close      . c-lineup-arglist)
                                          (substatement-open . 0)
                                          (case-label        . 4)
                                          (block-open        . 0)
                                          (knr-argdecl-intro . -)))
   )
 "EMBOSS C Programming Style")

;; Customisations for all of c-mode, c++-mode, and objc-mode
(defun emboss-c-mode-common-hook ()
  ;; add EMBOSS style and set it for the current buffer
```

```
  (c-add-style "EMBOSS" emboss-c-style t)
  ;; offset customisations not in emboss-c-style
  (c-set-offset 'member-init-intro '++)
  ;; other customisations
  (setq tab-width 8
        ;; this will make sure spaces are used instead of tabs
        indent-tabs-mode nil)
  ;; auto-newline [for hungry-delete use (c-toggle-auto-hungry-state 1)]
  (c-toggle-auto-state 1)

  ;; keybindings for C, C++, and Objective-C. Can put these in
  ;; c-mode-map because c++-mode-map and objc-mode-map inherit it
  (define-key c-mode-map "\C-m" 'newline-and-indent)
  )

(add-hook 'c-mode-common-hook 'emboss-c-mode-common-hook)
```

This will also let you reformat existing code (with some manual intervention) using the **emacs** command **`M-x indent-region`**.

C.3 ANSI C standard

EMBOSS C source code should follow the ANSI C standard (ANSI X3.159-1989 'Programming Language C'). This version of the language is often referred to as 'ANSI C', 'Standard C' or 'C89'. The semantics of this are given in the book:

The C Programming Language, 2nd edn
Brian Kernighan and Dennis Ritchie
Prentice Hall 1988; ISBN 0-13-110362-8

For a definitive reference manual, see:

C: A Reference Manual, 5th edn
Samuel P. Harbison and Guy L. Steele
Prentice Hall 2002; ISBN 0-13-089592-X.

This standard introduced:

- Function prototypes
- C++ style parameter declarations (argument types are specified in the function declarations)
- `void` pointers
- Support for international character sets and locales
- A more capable preprocessor
- Enumerated types (`enum`)
- Ability to pass structures to functions and have structures as return values.

Most of these are welcome additions. In EMBOSS, passing whole structures around is deprecated because it can be wasteful of stack memory and rather slow. Each EMBOSS object (C data structure) has a pointer defined for it which should be used for referencing (see below).

If you need to write system-dependent code then try to put such functions in a separate header file to isolate them as much as possible. The AJAX library file `ajsys.c` may well be suitable for such code. Remember that your code will need to work under many operating systems, e.g. Linux, Windows, MacOSX, etc.

C.4 Use of the pre-processor

The contents of every header (`*.h`) file must be enclosed in the pre-processor directives shown above.

The pre-processor can be used for:

- System-dependent code (but see the `ajsys` corollary above.)
- Commenting out code with `#if 0`
- Conditionals arising from system limits
- Definitions (`#define`); but consider `enum` or `static const` instead (see below).

An example of a conditional arising from system limits:

```
#include <limits.h>

#if INT_MAX > UCHAR_MAX
    ...
#else
#error "need int wider than char"
#endif
```

C.4.1 Constants

For constants, consider using `enum` and variables instead of `#define`, which is rarely visible in debuggers. For example:

```
enum { Red = 0xF00, Blue = 0x0F0, Green = 0x00F };
static const float pi = 3.14159265358;
```

C.4.2 Macros

Macros should avoid side effects. If possible, mention each argument exactly once. Put parentheses around all arguments. When the macro is an expression, put parentheses around the whole macro body. If the macro is the inline expansion of some function then capitalise the name and precede it with `M`:

```
#define MAJFUBAR(X) \
    (save = (X),                  \
```

```
    dosomethingwith(X),                    \
    (X) = save);
```

Try to write macros so that they are syntactically expressions, i.e. so you can put a semicolon at the end of their use.

C.5 Comments

Avoid fancy layout or decoration in comments. The styles used are as follows:

```
/* single line comments look like this */

/*
** Important single line comments look like multi-line comments.
*/

/*
** Multi-line comments look like this. Put the opening and closing
** comment sequences on lines by themselves. Use complete sentences
** with proper English grammar, capitalisation and punctuation.
*/

/* but you don't need to punctuate or capitalise one-liners */
```

The opening forward slash (/) of all comments should be indented to the same level as the code to which it applies, for example:

```
if (fubar())
{
     /* Print a message before exiting. */
     ajFatal("Program is FUBAR");
}
```

Avoid placing comments on the same line as the code itself. If, for tidiness, they are deemed essential, then set it off from the code using spaces. Try to indent such comments to the same column and don't continue them across multiple lines:

```
func1(); /* useless comment */
func2(); /* useless comment */
func3(); /* useless comment */
```

See Section 3.1, 'EMBOSS programming' for guidelines on when and how to use comments.

C.6 Variables

C.6.1 Global variables

Do not use them without a very compelling reason because they tend to make code very difficult to understand. If used sparingly and judiciously in some circumstances however they can greatly simplify the code. Use your discretion.

C.6.2 Static variables in functions

Such functions are not re-entrant so do not use them at all if you'd like the code to work in multi-threaded contexts. Occasionally there is very strong reason to use them but avoid them otherwise.

C.6.3 Variable declarations

Declare all variables at the top of each function. Declare one variable per line. It is easier to read if both datatypes and variables line up. This is tidy:

```
{
    ajint   a;
    ajint   b;
    ajlong  c;
...
```

whereas this is messy:

```
{
     ajint a,b;
     ajlong c;
...
```

This following code would conform to the latest C standard (C99) but is unclear (and C99 isn't supported by all the platforms on which EMBOSS should work):

```
{
.
.
    for (ajint a=0; a<b; a++)
.
.
```

The recommended layout is:

```
{
   ajint a;
   ajint b;

   ...

   for (a=0; a<b; a++)
.
.
```

C.6.4 Variable initialisation

Always initialise object pointer variables to NULL. There are circumstances in EMBOSS where it can be risky not to (see Section 5.3, 'Objects (C data structures)'). Initialise other datatypes as appropriate in the code. Align initialisation statements for ease of reading:

```
AjPStr  seqsubstring = NULL;
AjPDouble xarray       = NULL;
```

Do not initialise declarations with functions. Never use for example:

```
AjPStr seqsubstring = ajStrNew();
```

C.6.5 Use of `int`, `ajint`, `long` and `ajlong`

To circumvent any problems arising from system-dependent sizes of variables (see Section 5.1, 'Basic datatypes'), EMBOSS assumes an `ajint` is typically 32 bits and an `ajlong` is typically 64 bits (see Section 5.1, 'Basic datatypes'). Use `ajint`, if 32 bits is enough, instead of `int`. Use `ajlong` instead of `long` or `long long`.

Match your datatype to what you need. If, for example, you are using an Alpha box then both your `int` and `long` variables will be 64 bits. In this case don't use only `ajlong` out of laziness as your code will run more slowly on other platforms.

Standard C `int` and `long` *should* be used in some circumstances, for example, as parameters to C system library functions.

C.7 Precedence of operators

Avoid confusion introduced by relying on operator precedence. This could be confusing:

```
a = b * c + 1;
```

Whereas this is clear:

```
a = (b * c) + 1;
```

C.8 Structures and unions

C.8.1 Declaration

All application datatypes and private datatypes defined in library source files should be declared `static`. Public datatypes declared in library header files should not include this keyword!

Always use the EMBOSS method of declaring structures and unions, i.e. using `#define` and `typedef` in the manner shown:

```
typedef struct AjSStr
{
    ajuint Res;
    ajuint Len;
    char  *Ptr;
    ajuint Use;
    ajint  Padding;
}   AjOStr;
#define AjPStr AjOStr*
typedef AjPStr* AjPPStr;
```

The declaration must contain a structure name (`AjSStr`), object name (`AjOStr`) and pointer name (`AjPStr`) even if these are not used. In the example above, a pointer to the object pointer is also defined (`AjPPStr`). This is optional and should only be included if needed.

C.8.2 Naming conventions

The naming convention for AJAX public (external) structures is:

```
Structure name          AjSName
Object name             AjOName
Pointer name            AjPName
```

The naming convention for NUCLEUS public (external) structures is:

```
Structure name          EmbSName
Object name             EmbOName
Pointer name            EmbPName
```

Use of this convention, i.e. the `P` prefix for pointers, avoids problems where a pointer could appear to be an object in its own right.

This, for example, is bad practice in naming:

```
typedef struct T *T;
...
T t;
```

It is not clear from the variable name that `t` is a pointer.

The naming convention for private structures drops the `Aj` and `Emb` prefix. For private structures whether in AJAX, NUCLEUS or application code use:

```
Structure name          SName
Object name             OName
Pointer name            PName
```

For example:

```
/* @datastatic PBlastType *****************************************************
**
** DbiBlast types
**
** @attr ExtT [const char*] Table filename extension
** @attr ExtH [const char*] Header filename extension
** @attr ExtS [const char*] Sequence filename extension
** @attr IsProtein [AjBool] true for protein
** @attr IsBlast2 [AjBool] blast2.x or blast 1.x
** @attr Type [ajint] enumerated type
** @attr Padding [char[4]] Padding to alignment boundary
******************************************************************************/

typedef struct SBlastType
{
   const char* ExtT;
   const char* ExtH;
   const char* ExtS;
   AjBool      IsProtein;
   AjBool      IsBlast2;
   ajint       Type;
   char        Padding[4];
} OBlastType;

#define PBlastType OBlastType*
```

C.8.3 Datatype documentation

All public objects must be documented in a standard way (Appendix D, *Code documentation standards*). Separate individual objects by suitable whitespace (four newlines).

C.9 Functions

C.9.1 Function prototypes

All application functions and private functions in the libraries should be declared `static`. Public functions in the library should not include this keyword.

All functions must be prototyped using the full ANSI C style and indented for clarity. For example:

```
/*
** Prototype definitions
*/

void        ajListPushAppend (AjPList list, AjPListNode* tail);
AjPList     ajListNewListref (const AjPList list);
void        ajListExit       (void);
AjBool      ajListPeekFirst  (const AjPList thys, void** x);
void        ajListFree       (AjPList *list);
```

C.9.2 Implicit declarations

Examples of these would be:

```
extern fubar(int x);
main(int argc, char **argv)
```

Do not use them. Always specify what a function should return. This is better:

```
extern int fubar(int x);
int main(int argc, char **argv)
```

C.9.3 Function and parameter names

Application function names should be:

- Prepended with the application name (lower-case) followed by an underscore: `applicationname_FunctionName`
- Meaningful
- As short as possible
- Given in 'Sentence case' where the name has more than one word
- Use alphabetic characters only (other than the single underscore).

for example:

```
static void   abiview_TextDisplay(AjPGraph graphs, ajint nstart,ajint nstop,
                                  const AjPStr nseq, float tmax, float scale);

static ajint abiview_getResColour(char B);
```

Library function and function parameter names follow all but the first of the above conventions, additionally:

- All AJAX public (external) functions should begin with `aj`
- All NUCLEUS public (external) functions should begin with `emb`.

for example:

```
AjBool ajStrAssignSubS(AjPStr* Pstr, const AjPStr str,
                       ajint pos1, ajint pos2);

double embIepPhFromHconc  (double H);
```

When functions are added to the existing AJAX and NUCLEUS library files, and the changes are to be incorporated into the distribution, the function and parameter names must adhere to rules (Appendix D, *Code documentation standards*) that are defined in the header file and marked up in the code with the tag `@namXrule`. .

C.9.4 Function layout

The recommended order for the layout of code in functions is as follows:

- Variable declarations
- Calls to constructor functions
- Body of function
- Calls to destructor functions and `return`.

Always explicitly state a `return` at the end of every function, even for `void` functions.
For example:

```
/* @func ajStrAssignK *******************************************************
**
** Copy a single character to a string.
**
** @param [w] Pstr [AjPStr*] Target string.
** @param [r] chr [char] Source text.
** @return [AjBool] ajTrue if string was reallocated
** @@
******************************************************************************/

AjBool ajStrAssignK(AjPStr* Pstr, char chr)
{
    AjBool ret = ajFalse;
    AjPStr thys;

    if(!*Pstr)
      *Pstr=ajStrNewRes(2);

    thys = *Pstr;

    if(thys->Use != 1 || thys->Res <= 1)
    {
      ret = ajStrSetRes(Pstr, 2);
      thys = *Pstr;
    }

    thys->Ptr[0] = chr;
    thys->Ptr[1] = '\0';
    thys->Len =1;

    return ret;
}
```

C.9.5 Function documentation

All public (external) functions must be documented in a standard way (Appendix D, *Code documentation standards*). Separate objects by suitable whitespace (four newlines).

C.10 Program flow

C.10.1 Loops

Do not use `continue` statements without good reason in large blocks of code as they can make it difficult to follow.

C.10.2 `goto`

Do not use `goto`, or indeed functions for non-local jumps, unless there is really no alternative. They make code very difficult to follow.

C.10.3 `switch` statements

Consider using `switch` statements to improve the readability of code where you have excessively long chains of `if else` statements. Where the same `switch` statement is duplicated throughout the code, however, this will be difficult to maintain. Consider changing your code (probably the data model) so that the `switch` is not needed.

C.11 Memory allocation

C.11.1 General memory management

Do not define large arrays in variable declarations within functions. They will go on the stack and can cause many problems.

Do not explicitly use `malloc()` or `calloc()`; instead use the AJAX macros, e.g. `AJALLOC`, `AJNEW`, `AJCNEW`, `AJNEW0`, `AJCNEW0`, etc. (see Section 5.3, 'Objects (C data structures)').

C.11.2 Object construction

Call constructor functions explicitly, i.e. like this:

```
int main(int argc, char **argv)
{
    AjPStr  tmpstr  =  NULL;

    tmpstr = ajStrNew();
    ajStrAssignC(&tmpstr,"Hello");
```

not like this where construction occurs (the default string constructor is invoked by `ajStrAssignC`) but is not clear from the code:

```
AjPStr tmpstr = NULL;

    ajStrAssignC(&tmpstr,"Hello");
```

If possible call all necessary constructors at the start of the function. They will act as a reminder to put destructor functions at the bottom of the function. Doing this avoids memory leaks.

C.11.3 Object destruction

Your code should not leak memory. Always call the destructor function on any objects as required. This applies within all functions including the `main()` one: do not rely on the operating system to free up process memory:

```
int main(int argc, char **argv)
{
    AjPStr  tmpstr = NULL;

    tmpstr = ajStrNew();
    ajStrAssignC(&tmpstr,"Hello");

    ajStrDel(&tmpstr);

    return 0;
}
```

Appendix D Code documentation standards

D.1 Introduction

The guidelines for code documentation cover:

- General comments documenting the source code
- Structured comments and licence information at the head of each application source file
- Structured comments and licence information at the head of each library source file
- Structured comments describing each function
- Structured comments describing each datatype (C structure definitions)

A particular style (see below) is recommended for general C-style comments within the code. Such comments are an informal but valuable way of documenting the code. For the main body of documentation, structured comments are used.

The approach taken is based broadly on JavaDoc. Embedded documentation is marked up with text tags which begin with @ followed by a code identifying the item, then the free documentation text. No text should be given before the @ although any leading asterisks (*) or spaces are stripped before parsing. For example:

```
** @source ApplicationName
```

Text following a tag on a line must not contain HTML markup (this would complicate parsers that convert the text to XML for use in generating the online documentation). The text after the tag continues until a new tag is reached or a token (*/) indicating the end of a C-style comment is found.

D.2 General comments

Comments can make a program much easier to understand, but used heavily or poorly placed they can render good code unreadable. Use them conservatively. Good places to put comments include:

- Major logical steps within a program or function
- Tricky steps within a program or function
- Key variables, but do not document individual housekeeping variables, for example loop counters

- Places where you do something out of the ordinary (this will help others to debug the code)
- Places where you do something particularly clever (the technique might be useful for others).

Properties of code should only be commented if they are not obvious. By contrast, even seemingly obvious global properties and invariants may need to be made explicit. However, this doesn't have to be through comments. The `assert()` macro is an excellent executable comment. There are a few guidelines when writing comments:

- If your code needs a comment to be understood then you should look for ways to rewrite it so that it's clearer.
- Comments should be comprehensible to other people, not just yourself.
- Never write comments that might become incorrect or out of date. An inaccurate or misleading comment is worse than no comment at all.
- The size of a comment should be proportional to the size of the code that it refers to.
- All comments should be correctly formatted (see Appendix C, *C coding standards*).

D.3 Application code documentation

The application code documentation includes:

- General comments (see above)
- Standard header including the GPL
- Documentation of `main()` function
- Structured comments for datatypes
- Structured comments for functions.

D.3.1 Standard header

The application source file must begin with a standard header that includes the General Public License (GPL) statement, which should be used where possible. The header is shown below. The first few lines are specific to each source file, the remainder are constant:

```
/*********************************************************************
** @source ApplicationName
**
** ShortDescription
**
** @author Copyright (C) Year AuthorName
** @version VersionNumber description of version
** @modified EditDate EditorName description of edit
** @@
**
** This program is free software; you can redistribute it and/or
** modify it under the terms of the GNU General Public License
** as published by the Free Software Foundation; either version 2
** of the License, or (at your option) any later version.
```

```
**
** This program is distributed in the hope that it will be useful,
** but WITHOUT ANY WARRANTY; without even the implied warranty of
** MERCHANTABILITY or FITNESS FOR A PARTICULAR PURPOSE. See the
** GNU General Public License for more details.
**
** You should have received a copy of the GNU General Public License
** along with this program; if not, write to the Free Software
** Foundation, Inc., 59 Temple Place - Suite 330, Boston, MA 02111-1307, USA.
*****************************************************************/
```

Where:

- *`ApplicationName`* is the name of the application.
- *`ShortDescription`* is a short description of the application, which should be identical to the text given after the `documentation:` attribute in the application ACD file and in the `main()` function documentation (see below).
- *`Year`* is the year of first release and `AuthorName` is the original author.
- *`VersionNumber`* is the version number and may be followed by an optional description.
- *`EditDate`* and *`EditorName`* are the date and author of the last significant edit and may be followed by an optional description of the revision. The header may contain as many of these lines as required.

For example:

```
/* @source water
**
** Smith-Waterman local alignment
**
** @author: Copyright (C) Alan Bleasby
** @@
**
** This program is free software; you can redistribute it and/or
** modify it under the terms of the GNU General Public License
** as published by the Free Software Foundation; either version 2
** of the License, or (at your option) any later version.
**
** This program is distributed in the hope that it will be useful,
** but WITHOUT ANY WARRANTY; without even the implied warranty of
** MERCHANTABILITY or FITNESS FOR A PARTICULAR PURPOSE. See the
** GNU General Public License for more details.
**
** You should have received a copy of the GNU General Public License
** along with this program; if not, write to the Free Software
** Foundation, Inc., 59 Temple Place - Suite 330, Boston, MA 02111-1307, USA.
******************************************************************
```

D.3.2 `main()` function

This should be preceded by a header block matching the following format:

```
/* @prog ApplicationName *******************************************
**
```

```
** ShortDescription
**
******************************************************************************/
```

Where

- *ApplicationName* is the name of the application.
- *ShortDescription* is a short description of the application, which should be identical to the text given after the `documentation:` attribute in the application ACD file and in the application code standard header (see above).

For example:

```
/* @prog water ***************************************************************
**
** Smith-Waterman local alignment
**
******************************************************************************/
```

D.3.3 Application datatypes and functions

Application functions and datatypes are documented in exactly the same way as library code (see below). Structured comments should (ideally) be used but are not strictly required. Otherwise, use general C-style comments.

D.4 Library code documentation

The library code documentation includes:

- General comments (see above)
- Standard header including the GPL
- Structured comments for all datatypes
- Structured comments for all functions
- Functional sections (library source file only).

D.4.1 Standard header

Each library source file (`*.c`) must begin with a standard header. This is essentially the same as the application header and includes the GPL:

```
/*****************************************************************************
** @source AJAX LibraryArea functions
**
** ShortDescription
**
** @author Copyright (C) Year AuthorName
** @version VersionNumber description of version
** @modified EditDate EditorName description of edit
```

```
** @@
**
** This library is free software; you can redistribute it and/or
** modify it under the terms of the GNU Library General Public
** License as published by the Free Software Foundation; either
** version 2 of the License, or (at your option) any later version.
**
** This library is distributed in the hope that it will be useful,
** but WITHOUT ANY WARRANTY; without even the implied warranty of
** MERCHANTABILITY or FITNESS FOR A PARTICULAR PURPOSE. See the GNU
** Library General Public License for more details.
**
** You should have received a copy of the GNU Library General Public
** License along with this library; if not, write to the
** Free Software Foundation, Inc., 59 Temple Place - Suite 330,
** Boston, MA 02111-1307, USA.
*******************************************************************/
```

The tags are the same as in the application header except that *`LibraryArea`* describes the general area of the library, e.g. `"string"`.

For example:

```
/*****************************************************************************
** @source AJAX string functions
**
** AjPStr objects are reference counted strings
** Any change will need a new string object if the use count
** is greater than 1, so the original AjPStr is provided as a pointer
** so that it can be reallocated in any routine where string modification
** may be needed.
**
** In many cases the text is always a copy, even of a constant original, so
** that it can be simply freed.
**
** @author Copyright (C) 1998 Peter Rice
** @version 1.0
** @modified Jun 25 pmr First version
** @@
**
** This library is free software; you can redistribute it and/or
** modify it under the terms of the GNU Library General Public
** License as published by the Free Software Foundation; either
** version 2 of the License, or (at your option) any later version.
**
** This library is distributed in the hope that it will be useful,
** but WITHOUT ANY WARRANTY; without even the implied warranty of
** MERCHANTABILITY or FITNESS FOR A PARTICULAR PURPOSE. See the GNU
** Library General Public License for more details.
**
** You should have received a copy of the GNU Library General Public
** License along with this library; if not, write to the
** Free Software Foundation, Inc., 59 Temple Place - Suite 330,
** Boston, MA 02111-1307, USA.
******************************************************************************/
```

D.4.2 Library datatypes and functions

Structured comments (see below) are required for all library functions and datatypes.

D.4.3 Functional sections

Functions in a library source file (`*.c`) are organised into sections which define groups of functions with related functionality. The sections are documented using structured comments which define rules for naming functions, for naming and datatyping function parameters and for defining the return datatype. This ensures that functions within a library file, and to an extent across library files, present an intuitive and consistent API.

Where new functions are added to an existing library file, they should be positioned in the file under the appropriate section. Where a new library file is created, it is recommended but not strictly required that sections are added to the file.

Sections are of three types. The start of the structured comment is indicated by the tag given:

- File section (`@filesection`)
- Data section (`@datasection`)
- Standard section (`@section`).

A single file section should be given before the functions themselves. This contains basic documentation and rules for the library as a whole. The file section contains a number of data sections, one for each major datatype processed by the functions.

A data section is typically given for all the public datatypes (identified by `@data`) defined in the header file. For example, `ajstr.h` defines three new datatypes, `AjPStr`, `AjIStr` and `AjPStrTok`, but there are also functions whose major parameter is a C-style (`char*`) string. Therefore there are four data sections in total listed under the file section.

Standard sections are given under the data sections. These define groups of functions with related functionality, similar names, and similar parameters. The data section under which a standard section is given depends on the datatype of the return value or major parameter. For example, the default string constructor has no parameters but returns an `AjPStr` string and therefore is listed under the `AjPStr` data section. What defines the 'major' parameter is to some extent subjective. For all functions with multiple parameters, where the major parameter is not determined by the return type, it should be the first parameter in the list.

Each `@filesection`, `@datasection` and `@section` contains standard text tags described below.

D.4.3.1 Function section documentation tags

The available tags are preceded by `@`.

Tag	Description
`@filesection`	Start of structured comment for the file section and followed by the library name, and any number of lines of free text with markup.

`@datasection`	Start of structured comment for a data section and followed by the datatype, a description of the datatype, and any number of lines of free text with markup.
`@section`	Start of structured comment for a general section and followed by the section name and any number of lines of free text with markup.
`@fnote`	Used for general comments about the section, for instance, its relations to other sections.
`@namXrule`	Defines a rule for naming functions. It is followed by a text token indicating a component (substring) of the function name and one or more lines of free text describing the functionality of functions with that named component. `X` is an integer number which defines the order of that component in the function name (see below).
`@suffix`	Defines a naming rule for differentiating functions with essentially the same function but which differ by a single parameter datatype only. For example, two string insertion functions which both insert into an EMBOSS (`AjPStr`) string, but which insert either an `AjPStr` or a C-type (`char *`) string. The permissible suffix that may appear in the function name after the components defined by the `@namXrule` tags (see below) is given first. This is followed by the datatype (in brackets) and free text describing the datatype.
`@fdata` (standard section only)	Appears in standard sections only, where it must be present. It defines the major datatype (in brackets) that the functions in the section operate on or return. The datatype must be the same as that given in the current data section. Functions are checked to ensure the data type is used and passed appropriately for the function category (see `@fcategory`).
`@argrule`	Defines a rule for naming and datatyping function parameters and (optionally) for defining the function name suffix. It is followed by a text token indicating the suffix of the function name, or '`*`' if none is indicated. A second text token gives the name of the parameter. The parameter datatype is then given (in brackets) followed by one or more lines of free text describing the parameter.
`@valrule`	Defines the function return type. It is followed by a text token indicating the suffix of the function names for which this return type applies. The return datatype follows (in brackets) and is in turn followed by one or more lines of free text describing the return value under various conditions.
`@fcategory` (standard section only)	Defines a standard name used for the functional category. This is used to validate the way in which the data type (see `@fdata`) is passed to functions. Values can be: `"new"`,`"delete"`, `"assign"`, `"modify"`, `"cast"`, `"derive"`, `"use"`, `"iterate"`, `"input"`, `"output"`, `"misc"` or `"internals"`.

D.4.3.2 Example

An excerpt from `ajstr.c` is shown below, indicating the first file section, data section and standard section in the file:

```
/* @filesection ajstr ******************************************************
**
** @nam1rule aj Function belongs to the AJAX library.
**
** @suffix Len [size_t] length
** @suffix C [char*] C character string
** @suffix S [AjPStr] string object
** @suffix K [char] single character
*/



/* @datasection [AjPStr] String ********************************************
**
** Functions for manipulating AJAX (AjPStr) strings
**
** @nam2rule Str Function is for manipulating strings
**
*/



/* @section comparison
**
** Functions for comparing strings
**
** @fdata [AjPStr]
**
** @nam3rule   Match          Compare two complete strings.
** @nam4rule   MatchCase      Case-insensitive comparison.
** @nam4rule   MatchWild      Comparison using wildcard characters.
** @nam5rule   MatchWild      Case Case-insensitive comparison
**                            using wildcard characters.
** @nam5rule   MatchWildWord  Case-sensitive wildcard comparison of
**                            first words within two strings.
** @nam6rule   MatchWildWordCase  Case-insensitive wildcard comparison of
**                            first words within two strings.
** @nam4rule   MatchWord      Comparison using whole words.
** @nam5rule   MatchWordOne   Comparison using whole words matching any one.
** @nam5rule   MatchWordAll   Comparison using whole words matching every one.

** @nam3rule   Prefix         Compare start of string to given prefix.
** @nam4rule   Prefix         Case Case-insensitive comparison.
** @nam3rule   Suffix         Compare end of string to given suffix.
** @nam4rule   Suffix         Case Case-insensitive comparison.
**
** @argrule * str [const AjPStr] String
** @argrule C txt2 [const char*] String
** @argrule S str2 [const AjPStr] String
**
** @valrule * [AjBool] True on success
**
** @fcategory use
*
```

Prototypes of functions in the ‘comparison’ section are below:

```
AjBool ajStrMatchC   (const AjPStr thys, const char* txt);
AjBool  ajStrMatchS     (const AjPStr thys, const AjPStr str);
AjBool  ajStrMatchCaseC (const AjPStr thys, const char* text);
AjBool  ajStrMatchCaseS (const AjPStr thys, const AjPStr str);
```

```
AjBool  ajStrMatchWildC (const AjPStr thys, const char* text);
AjBool  ajStrMatchWildS (const AjPStr thys, const AjPStr wild);
AjBool  ajStrMatchWildCaseC  (const AjPStr thys, const char* text);
AjBool  ajStrMatchWildCaseS  (const AjPStr thys, const AjPStr wild);
AjBool  ajStrMatchWildWordC (const AjPStr str, const char* text);
AjBool  ajStrMatchWildWordS (const AjPStr str, const AjPStr text);
AjBool  ajStrMatchWildWordCaseC (const AjPStr str, const char* text);
AjBool  ajStrMatchWildWordCaseS (const AjPStr str, const AjPStr text);
AjBool  ajStrMatchWordAllS(const AjPStr str, const AjPStr str2);
AjBool  ajStrMatchWordOneS(const AjPStr str, const AjPStr str2);
AjBool  ajStrPrefixC(const AjPStr str, const char* txt2);
AjBool  ajStrPrefixS(const AjPStr str, const AjPStr str2);
AjBool  ajStrPrefixCaseC (const AjPStr str, const char* pref);
AjBool  ajStrPrefixCaseS (const AjPStr str, const AjPStr pref);
AjBool  ajStrSuffixC (const AjPStr thys, const char* suff);
AjBool  ajStrSuffixS (const AjPStr thys, const AjPStr suff);
AjBool  ajStrSuffixCaseC (const AjPStr str, const char* pref);
AjBool  ajStrSuffixCaseS (const AjPStr str, const AjPStr pref);
```

Let us examine how the name rules work for the function `ajStrMatchWildWordCaseS`. This is the most difficult example you are likely to find:

1. `@nam1rule aj Function belongs to the AJAX library.`

This rule in the file section defines the first function name component to be `aj`. All functions in the AJAX library have this prefix.

2. `@nam2rule Str Function is for manipulating strings.`

This rule in the data section for AJAX (`AjPStr`) strings defines the second function name component to be `Str`. All functions in the AJAX string library (`ajstr.h/c`) that operate on AJAX (`AjPStr`) strings have the prefix `ajStr`.

3. `@nam3rule Match Compare two complete strings.`

This rule in the standard section for 'comparison' functions defines the third function name component to be `Match`. The free text indicates that all functions with the prefix `ajStrMatch*` are for comparing two complete strings.

4. `@nam4rule MatchWild Comparison using wildcard characters.`

This rule defines the fourth function name component to be `Wild` and indicates that functions with the prefix `ajStrMatchWild*` compare two (complete) strings using wildcard characters.

5. `@nam5rule MatchWildWord Case-sensitive wildcard comparison of first words within two strings.`

This rule defines the fifth function name component to be `Word` and indicates that functions with the prefix `ajStrMatchWildWord` compare two complete strings using wildcard characters, using only the first words within the two strings.

6. `@nam6rule MatchWildWordCase Case - insensitive wildcard comparison of first words within two strings.`

This rule defines the sixth function name component to be `Case` and indicates that functions with the prefix `ajStrMatchWildWordCase` compare two complete strings using wildcard characters, using only the first words within the two strings, with case-insensitivity.

7. `@suffix C [char*] C character string.`

This rule defined in the file section indicates that the functions with the suffix `C` take a C-type (char *) string argument.

Important

The naming rules should only be defined where they are strictly necessary to differentiate existing functions. They should not be used in cases where one *could* break down the functionality semantically, but there is no justification to do so because there are no other similar functions. In such cases the function should have a single component after the basic prefix (e.g. `ajStr`). This component is typically one capitalised word, but where clarity is genuinely improved by using more than one word then **`SentenceCase`** should be used.

D.5 Datatype documentation

Structured comments appearing before C data structures have the following general format:

```
/* @data PointerName *********************************************
**
** DatatypeDescription
** DatatypeDescription cont.
**
** @alias StructName
** @alias ObjectName
**
** @new    Description
** @delete Description
** @assign Description
**
** @attr Element1 [Datatype1] Description
** @attr Element2 [Datatype2] Description
**
** @@
** Comments
** Comments cont.
**
******************************************************************/

typedef struct StructName
{
    Datatype1 Element1;
    Datatype2 Element2;
} ObjectName

#define PointerName ObjectName*
```

Where:

- *`PointerName`* is the pointer datatype, e.g. `AjPStr`.
- *`StructName`* is the structure name, e.g. `AjSStr`.

- `ObjectName` is the object datatype, e.g. `AjOStr`.
- `DatatypeDescription` is a description of the datatype that appears in the online documentation. It may span multiple lines.
- `Element*` is a variable name of an element of the structure.
- `Datatype*` is the datatype of an element of the structure.
- `Description` is a description of an element of the structure or of a function that operates on it (see below for use of tags).
- `Comments` are optional comments that will *not* appear in the online documentation.

In the example below there is an additional pointer datatype (`AjPPStr`). This is not required for most datatypes:

```
/* @data AjPStr ************************************************************
**
** Ajax string object.
**
** Holds a null terminated character string with additional data.
** The length is known and held internally.
** The reserved memory size is known and held internally.
** The reference count is known and held internally.
** New pointers can refer to the same string without needing
** to duplicate the character data.
**
** If a string has multiple references it cannot be changed. Any
** instance to be changed is first copied to a new string. This
** means that any function which can change the character data must
** pass a pointer to the string so that the string can be moved.
**
** A default null string is provided. New strings are by default
** implemented as pointers to this with increased reference counters.
**
** AjPStr is implemented as a pointer to a C data structure.
**
** @alias AjPPStr
** @alias AjSStr
** @alias AjOStr
** @iterator AjIStr
**
** @attr Res [ajuint] Reserved bytes (usable for expanding in place)
** @attr Len [ajuint] Length of current string, excluding NULL at end
** @attr Ptr [char*] The string, as a NULL-terminated C string.
** @attr Use [ajuint] Use count: 1 for single reference, more if several
**                    pointers share the same string.
**                    Must drop to 0 before deleting. Modifying means making
**                    a new string if not 1.
** @attr Padding [ajint] Padding to alignment boundary
** @@
****************************************************************************/

typedef struct AjSStr
{
  ajuint Res;
  ajuint Len;
  char *Ptr;
  ajuint Use;
  ajint Padding;
```

```
} AjOStr;
#define AjPStr AjOStr*
typedef AjPStr* AjPPStr;
```

D.5.1 Datatype documentation tags

The available tags are preceded by @. Some of the tags refer to functions which act upon the data structure.

Tag	Description
@data	Start of structured comment. Gives the datatype for the pointer to the data structure (e.g. AjPStr) and is followed by any number of lines of free text with markup. Use @datastatic if the data structure is not publicly visible.
@datastatic	Same as @data, but for static (internal) datatypes.
@alias	Aliases for related datatypes and names. The standard practice is to use the pointer name (AjPStr, etc.) as the main name, and the object name (AjOStr, etc.) and the structure name (AjSStr, etc.) as aliases. These are used to markup references through aliases in the source code.
@iterator	Name of the iterator datatype (e.g. AjIStr), if any, associated with this datatype.
@other	Name of any other datatype associated with this datatype.
@new	Constructor. Gives the function name, followed by a copy of the function description.
@delete	Destructor. Gives the function name, followed by a copy of the function description.
@assign	Assignment (overwrites the value). Gives the function name, followed by a copy of the function description.
@modify	Modifier (updates the value). Gives the function name, followed by a copy of the function description.
@cast	Casts the value into something else as the return type or another parameter. Gives the function name, followed by a copy of the function description.
@use	Uses the value but does not modify it or return any part. Gives the function name, followed by a copy of the function description.
@input	Input function (reads an input source and populates the datatype). Gives the function name, followed by a copy of the function description.
@output	Output function (writes the values to an output file). Gives the function name, followed by a copy of the function description.
@iterate	Iteration function (iterates over value or array of values). Gives the function name, followed by a copy of the function description.
@attr	Attribute name, type and description. Attributes are defined in the same order as they appear in the structure. The name and type must match the true definition.
@cc	Comment inserted between groups of attributes. No comments are allowed in the structure itself (to provide clean parsing) so comments are inserted in the header.

D.6 Function documentation

Structured comments for static functions begin with `@funcstatic` whereas exported functions are labelled `@func`. For exported functions a prototype is declared in an associated header file. Structured comments before each function must have the following general format:

```
/* @func FunctionName *********************************************
**
** FunctionDescription
** FunctionDescription cont.
**
** @param [ParameterCode1] ParameterName1 [Datatype1] Description
** @param [ParameterCode2] ParameterName2 [Datatype2] Description
** @return [Datatype] Description
** @cre Description
** @ure Description
** @exception ExceptName Description
** @see RelatedFunctionName
** @@
**
** Comments
** Comments cont.
**
******************************************************************/

Datatype FunctionName (Datatype ParameterName1, Datatype ParameterName2)
{
    /* Function C code goes here */
}
```

Where:

- *FunctionName* is the name of the function.
- *FunctionDescription* is a description of the function that appears in the online documentation. It may span multiple lines.
- *ParameterCode** is a parameter code (see below) describing a parameter and is followed by a short description of the parameter.
- *Datatype* is the datatype of a parameter or the datatype returned by the function (or `void` for void functions).
- *RelatedFunctionName* is the name of a function related to the function in question.
- *Comments* are optional comments that will *not* appear in the online documentation.

For example:

```
/* @func ajStrMatchC **********************************************************
**
** Simple test for matching a string and a text string.
```

```
**
** @param [r] str [const AjPStr] String
** @param [r] txt2 [const char*] Text
** @return [AjBool] ajTrue if two complete strings are the same
** @@
******************************************************************************/

AjBool ajStrMatchC(const AjPStr str, const char* txt2)
{
    if(!str || !txt2)
        return ajFalse;

    if(!strcmp(str->Ptr, txt2))
        return ajTrue;

    return ajFalse;
}
```

D.6.1 Function documentation tags

The available tags are preceded by `@` and are based on the set defined in JavaDoc.

Tag	Description
`@func`	Start of structured comment and followed by the function name and any number of lines of free text with markup. Use `@funcstatic` if the function is not publicly visible.
`@funcstatic`	Same as `@func`, but for `static` (internal) functions.
`@param`	Used once for each parameter. The parameter name (*`ParameterName`*) and type (*`Datatype`*) is as defined in the function. A parameter code (*`ParameterCode`*) is also given (see below).
`@cre`	Description of checked run-time error(s) for a parameter.
`@ure`	Description of *unchecked* run-time error(s) for a parameter.
`@return`	Datatype and description of the return value for successful execution.
`@error`	Description of the return value(s) for errors condition(s). Can be used more than once if appropriate.
`@exception`	Exception(s) thrown by this function. Can be used more than once if appropriate.
`@see`	Cross-reference to related functions. A function name (*`RelatedFunctionName`*) must be given that is identical to that given after the `@func` tag either in the same or a different source file.
`@@`	Effectively ends the structured comment. Any following text is ignored by the HTML conversion parser and will not appear in the online documentation.

D.6.2 Parameter codes

The code consists of lower-case letter(s) for the basic type and upper-case modifiers.

Code	Mnemonic	Description
Parameter codes (basic types)		
`r`	Read	Parameter value is used but not changed in any way.
`w`	Write	Parameter value is ignored. A new value is always written unless specific conditions (which should be documented) apply.
`u`	Update	Parameter value is used, and a new value is (typically) written.
`d`	Delete	Parameter value is a pointer to memory to be freed. The pointer will be set to `NULL`.
`f`	Pointer to function	Parameter value is a pointer to a function. The description should indicate the function type.
`v`	Vararg	Parameter value is a variable-length argument list of some form.
Parameter Codes (modifiers)		
`C`	Checked	The value is checked. See the `@cre` tag for checked runtime errors.
`E`	Empty	An empty value will be accepted, for example a string of length zero.
`N`	Null	A `NULL` value will be accepted.
`P`	Pointer	Use this if a pointer itself is potentially written to rather than just the data it points to.

Appendix E Resources

E.1 EMBOSS programmatic interfaces

Programmatic interfaces to EMBOSS are described below.

> **Note**
>
> The list of interfaces is by no means comprehensive. If you know of other relevant interfaces to EMBOSS please email emboss-bug@emboss.open-bio.org with details.

The available programmatic interfaces include:

Biopython	http://www.biopython.org Various modules utilise the EMBOSS programs.
Bio-Emboss	http://search.cpan.org/dist/Bio-Emboss/ A Perl interface to the EMBOSS libraries allowing you to write Perl programs that look and feel like a typical EMBOSS program.
BioPerl class for EMBOSS Applications `Bio::Tools::Run::EMBOSS Application`	http://doc.bioperl.org/releases/bioperl-1.0.2/Bio/Tools/Run/EMBOSSApplication.html The `EMBOSSApplication` class can represent any EMBOSS application. It is created by a `Bio::Factory::EMBOSS` object which primes it by reading an application ADC file.
BioPerl EMBOSS application factory class `Bio::Factory::EMBOSS`	http://doc.bioperl.org/bioperl-run/Bio/Factory/EMBOSS.html The EMBOSS `factory` class encapsulates access to existing EMBOSS programs.
BioPerl class for EMBOSS application qualifiers `Bio::Tools::Run::EMBOSS acd`	http://search.cpan.org/~birney/bioperl-run-1.4/Bio/Tools/Run/EMBOSSacd.pm The `EMBOSSacd` class represents all the possible command line arguments that can be given to an EMBOSS application.
BioMoby	http://biomoby.open-bio.org/ MOBY is a ontological system for interoperability between biological service providers whether they

	be data hosts or analytical services. The EMBOSS applications have been incorporated.
SOAPLAB	http://www.ebi.ac.uk/Tools/webservices/soaplab/overview **SoapLab** generates Web Services for command line programs. Two types of Web Services interface (*generic interface* and *typed interface*) are possible. The generic interface allows input data to be sent as weakly typed name:value pairs. This allows diverse applications to be integrated under a single programming interface. The typed Web Services interface has input/output type descriptions defined at WSDL/XSD level which allows Web Service clients to check input/output data types using standard means.

Index

Symbols